P9-DEY-786

Volume VI: Triumph and Tragedy

Winston Churchill (1874–1965) was the elder son of Lord Randolph Churchill and his American wife, Jennie Jerome. In 1908, he married Clementine Ogilvy, who gave him life-long support, and they had four daughters and one son.

Churchill entered the army in 1895, served in Cuba, India, Egypt, and the Sudan and his first publications were *The Story of the Malakand Field Force* (1898), *The River War* (1899) and *Savrola* (1900), his only novel. On a special commission for the *Morning Post*, he became involved in the Boer War, was taken prisoner and escaped. His experiences led to the writing of *London to Ladysmith, via Pretoria* and *Ian Hamilton's March*, both published in 1900.

He began his erratic political career in October 1900, when he was elected Conservative M.P. for Oldham. Four years later, however, he joined the Liberal party. In 1906, he became Under-Secretary of State for the colonies and showed his desire for reform in such writings as *My African Journey* (1908). He became President of the Board of Trade in 1908 and Home Secretary in 1910 and, together with Lloyd George, introduced social legislation which helped form much of the basis for modern Britain. Because he foresaw the possibilities of war with Germany after the Agadir crisis, he was made First Lord of the Admiralty in October 1911. He achieved major changes, including that of modernising and preparing the Royal Navy for war, despite unpopularity in the Cabinet because of the cost involved. In May 1915, however, pressurised by the Opposition, he left the Admiralty and served for a time as Lieutenant-Colonel in France. Lloyd George appointed him Minister of Munitions in July 1917 and Secretary of State for War and Air the following year. In 1924, he rejoined the Conservative party and was made Chancellor of the Exchequer by Baldwin. He resigned in January 1931 and, during the 1930s, wrote numerous books, amongst which were *My Early Life* (1930), *Thoughts and Adventures* (1932) and *Great Contemporaries* (1937). Churchill was again asked to take office in September 1939, after the German invasion of Poland and, when Chamberlain was forced to retire because of the Labour party's refusal to serve under him, Churchill became Prime Minister (May 1940–May 1945). From 1945, he spent most of his time writing *The Second World War* and returned to office in 1951. In 1953, he accepted the garter and also won the Nobel prize for literature. In April 1955, however, owing to increasing illness, he resigned as Prime Minister, although he continued to write. *A History of the English-speaking Peoples* (1956–8) is his major work of this time. He died at the age of ninety.

WINSTON S. CHURCHILL

THE SECOND WORLD WAR

VOLUME VI

TRIUMPH AND TRAGEDY

HOUGHTON MIFFLIN COMPANY BOSTON

ISBN 0-395-41060-6 (pbk.)
ISBN 978-0-395-41060-8 (pbk.)

Printed in the United States of America

DOC   30   29   28   27   26   25   24   23   22   21

4500615860

The quotation from When Greek Meets Greek by Reginald Leeper
is reprinted by courtesy of chatto and Windus.
The quotation from Invasion:1944 by Hans Speidel
is reprinted by courtesy of Henry Regnery Company.

Moral of the Work

───── ∞∞∞ ─────

IN WAR: RESOLUTION

IN DEFEAT: DEFIANCE

IN VICTORY: MAGNANIMITY

IN PEACE: GOODWILL

ACKNOWLEDGMENTS

I MUST again acknowledge the assistance of those who helped me with the previous volumes, namely, Lieutenant-General Sir Henry Pownall, Commodore G. R. G. Allen, Colonel F. W. Deakin, the late Sir Edward Marsh, Mr. Denis Kelly, and Mr. C. C. Wood. I have also to thank the very large number of others who have kindly read these pages and commented upon them.

I am obliged to Air Chief Marshal Sir Guy Garrod for his help in presenting the Air aspect.

Lord Ismay has continued to give me his aid, as have my other friends.

I record my obligation to Her Majesty's Government for permission to reproduce the text of certain official documents of which the Crown Copyright is legally vested in the Controller of Her Majesty's Stationery Office. At the request of Her Majesty's Government, on security grounds, I have paraphrased some of the telegrams published in this volume. These changes have not altered in any way the sense or substance of the telegrams.

I am indebted to the Roosevelt Trust for the use they have permitted of the President's telegrams quoted here, and also to others who have allowed their private letters to be published.

INTRODUCTION

WINSTON CHURCHILL began to write the first of what were to be the six volumes of *The Second World War* in 1946. It was a work he had expected to postpone to a later stage of his life, since he had looked forward in 1945 to extending his wartime leadership into the peace. The rejection of his party by the electorate was a heavy blow, which might have dulled his urge to write. But resilience was perhaps the most pronounced of his traits of character, and he had already written the history of another great war in which he had been a principal actor. Once committed to the task, he attacked it with an energy, enthusiasm and power of organisation which would have been remarkable in a professional historian of half his age.

His five-volume history of the First World War, *The World Crisis*, had drawn heavily on the evidence he had submitted to the Dardanelles Committee and on episodic accounts written for newspapers. Its origins were therefore in political debate and in journalism. He set about composing *The Second World War* in an entirely different manner—different, too, from the way in which he had written his great life of Marlborough. Then his technique had been to dictate long passages of narrative, later correcting points of detail in consultation with experts. Now he began by assembling a team of advisers and collecting the documents on which the writing was to be based. The documents were set up in print by the publishers, Cassell, while the advisers worked on the chronologies into which they would fit. Churchill meanwhile prepared himself by dictating recollections of what he had identified as key episodes. They consisted partly of firm impressions and partly of queries to his team about dates, times, places, and personalities. He also wrote copiously to fellow-actors, begging of them their own papers and recollections, and inviting their comments on what he proposed to say. When documents, chronologies, corrections, and comments were collated, he began to write. The bulk of the writing, which was completed in 1953, was done by his normal method of dictation; however, long passages of the first volume, which is very much an *apologia pro vita sua*, were composed in his own hand.

Churchill was not, did not aspire to be, and would very probably

ix

have despised the label of, a scientific historian. Like Clarendon and Macaulay, he saw history as a branch of moral philosophy. Indeed, he gave his history a Moral. Its phrases have become some of the most famous words he pronounced—"In War: Resolution; In Defeat: Defiance; In Victory: Magnanimity; In Peace: Goodwill." Each of the component volumes was also given a Theme—"How the English-speaking peoples through their unwisdom, carelessness and good nature allowed the wicked to rearm" is that of the first— which the author believed encapsulated the period with which the volume dealt, but which he also organised his material to illustrate. He justified this method by comparing it to that of Defoe's *Memoirs of a Cavalier*, "in which the author hangs the chronicle and discussion of great military and political events upon the thread of the personal experience of an individual."

The history is, indeed, intensely personal. Explicitly so, because Churchill asks the reader to regard it as a continuation of *The World Crisis*, the two together forming both "an account of another Thirty Years War" and an expression of his "life-effort" on which he was "content to be judged." Implicitly so, because he related many of the major episodes of the war autobiographically. An excellent example is his account of the air fighting on September 15, 1940, which is regarded as the crisis of the Battle of Britain. He was lunching at Chequers and decided, since the weather seemed to favour a German attack, to spend the afternoon at the Headquarters of the R.A.F. No. 11 Group. He and his wife at once drove there, were given seats in the command room from which the British fighters were controlled, and watched the development of the action:

Presently the red bulbs showed that the majority of our squadrons were engaged. A subdued hum arose from the floor, where the busy plotters pushed their discs to and fro in accordance with the swiftly-changing situation. . . . In a little while all our squadrons were fighting, and some had already begun to return for fuel. . . . I became conscious of the anxiety of the Commander. Hitherto I had watched in silence. I now asked, "What other reserves have we?" "There are none," said Air Marshal Park. In an account which he wrote afterwards he said that at this I "looked grave." Well I might. The odds were great; our margins small; the stakes infinite. . . . Then it appeared that the enemy were going home. No new attack appeared. In another ten minutes the action was ended. We climbed again the stairways that led to the surface, and almost as we emerged the "All Clear" sounded. . . . It was 4.30 p.m. before I got back to Chequers, and I immediately went to bed [an unvarying wartime habit]. I did not wake till eight. When I rang my Principal Private Secretary came in with the evening budget of news from all over the world. It was repellent. . . . "However," he said, as he finished his account, "all is redeemed by the air. We have shot down one hundred and eighty-three for the loss of under forty."

This account is both unique—neither Roosevelt, Stalin, nor Hitler left any first-hand narrative of their involvement in the direction of the war—and acutely revealing. Churchill was fascinated by military operations and followed their progress very closely. But he forbore absolutely to intervene in their control at the hour-by-hour and unit-by-unit level adopted by Hitler. He warned and advised, encouraged and occasionally excoriated. He appointed and removed commanders. But he did not presume to do their job. Another chapter conveys the extent of his forbearance. It comes in Volume IV and concerns the fall of Singapore in February 1942. Very properly, Churchill was not merely disheartened but outraged by the failure of the Malaya garrison, under its commander, General Percival, to halt a Japanese invading force which it outnumbered. When it became clear that Percival was about to be defeated, outrage mingled with desperation and disbelief. Breaking a rule, he signalled Wavell, the Supreme Commander, to urge that the newly arrived 18th Division fight "to the bitter end" and that "commanders and senior officers should die with their troops." In the event, the 18th Division was captured by the Japanese almost intact and General Percival marched into enemy lines under a white flag. By not one immoderate word does the author convey in his narrative how deeply he—and, he felt, his country—were wounded by this humiliating and disastrous episode.

The restraint shown in the Singapore chapter was determined by another principle which he had adopted: that of "never criticising any measure of war or policy after the event unless I had before expressed publicly or formally my opinion or warning about it." The effect is to invest the whole history with those qualities of magnanimity and good will by which he set such store, and the more so as it deals with personalities. The volumes are not only a chronicle of events. They are a record of meetings, debate, and disagreements with a world of people. Some were friends with whom he was forced to differ. Some were with opponents or future enemies with whom he nevertheless succeeded in making common cause, Stalin foremost among them. The descriptions of his personal relationships with these men would alone assure the permanent value of this history to our understanding of the Second World War.

But the value of these volumes is assured in a host of other ways. They have their defects: they take no account, because they could not, of the then still secret Ultra intelligence available throughout the war to the Prime Minister (though a longer cryptic reference on page 295 of Volume I alludes to its significance); and the first volume, in particular, may be judged excessively personal in its interpretation of the policies of the author's opponents. But these deficiencies do not detract from the history's monumental quality. It is an extraordinary achievement, extraordinary in its sweep and comprehensiveness, balance and literary effect; extraordinary in the singularity of its point of view; extraordinary as the labour of a man, already old, who still had ahead of him a career large enough to crown most other statesmen's lives; extraordinary as a contribution to the memorabilia of the English-speaking peoples. It is a great history and will continue to be read as long as Churchill and the Second World War are remembered.

JOHN KEEGAN

PREFACE

THIS volume concludes my personal narrative of the Second World War. Between the Anglo-American landings in Normandy on June 6, 1944, and the surrender of all our enemies fourteen months later tremendous events struck the civilised world. Nazi Germany was crushed, partitioned, and occupied; Soviet Russia established herself in the heart of Western Europe; Japan was defeated; the first atomic bombs were cast.

In this, as in earlier volumes, I have told the story as I knew and experienced it as Prime Minister and Minister of Defence of Great Britain. I have relied, as before, on the documents and speeches composed under the daily ordeal, in the belief that these give a truer picture of what happened at the time than could any afterthoughts. The original text was completed nearly two years ago. Other duties have since confined me to general supervision of the processes of checking the statements of fact contained in these pages and obtaining the necessary consents to the publication of the original documents.

I have called this volume *Triumph and Tragedy* because the overwhelming victory of the Grand Alliance has failed so far to bring general peace to our anxious world.

WINSTON S. CHURCHILL

Chartwell,
   Westerham,
      Kent
September 30, 1953

Theme of the Volume

HOW THE GREAT DEMOCRACIES
TRIUMPHED,
AND SO WERE ABLE TO RESUME
THE FOLLIES
WHICH HAD SO NEARLY
COST THEM THEIR
LIFE

TRIUMPH AND TRAGEDY

———————•❧❖❧•———————

TABLE OF CONTENTS

BOOK I

THE TIDE OF VICTORY

BOOK II

THE IRON CURTAIN

TABLE OF CONTENTS

MAPS AND DIAGRAMS

xxi

BOOK I

THE TIDE OF VICTORY

CHAPTER I

D DAY

The Normandy Landings – My Report to the House of Commons, June 6 – Important News from Stalin – His Telegram of June 11 – Enemy Dispositions on the Atlantic Wall – The German Warning System is Paralysed – Rundstedt's Mistake – I Visit the Beaches and Lunch with Montgomery, June 10 – Cruise in H.M.S. "Kelvin" – General Marshall's Message – My Telegrams to Stalin and Roosevelt, June 14.

OUR long months of preparation and planning for the greatest amphibious operation in history ended on D Day, June 6, 1944. During the preceding night the great armadas of convoys and their escorts sailed, unknown to the enemy, along the swept channels from the Isle of Wight to the Normandy coast. Heavy bombers of the Royal Air Force attacked enemy coast-defence guns in their concrete emplacements, dropping 5,200 tons of bombs. When dawn broke the United States Air Force came on the scene to deal with other shore defences, followed by medium and fighter-bombers. In the twenty-four hours of June 6 the Allies flew over 14,600 sorties. So great was our superiority in the air that all the enemy could put up during daylight over the invasion beaches was a mere hundred sorties. From midnight three airborne divisions were alighting, the British 6th Airborne Division north-east of Caen to seize bridgeheads over the river between the town and the sea, and two American airborne divisions north of Carentan to assist the seaborne assault on the beaches and to check the movement of enemy reserves into the Cotentin peninsula. Although in places the airborne divisions were more widely scattered than had been intended, the object was in every case achieved.

3

As dawn came and the ships, great and small, began to file into their prearranged positions for the assault the scene might almost have been a review. Immediate opposition was limited to an attack by torpedo-boats, which sank a Norwegian destroyer. Even when the naval bombardment began the reply from the coastal batteries was desultory and ineffective. There was no doubt that we had achieved a tactical surprise. Landing and support craft with infantry, with tanks, with self-propelled artillery, and a great variety of weapons, and engineer demolition teams to deal with the beach obstacles, all formed up into groups and moved towards the beaches. Among them were the D.D. ("swimming") tanks, which made their first large-scale appearance in battle. It was still very rough from the bad weather of the day before, and a good many of the "swimming" tanks foundered on the way.

Destroyers and gun and rocket batteries mounted on landing-craft pounded the beach defences, while farther to seaward battle-ships and cruisers kept down the fire of the defending batteries. Ground opposition was slight until the first landing-craft were a mile from the shore, but then mortar and machine-gun fire grew. Surf and the partly submerged obstacles and mines made the landings hazardous, and many craft were wrecked after setting down their troops, but the advance went on.

As soon as the foremost infantry got ashore they dashed forward towards their objectives, and in every case except one made good progress. On "Omaha" beach, north-west of Bayeux, the Vth American Corps ran into severe resistance. By an unlucky chance the enemy defences in this sector had recently been taken over by a German division in full strength and on the alert. Our Allies had a very stiff fight all day to make any lodgment at all, and it was not until the 7th that, after losing several thousand men, they were able to force their way inland. Although we did not gain all we sought, and in particular Caen remained firmly in enemy hands, the progress made on the first two days of the assault was judged very satisfactory.

From the Biscay ports a stream of U-boats, facing all risks and moving on the surface at high speed, sought to break up the invasion. We were well prepared. The western approaches to the Channel were guarded by numerous aircraft, forming our first line of defence. Behind them were the naval forces covering the landings. Meeting the full blast of our defence, the U-boats fared

4

badly. In the first crucial four days six were sunk by air attack and a similar number damaged. They were not able to make any impression on the invasion convoys, which continued to move to their objectives with trifling loss. Thereafter they were more cautious, but no more successful.

*　　*　　*　　*　　*

At noon on June 6 I asked the House of Commons to "take formal cognisance of the liberation of Rome by the Allied Armies under the command of General Alexander", the news of which had been released the night before. There was intense excitement about the landings in France, which everyone knew were in progress at the moment. Nevertheless I devoted ten minutes to the campaign in Italy and in paying my tribute to the Allied Armies there. After thus keeping them on tenterhooks for a little I said:

I have also to announce to the House that during the night and the early hours of this morning the first of the series of landings in force upon the European continent has taken place. In this case the liberating assault fell upon the coast of France. An immense armada of upwards of 4,000 ships, together with several thousand smaller craft, crossed the Channel. Massed airborne landings have been successfully effected behind the enemy lines, and landings on the beaches are proceeding at various points at the present time. The fire of the shore batteries has been largely quelled. The obstacles that were constructed in the sea have not proved so difficult as was apprehended. The Anglo–American Allies are sustained by about 11,000 first-line aircraft, which can be drawn upon as may be needed for the purposes of the battle. I cannot of course commit myself to any particular details. Reports are coming in in rapid succession. So far the commanders who are engaged report that everything is proceeding according to plan. And what a plan! This vast operation is undoubtedly the most complicated and difficult that has ever taken place. It involves tides, winds, waves, visibility, both from the air and the sea standpoint, and the combined employment of land, air, and sea forces in the highest degree of intimacy and in contact with conditions which could not and cannot be fully foreseen.

There are already hopes that actual tactical surprise has been attained, and we hope to furnish the enemy with a succession of surprises during the course of the fighting. The battle that has now begun will grow constantly in scale and in intensity for many weeks to come, and I shall not attempt to speculate upon its course. This I may say however.

Complete unity prevails throughout the Allied Armies. There is a brotherhood in arms between us and our friends of the United States. There is complete confidence in the Supreme Commander, General Eisenhower, and his lieutenants, and also in the commander of the Expeditionary Force, General Montgomery. The ardour and spirit of the troops, as I saw myself, embarking in these last few days was splendid to witness. Nothing that equipment, science, or forethought could do has been neglected, and the whole process of opening this great new front will be pursued with the utmost resolution both by the commanders and by the United States and British Governments whom they serve.

By the afternoon I felt justified in reporting to Stalin.

6 June 44

Everything has started well. The mines, obstacles, and land batteries have been largely overcome. The air landings were very successful, and on a large scale. Infantry landings are proceeding rapidly, and many tanks and self-propelled guns are already ashore. Weather outlook moderate to good.

His answer was prompt, and contained welcome news of the highest importance.

Marshal Stalin to Prime Minister   6 June 44

I have received your communication about the success of the beginning of the "Overlord" operations. It gives joy to us all and hope of further successes.

The summer offensive of the Soviet forces, organised in accordance with the agreement at the Teheran Conference, will begin towards the middle of June on one of the important sectors of the front. The general offensive of the Soviet forces will develop by stages by means of the successive bringing of armies into offensive operations. At the end of June and during July offensive operations will become a general offensive of the Soviet forces.

I shall not fail to inform you in due course of the progress of the offensive operations.

I was actually sending Stalin a fuller account of our progress when his telegram arrived.

Prime Minister to Marshal Stalin   7 June 44

I am well satisfied with the situation up to noon to-day, 7th. Only at one American beach has there been serious difficulty, and that has now been cleared up. 20,000 airborne troops are safely landed behind the flanks of the enemy's lines, and have made contact in each case with the American and British seaborne forces. We got across with small losses. We had expected to lose about 10,000 men. By to-night we hope to have the best part of a quarter of a million men ashore, including a considerable quantity of armour (tanks), all landed from special ships or swimming ashore by themselves. In this latter class of tanks there have been a good many casualties, especially on the American front, owing to the waves overturning the swimming tanks. We must now expect heavy counter-attacks, but we expect to be stronger in armour, and of course overwhelming in the air whenever the clouds lift.

2. There was a tank engagement of our newly landed armour with fifty enemy tanks of the 21st Panzer-Grenadier Division late last night towards Caen, as the result of which the enemy quitted the field. The British 7th Armoured Division is now going in, and should give us superiority for a few days. The question is, how many can they bring against us in the next week? The weather outlook in the Channel does not seem to impose any prohibition on our continued landings. Indeed, it seems more promising than before. All the commanders are satisfied that in the actual landing things have gone better than we expected.

3. Most especially secret. We are planning to construct very quickly two large synthetic harbours on the beaches of this wide, sandy bay of the Seine estuary. Nothing like these has ever been seen before. Great ocean liners will be able to discharge and run by numerous piers supplies to the fighting troops. This must be quite unexpected by the enemy, and will enable the build-up to proceed with very great independence of weather conditions. We hope to get Cherbourg at an early point in the operations.

4. On the other hand, the enemy will concentrate rapidly and heavily and the fighting will be continuous and increasing in scale. Still, we hope to have by D plus 30 about twenty-five divisions deployed, with all their corps troops, with both flanks of the second front resting on the sea and possessed of at least three good harbours—Cherbourg and the two synthetic harbours. This front will be constantly nourished and expanded, and we hope to include later the Brest peninsula. But all this waits on the hazards of war, which, Marshal Stalin, you know so well.

5. We hope that this successful landing and the victory of Rome, of which the fruits have still to be gathered from the cut-off Hun divisions, will cheer your valiant soldiers after all the weight they have had to bear, which no one outside your country has felt more definitely than I.

6. Since dictating the above I have received your message about the successful beginning of "Overlord", in which you speak of the summer offensive of the Soviet forces. I thank you cordially for this. I hope

7

you will observe that we have never asked you a single question, because of our full confidence in you, your nation, and your armies.

He replied:

9 June 44

I have received your message of June 7 with the information of the successful development of the operation "Overlord". We all greet you and the valiant British and American Armies and warmly wish you further successes.

The preparation of the summer offensive of the Soviet armies is concluding. To-morrow, June 10, the first stage will open in our summer offensive on the Leningrad front.

I repeated this at once to Roosevelt.

Stalin telegraphed again on June 11:

As is evident, the landing, conceived on a grandiose scale, has succeeded completely. My colleagues and I cannot but admit that the history of warfare knows no other like undertaking from the point of view of its scale, its vast conception, and its masterly execution. As is well known, Napoleon in his time failed ignominiously in his plan to force the Channel. The hysterical Hitler, who boasted for two years that he would effect a forcing of the Channel, was unable to make up his mind even to hint at attempting to carry out his threat. Only our Allies have succeeded in realising with honour the grandiose plan of the forcing of the Channel. History will record this deed as an achievement of the highest order.

The word "grandiose" is the translation from the Russian text which was given me. I think "majestic" was probably what Stalin meant. At any rate, harmony was complete.

*   *   *   *   *

Let us survey the enemy's dispositions and plans as we now know them. Marshal Rundstedt, with sixty divisions, was in command of the whole Atlantic Wall, from the Low Countries to the Bay of Biscay, and from Marseilles along the southern French shore. Under him Rommel held the coast from Holland to the Loire. His Fifteenth Army with nineteen divisions held the sector about Calais and Boulogne, and his Seventh Army had nine infantry and one Panzer division at hand in Normandy. The ten Panzer divisions on the whole Western Front were spreadeagled from Belgium to Bordeaux. How strange that the Germans, now on the defensive, made the same mistake as the French in 1940 and dispersed their most powerful weapon of counter-attack!

When Rommel took up his command in late January he had been displeased with the defences he found, and his energy improved them greatly. Along the coast there was a line of concrete works with all-round defence, many mines and difficult obstacles of various patterns, especially below high-water mark. Fixed guns pointed seawards, and field artillery covered the beaches. While there was no complete second line of defence, villages in rear were strongly fortified. Rommel was not content with the progress made, and had more time been left him our task would have been harder. Our opening bombardment by sea and air did not destroy many of the concrete works, but by stunning their defenders reduced their fire and also upset their Radar.

The German warning system had been completely paralysed. From Calais to Guernsey the Germans had no fewer than one hundred and twenty major pieces of Radar equipment for finding our convoys and directing the fire of their shore batteries. These were grouped in forty-seven stations. We discovered them all, and attacked them so successfully with rocket-firing aircraft that on the night before D Day not one in six was working. The serviceable ones were deceived by the device of tin-foil strips known as "Window",* which simulated a convoy heading east of Fécamp, and they thus failed to detect the real landings. One piece of equipment near Caen managed to keep going and discovered the approach of the British force, but its reports were ignored by the plotting centre as they were not corroborated by any of the other stations. Nor was this the only menace which was overcome. Encouraged by their success two years before in concealing the passage up the Channel of the *Scharnhorst* and *Gneisenau*, the enemy had built many more jamming stations for thwarting both the ships which directed our night fighters and the Radar beams upon which many of our forces depended for an accurate landfall. But they too were discovered, and Bomber Command made some highly concentrated raids upon them. All were obliterated, and our radio and Radar aids were secure. It may be mentioned that all the Allied effort in the radio war for D Day was British.

It is indeed remarkable that the vast, long-planned assault fell on the enemy as a surprise both in time and place. The German High Command was told that the weather would be too rough

* See Vols. IV, pp. 257-9, and V, p. 459 (Cassell's editions)

that day for amphibious operations, and had received no recent air reports of the assembly of our thousands of ships along the English shore. Early on June 5 Rommel left his headquarters to visit Hitler at Berchtesgaden, and was in Germany when the blow fell. There had been much argument about which front the Allies would attack. Rundstedt had consistently believed that our main blow would be launched across the Straits of Dover, as that was the shortest sea route and gave the best access to the heart of Germany. Rommel for long agreed with him. Hitler and his staff however appear to have had reports indicating that Normandy would be the principal battleground.* Even after we had landed uncertainties continued. Hitler lost a whole critical day in making up his mind to release the two nearest Panzer divisions to reinforce the front. The German Intelligence Service grossly overestimated the number of divisions and the amount of suitable shipping available in England. On their showing there were ample resources for a second big landing, so Normandy might be only a preliminary and subsidiary one. On June 19 Rommel reported to von Rundstedt, ". . . a large-scale landing is to be expected on the Channel front on both sides of Cap Gris Nez or between the Somme and Le Havre,"† and he repeated the warning a week later. Thus it was not until the third week in July, six weeks after D Day, that reserves from the Fifteenth Army were sent south from the Pas de Calais to join the battle. Our deception measures both before and after D Day had aimed at creating this confused thinking. Their success was admirable and had far-reaching results on the battle.

*   *   *   *   *

On June 10 General Montgomery reported that he was sufficiently established ashore to receive a visit. I therefore set off in my train to Portsmouth, with Smuts, Brooke, General Marshall, and Admiral King. All three American Chiefs of Staff had flown to the United Kingdom on June 8 in case any vital military decision had to be taken at short notice. A British and an American destroyer awaited us. Smuts, Brooke, and I embarked in the former, and General Marshall and Admiral King, with their staffs, in the latter, and we crossed the Channel without incident to our respective fronts. Montgomery, smiling and confident, met me

* Blumentritt, *Von Rundstedt*, pp. 218, 219.
† Chester Wilmot, *The Struggle for Europe*, p. 318.

at the beach as we scrambled out of our landing-craft. His army had already penetrated seven or eight miles inland. There was very little firing or activity. The weather was brilliant. We drove through our limited but fertile domain in Normandy. It was pleasant to see the prosperity of the countryside. The fields were full of lovely red and white cows basking or parading in the sunshine. The inhabitants seemed quite buoyant and well nourished and waved enthusiastically. Montgomery's headquarters, about five miles inland, were in a château with lawns and lakes around it. We lunched in a tent looking towards the enemy. The General was in the highest spirits. I asked him how far away was the actual front. He said about three miles. I asked him if he had a continuous line. He said, "No." "What is there then to prevent an incursion of German armour breaking up our luncheon?" He said he did not think they would come. The staff told me the château had been heavily bombed the night before, and certainly there were a good many craters around it. I told him he was taking too much of a risk if he made a habit of such proceedings. Anything can be done once or for a short time, but custom, repetition, prolongation, is always to be avoided when possible in war. He did in fact move two days later, though not till he and his staff had had another dose.

It continued fine, and apart from occasional air alarms and anti-aircraft fire there seemed to be no fighting. We made a considerable inspection of our limited bridgehead. I was particularly interested to see the local ports of Port-en-Bessin, Courseulles, and Ouistreham. We had not counted much on these little harbours in any of the plans we had made for the great descent. They proved a most valuable acquisition, and soon were discharging about two thousand tons a day. I dwelt on these agreeable facts as we drove or walked round our interesting but severely restricted conquest.

Smuts, Brooke, and I went home in the destroyer *Kelvin*. Admiral Vian, who now commanded all the flotillas and light craft protecting the Arromanches harbour, was on board. He proposed that we should go and watch the bombardment of the German position by the battleships and cruisers protecting the British left flank. Accordingly we passed between the two battleships, which were firing at twenty thousand yards, and through the cruiser squadron, firing at about fourteen thousand yards, and soon we were within seven or eight thousand yards of the shore, which was thickly wooded. The bombardment was leisurely and continuous, but there was no reply from the enemy. As we were about to turn I said to Vian, "Since we are so near, why shouldn't we have a plug at them ourselves before we go home?" He said, "Certainly," and in a minute or two all our guns fired on the silent coast. We were of course well within the range of their artillery, and the moment we had fired Vian made the destroyer turn about and depart at the highest speed. We were soon out of danger and passed through the cruiser and battleship lines. This is the only time I have ever been on board a naval vessel when she fired "in anger"—if it can be so called. I admired the Admiral's sporting spirit. Smuts too was delighted. I slept soundly on the four-hour voyage to Portsmouth. Altogether it had been a most interesting and enjoyable day.

* * * * *

At our train we found the three American Chiefs of Staff. They were highly pleased with all they had seen on the American beaches, and full of confidence in the execution of our long-cherished design. We dined together in a happy mood. During the dinner I noticed General Marshall writing industriously, and presently he handed me a message he had written to Admiral Mountbatten, which he suggested we should all sign.

10 June 44

To-day we visited the British and American armies on the soil of France. We sailed through vast fleets of ships, with landing-craft of many types pouring more and more men, vehicles, and stores ashore. We saw clearly the manoeuvre in process of rapid development. We have shared our secrets in common and helped each other all we could. We wish to tell you at this moment in your arduous campaign that we realise that much of this remarkable technique, and therefore the success of the venture, has its origin in developments effected by you and your staff of Combined Operations.

ARNOLD, BROOKE, CHURCHILL, KING,
MARSHALL, SMUTS

Mountbatten must indeed have valued this tribute. The vast, intricate operation, with all its novel and ingenious devices, could not have been achieved without the devoted efforts of the staff of Combined Operations, the organisation which had been created in 1940 under Admiral Keyes, and had been carried by his successor to full fruition.

*   *   *   *   *

When time permitted I reported again to my two great companions.

Prime Minister to Marshal Stalin   14 June 44

I visited the British sector of the front on Monday, as you may have seen from the newspapers. The fighting is continuous, and at that time we had fourteen divisions operating on a front of about seventy miles. Against this the enemy have thirteen divisions, not nearly so strong as ours. Reinforcements are hurrying up from their rear, but we think we can pour them in much quicker from the sea. It is a wonderful sight to see this city of ships stretching along the coast for nearly fifty miles and apparently secure from the air and the U-boats which are so near. We hope to encircle Caen, and perhaps to make a capture there of prisoners. Two days ago the number of prisoners was 13,000, which is more than all the killed and wounded we had lost up to that time. Therefore it may be said that the enemy have lost nearly double what we have, although we have been continuously on the offensive. During yesterday the advances were quite good, though the enemy resistance is stiffening as his strategic reserves are thrown into the battle. I should think it quite likely that we should work up to a battle of about a million a side, lasting through June and July. We plan to have about two million there by mid-August.

Every good wish for your successes in Karelia.

To the President I wrote on the same day about various questions, including the visit of de Gaulle to France, which I had arranged without consulting Roosevelt beforehand. I added:

I had a jolly day on Monday on the beaches and inland. There is a great mass of shipping extended more than fifty miles along the coast. It is being increasingly protected against weather by the artificial harbours, nearly every element of which has been a success, and will soon have effective shelter against bad weather. The power of our air and of our anti-U-boat forces seems to ensure it a very great measure of protection. After doing much laborious duty we went and had a plug at the Hun from our destroyer, but although the range was 6,000 yards he did not honour us with a reply.

Marshall and King came back in my train. They were greatly reassured by all they saw on the American side, and Marshall wrote out a charming telegram to Mountbatten, saying how many of these new craft had been produced under his organisation and what a help they had been. You used the word "stupendous" in one of your early telegrams to me. I must admit that what I saw could only be described by that word, and I think your officers would agree as well. The marvellous efficiency of the transportation exceeds anything that has ever been known in war. A great deal more has to be done, and I think more troops are needed. We are working up to a battle which may well be a million a side. The Chiefs of Staff are searching about for the best solution of these problems as between the Mediterranean and "Overlord".

How I wish you were here!

CHAPTER II

NORMANDY TO PARIS

The Struggle for Caen – Effect of Our Air Offensive on the Enemy's Communications – The Allies Form a Continuous Front – The Flying Bomb Attack on London Begins – Hitler's Conference near Soissons, June 17 – Our Build-up Across the Beaches – The "Mulberry" Harbours and "Pluto" – Correspondence with Stalin – The British Attack on Caen, July 8 – Caen Captured – Congratulations from Smuts and Stalin – Rommel is Wounded and Rundstedt is Replaced – Montgomery's General Offensive, July 18 – I Fly to Cherbourg, July 20 – The Wonderful D.U.K.W.s – Visit to Montgomery – Another Attempt on Hitler's Life – The American Break-out, July 25 – Canadian Attack Down the Falaise Road – Vire Taken – Correspondence with Montgomery – Another Visit to Montgomery – Eisenhower Arrives – Patton's Drive Across Brittany – The Fall of Brest, September 19 – The Falaise Pocket – Eight German Divisions Annihilated – The Liberation of Paris, August 25.

ONCE ashore the first need of the Allies was to consolidate the immediate defence of their beaches and form a continuous front by expanding from them. The enemy fought stubbornly and were not easily overcome. In the American sector the marshes near Carentan and at the mouth of the river Vire hampered our movements, and everywhere the country was suited to infantry defence. The *bocage* which covers much of Normandy consists of a multitude of small fields divided by banks, with ditches and very high hedges. Artillery support for an attack is thus hindered by lack of good observation and it was extremely difficult to use tanks. It was infantry fighting all the way, with every little field a potential strong-point. Nevertheless good progress was made, except for the failure to capture Caen. This small but famous town was to be the scene of bitter struggles over many days. To us it was important, because, apart from the fact that there was good ground to the east for constructing air-strips, it was the hinge on which our whole plan turned. Montgomery's intention was to make a great left wheel by the American forces, with Caen as their left-hand pivot. It was equally important for the Germans. If their lines were pierced there the whole of their Seventh Army would be forced south-eastwards towards the Loire, opening a gap between it and the Fifteenth Army in the north. The way to Paris would then be open. Thus in the following weeks Caen became the scene of ceaseless attacks and the most stubborn defence, drawing towards it a great part of the German divisions, and especially their armour. This was a help as well as a hindrance.

The Germans, though the reserve divisions of their Fifteenth Army were still held intact north of the Seine, had of course been reinforced from elsewhere, and by June 12 twelve divisions were in action, four of them Panzers. This was less than we had expected. The tremendous air offensive had hampered all the enemy's communications. Every bridge across the Seine below Paris and the principal bridges across the river Loire were by now destroyed. Most of the reinforcing troops had to use the roads and railways running through the gap between Paris and Orleans, and were subjected to continuous and damaging attacks by day and night from our air forces. A German report of July 8 said, "From Paris to the west and south-west all rail communications are broken." Not only were the enemy unable to reinforce quickly, but their divisions arrived piecemeal, short of equipment, and fatigued by long night marches, and they were thrown into the line as they came. The German command had no chance to form a striking force behind the battle for a powerful, well-concerted counter-offensive.

By June 11 the Allies had formed a continuous front inland, and our fighters were operating from half a dozen forward air-strips. The next task was to secure a lodgment area big enough to hold sufficient forces for the decisive break-out. The Americans thrust westward across the Cherbourg peninsula towards Barneville, on the western coast, which they reached on June 17. Simultaneously they advanced northwards, and after sharp fighting stood before the outer defences of Cherbourg on the 22nd. The enemy resisted stoutly till the 26th in order to carry out demolitions These were so thorough that heavy loads could not be brought in through the port till the end of August.

* * * * *

Beyond the battlefield other events influenced the future. On the night of June 12–13 the first flying bombs fell on London. They were launched in Northern France from places remote from our landed armies. Their early conquest would bring relief to our civil population, once again under bombardment. Part of the Strategic Air Force renewed attacks on these sites, but there could of course be no question of distorting the land battle on this account. As I said in Parliament, the people at home could feel they were sharing the perils of their soldiers.

On June 17, at Margival, near Soissons, Hitler held a conference with Rundstedt and Rommel. His two generals pressed on him strongly the folly of bleeding the German Army to death in Normandy. They urged that before it was destroyed the Seventh Army should make an orderly withdrawal towards the Seine, where, together with the Fifteenth Army, it could fight a defensive but mobile battle with at least some hope of success. But Hitler would not agree. Here, as in Russia and Italy, he demanded that no ground should be given up and all should fight where they stood. The generals were of course right. Hitler's method of fighting to the death at once on all fronts lacked the important element of selection.

In the battle area along the coast our consolidation was making headway. Bombarding ships of all types, including battleships, continued to support the armies on shore, particularly in the eastern sector, where the enemy concentrated the bulk of his armour and where his batteries were most troublesome. U-boats and light surface vessels tried to attack, though with little success, but sea mines, which were mostly laid by aircraft, took a serious toll of Allied shipping and delayed our build-up. Attacks from enemy bases to the eastward, particularly from Havre, were warded off, and in the west an Allied naval bombarding squadron later co-operated with the American Army in the capture of Cherbourg. Across the beaches progress was good. In the first six days 326,000 men, 54,000 vehicles, and 104,000 tons of stores were landed. In spite of serious losses among landing-craft an immense supply organisation was rapidly taking shape. An average of more than two hundred vessels and craft of all types was arriving daily with supplies. The gigantic problem of handling such a volume of shipping was aggravated by bad weather. Nevertheless remarkable progress was made. The Merchant Navy played an outstanding part. Their seamen cheerfully accepted all the risks of war and weather, and their staunchness and fidelity played an impressive part in the vast enterprise.

By June 19 the two "Mulberry" harbours, one at Arromanches, the other ten miles farther west, in the American sector, were making good progress. The submarine pipe-lines ("Pluto") were to come into action later, but meanwhile Port-en-Bessin was being developed as the main supply port for petrol.* But then a four-day gale began which almost entirely prevented the landing of men and material, and did great damage to the newly sunk breakwaters. Many floating bombardons which were not designed for such conditions broke from their moorings and crashed into other breakwaters and the anchored shipping. The harbour in the American sector was ruined, and its serviceable parts were used to repair Arromanches. This gale, the like of which had not been known in June for forty years, was a severe misfortune. We were already behind our programme of unloading. The break-out was equally delayed, and on June 23 we stood only on the line we had prescribed for the 11th.

* * * * *

The Soviet offensive had now begun, and I kept Stalin constantly informed of our fortunes.

Prime Minister to Marshal Stalin   25 June 44

We now rejoice in the opening results of your immense operations, and will not cease by every human means to broaden our fronts engaged with the enemy and to have the fighting kept at the utmost intensity.

2. The Americans hope to take Cherbourg in a few days. The fall of Cherbourg will soon set three American divisions free to reinforce our attack southwards, and it may be 25,000 prisoners will fall into our hands at Cherbourg.

3. We have had three or four days of gale—most unusual in June—which has delayed the build-up and done much injury to our synthetic

---

* The "Pluto" project included first the laying of pipe-lines in the assault area through which seagoing tankers could discharge petrol direct to the shore. Submarine pipe-lines across the Channel were laid later from the Isle of Wight to Cherbourg and from Dungeness to Boulogne.

harbours in their incomplete condition. We have provided the means to repair and strengthen them. The roads leading inland from the two synthetic harbours are being made with great speed by bulldozers and steel networks unrolled. Thus, with Cherbourg, a large base will be established from which very considerable armies can be operated irrespective of weather.

4. We have had bitter fighting on the British front, where four out of the five Panzer divisions are engaged. The new British onslaught there has been delayed a few days by the bad weather, which delayed the completion of several divisions. The attack will begin to-morrow.

5. The advance in Italy goes forward with great rapidity, and we hope to be in possession of Florence in June and in contact with the Pisa–Rimini line by the middle or end of July. I shall send you a telegram presently about the various strategic possibilities which are open in this quarter. The overriding principle which, in my opinion, we should follow is the continuous engagement of the largest possible number of Hitlerites on the broadest and most effective fronts. It is only by hard fighting that we can take some of the weight off you.

6. You may safely disregard all the German rubbish about the results of their flying bomb. It has had no appreciable effect upon the production or life of London. Casualties during the seven days it has been used are between ten and eleven thousand. The streets and parks remain full of people enjoying the sunshine when off work or duty. Parliament debates continually throughout the alarms. The rocket development may be more formidable when it comes. The people are proud to share in a small way the perils of our own soldiers and of your soldiers, who are so highly admired in Britain. May all good fortune attend your new onfall.

Stalin sent me his congratulations on the fall of Cherbourg, and gave further information about his own gigantic operation.

Marshal Stalin to Prime Minister   27 June 44

The Allied forces have liberated Cherbourg, thus crowning their efforts in Normandy with another great victory. I greet the increasing successes of the brave British and American forces, who have developed their operations both in Northern France and Italy.

If the scale of military operations in Northern France is becoming increasingly powerful and dangerous for Hitler, the successful development of the Allies' offensive in Italy is also worthy of every attention and applause. We wish you new successes.

Concerning our offensive, it can be said that we shall not give the Germans a breathing-space, but shall continue to widen the front of our offensive operations by increasing the strength of our onslaught against the German armies. You will of course agree with me that this is indispensable for our common cause.

As regards the Hitlerite flying bombs, this expedient, it is clear, can have no serious importance either for operations in Normandy or for the population of London, whose bravery is known to all.

I replied:

Prime Minister to Marshal Stalin   1 July 44

This is the moment for me to tell you how immensely we are all here impressed with the magnificent advances of the Russian armies, which seem, as they grow in momentum, to be pulverising the German armies which stand between you and Warsaw, and afterwards Berlin. Every victory that you gain is watched with eager attention here. I realise vividly that all this is the second round you have fought since Teheran, the first of which regained Sebastopol, Odessa, and the Crimea and carried your vanguards to the Carpathians, Sereth, and Pruth.

The battle is hot in Normandy. The June weather has been tiresome. Not only did we have a gale on the beaches worse than any in the summer-time records of many years, but there has been a great deal of cloud. This denies us the full use of our overwhelming air superiority, and also helps the flying bombs to get through to London. However, I hope that July will show an improvement. Meanwhile the hard fighting goes in our favour, and although eight Panzer divisions are in action against the British sector we still have a good majority of tanks. We have well over three-quarters of a million British and Americans ashore, half and half. The enemy is burning and bleeding on every front at once, and I agree with you that this must go on to the end.

*   *   *   *   *

In the last week of June the British established a bridgehead across the river Odon south of Caen. Efforts to extend it southward and eastward across the river Orne were repelled. The southern sector of the British front was twice attacked by several Panzer divisions. In violent conflicts the Germans were severely defeated, with heavy losses from our air and powerful artillery.* It was now our turn to strike, and on July 8 a strong attack on Caen was launched from the north and north-west. The first of the tactical bombardments by Allied heavy bombers, which henceforward were a marked feature, prepared the way. Royal Air Force heavy bombers dropped more than 2,000 tons on the German defences, and at dawn British infantry, hampered un-

---

* These attacks were the result of Hitler's instructions at the Soissons conference. On July 1 Keitel telephoned Rundstedt and asked, "What shall we do?" Rundstedt answered, "Make peace, you idiots. What else can you do?"

avoidably by the bomb-craters and the rubble of fallen buildings, made good progress. By July 10 all of Caen on our side of the river was gained and I could say to Montgomery, "Many congratulations on your capture of Caen." He replied:

General Montgomery to Prime Minister   11 July 44

Thank you for your message. We wanted Caen badly. We used a great weight of air-power to ensure quick success, and the whole battle area leading up to Caen is a scene of great destruction. The town itself also suffered heavily. All to-day the 9th and 10th Panzer Divisions have been attacking furiously to retake Pt. 112, to the [north-east] of Évrecy, and another division has been assaulting the 30th U.S. Division to the north-west of St. Lô. Very heavy losses have been inflicted on all three divisions, and the more they will attack us in this way the better. All goes well.

Smuts, who had now returned to South Africa, sent a prescient and suggestive telegram.

        10 July 44

In view of the spectacular Russian advance, and of the capture of Caen, which forms a welcome pendant, the Germans cannot, as things are now developing, face both fronts. They will soon have to decide whether to throw their main weight against the attack from the east or that from the west. *Knowing what to expect from a Russian invasion, it is likely that they will decide for concentrating on the Russian front. This will help to ease our task in the west.**

Having broken through at Caen, it is essential that we should maintain the initiative and offensive without pause, and that we should advance to the rear of the German flying bomb bases as soon as possible.

I must express my regret at the decision affecting Alexander's advance.† Considering however your success in coping with similar obstructions in the past, I continue to hope that in the end your strategy will again prove successful, backed as it is by every sound military as well as political consideration.

Stalin, who followed our fortunes with daily attention, also sent his "congratulations on the occasion of the splendid new victory of the British forces in the liberation of the town of Caen".

By the middle of July thirty Allied divisions were ashore. Half were American and half British and Canadian. Against these the Germans had gathered twenty-seven divisions. But they had already suffered 160,000 casualties, and General Eisenhower estimated their fighting value as no higher than sixteen divisions.

---

* Author's italics.
† The decision to make a landing in Southern France.

An important event now occurred. On July 17 Rommel was severely wounded. His car was attacked by our low-flying fighters, and he was carried to hospital in what was thought a dying condition. He made a wonderful recovery, in time to meet his death later on at Hitler's orders. In early July Rundstedt was replaced in the overall command of the Western Front by von Kluge, a general who had won distinction in Russia.

*   *   *   *   *

Montgomery's general offensive, planned for July 18, now approached. "God with you," I said. He replied:

17 July 44

Thank you for your message. General conditions for big attack to-morrow now very favourable, as main enemy weight has moved to west of Orne, as was intended, to oppose my attacks in Évrecy area, and these attacks will be continued to-day and to-night.

For complete success to-morrow good flying weather essential. Am determined to loose the armoured divisions to-morrow if in any way possible, and will delay zero hour up to 3 p.m. if necessary.

The British Army attacked with three corps, with the aim of enlarging their bridgeheads and carrying them well beyond the river Orne. The operation was preceded by an even greater bombardment by the Allied air. The German Air Force was totally prevented from interfering. Good progress was made to the east of Caen, until clouded skies began to hamper our planes and led to a week's delay in launching the break-out from the American sector. I thought this was an opportunity to visit Cherbourg and to spend a few days in the "Mulberry" harbour. On the 20th I flew direct in an American Army Dakota to their landing-ground on the Cherbourg peninsula, and was taken all round the harbour by the United States commander. Here I saw for the first time a flying bomb launching-point. It was a very elaborate affair. I was shocked at the damage the Germans had done to the town, and shared the staff disappointment at the inevitable delay in getting the port to work. The basins of the harbour were thickly sown with contact mines. A handful of devoted British divers were at work day and night disconnecting these at their mortal peril. Warm tributes were paid to them by their American comrades. After a long and dangerous drive to the United States beach-head known as Utah Beach I went aboard a British motor torpedo-boat, and thence had a rough passage to

NORTH-WEST EUROPE

Arromanches. As one gets older sea-sickness retreats. I did not succumb, but slept soundly till we were in the calm waters of our synthetic lagoon. I went aboard the cruiser *Enterprise*, where I remained for three days, making myself thoroughly acquainted with the whole working of the harbour, on which all the armies now almost entirely depended, and at the same time transacting my London business.

The nights were very noisy, there being repeated raids by single aircraft, and more numerous alarms. By day I studied the whole process of the landing of supplies and troops, both at the piers, in which I had so long been interested, and on the beaches. On one occasion six tank landing-craft came to the beach in line. When their prows grounded their drawbridges fell forward and out came the tanks, three or four from each, and splashed ashore. In less than eight minutes by my stop-watch the tanks stood in column of route on the highroad ready to move into action. This was an impressive performance, and typical of the rate of discharge which had now been achieved. I was fascinated to see the D.U.K.W.s swimming through the harbour, waddling ashore, and then hurrying up the hill to the great dump where the lorries were waiting to take their supplies to the various units. Upon the wonderful efficiency of this system, now yielding results far greater than we had ever planned, depended the hopes of a speedy and victorious action.

On the first night when I visited the wardroom the officers were singing songs. At the end they sang the chorus of "Rule, Britannia". I asked them what were the words. Nobody knew them. So I recited some of Thomson's noble lines myself, and for the benefit and the instruction of the reader (if he needs any) I reprint them here:

> The nations not so blest as thee
>   Must in their turn to tyrants fall:
> While thou shalt flourish great and free,
>   The dread and envy of them all.

> The Muses still, with freedom found,
>   Shall to thy happy coasts repair;
> Blest isle, with matchless beauty crowned,
>   And manly hearts to guard the fair.

*   *   *   *   *

23

On my last day at Arromanches I visited Montgomery's head-quarters, a few miles inland. The Commander-in-Chief was in the best of spirits on the eve of his largest operation, which he explained to me in all detail. He took me into the ruins of Caen and across the river, and we also visited other parts of the British front. Then he placed at my disposal his captured Storch aeroplane, and the Air Commander himself piloted me all over the British positions. This aircraft could land at a pinch almost anywhere, and consequently one could fly at a few hundred feet from the ground, gaining a far better view and knowledge of the scene than by any other method. I also visited several of the air stations, and said a few words to gatherings of officers and men. Finally I went to the field hospital, where, though it was a quiet day, a trickle of casualties was coming in. One poor man was to have a serious operation, and was actually on the table about to take the anæsthetic. I was slipping away when he said he wanted me. He smiled wanly and kissed my hand. I was deeply moved, and very glad to learn later on that the operation had been entirely successful.

I flew back home that evening, July 23, and arrived before dark. To Captain Hickling, the naval officer in charge of Arromanches, I paid the tribute that was due.

25 July 44

I send you and all under your command my warmest congratulations on the splendid work that has been done at Arromanches. This miraculous port has played, and will continue to play, a most important part in the liberation of Europe. I hope to pay you another visit before long.

The above message should be promulgated to all concerned, in such a way that it does not become known to the enemy, who are as yet ignorant of the capacity and potentialities of Arromanches.

They wanted to call the harbour "Port Churchill". But this for various reasons I forbade.

*   *   *   *   *

At this time the orders which had held the German Fifteenth Army behind the Seine were cancelled, and several fresh divisions were sent to reinforce the hard-pressed Seventh Army. Their transference, by rail or road, or across the Seine by the ferry system which had replaced the broken bridges, was greatly delayed and injured by our air forces. The long-withheld aid reached the field too late to turn the scale.

During the pause in the fighting in Normandy there took place on July 20 a renewed, unsuccessful attempt on Hitler's life. According to the most trustworthy story, Colonel von Stauffenberg had placed under Hitler's table, at a staff meeting, a small case containing a time-bomb. Hitler was spared from the full effect of the explosion by the heavy table-top and its supporting cross-pieces, and also by the light struc.ure of the building itself, which allowed an instantaneous dispe: sal of the pressures. Several officers present were killed, but the Fuehrer, though badly shaken and wounded, arose exclaiming, "Who says I am not under the special protection of God?" All the fury of his nature was aroused by this plot, and the vengeance which he inflicted on all suspected of being in it makes a terrible tale.

*　*　*　*　*

The hour of the great American break-out under General Omar Bradley came at last. On July 25 their VIIth Corps struck southwards from St. Lô, and the next day the VIIIth Corps, on their right, joined the battle. The bombardment by the United States Air Force had been devastating, and the infantry assault prospered. Then the armour leaped through and swept on to the key point of Coutances. The German escape route down that coast of Normandy was cut, and the whole German defence west of the Vire was in jeopardy and chaos. The roads were jammed with retreating troops, and the Allied bombers and fighter-bombers took a destructive toll of men and vehicles. The advance drove forward. Avranches was taken on July 31, and soon afterwards the sea corner, opening the way to the Brittany peninsula, was turned. The Canadians, under General Crerar, made a simultaneous attack from Caen down the Falaise road. This was effectively opposed by four Panzer divisions. Montgomery, who still commanded the whole battle line, thereupon transferred the weight of the British attack to the other front, and gave orders to the British Second Army, under General Dempsey, for a new thrust from Caumont to Vire. Preceded again by heavy air bombing, this started on July 30, and Vire was reached a few days later.

*　*　*　*　*

When the main American offensive began and the Canadian Corps was checked on the Falaise road some invidious comparisons were made at our expense.

Prime Minister to General Montgomery   27 July 44

It was announced from S.H.A.E.F. last night that the British had sustained "quite a serious setback". I am not aware of any facts that justify such a statement. It seems to me that only minor retirements of, say, a mile have taken place on the right wing of your recent attack, and that there is no justification for using such an expression. Naturally this has created a good deal of talk here. I should like to know exactly what the position is, in order to maintain confidence among wobblers or critics in high places.

2. For my own most secret information, I should like to know whether the attacks you spoke of to me, or variants of them, are going to come off. It certainly seems very important for the British Army to strike hard and win through; otherwise there will grow comparisons between the two armies which will lead to dangerous recrimination and affect the fighting value of the Allied organisation. As you know, I have the fullest confidence in you and you may count on me.

Montgomery replied:

27 July 44

I know of no "serious setback". Enemy has massed great strength in area south of Caen to oppose our advance in that quarter. Very heavy fighting took place yesterday and the day before, and as a result the troops of Canadian Corps were forced back 1,000 yards from farthest positions they had reached. . . .

My policy since the beginning has been to draw the main enemy armoured strength on to my eastern flank and to fight it there, so that our affairs on western flank could proceed the easier. In this policy I have succeeded; the main enemy armoured strength is now deployed on my eastern flank, to east of the river Odon, and my affairs in the west are proceeding the easier and the Americans are going great guns.

As regards my future plans. The enemy strength south of Caen astride the Falaise road is now very great, and greater than anywhere else on the whole Allied front. I therefore do not intend to attack him there. Instead I am planning to keep the enemy forces tied to that area and to put in a very heavy blow with six divisions from Caumont area, where the enemy is weaker. This blow will tend to make the American progress quicker.

Montgomery's optimism was justified by events, and on August 3 I telegraphed:

Prime Minister to General Montgomery   3 Aug 44

I am delighted that the unfolding of your plan, which you explained to me, has proceeded so well. It is clear that the enemy will hold on to his eastern flank and hinge with desperate vigour. I am inclined to feel that the Brest peninsula will mop up pretty cheaply. I rejoice that our armour and forward troops have taken Vire. It looks on the map as if you ought to have several quite substantial cops. Naturally I earnestly desire to see the Second Army armour, which cannot be far short of 2,500, loose on the broad plains. In this war by-passing has become a brand-new thing on land as well as at sea. I may come to you for a day in the course of the next week, before I go to Italy. Every good wish.

General Montgomery to Prime Minister   1 Aug 44
Thank you for your message.

2. I fancy we will now have some heavy fighting on eastern flank, and especially on that part from Villers-Bocage to Vire which faces due east. The enemy has moved considerable strength to that part from area south and south-east of Caen.

3. I am therefore planning to launch a heavy attack with five divisions from Caen area directed towards Falaise. Am trying to get this attack launched on August 7.

4. I have turned only one American corps westward into Brittany, as I feel that will be enough.

The other corps of Third United States Army will be directed on Laval and Angers. The whole weight of First United States Army will be put into the swing round the south flank of Second Army and directed against Domfront and Alençon.

5. Delighted to welcome you here next week or at any time.

Prime Minister to General Montgomery   6 Aug 44
I was sorry I could not reach you yesterday. If possible I will come to-morrow, Monday. Please make no special arrangements on my account or inconvenience yourself in any way. Eisenhower, with whom I spent yesterday afternoon, suggested I should also visit Bradley at his headquarters, which I should like to do in the afternoon if you see no objection. The party would consist of self, General Hollis, and Tommy only.

Accordingly, on the 7th I went again to Montgomery's head-quarters by air, and after he had given me a vivid account with his maps an American colonel arrived to take me to General Bradley. The route had been carefully planned to show me the frightful devastation of the towns and villages through which the United States troops had fought their way. All the buildings were pulverised by air bombing. We reached Bradley's headquarters about four o'clock. The General welcomed me cordially, but I could feel there was great tension, as the battle was at its height

and every few minutes messages arrived. I therefore cut my visit
short and motored back to my aeroplane, which awaited me. I
was about to go on board when, to my surprise, Eisenhower
arrived. He had flown from London to his advanced headquarters,
and, hearing of my movements, intercepted me. He had not yet
taken over the actual command of the army in the field from
Montgomery; but he supervised everything with a vigilant eye,
and no one knew better than he how to stand close to a tremendous event without impairing the authority he had delegated to
others.   ✱   ✱   ✱   ✱   ✱

The Third United States Army, under General Patton, had now
been formed and was in action. He detached two armoured and
three infantry divisions for the westward and southerly drive to
clear the Brittany peninsula. The cut-off enemy at once retreated
towards their fortified ports. The French Resistance Movement,
which here numbered 30,000 men, played a notable part, and the
peninsula was quickly overrun. By the end of the first week in
August the Germans, amounting to 45,000 garrison troops and
remnants of four divisions, had been pressed into their defensive
perimeters at St. Malo, Brest, Lorient, and St. Nazaire. Here they
could be penned and left to wither, thus saving the unnecessary losses which immediate assaults would have required. The
damage done to Cherbourg had been enormous, and it was
certain that when the Brittany ports were captured they would
take a long time to repair. The fertility of the "Mulberry"
at Arromanches, the sheltered anchorages, and the unforeseen
development of smaller harbours on the Normandy coast had
lessened the urgency of capturing the Brittany ports, which had
been so prominent in our early plans. Moreover, with things
going so well we could count on gaining soon the far better
French ports from Havre to the north. Brest, however, which
held a large garrison, under an active commander, was dangerous,
and had to be eliminated. It surrendered on September 19 to
violent attacks by three U.S. divisions.

     ✱   ✱   ✱   ✱   ✱

While Brittany was thus being cleared or cooped the rest of
Patton's Third Army drove eastward in the "long hook" which
was to carry them to the gap between the Loire and Paris and
down the Seine towards Rouen. The town of Laval was entered

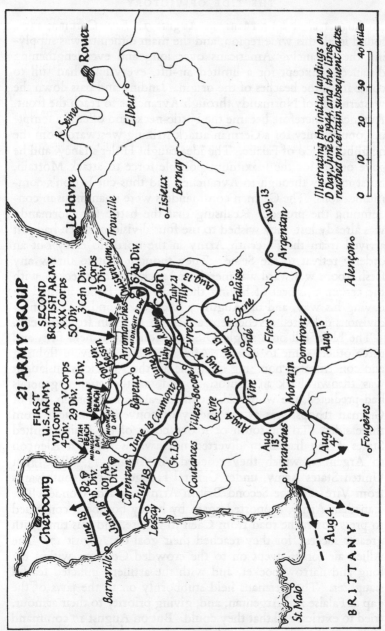

Illustrating the initial landings on
D Day, June 6, 1944, and the lines
reached on certain subsequent dates

0   10   20   30   40 Miles

21 ARMY GROUP

SECOND
BRITISH ARMY
XXX Corps
50 Div.

FIRST
U.S. ARMY
VII Corps   V Corps
4 Div.   29 Div.   1 Div.

3 Cdn Div.

1 Corps
3 Div.

6 Ab. Div.

R. Seine   Rouen

Elbeuf

Le Havre

Lisieux   Bernay

Trouville

Aug. 13

Argentan

Caen   Aug. 21

July 21
Tilly   Falaise

MIDNIGHT D DAY

Ouistreham

Aug. 13

R. Orne

Evrecy

R. Odon   July 9

Villers-Bocage   Aug. 4

Condé   Flers

July 18
Caumont

Aug. 13
Vire

Aug. 13
Mortain

Aug. 13
Domfront

Alençon

Arromanches   Aug. 4

Port-en-Bessin
Port

Bayeux

MIDNIGHT D DAY

June 18

OMAHA
BEACH
MIDNIGHT
D DAY

UTAH
BEACH
MIDNIGHT
D DAY

82 Ab. Div.

101 Ab.
Div.

Carentan
July 13

Coutances

St. Lô

R. Vire

June 18   Cherbourg

June 18

June 18

Barneville

Lessay

Aug. 4   Fougères

Aug. 4

BRITTANY   Avranches

St. Malo

N O R M A N D Y

on August 6, and Le Mans on August 9. Few Germans were found in all this wide region, and the main difficulty was supplying the advancing Americans over long and ever-lengthening distances. Except for a limited air-lift, everything had still to come from the beaches of the original landing and pass down the western side of Normandy through Avranches to reach the front. Avranches therefore became the bottle-neck, and offered a tempting opportunity for a German attack striking westward from the neighbourhood of Falaise. The idea caught Hitler's fancy, and he gave orders for the maximum possible force to attack Mortain, burst its way through to Avranches, and thus cut Patton's communications. The German commanders were unanimous in condemning the project. Realising that the battle for Normandy was already lost, they wished to use four divisions which had just arrived from the Fifteenth Army in the north to carry out an orderly retreat to the Seine. They thought that to throw any fresh troops westward was merely to "stick out their necks", with the certain prospect of having them severed. Hitler insisted on having his way, and on August 7 five Panzer and two infantry divisions delivered a vehement attack on Mortain from the east.

The blow fell on a single U.S. division, but it held firm and three others came to its aid. After five days of severe fighting and concentrated bombing from the air the audacious onslaught was thrown back in confusion, and, as the enemy generals had predicted, the whole salient from Falaise to Mortain, full of German troops, was at the mercy of converging attacks from three sides. To the south of it one corps of the Third United States Army had been diverted northwards through Alençon to Argentan, which they reached on August 13. The First United States Army, under General Hodges, thrust southwards from Vire, and the Second British Army towards Condé. The Canadian Army, supported again by heavy bombers, continued to press down the road from Caen to Falaise, and this time with greater success, for they reached their goal on August 17. The Allied air forces swept on to the crowded Germans within the long and narrow pocket, and with the artillery inflicted fearful slaughter. The Germans held stubbornly on to the jaws of the gap at Falaise and Argentan, and, giving priority to their armour, tried to extricate all that they could. But on August 17 command and control broke down and the scene became a shambles. The jaws closed on August 20, and although by then a considerable part of the enemy had been able to scramble eastwards no fewer than eight German divisions were annihilated. What had been the Falaise pocket was their grave. Von Kluge reported to Hitler: "The enemy air superiority is terrific and smothers almost all our movements. Every movement of the enemy however is prepared and protected by his air forces. Losses in men and material are extraordinary. The morale of the troops has suffered very heavily under constant murderous enemy fire."

* * * * *

The Third United States Army, besides clearing the Brittany peninsula and contributing with their "short hook" to the culminating victory at Falaise, thrust three corps eastwards and north-eastwards from Le Mans. On August 17 they reached Orleans, Chartres, and Dreux. Thence they drove north-westwards down the left bank of the river to meet the British advancing on Rouen. Our Second Army had experienced some delay. They had to reorganise after the Falaise battle, and the enemy found means to improvise rearguard positions. However, the pursuit was pressed hotly, and all the Germans south of the Seine were soon seeking desperately to retreat across it, under destructive air attacks. None of the bridges destroyed by previous air bombardments had been repaired, but there were a few pontoon bridges and a fairly adequate ferry service. Very few vehicles could be saved. South of Rouen immense quantities of transport were abandoned. Such troops as escaped were in no condition to resist on the farther bank of the river.

Eisenhower, who had taken up supreme command, was determined to avoid a battle for Paris. Stalingrad and Warsaw had proved the horrors of frontal assaults and patriotic risings, and he therefore resolved to encircle the capital and force the garrison to surrender or flee. By August 20 the time for action had come. Patton had crossed the Seine near Mantes, and his right flank had reached Fontainebleau. The French Underground had revolted. The police were on strike. The Prefecture was in Patriot hands. An officer of the Resistance reached Patton's headquarters with vital reports, and on the morning of Wednesday, August 23, these were delivered to Eisenhower at Le Mans.

Attached to Patton was the French 2nd Armoured Division,

under General Leclerc, which had landed in Normandy on August 1, and played an honourable part in the advance.* De Gaulle arrived the same day, and was assured by the Allied Supreme Commander that when the time came—and as had been long agreed—Leclerc's troops would be the first in Paris. That evening the news of street-fighting in the capital decided Eisenhower to act, and Leclerc was told to march. At 7.15 p.m. General Bradley delivered these instructions to the French commander, whose division was then quartered in the region of Argentan. The operation orders, dated August 23, began with the words, "Mission (1) s'emparer de Paris . . ."

Leclerc wrote to de Gaulle: "I have had the impression . . . of living over again the situation of 1940 in reverse—complete disorder on the enemy side, their columns completely surprised." He decided to act boldly and evade rather than reduce the German concentrations. On August 24 the first detachments moved on the city from Rambouillet, where they had arrived from Normandy the day before. The main thrust, led by Colonel Billotte, son of the commander of the First French Army Group, who was killed in May 1940, moved up from Orleans. That night a vanguard of tanks reached the Porte d'Orléans, and at 9.22 precisely entered the square in front of the Hôtel de Ville. The main body of the division got ready to enter the capital on the following day. Early next morning Billotte's armoured columns held both banks of the Seine opposite the Cité. By the afternoon the headquarters of the German commander, General von Cholitz, in the Hôtel Meurice, had been surrounded, and Cholitz surrendered to a French lieutenant, who brought him to Billotte. Leclerc had meanwhile arrived and established himself at the Gare Montparnasse, moving down in the afternoon to the Prefecture of Police. About four o'clock von Cholitz was taken before him. This was the end of the road from Dunkirk to Lake Chad and home again. In a low voice Leclerc spoke his thoughts aloud: "Maintenant, ça y est", and then in German he introduced himself to the vanquished. After a brief and brusque discussion the capitulation of the garrison was signed, and one by one their remaining strong-points were occupied by the Resistance and the regular troops.

The city was given over to a rapturous demonstration. German

---

* See Vol. V (Cassell's edition) pp. 544–6.

prisoners were spat at, collaborators dragged through the streets, and the liberating troops fêted. On this scene of long-delayed triumph there arrived General de Gaulle. At 5 p.m. he reached the Rue St. Dominique, and set up his headquarters in the Ministry of War. Two hours later at the Hôtel de Ville he appeared for the first time as the leader of Free France before the jubilant population in company with the main figures of the Resistance and Generals Leclerc and Juin. There was a spontaneous burst of wild enthusiasm. Next afternoon, on August 26, de Gaulle made his formal entry on foot down the Champs Élysées to the Place de la Concorde, and then in a file of cars to Notre Dame. There was some firing from inside and outside the cathedral by hidden collaborators. The crowd scattered, but after a short moment of panic the solemn dedication of the liberation of Paris proceeded to its end.

*　　*　　*　　*　　*

By August 30 our troops were crossing the Seine at many points. Enemy losses had been tremendous: 400,000 men, half of them prisoners, 1,300 tanks, 20,000 vehicles, 1,500 field-guns. The German Seventh Army, and all divisions that had been sent to reinforce it, were torn to shreds. The Allied break-out from the beach-head had been delayed by bad weather and Hitler's mistaken resolve. But once that battle was over everything went with a run, and the Seine was reached six days ahead of the planned time. There has been criticism of slowness on the British front in Normandy, and the splendid American advances of the later stages seemed to indicate greater success on their part than on ours. It is therefore necessary to emphasise again that the whole plan of campaign was to pivot on the British front and draw the enemy's reserves in that direction in order to help the American turning movement. The object of the Second British Army was described in its original plan as "to protect the flank of the U.S. armies while the latter captured Cherbourg, Angers, Nantes, and the Brittany ports". By determination and hard fighting this was achieved. General Eisenhower, who fully comprehended the work of his British comrades, wrote in his official report: "Without the great sacrifices made by the Anglo-Canadian armies in the brutal, slugging battles for Caen and Falaise the spectacular advances made elsewhere by the Allied forces could never have come about."

33

CHAPTER III

THE PILOTLESS BOMBARDMENT

The Attack on London Begins, June 13 – The Construction and Performance of the Flying Bomb – The Destruction of the Guards Chapel, June 18 – Damage and Casualties – Allied Counter-Measures – I Appoint a Small Committee, June 22 – My Speech to the Commons, July 6 – Bomber Command Find New Targets – The Re-deployment of the Anti-Aircraft Batteries Along the Coast – The Flying Bomb is Mastered – Credit for All – The Long-Range Rocket – Controversy About its Size – The Swedish Rocket – The Scientific Intelligence Report of August 26 – An Impressive Technical Achievement – The First Rockets Fall on London, September 8 – An Opinion by Speer – The Failure of the V3 – The Sufferings of Belgium – Duncan Sandys' Report to the War Cabinet on Guided Missiles.

*T*HE long-studied assault on England by unmanned missiles now began. The target was Greater London. For more than a year we had argued among ourselves about the character and scale of the attack, and every preparation which our wits could devise and our resources permit had been made in good time.

In the early hours of June 13, exactly a week after D Day, four pilotless aircraft crossed our coast. They were the premature result of a German order, sent urgently on D Day in reaction to our successes in Normandy. One reached Bethnal Green, where it killed six people and injured nine; the others caused no casualties. Nothing further happened until late on June 15, when the Germans started their campaign of "Retaliation" (*Vergeltung*) in earnest. More than two hundred of the missiles came against us within twenty-four hours, and over three thousand were to follow in the next five weeks.

The flying bomb, as we came to call it, was named V1 by

Hitler, since he hoped—with some reason—that it was only the first of a series of terror weapons which German research would provide. To Londoners the new weapon was soon known as the "doodle-bug", or "buzz bomb", from the strident sound of its engine, which was a jet of new and ingenious design. The bomb flew at speeds up to four hundred miles an hour, and at heights around three thousand feet, and it carried about a ton of explosive. It was steered by a magnetic compass, and its range was governed by a small propeller, which was driven round by the passage of the bomb through the air. When the propeller had revolved a number of times which corresponded to the distance of London from the launching site the controls of the missile were tripped to make it dive to earth. The blast damage was all the more vicious because the bomb usually exploded before penetrating the ground.

This new form of attack imposed upon the people of London a burden perhaps even heavier than the air raids of 1940 and 1941. Suspense and strain were more prolonged. Dawn brought no relief, and cloud no comfort. The man going home in the evening never knew what he would find; his wife, alone all day or with the children, could not be certain of his safe return. The blind, impersonal nature of the missile made the individual on the ground feel helpless. There was little that he could do, no human enemy that he could see shot down.

*　*　*　*　*

My daughter Mary was still serving in the Hyde Park Anti-Aircraft Battery. On the morning of Sunday, June 18, when I was at Chequers, Mrs. Churchill told me she would pay the battery a visit. She found it in action. One bomb had passed over it and demolished a house in the Bayswater Road. While my wife and daughter were standing together on the grass they saw a tiny black object dive out of the clouds, which looked as if it would fall very near Downing Street. My car had gone to collect the letters, and the driver was astonished to see all the passers-by in Parliament Square fall flat on their faces. There was a dull explosion near by and everyone went on their business. The bomb had fallen on the Guards Chapel at Wellington Barracks. A special service for which a large number of members of the Brigade, active and retired, had gathered was going on. There was a direct hit. The whole building was demolished in a second, and nearly two hundred Guardsmen, including many distinguished officers, and their relations and friends were left killed or maimed under the ruins. This was a tragic event. I was still in bed working at my boxes when my wife returned. "The battery has been in action," she said, "and the Guards Chapel is destroyed."

I gave directions at once that the Commons should retire again to the Church House, whose modern steel structure offered somewhat more protection than the Palace of Westminster. This involved a lot of messages and rearrangement. We had a brief interlude in Secret Session, and a Member indignantly asked, "Why have we come back here?" Before I could reply another Member intervened. "If the hon. gentleman will walk a few hundred yards to Birdcage Walk he will see the reason." There was a long silence and the matter dropped.

As the days passed, every borough in London was hit. The worst damage lay in a belt extending from Stepney and Poplar south-westwards to Wandsworth and Mitcham. Of individual boroughs Croydon suffered most hits, including eight bombs in a single day, followed by Wandsworth, Lewisham, Camberwell, Woolwich and Greenwich, Beckenham, Lambeth, Orpington, Coulsdon and Purley, West Ham, Chislehurst, and Mitcham.* About three-quarters of a million houses were damaged, twenty-three thousand of them beyond repair. But although London was the worst sufferer the casualties and the damage spread well outside its bounds. Parts of Sussex and Kent, popularly known as "Bomb Alley" because they lay on the line of route, paid a heavy toll; and bombs, although all were aimed at Tower Bridge, fell far and wide over the countryside from Hampshire to Suffolk. One landed near my home at Westerham, killing, by a cruel mischance, twenty-two homeless children and five grown-ups collected in a refuge made for them in the woods.

*   *   *   *   *

Our Intelligence had accurately foretold six months before how the missiles would perform, but we had not found it easy to prepare fighter and gun defences of adequate quality. Hitler had in fact believed, from trials he had witnessed of a captured Spitfire

* In order of intensity—*i.e.*, bombs per 100 acres—the order was different: first the City of London area, and then Penge, Bermondsey, Deptford, Greenwich, Camberwell, Lewisham, Stepney, Poplar, Lambeth, Battersea, Mitcham, and Wandsworth.

against a flying bomb, that our fighters would be useless. Our timely warning enabled us to disappoint him, but only by a narrow margin. Our fastest fighters, specially stripped and with added power, could barely overtake the speediest missiles. Many bombs did not fly as fast as their makers intended, but even so it was often difficult for our fighters to catch them in time. To make things worse, the enemy fired the bombs in salvoes, in the hope of saturating our defences. Our normal procedure of "scrambling" was too slow, and so the fighters had to fly standing patrols, finding and chasing their quarry with the help of instructions and running commentaries from Radar stations and Observer Corps posts on the ground. The flying bombs were much smaller than normal aircraft, and so they were difficult either to see or hit. There were poor chances of a "kill" from much more than three hundred yards; but it was dangerous to open fire from less than two hundred yards, because the exploding bomb might destroy the attacking fighter.

The red flame of their exhausts made the bombs easier to see in the dark, and during the first two nights our anti-aircraft batteries in London fired on them and claimed to have brought many down. This tended to serve the enemy's purpose, since some of the missiles might otherwise have fallen in open country beyond the capital. Firing in the Metropolitan area was therefore stopped, and by June 21 the guns had moved to the advanced line on the North Downs. Many of the bombs flew at heights which we at first thought would be awkward for the guns, rather too low for the heavies and too high for the others; but fortunately it proved possible to use the heavies against lower targets than we had previously thought. We had realised of course that some bombs would escape both fighters and guns, and these we tried to parry by a vast balloon barrage deployed to the south and south-east of London. In the course of the campaign the barrage did in fact catch 232 bombs, each of which would almost inevitably have fallen somewhere in the London area.

Nor had we been content with defensive measures. The original "ski sites", ninety-six in number, from which the bombs were to have been launched in France had been heavily attacked by our bombers from December 1943 onwards and substantially eliminated.* But, despite all our efforts, the enemy had succeeded

* See Vol. V (Cassell's edition), pp. 210, 212–13.

in launching the assault from new and less pretentious sites, and bombs were penetrating our defences in numbers which, although far smaller than the enemy had originally hoped, were presenting us with many problems. For the first week of the bombardment I kept the control in my own hands; but on June 20 I passed it to an Inter-Service Committee under Duncan Sandys which was known by the code-name of "Crossbow".

Prime Minister to Home Secretary, Sir Edward Bridges,   22 June 44
and to General Ismay, for C.O.S. Committee

Now that we can see our way a little clearer, and after consultation with the Chiefs of Staff, I have decided that the "Crossbow" Committee, over which I have hitherto presided, should consist of a smaller group charged with the responsibility for reporting upon the effects of the flying bomb and the flying rocket and the progress of countermeasures and precautions to meet it. The Joint Parliamentary Secretary, Ministry of Supply [Mr. Duncan Sandys], will be chairman, and the membership should be kept as small as possible. . . .

This Committee will report daily, or as often as may be necessary, to me, the Home Secretary, the Secretary of State for Air, and the Chiefs of Staff.

The Home Secretary, the Secretary of State for Air, and I will attend together should occasion arise.

The committee included Air Marshal Bottomley, Deputy Chief of the Air Staff, Air Marshal Hill, Air Officer Commanding Air Defence of Great Britain, and General Pile, General Officer Commanding Anti-Aircraft Command.

*   *   *   *   *

On July 6 I unfolded to the House of Commons, many of whose constituencies were feeling the strain of the attack, the preparation and action the Government had taken since early in 1943. At any rate, no one could say that we had been caught by surprise. There was no complaint. Everyone saw we just had to lump it, an ordeal made easier by our hopes of a successful advance in Normandy. My account was detailed.

The total weight of bombs so far dropped by us on flying bomb and rocket targets in France and Germany, including Peenemünde, has now reached about fifty thousand tons, and the number of reconnaissance flights totals many thousands. The scrutiny and interpretation of the tens of thousands of air photographs obtained for this purpose has alone been a stupendous task, discharged by the Air Reconnaissance and

Photographic Interpretation units of the Royal Air Force. These efforts have been exacting to both sides, friends and foes. Quite a considerable proportion of our flying power has been diverted for months past from other forms of offensive activity. The Germans for their part have sacrificed a great deal of manufacturing strength which would have increased their fighter and bomber forces working in conjunction with their hard-pressed armies on every front. It has yet to be decided who has suffered and will suffer most in the process. There has in fact been in progress for a year past an unseen battle into which great resources have been poured by both sides. This invisible battle has now flashed into the open, and we shall be able, and indeed obliged, to watch its progress at fairly close quarters. . . .

We must neither underrate nor exaggerate. In all up to six o'clock this morning about two thousand seven hundred and fifty flying bombs have been discharged from the launching stations along the French coast. A very large proportion of these have either failed to cross the Channel or have been shot down and destroyed by various methods. . . . The weather however during June has been very unfavourable to us for every purpose. In Normandy it has robbed us in great part of the use of our immense superiority. . . . In Britain it has made more difficult the work and combination of the batteries and aircraft. It has also reduced the blows we strike at every favourable opportunity at the launching sites and suspected points on the other side of the Channel. Nevertheless the House will, I think, be favourably surprised to learn that the total number of flying bombs launched from the enemy's stations has killed exactly one person per bomb. . . . Actually the latest figures up to six o'clock this morning are 2,754 flying bombs launched and 2,752 fatal casualties sustained. . . . A very high proportion of the casualties, somewhere around 10,000, not always severe or mortal, has fallen upon London, which presents to the enemy a target eighteen miles wide by over twenty miles deep. It is therefore the unique target of the world for the use of a weapon of such proved inaccuracy. The flying bomb is a weapon literally and essentially indiscriminate in its nature, purpose, and effect. The introduction by the Germans of such a weapon obviously raises some grave questions, upon which I do not propose to trench to-day.

* * * * *

Arrangements had been made to evacuate mothers and children and to open the deep shelters which had hitherto been held in reserve, and I explained that everything in human power would be done to defeat this novel onslaught; but I ended on a note which seemed appropriate to the mood of the hour.

We shall not allow the battle operations in Normandy nor the attacks we are making against special targets in Germany to suffer. They come first, and we must fit our own domestic arrangements into the general scheme of war operations. There can be no question of allowing the slightest weakening of the battle in order to diminish in scale injuries which, though they may inflict grievous suffering on many people and change to some extent the normal, regular life and industry of London, will never stand between the British nation and their duty in the van of a victorious and avenging world. It may be a comfort to some to feel that they are sharing in no small degree the perils of our soldiers overseas, and that the blows which fall on them diminish those which in other forms would have smitten our fighting men and their allies. But I am sure of one thing, that London will never be conquered and will never fail, and that her renown, triumphing over every ordeal, will long shine among men.

We now know that Hitler had thought that the new weapon would be "decisive" in fashioning his own distorted version of peace. Even his military advisers, who were less obsessed than their master, hoped that London's agony would cause some of our armies to be diverted to a disastrous landing in the Pas de Calais in an attempt to capture the launching sites. But neither London nor the Government flinched, and I had been able to assure General Eisenhower on June 18 that we would bear the ordeal to the end, asking for no change in his strategy in France.

*   *   *   *   *

Our bombing attacks on launching sites went on for a time, but it was clear before the end of June that these were now poor targets. Bomber Command, anxious to share more effectively in relieving London, sought better ones; and they were soon found. The main storage depots for the flying bombs in France now lay in a few large natural caverns around Paris, long exploited by French mushroom-growers. One of these caverns, at St. Leu d'Esserent, in the Oise valley, was rated by the Germans to store two thousand bombs, and it had supplied 70 per cent. of all the bombs fired in June. Early in July it was largely destroyed by Bomber Command, using some of their heaviest bombs to crush the roof in. Another, rated to hold one thousand, was smashed by American bombers. We know that at least three hundred flying bombs were irretrievably buried in this one cavern. London was spared all these, and the Germans were forced to use bombs of a type which they had previously condemned as unsatisfactory.

Our bombers did not achieve their success without loss. Of all our forces they were the earliest engaged against the flying bombs. They had attacked research centres and factories in Germany, and launching sites and supply depots in France. By the end of the campaign nearly two thousand airmen of British and Allied bombers had died in London's defence.

* * * * *

At the headquarters of Air Defence of Great Britain much thought had been given to the rôles of fighters and guns. Our dispositions had seemed sensible enough: fighters ranging out over the sea and over most of Kent and Sussex, where the bombs were dispersed, and guns concentrated in a belt nearer London, where the bombs drew into a more compact front as they approached their target. This seemed to give each method of defence its best chance, and it was no surprise that in the first few weeks of the campaign, as indeed in all other campaigns previously, the fighters had much more success than the guns. By the second week of July however General Pile and some discerning experts came to the conclusion that the guns could do very much better without undue prejudice to the success of the fighters if the batteries were moved on to the coast. Their Radar for fire control would have more scope, and it would be safer to use the proximity-fuzed shells which were now arriving from America.* We had not been sure if the guns could use their Radar on the coast, owing to the danger of enemy jamming, but so good had been our Intelligence, and so accurate our bombing, that by D Day we had put all the German jamming stations out of action. It was nevertheless a grave decision to uproot the enormous Anti-Aircraft organisation from the North Downs and to redeploy it on the coast, knowing that this might spoil the success of the fighters.

On July 17 Duncan Sandys, who had pressed strongly for this change, reported to the War Cabinet:

The lay-out of our defences against the flying bomb has been reviewed in the light of the results obtained during the past few weeks. Experience has shown that under the original plan fighters and guns

---

* These shells, which were designed to explode as they passed near the target, were dangerous to use over land, since if they missed the target badly they did not explode until they fell to earth.

frequently interfered with one another and that an unnecessarily large proportion of the flying bombs destroyed were brought down over land. It has accordingly been decided to re-deploy our defences in four distinct belts, as follows:

(i) *Fighter Belt at Sea*

Fighter aircraft will operate under close radio control at a distance of not less than 10,000 yards from the shore.

(ii) *Coastal Gun Belt*

All anti-aircraft guns allotted for defence against the flying bomb will be deployed in a narrow strip 5,000 yards in width, extending along the coast from Beachy Head to St. Margaret's Bay. These guns will fire out to sea up to the 10,000-yard limit.

(iii) *Inland Fighter Belt*

Inland, between the coastal gun zone and the balloon barrage, there will be a second fighter belt in which aircraft will operate under running commentary control. The bursts of anti-aircraft fire in the gun belt should be a great help to pilots in spotting the line of flight of approaching bombs. By night they will have the additional assistance of searchlights over the whole of the inland fighter zone.

(iv) *Balloon Belt*

There will be no important changes in the boundaries of the balloon barrage.

The re-deployment of the anti-aircraft guns on to their new sites along the coast was carried out over the week-end, and the new defence plan came into operation at six o'clock this morning.

The new deployment was a vast undertaking, and it was executed with the most praiseworthy speed. Nearly four hundred heavy and six hundred Bofors guns had to be moved and re-sited. Three thousand miles of telephone cable were laid. Twenty-three thousand men and women were moved, and the vehicles of Anti-Aircraft Command travelled two and three-quarter million miles in a week. In four days the move to the coast was completed.

This whole operation was decided upon and carried out on their own responsibility by Air Marshal Hill and General Pile, with the approval of Duncan Sandys. For a few days after the re-deployment our combined defences destroyed far fewer bombs, mainly because the fighters were much hampered by the new restrictions on their movement. But this setback did not last long. The guns soon got their grip, and the results improved rapidly. With the new Radar and predicting equipment, and above all with the new proximity fuzes, all of which we had asked for from America six months before, the performance of the gunners exceeded all our hopes. By the end of August not more than one bomb in seven got through to the London area. The record "bag" was on August 28, when ninety-four bombs approached our coast and all but four were destroyed. The balloons caught two, the fighters twenty-three, and the guns sixty-five. The V1 had been mastered.

The Germans, who keenly watched the performance of our guns from across the Channel, were completely bewildered by the success of our artillery. They had still not solved the mystery when their launching sites were overwhelmed in the first week of September by the victorious and rapid advance of the British and Canadian armies from Normandy to Antwerp. The success of the armies released London and its defenders from the intense strain of the previous three months, and on September 6 Mr. Herbert Morrison, Home Secretary and Minister of Home Security, was able to announce, "The Battle of London is won". Although the Germans thereafter irritated us from time to time with flying bombs launched from aircraft, and with a few long-range bombs from Holland, the threat was thenceforward insignificant. In all about eight thousand bombs were launched against London, and about two thousand four hundred got through.* Our total civilian casualties were 6,184 killed and 17,981 seriously injured. These figures do not tell the whole story. Many people, though wounded, did not have to stay in hospital, and their sufferings have gone unrecorded.

Our Intelligence had played a vital part. The size and performance of the weapon, and the intended scale of attack, were known to us in excellent time. This enabled our fighters to be made ready. The launching sites and the storage caverns were found, enabling our bombers to delay the attack and mitigate its violence. Every known means of getting information was employed, and it was pieced together with great skill. To all our sources, many of whom worked amid deadly danger, and some of whom will be for ever unknown to us, I pay my tribute.

But good Intelligence alone would have been useless. Fighters, bombers, guns, balloons, scientists, Civil Defence, and all the organisation that lay behind them, had each played their parts to

---

* The exact German figure for flying bombs launched against London from sites in France was 8,564, of which 1,006 crashed soon after launching.

the full. It was a great and concerted defence, made absolute by the victory of our armies in France.

*   *   *   *   *

A second threat drew near. This was the long-range rocket, or V2, with which we had been so preoccupied twelve months before. The Germans however had found it difficult to perfect, and in the meantime it had been overtaken by the flying bomb. But almost as soon as the bombs began to hit us the signs appeared that a rocket assault was also approaching. The weight of the rocket and its war-head became subjects of high dispute. Certain early but doubtful Intelligence reports had suggested war-heads of five to ten tons, and these were seized upon by those of our experts who believed on other grounds that such weights were reasonable. Some believed that the rocket would weigh eighty tons, with a ten-ton war-head. Lord Cherwell, now strongly vindicated in his stand for the flying bomb in June 1943,* even before there were any indications of it from Intelligence was very doubtful whether we should ever see the rocket in operation at all, and certainly not the monster of eighty tons. Between the extremes there were a few Intelligence reports which suggested a much lighter rocket than eighty tons; but, with all the controversy, anxiety remained acute.

We knew that work was continuing at Peenemünde, and sparse reports from the Continent renewed our concern about the scale and imminence of the attack. On July 18 Dr. Jones informed the Crossbow Committee that there might well be a thousand rockets already in existence. On July 24 Sandys reported to the Cabinet: "Although we have as yet no reliable information about the movement of projectiles westwards from Germany, it would be unwise to assume from this negative evidence that a rocket is not imminent." In a minute to me the following day the Chiefs of Staff wrote: "The Air Staff agree with this statement, and the Chiefs of Staff consider that the War Cabinet should be warned." The situation was discussed by the Cabinet on July 27, and we considered proposals by Mr. Herbert Morrison which would have involved evacuating about a million people from London.

Every effort was made to complete the remaining gaps in our knowledge about the size, performance, and characteristics of the rocket. Fragmentary evidence from many sources was pieced

* See Vol. V (Cassell's edition), pp 206, 212–13.

together by our Intelligence services and presented to the Crossbow Committee. From this it was deduced that the rocket weighed twelve tons, with a one-ton war-head. This light weight explained many things that had puzzled us, such as the absence of elaborate launching arrangements. These calculations were confirmed when the Royal Aircraft Establishment had the opportunity to examine the wreckage of an actual rocket. It came into our hands as the result of a lucky and freak error in the trials at Peenemünde on June 13, and according to a prisoner the explanation was as follows. For some time the Germans had been using glider bombs against our shipping. These were launched from aircraft and guided to the target by radio. It was now decided to see whether a rocket could be steered in the same way. An expert operator was obtained, and placed in a good position to watch the missile from the start. The Peenemünde experimenters were well accustomed to seeing a rocket rise, and it had not occurred to them that the glider-bomb expert would be surprised by the spectacle. But surprised he was, so much so that he forgot his own part in the procedure. In his astonishment he pushed the control lever well off to the left and held it there. The rocket obediently kept turning to the left, and by the time the operator had pulled himself together it was out of control range and heading for Sweden. There it fell. We soon heard about it, and after some negotiations the remains were brought to Farnborough, where our experts sorted out the battered fragments with noteworthy success.

Before the end of August we knew exactly what to expect. This is shown by the following tables, which compare figures given in a Scientific Intelligence report dated August 26 with those discovered after the war in German records.

—	British Estimate on August 26, 1944	German Figures
Total weight	$11\frac{1}{2}$–14 tons Probably 12–13	12.65 tons
Weight of war-head—i.e., amount of explosive	1 ton	1 ton (sometimes .97 ton)
Range	200–210 miles	207 miles

TOTAL STOCKS AND MONTHLY PRODUCTION

—	British Estimate on August 26, 1944	German Figures
Total stocks	Perhaps 2,000	1,800
Monthly production	About 500	300 in May 1944. Average from Sept. 1944 to March 1945, 618

The rocket was an impressive technical achievement. Its thrust was developed in a jet from the combustion of alcohol and liquid oxygen, nearly four tons of the former and five of the latter being consumed in about a minute. To force these fuels into the jet chamber at the required rate needed a special pump of nearly a thousand horse-power. The pump in its turn was worked by a turbine driven by hydrogen peroxide. The rocket was controlled by gyroscopes or by radio signals operating on large graphite vanes placed behind the jet to deflect the exhaust gases and so steer the rocket. The rocket first rose vertically for six miles or so, and automatic controls then turned it over to climb with increasing speed at about forty-five degrees. When the speed was sufficient for the desired range further controls cut off the fuels from the jet, and the missile then flew in a gigantic parabola, reaching a height of about fifty miles and falling about two hundred miles away from the launching point. Its maximum speed was about four thousand miles an hour, and the whole flight took no more than three or four minutes.

*   *   *   *   *

At the end of August it seemed that our armies might expel the enemy from all territory within the two-hundred-mile range of the rocket from London, but he managed to hold Walcheren and, The Hague. On September 8, a week after the main V1 bombardment ceased, the Germans launched their first two rockets against London.* The first V2 fell at Chiswick at seventeen minutes to seven in the evening, the other at Epping sixteen

* The first long-range rocket to be successfully fired in war had been launched about ten hours earlier, against Paris, but this, as it turned out, was of minor consequence.

seconds later. About thirteen hundred* were fired against England in the seven months before our armies could liberate The Hague, whence most rockets were launched. Many fell short, but about five hundred hit London. The total casualties caused by the V2 in England were 2,724 killed and 6,476 seriously injured. On the average each rocket caused about twice as many casualties as a flying bomb. Although the war-heads were of much the same size, the strident engine of the flying bomb warned people to take cover. The rocket approached in silence.

Many counter-measures were tried, and still more explored. The raid on Peenemünde over a year before did more than everything else to alleviate the threat. The V2 attack would otherwise have started at least as early as the V1 attack, and it would have been from a shorter range, and therefore more accurate in June than it was in September and after. The United States Air Force continued to bomb Peenemünde in July and August, and both they and Bomber Command attacked factories making rocket components. We owe it to our armies that they had pushed the rocket back to the limit of its range before the Germans were at last ready to open fire. Our fighters and tactical bombers continually worried the launching points near The Hague. We made ready to jam the radio control of the rockets, should the Germans use it, and we even considered attempting to burst the rockets in the air by gunfire as they fell.

Our efforts confined the attack to four or five hundred rockets a month, shared between London and the Continent, compared with an intended rate of nine hundred. Thus, although we could do little against the rocket once it was launched, we postponed and substantially reduced the weight of the onslaught. About two hundred rockets a month were aimed against London, most of the rest against Antwerp, and a few against other Continental targets. The enemy made no mention of his new missiles until November 8, and I did not feel the need for a public statement until November 10. I was then able to assure the House that the scale and effects of the attack had not hitherto been serious. This fortunately continued to be true throughout the remaining months of the war.

Despite the great technical achievements, Speer, the highly

---

* The German records show that 1,190 rockets were successfully launched against London, out of 1,359 attempts.

competent German Minister of Munitions, deplored the effort that had been put into making rockets. He asserted that each one took as long to produce as six or seven fighters, which would have been far more useful, and that twenty flying bombs could have been made for the cost of one rocket. This post-war information confirms the views Lord Cherwell had so often expressed before the event.

It was fortunate that the Germans spent so much effort on rockets instead of on bombers. Even our Mosquitoes, each of which was probably no dearer than a rocket, dropped on the average 125 tons of bombs per aircraft within one mile of the target during their life, whereas the rocket dropped one ton only, and that with an average error of fifteen miles.

* * * * *

Hitler had hoped to have yet another "V" weapon. This was to have been a multi-barrel long-range gun installation dug into the ground near the village of Mimoyecques, in the Pas de Calais. Each of the fifty smooth-bore barrels was about four hundred feet long, and it was to fire a shell about six inches in diameter, and stabilised, not by spin, but by fins like a dart. Explosive charges were placed in side-tubes at frequent intervals up the barrel, and were ignited in succession as the projectile accelerated. The shell was intended to emerge from the barrel with a speed of at least five thousand feet per second, and with so many barrels the designers hoped to fire a shell at London every few minutes. This time however Hitler's hopes were completely disappointed: all the trial projectiles "toppled" in flight, and range and accuracy were therefore very poor. A hundred scientists, technicians, and serving officers met in Berlin on May 4, 1944, and came to the unpleasant conclusion that the Fuehrer would have to be told of the failure. We did not know this until afterwards, and as a precaution our bombers repeatedly smashed the concrete structure at Mimoyecques, which five thousand workmen laboured as repeatedly to repair.

* * * * *

While I have recorded the story of Hitler's "Retaliation" campaign against England, we must not forget that Belgium suffered with equal bitterness when the Germans attempted to use the same vindictive weapons against her liberated cities. We did not of course allow the German attack to go unparried. Our bombing of German production centres, and other targets, happily reduced the scale of effort against Belgium as much as against ourselves; but it was not easy to re-deploy fighter and gun defences, with all their elaborate control, in the newly won territories. German records show that by the end of the war Antwerp had been the target for 8,696 flying bombs and 1,610 rockets. 5,960 of all these fell within eight miles of the city centre, and between them they killed 3,470 Belgian citizens and 682 Allied Service-men. A further 3,141 flying bombs were aimed against Liége, and 151 rockets against Brussels. The people of Belgium bore this senseless bombardment in a spirit equal to our own.

*   *   *   *   *

The German "V" weapons, though in the event unsuccessful, impressed us with the potentialities of these new methods. In a report to the Cabinet Duncan Sandys emphasised the decisive importance which guided missiles might have in future wars, and pointed out the need for devoting substantial resources to their development. The following extract may be deemed significant:

The advent of the long-range, radio-controlled, jet-propelled projectile has opened up vast new possibilities in the conduct of military operations. In future the possession of superiority in long-distance rocket artillery may well count for as much as superiority in naval or air power. High-grade scientific and engineering staff, together with extensive research facilities, will have to be maintained as a permanent part of our peace-time military organisation.

We began to design our own guided missiles, and by the end of the war we had founded a permanent organisation for this purpose.

*   *   *   *   *

Such is the tale of the new weapons on which Hitler pinned his stubborn hopes for many months, and of their defeat by the foresight of the British Administration, the skill of the Services, and the fortitude of the people who, by their conduct for the second time in this war, gave "Greater London" a prouder meaning.

CHAPTER IV

ATTACK ON THE SOUTH OF FRANCE?

The Strategic Decisions of the Teheran Conference – The Plan to Land in the South of France – Delay in the Capture of Rome – General Marshall's Visit to England and the Mediterranean – "Overlord's" Need for More Ports in the South or West – A Telegram from Smuts, June 23 – Opposing Views of the British and American Chiefs of Staff – Correspondence with President Roosevelt – General Wilson is Ordered to Attack the French Riviera – My Plan for a Landing on the Atlantic Coast – A Visit to Eisenhower and a Conference at Portsmouth, August 7 – Mr. Roosevelt's Adverse Telegram.

LIBERATING Normandy was a supreme event in the European campaign of 1944, but it was only one of several concentric strokes upon Nazi Germany. In the east the Russians were flooding into Poland and the Balkans, and in the south Alexander's armies in Italy were pressing towards the river Po. Decisions had now to be taken about our next move in the Mediterranean, and it must be recorded with regret that these occasioned the first important divergence on high strategy between ourselves and our American friends.

The design for final victory in Europe had been outlined in prolonged discussion at the Teheran Conference in November 1943. Its decisions still governed our plans, and it would be well to recall them. First and foremost we had promised to carry out "Overlord". Here was the dominating task, and no one disputed that here lay our prime duty. But we still wielded powerful forces in the Mediterranean, and the question had remained, what should they do? We had resolved that they should capture Rome, whose near-by airfields were needed for bombing Southern Germany. This accomplished, we planned to advance up the peninsula as far as the Pisa–Rimini line, and there hold as many enemy divisions as possible in Northern Italy. This however was not all. A third operation was also agreed upon, namely, an amphibious landing in the south of France, and it was on this project that controversy was about to descend. It was originally conceived as a feint or threat to keep German troops on the Riviera and stop them joining the battles in Normandy, but in Cairo the Americans had pressed for a real attack by ten divisions, and at Teheran Stalin had supported them. I accepted the change, largely to prevent undue diversions to Burma, although I contemplated other ways of exploiting success in Italy, and the plan had been given the code-name "Anvil".

One thing was plain: it was no use landing in the south of France unless we did so at the right time. The mere threat of an assault would suffice to keep German troops in the region; a real invasion might induce the enemy to reinforce them; but once we joined battle in Normandy "Anvil's" value was much reduced, because Hitler was not likely to detach troops from the main struggle in the north for the sake of keeping his hold on Provence. If we invaded the Riviera at all we must do so at the same time as or just before we landed in Normandy, and this was what we intended to do when we made our plans at Teheran.

A second condition also governed "Anvil's" usefulness. Many of the forces needed for the operation—that is to say, for the full-scale invasion as opposed to a feint or threat—would have to come from our armies in Italy. But these had first to accomplish the arduous and important task of seizing Rome and the airfields. Until this was done little could be spared or taken from Alexander's forces. Rome must fall before "Anvil" could start.

All turned on the capture of Rome. If we could seize it quickly all would be well. Troops could then be withdrawn from the Italian front and "Anvil" launched in good time. If not, a feint landing would suffice. If we landed in earnest, but after "Overlord" had started, our forces would have a long way to go before they could reach Eisenhower's armies, and by then the battle of the beaches would be over. They would be too late to help. This in fact was what happened, and indeed already seemed likely to happen early in 1944.

At Teheran we had confidently expected to reach Rome early in the spring, but this had proved impossible. The important descent at Anzio to accelerate the capture of Rome had drawn eight or ten German divisions away from the vital theatre, or more than was expected to be attracted to the Riviera by "Anvil". This in effect superseded it by achieving its object. Nevertheless the Riviera project went forward as if nothing had happened.

Apart from "Anvil" hanging somewhat vaguely in the future, some of the finest divisions of the armies in Italy had rightly been assigned to the main operation of "Overlord" and had sailed for England before the end of 1943. Alexander had thus been weakened and Kesselring had been strengthened. The Germans had sent reinforcements to Italy, had parried the Anzio swoop, and had stopped us entering Rome until just before D Day. The hard fighting had of course engulfed important enemy reserves which might otherwise have gone to France, and it certainly helped "Overlord" in its critical early stages, but none the less our advance in the Mediterranean had been gravely upset. Landing-craft were another obstacle. Many of them had been sent to "Overlord". "Anvil" could not be mounted until they came back, and this depended on events in Normandy. These facts had been long foreseen, and as far back as March 21 General Maitland Wilson, the Supreme Commander in the Mediterranean, reported that "Anvil" could not be launched before the end of July. Later he put it at mid-August, and declared that the best way to help "Overlord" was to abandon any attack on the Riviera and concentrate on Italy.

When Rome fell on June 4 the problem had to be reviewed. Should we go on with "Anvil" or should we make a new plan?

General Eisenhower naturally wanted to strengthen his attack in North-West Europe by all available means. Strategic possibilities in Northern Italy did not attract him, but he consented to return the landing-craft as soon as possible if this would lead to a speedy "Anvil". The American Chiefs of Staff agreed with Eisenhower, holding rigidly to the maxim of concentration at the decisive point, which in their eyes meant only North-West Europe. They were supported by the President, who was mindful of the agreements made with Stalin many months before at Teheran. Yet all was changed by the delay in Italy.

*   *   *   *   *

Soon after D Day General Marshall came to England and expressed his concern about yet another problem. Enormous forces were accumulating in the United States, and should join the battle as soon as possible. This they could do either by sailing direct to France or by coming through the United Kingdom, and arrangements had been made accordingly, but so great was the promised influx that Marshall doubted whether our ports would suffice. At this period we held only a few harbours along the French coast of the Channel, and although Eisenhower intended to capture Brest, and other landing-places in the Bay of Biscay might also fall to us if things went well, we could not be sure of seizing them, and still less of clearing them, in sufficient time. Yet a full and speedy build-up was vital to the success of "Overlord". The solution which General Marshall proposed was to capture entirely new bases in either the west or the south of France, and preferably in the west because this was the more quickly reached from America.

I was fully alive to all this, and had for some time contemplated a descent on the Biscay coast from North Africa, even though this could not be achieved before the end of July or early in August. But I was equally anxious not to wreck Alexander's victory in Italy. I considered that the options might still remain open and all preparations should be made to move in whatever direction seemed best.

On June 14 the Combined Chiefs of Staff decided to prepare an amphibious operation in the Mediterranean which might strike either in the south of France or in the Bay of Biscay or at the head of the Adriatic. Its destination could be left open for the moment. Three days later General Marshall visited the Mediterranean to confer with the commanders. General Wilson was impressed with "Overlord's" need for more ports, of which he then learnt for the first time, but he did not alter his judgment against "Anvil", and on June 19 told the Combined Chiefs of Staff that he still thought his best contribution to the common end would be to press forward with all his resources into the Po valley. Thereafter, with the help of an amphibious operation against the Istrian peninsula, at the head of the Adriatic, which is dominated by and runs south from Trieste, there would be attractive prospects of advancing through the Ljubljana Gap into Austria and Hungary and striking at the heart of Germany from another direction. Alexander agreed.

Smuts was in Italy at the time and telegraphed to me:

Field-Marshal Smuts to Prime Minister 23 June 44

I have discussed with Wilson and Alexander the future employment of latter's forces, and summarise results for your information. Neither of them favours any of present "Anvil" proposals, as their results will be at least doubtful in directly helping Eisenhower, and in any case would involve very serious loss of time when time is so important to us. The success already achieved by Alexander and present high morale of his army tells strongly against any break-up of his forces and interruption of their victorious advance. With the reinforcements forthcoming for Eisenhower he should be able not only to hold his own and extend his right flank to the Loire, but also to advance eastwards towards or beyond Paris. The extension of his left flank is a matter for Staff consideration and report, but this ought not to delay a decision on present question of the switch-over, which is both most urgent and critical.

As regards plan for Alexander's advance, he and Wilson agree that there will be no difficulty in his break-through to the Po and thereafter swinging east towards Istria, Ljubljana, and so on to Austria. Alexander favours an advance both by land and sea, while Wilson favours the latter and thinks three seaborne divisions with one or two airborne divisions would suffice and make possible capture of Trieste by beginning of September. Thereafter the advance will reopen eastward, gathering large Partisan support and perhaps forcing the enemy out of the Balkans. The co-operation between our and the Russian advance towards Austria and Germany would constitute as serious a threat to the enemy as Eisenhower's advance from the west, and the three combined are most likely to produce early enemy collapse.

Alexander, who has just held a meeting with his commanders, is sending C.I.G.S. separately his views. I would only add that considered views of two such competent and experienced leaders as Wilson and Alexander weigh heavily with me, and should not lightly be set aside by Combined Chiefs of Staff, whose planning in any case does not exclude alternative now pressed by both of them. Both have impressed on me urgency of a decision on many grounds, if possible before end of next week.

*   *   *   *   *

On June 23 General Eisenhower advised the Combined Chiefs of Staff to concentrate our forces in direct support of the decisive battle in Northern France. He admitted that an advance through the Ljubljana Gap might contain German troops, but it would not draw any of their divisions from France. As for a descent in the Bay of Biscay, he agreed that Bordeaux was closer to the United

States than Marseilles, but maintained that the latter could be captured more quickly by forces already in the Mediterranean, and would furnish a direct route northwards to join in the battle for the Ruhr. He therefore urged that "Anvil" should be undertaken, at the expense of course of our armies in Italy, since "in my view the resources of Great Britain and the United States will not permit us to maintain two major theatres in the European war, each with decisive missions".

We were all agreed that "Overlord" took priority; the point was how the armies in the second theatre, Italy, could best help to overthrow Germany. The American Chiefs of Staff strongly supported Eisenhower. They condemned what they called the "commitment of Mediterranean resources to large-scale operations in Northern Italy and into the Balkans". Our own Chiefs of Staff took the opposite view. On June 26 they declared that the Allied forces in the Mediterranean could best help "Overlord" by destroying the Germans who faced them. In order to launch "Anvil" on August 15 withdrawals from the Italian front would have to begin at once, and rather than land on the Riviera they preferred to send troops by sea direct to Eisenhower. With much prescience they remarked: "We think that the mounting of 'Anvil' on a scale likely to achieve success would hamstring General Alexander's remaining forces to such an extent that any further activity would be limited to something very modest."

They urged that Alexander should develop his offensive in Italy so as to engage and destroy all the German forces opposed to him, that General Wilson should do all he could to emphasise *the threat of an assault** on the south of France, and that Wilson should prepare to send Eisenhower one or more American divisions and/or all the French divisions which he was capable of receiving and which our shipping resources would permit.

This direct conflict of opinions, honestly held and warmly argued by either side, could only be settled, if at all, between the President and myself, and an interchange of telegrams now took place.†

"The deadlock," I said on June 28, "between our Chiefs of Staff raises most serious issues. Our first wish is to help General Eisenhower in the most speedy and effective manner. But we do

* Author's italics.
† The complete texts of the more important documents can be studied in Appendix D.

not think this necessarily involves the complete ruin of all our great affairs in the Mediterranean, and we take it hard that this should be demanded of us. . . . I most earnestly beg you to examine this matter in detail for yourself. . . . Please remember how you spoke to me at Teheran about Istria, and how I introduced it at the full Conference. This has sunk very deeply into my mind, although it is not by any means the immediate issue we have to decide."

Later I summed up my conclusions to Mr. Roosevelt.

(a) Let us reinforce "Overlord" directly, to the utmost limits of landings from the west.

(b) Let us next do justice to the great opportunities of the Mediterranean commanders, and confine ourselves to minor diversions and threats to hold the enemy around the Gulf of Lions.

(c) Let us leave General Eisenhower all his landing-craft as long as he needs them to magnify his landing capacity.

(d) Let us make sure of increasing to the maximum extent the port capacity in the "Overlord" battle area.

(e) Let us resolve not to wreck one great campaign for the sake of another. Both can be won.

The President's reply was prompt and adverse. He was resolved to carry out what he called "the grand strategy" of Teheran, namely, exploiting "Overlord" to the full, "victorious advances in Italy, and an early assault on Southern France". Political objects might be important, but military operations to achieve them must be subordinated to striking at the heart of Germany by a campaign in Europe. Stalin himself had favoured "Anvil", and had classified all other operations in the Mediterranean as of lesser importance, and Mr. Roosevelt declared he could not abandon it without consulting him. The President continued:

My interest and hopes centre on defeating the Germans in front of Eisenhower and driving on into Germany, *rather than on limiting this action for the purpose of staging a full major effort in Italy.** I am convinced we will have sufficient forces in Italy, with "Anvil" forces withdrawn, to chase Kesselring north of Pisa–Rimini and maintain heavy pressure against his army at the very least to the extent necessary to contain his present force. I cannot conceive of the Germans paying the price of ten additional divisions, estimated by General Wilson, in order to keep us out of Northern Italy.

* Author's subsequent italics throughout.

We can—and Wilson confirms this—immediately withdraw five divisions (three United States and two French) from Italy for "Anvil". *The remaining twenty-one divisions, plus numerous separate brigades, will certainly provide Alexander with adequate ground superiority. . . .*

Mr. Roosevelt contended that a landing in the Bay of Biscay would be a waste of shipping. If Eisenhower wanted more troops they were ready in the United States and he had only to ask for them. But it was his objections to a descent on the Istrian peninsula and a thrust against Vienna through the Ljubljana Gap that revealed both the rigidity of the American military plans and his own suspicion of what he called a campaign "in the Balkans". He claimed that Alexander and Smuts, "for several natural and very human reasons", were inclined to disregard two vital considerations. First, the operation infringed "the grand strategy". Secondly, it would take too long and we could probably not deploy more than six divisions. "I cannot agree," he wrote, "to the employment of United States troops against Istria *and into the Balkans*, nor can I see the French agreeing to such use of French troops. . . . For purely political considerations over here, I should never survive even a slight setback in 'Overlord' *if it were known that fairly large forces had been diverted to the Balkans.*"*

No one involved in these discussions had ever thought of moving armies *into the Balkans*; but Istria and Trieste were strategic and political positions, which, as he saw very clearly, might exercise profound and widespread reactions, especially after the Russian advances.

The President suggested at one point that we should lay our respective cases before Stalin. I said I did not know what he would say if the issue were put to him to decide. On military grounds he might have been greatly interested in the eastward movement of Alexander's army, which, without entering the Balkans, would profoundly affect all the forces there, and which, in conjunction with any attacks Stalin might make upon Roumania or with Roumania against Transylvania, might produce the most far-reaching results. On a long-term political view he might prefer that the British and Americans should do their share in France in the very hard fighting that was to come, and that East, Middle, and Southern Europe should fall naturally into his control. But I felt it was better to settle the matter for ourselves and between

* The full text of the President's telegram will be found in Appendix D.

ourselves. I was sure that if we could have met, as I so frequently proposed, we should have reached a happy agreement.

On July 2 the President declared that he and his Chiefs of Staff were still convinced that "Anvil" should be launched at the earliest possible date, and he asked us to direct General Wilson accordingly. He said that at Teheran he had only contemplated a series of raids in force in Istria if the Germans started a general retirement from the Dodecanese and Greece. But this had not happened yet.

"Therefore," he concluded, "I am compelled by the logic of not dispersing our main efforts to a new theatre to agree with my Chiefs of Staff.

"I honestly believe that God will be with us as He has in 'Overlord' and in Italy and in North Africa. I always think of my early geometry—'a straight line is the shortest distance between two points'."

For the time being I resigned myself, and the same day General Wilson was ordered to attack the south of France on August 15. Preparations began at once, but the reader should note that "Anvil" was renamed "Dragoon". This was done in case the enemy had learnt the meaning of the original code-word.

* * * * *

By early August however a marked change had come over the battlefield in Normandy and great developments impended. On the 4th I reopened with the President the question of switching "Dragoon" to the west.

Prime Minister to President Roosevelt   4 Aug 44
The course of events in Normandy and Brittany, and especially the brilliant operations of the United States Army, give good prospects that the whole Brittany peninsula will be in our hands within a reasonable time. I beg you will consider the possibility of switching "Dragoon" into the main and vital theatre, where it can immediately play its part at close quarters in the great and victorious battle in which we are now engaged.

2. I cannot pretend to have worked out the details, but the opinion here is that they are capable of solution. Instead of having to force a landing against strong enemy defences we might easily find welcoming American troops at some point or other from St. Nazaire north-westward along the Brittany peninsula. I feel that we are fully entitled to use the extraordinary flexibility of sea- and air-power to move with the

moving scene. The arrival of the ten divisions assigned to "Dragoon", with their L.S.T.s, might be achieved rapidly, and if this came off it would be decisive for Eisenhower's victorious advance by the shortest route right across France.

3. I most earnestly ask you to instruct your Chiefs of Staff to study this proposal, on which our people here are already at work.

I also hoped that Hopkins might be able to help.

Prime Minister to Mr. Harry Hopkins   6 Aug 44

I am grieved to find that even splendid victories and widening opportunities do not bring us together on strategy. The brilliant operations of the American Army have not only cut off the Brest peninsula, but in my opinion have to a large extent demoralised the scattered Germans who remain there. St. Nazaire and Nantes, one of your major disembarkation ports in the last war, may be in our hands at any time. Quiberon Bay, Lorient, and Brest will also soon fall into our hands. It is my belief that the German troops on the Atlantic shore south of the Cherbourg peninsula are in a state of weakness and disorder and that Bordeaux could be obtained easily, cheaply, and swiftly. The possession of these Atlantic ports, together with those we have now, will open the way for the fullest importation of the great armies of the United States still awaiting their opportunity. In addition the ten divisions now mounted for "Dragoon" could be switched into St. Nazaire as soon as it is in Allied possession, in this case American possession. Thus Eisenhower might speedily be presented with a new great port, as well as with a new army to operate on his right flank in the march towards the Seine.

2. I repeat that the above is additional to anything that has been foreshadowed in the schedules of transportation either from Great Britain or the United States. Instead of this we are to be forced to make a heavy attack from the sea on the well-fortified Riviera coast and to march westward to capture the two fortresses of Toulon and Marseilles, thus opening a new theatre where the enemy will at the outset be much stronger than we are, and where our advance runs cross-grained to the country, which abounds in most formidable rocky positions, ridges, and gullies.

3. Even after taking the two fortresses of Toulon and Marseilles we have before us the lengthy advance up the Rhone valley before we even get to Lyons. None of this operation can influence Eisenhower's battle for probably ninety days after the landings.* We start 500 miles away from the main battlefield instead of almost upon it at St. Nazaire. There is no correlation possible between our armies in the Brest and

* The first major operations in which the "Dragoon" armies took part after their junction with General Eisenhower's forces were in mid-November.

Cherbourg peninsulas and the troops operating against Toulon and Marseilles. When Marseilles is gained the turn-round from the United States is about fourteen days longer than the straight run across the Atlantic.

4. Of course we are going to win anyway, but these are very hard facts. When "Anvil" was raised at Teheran it was to be a diversionary or holding operation a week before or a week later than "Overlord" D Day, in the hope of drawing about eight German divisions away from the main battle. The decision to undertake Anzio and the delays at Cassino forced us to continue putting off "Anvil", until its successor "Dragoon" bears no relation to the original conception. However, out of evil came good, and the operations in Italy being persevered in drew not fewer than twelve divisions from the German reserves in North Italy and elsewhere, and they have been largely destroyed. The coincidence that the defeat of Kesselring's army and the capture of Rome occurred at the exact time of launching "Overlord" more than achieved all that was ever foreseen from "Anvil", and, to those who do not know the inner history, wears the aspect of a great design. Thus I contend that what "Anvil" was meant for is already gained.

5. Bowing to the United States Chiefs of Staff under recorded protest and the overriding of our views, we have done everything in human power, including the provision of nearly one-half the naval forces about to be engaged. If nothing can be done to save the situation I earnestly pray the American view may be right. But now an entirely new situation has developed through the victories that have been won in France and the greater victories that seem possible. It is in these circumstances that I have thought it right, on the recommendation of the British Chiefs of Staff, to reopen the question. There are still three or four days in which the decision to send to St. Nazaire the forces now destined and largely loaded for "Dragoon" could be reconsidered. I admit the arguments against late changes in plans, but they ought to be fairly weighed against what seems to us to be the overwhelming case for strengthening the main battle, and thus possibly finishing up Hitler this year.

6. You know the great respect and regard which I have for Marshall, and if you feel able to embroil yourself in these matters I should be glad if you would bring my views before him, especially the later paragraphs, which are my reply to any complaint he may have made that I supported "Anvil" at Teheran and have turned against it since.

7. Let me know also whether my last speech was satisfactory from the American military standpoint and whether there were any points which you would rather I had stated differently. I set the good relations of our Armies above everything else.

Kindest regards.

The reply was far from comforting

Mr. Harry Hopkins to Prime Minister   7 Aug 44
Your wire received. While there has been no reply as yet from the President to your message relative to the same matter, I am sure his answer will be in the negative. While I have seen no analysis of logistics involved, I am absolutely certain you will find the supply problem insurmountable. Divisions are already available for Eisenhower's immediate build-up which will tax the ports to the limit. Then, too, no one knows the condition of the Brittany ports. It seems to me that our tactical position to-day in "Overlord" is precisely as planned and as we anticipated it would be when "Anvil" was laid on. To change the strategy now would be a great mistake, and I believe would delay rather than aid in our sure conquest of France. I believe too the movement north from "Anvil" will be much more rapid than you anticipate. They have nothing to stop us. The French will rise and abyssiniate large numbers of Germans, including, I trust, Monsieur Laval. A tremendous victory is in store for us.

*   *   *   *   *

That day I visited Eisenhower at his headquarters near Portsmouth and unfolded to him my last hope of stopping the "Dragoon" operation. After an agreeable luncheon we had a long and serious conversation. Eisenhower had with him Bedell Smith and Admiral Ramsay. I had brought the First Sea Lord, as the movement of shipping was the key. Briefly, what I proposed was to continue loading the "Dragoon" expedition, but when the troops were in the ships to send them through the Straits of Gibraltar and enter France at Bordeaux. The matter had been long considered by the British Chiefs of Staff, and the operation was considered feasible. I showed Eisenhower the telegram I had sent to the President, whose reply I had not yet received, and did my best to convince him. The First Sea Lord strongly supported me. Admiral Ramsay argued against any change of plan. Bedell Smith, on the contrary, declared himself strongly in favour of this sudden deflection of the attack, which would have all the surprise that sea-power can bestow. Eisenhower in no way resented the views of his Chief of Staff. He always encouraged free expression of opinion in council at the summit, though of course whatever was settled would receive every loyalty in execution.

However, I was quite unable to move him, and next day I received the President's reply.

President Roosevelt to Prime Minister   8 Aug 44

I have consulted by telegraph with my Chiefs of Staff, and am unable to agree that the resources allocated to "Dragoon" should be considered available for a move into France via ports on the coast of Brittany.

On the contrary, it is my considered opinion that "Dragoon" should be launched as planned at the earliest practicable date, and I have full confidence that it will be successful and of great assistance to Eisenhower in driving the Huns from France.

There was no more to be done about it. It is worth noting that we had now passed the day in July when for the first time in the war the movement of the great American armies into Europe and their growth in the Far East made their numbers in action greater than our own. Influence on Allied operations is usually increased by large reinforcements. It must also be remembered that had the British views on this strategic issue been accepted the tactical preparations might well have caused some delay, which again would have reacted on the general argument.

Prime Minister to President Roosevelt   8 Aug 44

I pray God that you may be right. We shall of course do everything in our power to help you achieve success.

CHAPTER V

BALKAN CONVULSIONS
THE RUSSIAN VICTORIES

The Need for Political Agreement with Russia in Central and Eastern Europe – Mr. Eden's Suggestion about Greece and Roumania, May 18 – My Telegram to the President, May 31 – Nervousness in the State Department – Mr. Roosevelt's Telegram of June 11, and My Reply – My Message to the President of June 23 – An Argument Between Friends – I Telegraph Stalin about Turkey, July 11 – His Non-committal Answer – The Russian Summer Campaign – The Finns Sue for an Armistice, August 25 – Advance to the Niemen – Twenty-five German Divisions Cease to Exist – The Red Army Crosses the Vistula – Revolution in Roumania.

T HE advance of the Soviet armies into Central and Eastern Europe in the summer of 1944 made it urgent to come to a political arrangement with the Russians about those regions. Post-war Europe seemed to be taking shape. Difficulties in Italy had already begun owing to Russian intrigues. We were striving to reach a balanced result in Yugoslav affairs by direct negotiation with Tito. But no progress had as yet been made with Moscow about Poland, Hungary, Roumania, and Bulgaria. The whole subject had been surveyed at the meeting of the Imperial Conference in London in May, and I had then minuted to the Foreign Secretary:

4 May 44

A paper should be drafted for the Cabinet, and possibly for the Imperial Conference, setting forth shortly—for that is essential—the brute issues between us and the Soviet Government which are developing in Italy, in Roumania, in Bulgaria, in Yugoslavia, and above all in Greece. It ought to be possible to get this on one page.

2. I cannot say there is much in Italy, but broadly speaking the issue

is: are we going to acquiesce in the Communisation of the Balkans and perhaps of Italy? Mr. Curtin touched upon this this morning, and I am of opinion on the whole that we ought to come to a definite conclusion about it, and that if our conclusion is that we resist the Communist infusion and invasion we should put it to them pretty plainly at the best moment that military events permit. We should of course have to consult the United States first.

And again on the same day:

Evidently we are approaching a showdown with the Russians about their Communist intrigues in Italy, Yugoslavia, and Greece. I think their attitude becomes more difficult every day.

On May 18 the Soviet Ambassador in London called at the Foreign Office to discuss a general suggestion which Mr. Eden had made that the U.S.S.R. should temporarily regard Roumanian affairs as mainly their concern under war conditions while leaving Greece to us. The Russians were prepared to accept this, but wished to know if we had consulted the United States. If so they would agree. I minuted on the record of this conversation: "I should like to telegraph to the President about this. He would like the idea, especially as we should keep in close touch with him."

On May 31 I accordingly sent a personal telegram to Mr. Roosevelt.

Prime Minister to President Roosevelt   31 May 44
There have recently been disquieting signs of a possible divergence of policy between ourselves and the Russians in regard to the Balkan countries, and in particular towards Greece. We therefore suggested to the Soviet Ambassador here that we should agree between ourselves as a practical matter that the Soviet Government would take the lead in Roumanian affairs, while we would take the lead in Greek affairs, each Government giving the other help in the respective countries. Such an arrangement would be a natural development of the existing military situation, since Roumania falls within the sphere of the Russian armies and Greece within the Allied command under General Wilson in the Mediterranean.

2. The Soviet Ambassador here told Eden on May 18 that the Soviet Government agreed with this suggestion, but before giving any final assurance in the matter they would like to know whether we had consulted the United States Government and whether the latter had also agreed to this arrangement.

3. I hope you may feel able to give this proposal your blessing. We do not of course wish to carve up the Balkans into spheres of influence, and in agreeing to the arrangement we should make it clear that it applied only to war conditions and did not affect the rights and responsibilities which each of the three Great Powers will have to exercise at the peace settlement and afterwards in regard to the whole of Europe. The arrangement would of course involve no change in the present collaboration between you and us in the formulation and execution of Allied policy towards these countries. We feel however that the arrangement now proposed would be a useful device for preventing any divergence of policy between ourselves and them in the Balkans.

4. Meanwhile Halifax has been asked to raise this matter with the State Department on the above lines.

The first reactions of the State Department were cool. Mr. Hull was nervous of any suggestion that "might appear to savour of the creation or acceptance of the idea of spheres of influence".

On June 8 I sent the following message to Lord Halifax in Washington:

Prime Minister to Lord Halifax (Washington)   8 June 44

There is no question of spheres of influence. We all have to act together, but someone must be playing the hand. It seems reasonable that the Russians should deal with the Roumanians and Bulgarians, upon whom their armies are impinging, and that we should deal with the Greeks, who are in our assigned theatre, who are our old allies, and for whom we sacrificed 40,000 men in 1941. I have reason to believe that the President is in entire agreement with the line I am taking about Greece. The same is true of Yugoslavia. I keep him constantly informed, but on the whole we, His Majesty's Government, are playing the hand, and have to be very careful to play it agreeably with the Russians. No fate could be worse for any country than to be subjected in these times to decisions reached by triangular or quadrangular telegraphing. By the time you have got one thing settled three others have gone astray. Moreover, events move very rapidly in these countries.

2. On the other hand, we follow the lead of the United States in South America as far as possible, as long as it is not a question of our beef and mutton. On this we naturally develop strong views on account of the little we get.

On June 11 Mr. Roosevelt cabled:

... Briefly, we acknowledge that the military responsible Government in any given territory will inevitably make decisions required by military developments, but are convinced that the natural tendency for such decisions to extend to other than military fields would be strengthened by an agreement of the type suggested. In our opinion, this would certainly result in the persistence of differences between you and the Soviets and in the division of the Balkan region into spheres of influence despite the declared intention to limit the arrangement to military matters.

We believe efforts should preferably be made to establish consultative machinery to dispel misunderstandings and restrain the tendency toward the development of exclusive spheres.

Prime Minister to President Roosevelt   11 June 44

I am much concerned to receive your message. Action is paralysed if everybody is to consult everybody else about everything before it is taken. Events will always outstrip the changing situations in these Balkan regions. Somebody must have the power to plan and act. A Consultative Committee would be a mere obstruction, always over-ridden in any case of emergency by direct interchanges between you and me, or either of us and Stalin.

2. See, now, what happened at Easter. We were able to cope with this mutiny of the Greek forces entirely in accordance with your own views. This was because I was able to give constant orders to the military commanders, who at the beginning advocated conciliation, and above all no use or even threat of force. Very little life was lost. The Greek situation has been immensely improved, and, if firmness is maintained, will be rescued from confusion and disaster. The Russians are ready to let us take the lead in the Greek business, which means the E.A.M. and all its malice can be controlled by the national forces of Greece. Otherwise civil war and ruin to the land you care about so much. I always reported to you, and I always will report to you. You shall see every telegram I send. I think you might trust me in this.

3. If in these difficulties we had had to consult other Powers and a set of triangular or quadrangular telegrams got started the only result would have been chaos or impotence.

4. It seems to me, considering the Russians are about to invade Roumania in great force and are going to help Roumania recapture part of Transylvania from Hungary, provided the Roumanians play, which they may, considering all that, it would be a good thing to follow the Soviet leadership, considering that neither you nor we have any troops there at all and that they will probably do what they like anyhow. Moreover, I thought their terms, apart from indemnity, very sensible, and even generous. The Roumanian Army has inflicted many injuries upon the Soviet troops, and went into the war against Russia with glee. I see no difficulty whatever in our addressing the Russians

at any time on any subject, but please let them go ahead upon the lines agreed as they are doing all the work.

5. Similarly with us in Greece. We are an old ally of Greece. We had 40,000 casualties in trying to defend Greece against Hitler, not counting Crete. The Greek King and the Greek Government have placed themselves under our protection. They are at present domiciled in Egypt. They may very likely move to the Lebanon, which would be a better atmosphere than Cairo. Not only did we lose the 40,000 men above mentioned in helping Greece, but a vast mass of shipping and warships, and by denuding Cyrenaica to help Greece we also lost the whole of Wavell's conquests in Cyrenaica. These were heavy blows to us in those days. Your telegrams to me in the recent crisis worked wonders. We were entirely agreed, and the result is entirely satisfactory. Why is all this effective direction to be broken up into a committee of mediocre officials such as we are littering about the world? Why can you and I not keep this in our own hands, considering how we see eye to eye about so much of it?

6. To sum up, I propose that we agree that the arrangements I set forth in my message of May 31 may have a trial of three months, after which it must be reviewed by the three Powers.

On June 13 the President agreed to this proposal, but added: "We must be careful to make it clear that we are not establishing any post-war spheres of influence."

I shared his view, and replied the next day:

I am deeply grateful to you for your telegram. I have asked the Foreign Secretary to convey the information to Molotov and to make it clear that the reason for the three months' limit is in order that we should not prejudge the question of establishing post-war spheres of influence.

I reported the situation to the War Cabinet that afternoon, and it was agreed that, subject to the time-limit of three months, the Foreign Secretary should inform the Soviet Government that we accepted this general division of responsibility. This was done on June 19. The President however was not happy about the way we had acted, and I received a pained message saying "we were disturbed that your people took this matter up with us only after it had been put up to the Russians." On June 23 accordingly I outlined to the President, in reply to his rebuke, the situation as I saw it from London.

Prime Minister to President Roosevelt   23 June 44

The Russians are the only Power that can do anything in Roumania, and I thought it was agreed between you and me that on the basis of their reasonable armistice terms, excepting indemnities, they should try to give coherent direction to what happened there. In point of fact we have all three co-operated closely in handling in Cairo the recent Roumanian peace-feelers. On the other hand, the Greek burden rests almost entirely upon us, and has done so since we lost 40,000 men in a vain endeavour to help them in 1941. Similarly, you have let us play the hand in Turkey, but we have always consulted you on policy, and I think we have been agreed on the line to be followed. It would be quite easy for me, on the general principle of slithering to the Left, which is so popular in foreign policy, to let things rip, when the King of Greece would probably be forced to abdicate and E.A.M. would work a reign of terror in Greece, forcing the villagers and many other classes to form Security Battalions under German auspices to prevent utter anarchy. The only way I can prevent this is by persuading the Russians to quit boosting E.A.M. and ramming it forward with all their force. Therefore I proposed to the Russians a temporary working arrangement for the better conduct of the war. This was only a proposal, and had to be referred to you for your agreement.

2. I cannot admit that I have done anything wrong in this matter. It would not be possible for three people in different parts of the world to work together effectively if no one of them may make any suggestion to either of the others without simultaneously keeping the third informed. A recent example of this is the message you have sent quite properly to Uncle Joe about your conversations with the Poles, of which as yet I have heard nothing from you. I am not complaining at all of this, because I know we are working for the general theme and purposes, and I hope you will feel that has been so in my conduct of the Greek affair.

3. I have also taken action to try to bring together a union of the Tito forces with those in Serbia, and with all adhering to the Royal Yugoslav Government, which we have both recognised. You have been informed at every stage of how we are bearing this heavy burden, which at present rests mainly on us. Here again nothing would be easier than to throw the King and the Royal Yugoslav Government to the wolves and let a civil war break out in Yugoslavia, to the joy of the Germans. I am struggling to bring order out of chaos in both cases and concentrate all efforts against the common foe. I am keeping you constantly informed, and I hope to have your confidence and help within the spheres of action in which initiative is assigned to us.

The President's reply of June 27 settled this argument between friends. "It appears," he said, "that both of us have inadvertently taken unilateral action in a direction that we both now agree to have been expedient for the time being. It is essential that we should always be in agreement in matters bearing on our Allied war effort."

I replied the same day: "You may be sure I shall always be looking to our agreement in all matters before, during, and after."

The difficulties however continued on a Governmental level. The Russians insisted on consulting the Americans direct.

* * * * *

Another issue also claimed our attention. The Russian armies were now poised on the borders of Roumania. Here was Turkey's last chance to enter the war on the Allied side, and her entry at this stage would have a potent influence on the future of South-Eastern Europe. She now offered to go as far as breaking off relations with the Axis.

I gave Stalin my views on these events.

Prime Minister to Marshal Stalin   11 July 44

Some weeks ago it was suggested by Eden to your Ambassador that the Soviet Government should take the lead in Roumania, and the British should do the same in Greece. This was only a working arrangement to avoid as much as possible the awful business of triangular telegrams, which paralyses action. Molotov then suggested very properly that I should tell the United States, which I did, and always meant to, and after some discussion the President agreed to a three months' trial being made. These may be three very important months, Marshal Stalin, July, August, and September. Now however I see that you find some difficulty in this. I would ask whether you should not tell us that the plan may be allowed to have its chance for three months. No one can say it affects the future of Europe or divides it into spheres; but we can get a clear-headed policy in each theatre, and we will all report to the others what we are doing. However, if you tell me it is hopeless I shall not take it amiss.

2. There is another matter I should like to put to you. Turkey is willing to break relations immediately with the Axis Powers. I agree with you that she ought to declare war, but I fear that if we tell her to do so she will defend herself by asking both for aircraft to protect her towns, which we shall find it hard to spare or put there at the present moment, and also for joint military operations in Bulgaria and the Ægean, for which we have not at present the means. And in addition to all this she will demand once again all sorts of munitions, which we cannot spare because the stocks we had ready for her at the beginning

of the year have been drawn off in other directions. It seems to me therefore wiser to take this breaking off relations with Germany as a first instalment. We can then push a few things in to help her against a vengeance attack from the air, and out of this, while we are together, her entry into the war might come. The Turkish alliance in the last war was very dear to the Germans, and the fact that Turkey had broken off relations would be a knell to the German soul. This seems to be a pretty good time to strike such a knell.

3. I am only putting to you my personal thoughts on these matters, which are also being transmitted by Eden to M. Molotov.

4. We have about a million and fifty thousand men in Normandy, with a vast mass of equipment, and rising by 25,000 a day. The fighting is very hard, and before the recent battles, for which casualties have not yet come in, we and the Americans had lost 64,000 men. However, there is every evidence that the enemy has lost at least as many, and we have besides 51,000 prisoners in the bag. Considering that we have been on the offensive and had the landing from the sea to manage, I consider that the enemy has been severely mauled. The front will continue to broaden and the fighting will be unceasing.

5. Alexander is pushing very hard in Italy also. He hopes to force the Pisa–Rimini line and break into the Po valley. This will either draw further German divisions on to him or yield up valuable strategic ground.

6. The Londoners are standing up well to the bombing, which has amounted to 22,000 casualties so far and looks like becoming chronic.

7. Once more, congratulations on your glorious advance to Vilna.

His reply was non-committal.

Marshal Stalin to Prime Minister   15 July 44
As regards the question of Roumania and Greece. . . . One thing is clear to me: it is that the American Government have some doubts regarding this question, and that it would be better to revert to this matter when we receive the American reply to our inquiry. As soon as the observations of the American Government are known I shall not fail to write to you further on this question.

2. The question of Turkey should be considered in the light of those facts which have been well known to the Governments of Great Britain, the Soviet Union, and the U.S.A. from the time of the last negotiations with the Turkish Government at the end of last year. You of course will remember how insistently the Governments of our three countries proposed to Turkey that she should enter the war against Hitlerite Germany on the side of the Allies as long ago as in November and December of 1943. Nothing came of this. As you know, on the

initiative of the Turkish Government in May-June of this year we again entered into negotiations with the Turkish Government, and twice we proposed to them the same thing that the three Allied Governments had proposed to them at the end of last year. Nothing came of this either. As regards these or other half-measures on the part of Turkey, at the present time I see no benefit in them for the Allies. In view of the evasive and vague attitude with regard to Germany adopted by the Turkish Government, it is better to leave Turkey in peace and to her own free will and not to exert fresh pressure on Turkey. This of course means that the claims of Turkey, who has evaded war with Germany, to special rights in post-war matters also lapse. . . .

We were thus unable to reach any final agreement about dividing responsibilities in the Balkan peninsula. Early in August the Russians dispatched from Italy by a subterfuge a mission to E.L.A.S. in Northern Greece. In the light of American official reluctance and of this instance of Russian bad faith, we abandoned our efforts to reach a major understanding until I met Stalin in Moscow two months later. By then much had happened on the Eastern Front.

*　　*　　*　　*　　*

The Russian summer campaign was a tale of sweeping success. I can but summarise it here.

The advance opened with a secondary offensive against the Finns. Between Lake Ladoga and the sea they had deepened and strengthened the original Mannerheim Line into a formidable defensive system. However, the Russian troops, very different in quality and armament from those who had fought here in 1940, broke through after twelve days of hard fighting and captured Viborg on June 21. Operations were begun the same day to clear the north shore of Lake Ladoga, and at the end of the month they had thrown their opponents back and reopened the railway from Leningrad to Murmansk, the terminal of our Arctic convoys. The Finns struggled on for a while, supported by German troops, but they had had enough, and on August 25 sued for an armistice.

The attack on the German front between Vitebsk and Gomel began on June 23. These, with Bobruisk, Mogilev, and many other towns and villages, had been turned into strong positions, with all-round defence, but they were successively surrounded and disposed of, while the Russian armies poured through the gaps between. Within a week they had broken through to a depth of eighty miles. Taking swift advantage of their success, they captured Minsk on July 6, and closed the retreating enemy along a hastily organised line running southwards from Vilna to the Pripet Marshes, from which the Germans were soon swept by the irresistible Russian flood. At the end of July the Red armies had reached the Niemen at Kovno and Grodno. Here, after an advance of 250 miles in five weeks, they were brought to a temporary halt to replenish. The German losses had been crushing. Twenty-five divisions had ceased to exist, and an equal number were cut off in Courland.* On July 17 alone 57,000 German prisoners were marched through Moscow—who knows whither?

South of the Pripet Marshes the Russian successes were no less magnificent. On July 13 a series of attacks were launched on the front between Kovel and Stanislav. In ten days the whole German front was broken and the Russians had reached Jaroslav, on the San river, 120 miles farther west. Stanislav, Lemberg, and Przemysl, isolated by this advance, were soon accounted for, and on July 30 the triumphant Russians crossed the Vistula south of Sandomir. Here supply imposed a halt. The crossing of the Vistula was taken by the Polish Resistance Movement in Warsaw as a signal for the ill-fated rising which is recorded in another chapter.

There was still a further far-reaching Russian success in this great campaign. To the southward of their victories lay Roumania. Till August was far advanced the German line from Cernowitz to the Black Sea barred the way to Roumania, the Ploesti oilfields, and the Balkans. It had been weakened by withdrawal of troops to sustain the sagging line farther north, and under violent attacks, beginning on August 22, it rapidly disintegrated. Aided by landings on the coast, the Russians made short work of the enemy. Sixteen German divisions were lost. On August 23 a *coup d'état* in Bucharest, organised by the young King Michael and his close advisers, led to a complete reversal of the whole military position. The Roumanian armies followed their King to a man. Within three days before the arrival of the Soviet troops the German forces had been disarmed or had retired over the northern frontiers. By September 1 Bucharest had been evacuated by the Germans. The Roumanian armies disintegrated

* Guderian, *Panzer Leader*, p. 352.

OPERATIONS ON THE RUSSIAN FRONT, JUNE 1944-JAN. 1945

73

and the country was overrun. The Roumanian Government capitulated. Bulgaria, after a last-minute attempt to declare war on Germany, was overwhelmed. Wheeling to the west, the Russian armies drove up the valley of the Danube and through the Transylvanian Alps to the Hungarian border, while their left flank, south of the Danube, lined up on the frontier of Yugoslavia. Here they prepared for the great westerly drive which in due time was to carry them to Vienna.

CHAPTER VI

ITALY AND THE RIVIERA LANDING

The Allied Pursuit Beyond Rome – The Toll of "Anvil" – The Gothic Line – The Fifth Army Reduced by 100,000 Men – The Advance to the Arno – I Fly to Naples and Meet Tito – Balkan Strategy and the Istrian Peninsula – Tito, Communism, and King Peter – Allied Military Government in Istria – My Second Meeting with Tito – I Report to the President – A Sunshine Holiday – I Fly to Corsica – The Landings on the French Riviera – My Telegram to the King, August 17 – And to General Eisenhower, August 18 – An Outline of the "Dragoon" Operation – My Summing Up of the "Anvil"-"Dragoon" Story – Correspondence with Smuts – The Vienna Hope.

A FTER Rome fell on June 4 Kesselring's broken armies streamed northwards in disorder, harassed and disorganised by continuous air attacks and closely pursued on land. General Clark's Fifth U.S. Army took the coastal roads towards Pisa, while our Eighth Army followed up astride the Tiber, heading for Lake Trasimene. The pace was hot.

Prime Minister to General Alexander   9 June 44

All our information here goes to reinforce your estimate of the ruin you have wrought on the German armies in Italy. Your whole advance is splendid, and I hope the remnants of what were once the German armies will be collected.

Alexander hoped greatly that the "Anvil" plan to land in the south of France would be put aside and that he would be allowed to keep intact his battle-trained troops, now flushed with victory. If so, he was confident of breaking through the Apennines into the valley of the Po and beyond within a few months. He failed by a very narrow margin, as this story will show, and it seems

75

certain that but for the deprivations and demands of "Anvil" the campaign in Italy could have been over by Christmas.

In any case there was hard fighting ahead. Nineteen German divisions had been involved in the battles of May and early June. Three of them had ceased to exist; most of the others were gravely stricken and hurrying northwards in confused retreat. But Kesselring was a good general, with a competent staff. His problem was to delay our advance until he had reorganised his troops and occupied his next prepared position, the so-called Gothic Line, which ran from the west coast beyond Pisa, curved along the mountains north of Florence, and then struck off to the Adriatic at Pesaro. The Germans had been working on this line for more than a year, but it was still unfinished. Kesselring had to fight for time to complete and man it and to receive the eight divisions which were being sent him from Northern Europe, the Balkans, Germany, and Russia.

After ten days of pursuit German resistance began to stiffen, and the Eighth Army had a hard fight to overcome a strong position on the famous shores of Lake Trasimene. It was not till June 28 that the enemy were ejected and fell back on Arezzo. On the west coast American troops of the Fifth Army took Cecina, not without difficulty, on July 1, and on their right the French Corps, also under General Clark's command, reached Siena soon afterwards. The enemy made a corresponding withdrawal on the Adriatic coast, enabling the Polish Corps swiftly to occupy Pescara and drive on towards Ancona. At this time too a French colonial division, transhipped from Corsica, took Elba, with two thousand prisoners, after a couple of days' sharp fighting, in which they received strong naval and air support.

In early July, as a result of the discussions which had been proceeding with the United States, Alexander was ordered to detach forces, amounting finally to seven divisions, for "Anvil". The Fifth Army alone was thereby reduced from nearly 250,000 men to 153,000. Despite this blow Alexander persevered with vigour in his pursuit and plan. The Germans, re-formed and rebuilt to the equivalent of fourteen full divisions, faced him on a line from Rosignano to Arezzo, and thence to the Adriatic south of Ancona. This was one of a succession of covering positions which the enemy were to hold with increasing obstinacy to stop us reaching their Gothic Line. Arezzo fell to the British on July 16, after heavy

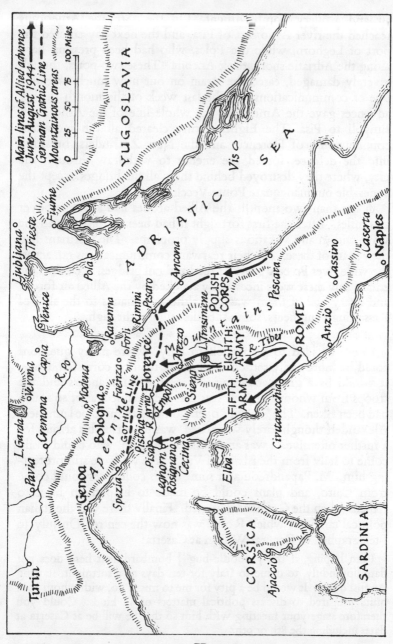

air and artillery bombardment. On the 18th the Americans reached the river Arno east of Pisa, and the next day entered the port of Leghorn, while the Poles, who had been pressing hard along the Adriatic shore, took Ancona. These two ports, though severely damaged, eased the strain on our now much-extended line of communications. In the last week of the month further advances gave the Americans the whole line of the Arno from Empoli to Pisa. The Eighth Army cleared all the mountain country south of Florence, and the New Zealanders, breaking into the defence, forced the enemy to withdraw through the city, where they destroyed behind them all the bridges except the venerable but inadequate Ponte Vecchio.

In less than two months the Allied armies had advanced over 250 miles. After the first fortnight it had been hard going all the way, with many vexatious supply problems. The Germans had their share of these. All their rearward communications ran across the wide river Po on a score of road and rail bridges. Towards the end of July these were incessantly attacked by the Allied air forces, and every one of them was cut, although, thanks to the skill of Kesselring's engineers, some supplies still got through.

*   *   *   *   *

I now decided to go myself to Italy, where many questions could be more easily settled on the spot than by correspondence. It would be a great advantage to see the commanders and the troops from whom so much was being demanded, after so much had been taken. The "Anvil" operation was about to be launched. Alexander, though sorely weakened, was preparing his armies for a further offensive. I was anxious to meet Tito, who could easily come to Italy from the island of Vis, where we were still protecting him. M. Papandreou and some of his colleagues could come from Cairo, and plans could be made to help them back to Athens when the Germans departed. Finally there was the Italian political tangle of which Rome was now the centre. On July 30 I telegraphed to General Wilson at Caserta:

I am hoping, if the ["doodle-bug"] bombardment here does not flare up unduly, to come to Italy for ten days or a fortnight, starting August 6 or 7. It would be a pity for me to miss Tito, with whom I am quite prepared to discuss political matters of all kinds. Could you therefore stage your meeting with him so that he will be at Caserta at dates including the 8th or 9th?

And on August 4 to General Alexander:

I thought it would be better for us to make our plans [for my visit] together when I arrive. Mind you do not let me get in your way. I do not want a heavy programme, nor to engage myself to see anybody except you, Wilson, and Tito. I have no doubt I shall find plenty to do when I am on the spot.

*   *   *   *   *

The days were so crowded with Cabinet business that my dates receded. On August 9 I telegraphed to Mr. Duff Cooper that I hoped to arrive at the Maison Blanche airfield, outside Algiers, about 6.30 a.m. on Friday, August 11, and would stay there for about three hours on my way to Naples. I added, "You may tell de Gaulle in case he wishes to see me at your house or the Admiral's villa. The visit is quite informal."

We arrived punctually. Duff Cooper met me, and took me to his house, which his wife had made most comfortable. He told me he had conveyed my invitation or suggestion to de Gaulle, and that the General had refused. He did not wish to intrude upon the repose I should need at this brief halt on my journey. I thought this needlessly haughty, considering all the business we had in hand and what I could have told him. He was however still offended by what had happened at "Overlord", and thought this was a good chance of marking his displeasure. I did not in fact see him again for several months.

I reached Naples that afternoon, and was installed in the palatial though somewhat dilapidated Villa Rivalta, with a glorious view of Vesuvius and the bay. Here General Wilson explained to me that all arrangements had been made for a conference next morning with Tito and Subašić, the new Yugoslav Prime Minister of King Peter's Government in London. They had already arrived in Naples, and would dine with us the next night.

On the morning of August 12 Marshal Tito came up to the villa. He wore a magnificent gold and blue uniform which was very tight under the collar and singularly unsuited to the blazing heat. The uniform had been given him by the Russians, and, as I was afterwards informed, the gold lace came from the United States. I joined him on the terrace of the villa, accompanied by Brigadier Maclean and an interpreter.

I suggested that the Marshal might first like to see General

Wilson's War Room, and we moved inside. The Marshal, who was attended by two ferocious-looking bodyguards, each carrying automatic pistols, wanted to bring them with him in case of treachery on our part. He was dissuaded from this with some difficulty, and proposed to bring them to guard him at dinner instead.

I led the way into a large room, where maps of the battle-fronts covered the walls. I began by displaying the Allied front in Normandy and indicating our broad strategic moves against the German armies in the West. I pointed out Hitler's obstinacy in refusing to yield an inch of territory, how numerous divisions were locked up in Norway and in the Baltic provinces, and said that his correct strategy would be to withdraw his troops from the Balkans and concentrate them on the main battle-fronts. Allied pressure in Italy and the Russian advance from the east might force him to go, but we must reckon on the possibility of his staying. As I talked I pointed on the map to the Istrian peninsula, and asked Tito where, if we were able to reach it from the east coast of Italy, his forces could be sent to co-operate with us. I explained that it would help if a small port could be opened on the Yugoslav coast so that we could send in war material by sea. In June and July we had sent nearly two thousand tons of stores to his forces by air, but could do much more if we had a port. Tito said that although German opposition had intensified lately, and Yugoslav losses increased, he was able to raise considerable forces in Croatia and Slovenia, and he would certainly favour an operation against the Istrian peninsula, in which Yugo-slav forces would join.

We now moved into a small sitting-room, and I began to question him about his relations with the Royal Yugoslav Government. He said that violent fighting still continued between the Partisans and Mihailović, whose power rested on German and Bulgar help. Reconciliation was unlikely. I replied that we had no desire to intervene in internal Yugoslav affairs, but wanted his country to be strong, united, and independent. Dr. Subašić was very loyal to this idea. Moreover, we ought not to let the King down. Tito said that he understood our obligation towards King Peter, but was not able to do anything about it until after the war, when the Yugoslav people themselves would decide.

I then turned to the future, and suggested that the right solution for Yugoslavia would be a democratic system based on the peasantry, and perhaps some gradual measure of agrarian reform where the holdings were too small. Tito assured me that, as he had stated publicly, he had no desire to introduce the Communist system into Yugoslavia, if only because most European countries after the war would probably be living under a democratic régime. Developments in small countries depended on relations between the Great Powers. Yugoslavia should be able to profit by the growing improvement in these relations and develop along democratic lines. The Russians had a mission with the Partisans, but its members, far from expressing any idea of introducing the Soviet system into Yugoslavia, had spoken against it.

I asked Tito if he would reaffirm his statement about Communism in public, but he did not wish to do this as it might seem to have been forced upon him. It was agreed that he should however discuss the suggestion with Dr. Subašić, whom he was meeting for the first time that afternoon.

We then lunched together, and arranged that if the talks with Dr. Subašić made favourable progress we should meet again the following evening. In the meantime I undertook to draft a memorandum on Yugoslav affairs, and the Marshal promised to send me a letter on certain specific matters about supplies.

    *   *   *   *   *

Early in the day Tito had met General Gammell, Chief of Staff to General Wilson, and been given an important memorandum on Allied projects in Istria and thereabouts. It read as follows:

In the event of Allied forces occupying Northern Italy, Austria, or Hungary it is the Supreme Allied Commander's intention to impose Allied military government in the area which was under Italian rule at the outbreak of war, which automatically suspends Italian sovereignty. The Military Governor will be the General Officer Commanding the Allied armies in the area. It is intended that the area shall remain under direct Allied administration until its disposition has been determined by negotiation between the Governments concerned.

2. This direct Allied military government is necessary in order to safeguard the bases and lines of communication of the Allied troops of occupation in Central Europe.

3. As the Allied forces of occupation will have to be supplied

through the port of Trieste, it will be necessary for them to have secure lines of communication protected by British troops on the route through Ljubljana–Maribor–Graz.

4. The Supreme Allied Commander looks to the Yugoslav authorities to co-operate with him in carrying out this policy, and he intends to maintain the closest liaison with them.

Tito had grumbled at these proposals in a letter to me, and when we met again on the afternoon of August 13, Mr. Stevenson, our Ambassador to Yugoslavia, and Dr. Subašić being present, I said that it was an operational question which needed careful study, and also close consultation with the American President. The status of Istria, which still remained Italian, could not be prejudged. It might be a good thing to remove it from Italian sovereignty, but this must be decided at the Peace Conference, or, if there were none, by a meeting of the principal Powers, at which Yugoslavia could state her claim. The United States Government was against territorial changes in time of war, and we ought not to discourage the Italians more than could be helped because they were now making a useful contribution to the war. The best solution might therefore be for the territory to be administered under Allied military government when it was freed from the Germans.

Tito said that he could not accept an Italian civil administration, and pointed out that his National Liberation Movement already controlled many of these areas, and should at least be associated in their administration. He and Subašić agreed to send us a joint memorandum on Istria, and there the matter rested for the moment.

We then discussed how to produce a united Yugoslav Navy, and how to send him light tanks, gunboats, and artillery. I said that we would do what we could, but I warned him that we should lose interest if the fighting in Yugoslavia developed into mere civil war and the struggle against the Germans became only a side issue.

I had referred to this in a note which I had sent to Tito on August 12. We now considered the wider implications of this document, which ran as follows:

The desire of His Majesty's Government is to see a united Yugoslav Government, in which all Yugoslavs resisting the enemy are represented, and a reconciliation between the Serbian people and the National Liberation Movement.

2. His Majesty's Government intend to continue, and if possible to increase, the supply of war material to Yugoslav forces now that an agreement has been reached between the Royal Yugoslav Government and the National Liberation Movement. They expect, in return, that Marshal Tito will make a positive contribution to the unification of Yugoslavia by including in the declaration which he has already agreed with the Yugoslav Prime Minister to make, not only a statement regarding his intention not to impose Communism on the country, but also a statement to the effect that he will not use the armed strength of the Movement to influence the free expression of the will of the people on the future régime of the country.

3. Another contribution which Marshal Tito could make to the common cause is to agree to meet King Peter, preferably on Yugoslav soil.

4. If it should turn out that any large quantities of ammunition sent by His Majesty's Government are used for fratricidal strife other than in self-defence, it would affect the whole question of Allied supplies, because we do not wish to be involved in Yugoslav political differences.

5. We should like to see the Royal Yugoslav Navy and Air Force working all out for national liberation, but this cannot be agreed unless first of all due consideration is paid to the King, the constitutional flag, and the closer unity of the Government and the Movement.

6. His Majesty's Government, while regarding Marshal Tito and his brave men with the utmost admiration, are not satisfied that sufficient recognition has been given to the power and rights of the Serbian people, or to the help which has been given, and will be continued, by His Majesty's Government.

The Yugoslavs objected to my suggestion that the Partisan movement was divorced from the Serbian people. I did not press this point, particularly as Tito had said that he was prepared later on to make a public statement about not introducing Communism into Yugoslavia after the war. We then discussed a possible meeting between him and King Peter. I said that democracy had flowered in England under constitutional monarchy, and thought that Yugoslavia's international position would be stronger under a king than as a republic. Tito said his country had had an unfortunate experience with her King, and it would take time for King Peter to live down his connection with Mihailović. He had no objection in principle to meeting the King, but thought that the moment had not yet come. We therefore agreed to leave it to him and Dr. Subašić to decide on the most opportune occasion.

*   *   *   *   *

Later I entertained Tito to dinner. He was still confined in his gold-lace strait-jacket. I was so glad to be wearing only a white duck suit.

I now reported the results of these talks to the President.

Prime Minister to President Roosevelt   14 Aug 44

I have had meetings during the last two days with Marshal Tito and the Yugoslav Prime Minister. I told both the Yugoslav leaders that we had no thought but that they should combine their resources so as to weld the Yugoslav people into one instrument in the struggle against the Germans. Our aim was to promote the establishment of a stable and independent Yugoslavia, and the creation of a united Yugoslav Government was a step towards this end.

2. The two leaders reached a satisfactory agreement on a number of practical questions. They agreed that all the Yugoslav naval forces will now be united in the struggle under a common flag. This agreement between the Yugoslav Prime Minister and Marshal Tito will enable us with more confidence to increase our supplies of war material to the Yugoslav forces.

3. They agreed between themselves to issue simultaneously a statement in a few days' time, which I hope will strengthen and intensify the Yugoslav war effort. They are going off together to-day to Vis to continue their discussions.

4. I am informing Marshal Stalin of the result of these meetings.

*   *   *   *   *

On all these three days at Naples I mingled pleasure with toil. Admiral Morse, who commanded the naval forces, took me each day in his barge on an expedition, of which the prime feature was a bathe. On the first we went to the island of Ischia, with its hot springs, and on the return we ran through an immense United States troop convoy sailing for the landing on the Riviera. All the ships were crowded with men, and as we passed along their lines they cheered enthusiastically. They did not know that if I had had my way they would be sailing in a different direction. However, I was proud to wave to these gallant soldiers. We also visited Capri. I had never seen the Blue Grotto before. It is indeed a miracle of transparent, sparkling water of a most intense, vivid blue. We bathed in a small, warm bay, and repaired to luncheon at a comfortable inn. I summoned up in my mind all I could remember about the Emperor Tiberius. Certainly in Capri he had chosen an agreeable headquarters from which to rule the world. These days, apart from business, were a sunshine holiday.

*　*　*　*　*

On the afternoon of August 14 I flew in General Wilson's Dakota to Corsica in order to see the landing at "Anvil" which I had tried so hard to stop, but to which I wished all success. We had a pleasant flight to Ajaccio, in the harbour of which General Wilson and Admiral Sir John Cunningham had posted themselves on board a British headquarters ship. The airfield was very small and not easily approached. The pilot was excellent. He had to come in between two bluffs, and his port-wing was scarcely fifteen feet from one of them. The General and the Admiral brought me aboard, and we spent a long evening on our affairs. I was to start at daylight in the British destroyer *Kimberley*. I took with me two members of the American Administration, General Somervell and Mr. Patterson, the Assistant Secretary of War, who were on the spot to see their venture. Captain Allen, whose help in these volumes I have acknowledged, was sent by the Admiral to see that we did not get into trouble. We were five hours sailing before we reached the line of battleships bombarding at about fifteen thousand yards. I now learned from Captain Allen that we were not supposed to go beyond the ten-thousand-yard limit for fear of mines. If I had known this when we passed the *Ramillies*, which was firing at intervals, I could have asked for a picket-boat and gone ashore. As it was we did not go nearer than about seven thousand yards. Here we saw the long rows of boats filled with American storm troops steaming in continuously to the Bay of St. Tropez. As far as I could see or hear not a shot was fired either at the approaching flotillas or on the beaches. The battleships had now stopped firing, as there seemed to be nobody there. We then returned to Ajaccio. I had at least done the civil to "Anvil", and indeed I thought it was a good thing I was near the scene to show the interest I took in it. On the way back I found a lively novel, *Grand Hotel*, in the captain's cabin, and this kept me in good temper till I got back to the Supreme Commander and the Naval Commander-in-Chief, who had passed an equally dull day sitting in the stern cabin.

85

On August 16 I got back to Naples, and rested there for the night before going up to meet Alexander at the front. I telegraphed to the King, from whom I had received a very kind message.

Prime Minister to the King   17 Aug 44

With humble duty.

From my distant view of the "Dragoon" operation, the landing seemed to be effected with the utmost smoothness. How much time will be taken in the advance first to Marseilles and then up the Rhone valley, and how these operations will relate themselves to the far greater and possibly decisive operations in the north [Normandy], are the questions that now arise.

I am proceeding to-day to General Alexander's headquarters. It is very important that we ensure that Alexander's army is not so mauled and milked that it cannot have a theme or plan of campaign. This will certainly require a conference on something like the "Quadrant" scale, and at the same place [Quebec].

My vigour has been greatly restored by the change and movement and the warm weather. I hope to see various people, including Mr. Papandreou, in Rome, where I expect to be on the 21st.

May I express to Your Majesty the pleasure and encouragement which Your Majesty's gracious message gave me.

And to General Eisenhower:

Prime Minister to General Eisenhower (France)   18 Aug 44

I am following with thrilled attention the magnificent developments of operations in Normandy and Anjou. I offer you again my sincere congratulations on the truly marvellous results achieved, and hope for surpassing victory. You have certainly among other things effected a very important diversion from our attack at "Dragoon". I watched this landing yesterday from afar. All I have learnt here makes me admire the perfect precision with which the landing was arranged and the intimate collaboration of British-American forces and organisations. I shall hope to come and see you and Montgomery before the end of the month. Much will have happened by then. It seems to me that the results might well eclipse all the Russian victories up to the present. All good wishes to you and Bedell.

*   *   *   *   *

This chapter may close with an outline of the "Anvil"-"Dragoon" operations themselves.

The Seventh Army, under General Patch, had been formed to carry out the attack. It consisted of seven French and three U.S.

divisions, together with a mixed American and British air borne division. The three American divisions comprised General Truscott's VIth Corps, which had formed an important part of General Clark's Fifth Army in Italy. In addition, up to four French divisions and a considerable part of the Allied air forces were withdrawn from Alexander's command.

The new expedition was mounted from both Italy and North Africa, Naples, Taranto, Brindisi, and Oran being used as the chief loading ports. Great preparations had been made throughout the year to convert Corsica into an advanced air base and to use Ajaccio as a staging port for landing-craft proceeding to the assault from Italy. All these arrangements now bore fruit. Under the Commander-in-Chief, Admiral Sir John Cunningham, the naval attack was entrusted to Vice-Admiral Hewitt, U.S.N., who had had much experience in similar operations in the Mediterranean. Lieut.-General Eaker, U.S.A.A.F., commanded the air forces, with Air Marshal Slessor as his Deputy.

Landing-craft restricted the first seaborne landing to three divisions, and the more experienced Americans led the van. Shore defences all along the coast were strong, but the enemy were weak in numbers and some were of poor quality. In June there had been fourteen German divisions in Southern France, but four of these were drawn away to the fighting in Normandy and no more than ten remained to guard the 200 miles of coastline. Only three of these lay near the beaches on which we landed. The enemy were also short of aircraft. Against our total of 5,000 in the Mediterranean, of which 2,000 were based in Corsica and Sardinia, they could muster a bare two hundred, and these were mauled in the days before the invasion. In the midst of the Germans in Southern France over 25,000 armed men of the Resistance were ready to revolt. We had sent them their weapons, and, as in so many other parts of France, they had been organised by some of that devoted band of men and women trained in Britain for the purpose during the past three years.

The strength of the enemy's defences demanded a heavy pre-liminary bombardment, which was provided from the air for the previous fortnight all along the coast, and, jointly with the Allied Navies, on the landing beaches immediately before our descent. No fewer than six battleships, twenty-one cruisers, and a hundred destroyers took part. The three U.S. divisions, with American and French Commandos on their left, landed early on August 15 between Cannes and Hyères. Thanks to the bombardment, successful deception plans, continuous fighter cover, and good staff work, our casualties were relatively few. During the previous night the airborne division had dropped around Le Muy, and soon joined hands with the seaborne attack.

By noon on the 16th the three American divisions were ashore. One of them moved northwards to Sisteron, and the other two struck north-west towards Avignon. The IInd French Corps landed immediately behind them and made for Toulon and Marseilles. Both places were strongly defended, and although the French were built up to a force of five divisions the ports were not fully occupied till the end of the month. The installations were severely damaged, but Port de Bouc had been captured intact with the aid of the Resistance, and supplies soon began to flow. This was a valuable contribution by the French forces under General de Lattre de Tassigny. In the meantime the Americans had been moving fast, and on August 28 were beyond Valence and Grenoble. The enemy made no serious attempt to stop the advance, except for a stiff fight at Montélimar by a Panzer division. The Allied Tactical Air Force was treating them roughly and destroying their transport. Eisenhower's pursuit from Normandy was cutting in behind them, having reached the Seine at Fontainebleau on August 20, and five days later it was well past Troyes. No wonder the surviving elements of the German Nineteenth Army, amounting to a nominal five divisions, were in full retreat, leaving 50,000 prisoners in our hands. Lyons was taken on September 3, Besançon on the 8th, and Dijon was liberated by the Resistance Movement on the 11th. On that day "Dragoon" and "Overlord" joined hands at Sombernon. In the triangle of South-West France, trapped by these concentric thrusts, were the isolated remnants of the German First Army, over 20,000 strong, who freely gave themselves up.

* * * * *

To sum up the "Anvil"-"Dragoon" story, the original proposal at Teheran in November 1943 was for a descent in the south of France to help take the weight off "Overlord". The timing was to be either in the week before or the week after D Day. All this was changed by what happened in the interval. The latent

"ANVIL"

threat from the Mediterranean sufficed in itself to keep ten German divisions on the Riviera. Anzio alone had meant that the equivalent of four enemy divisions was lost to other fronts. When, with the help of Anzio, our whole battle line advanced, captured Rome and threatened the Gothic Line, the Germans hurried a further eight divisions to Italy. Delay in the capture of Rome and the dispatch of landing-craft from the Mediterranean to help "Overlord" caused the postponement of "Anvil"-"Dragoon" till mid-August, or two months later than had been proposed. `It therefore did not in any way affect "Overlord". When it was belatedly launched it drew no enemy down from the Normandy battle theatre. Therefore none of the reasons present in our minds at Teheran had any relation to what was done and "Dragoon" caused no diversion from the forces opposing General Eisenhower. In fact, instead of helping him, he helped it by threatening the rear of the Germans retiring up the Rhone valley. This is not to deny that the operation as carried out eventually brought important assistance to General Eisenhower by the arrival of another army on his right flank and the opening of another line of communications thither. For this a heavy price was paid. The army of Italy was deprived of its opportunity to strike a most formidable blow at the Germans, and very possibly to reach Vienna before the Russians, with all that might have followed therefrom. But once the final decision was reached I of course gave "Dragoon" my full support, though I had done my best to constrain or deflect it.

* * * * *

At this time I received some pregnant messages from Smuts, now back at the Cape. He had always agreed wholeheartedly with my views on "Dragoon", "but," he now wrote (August 30) "please do not let strategy absorb all your attention to the damage of the greater issue now looming up.

"*From now on it would be wise to keep a very close eye on all matters bearing on the future settlement of Europe. This is the crucial issue on which the future of the world for generations will depend. In its solution your vision, experience, and great influence may prove a main factor.*" *

I have been taxed in the years since the war with pressing after Teheran, and particularly during these weeks under review, for

My italics.—W. S. C.

a large-scale Allied invasion of the Balkans in defiance of American thinking on the grand strategy of the war. The essence of my oft-repeated view is contained in the following reply to these messages from Smuts:

Prime Minister to Field-Marshal Smuts   31 Aug 44
Local success of "Dragoon" has quite delighted Americans, who intend to use this route to thrust in every reinforcement. Of course 45,000 prisoners have been taken, and there will be many more. The idea now, from which nothing will turn them, is to work in a whole Army Group through the captured ports instead of using the much easier ports on the Atlantic.

"My object now," I said, "is to keep what we have got in Italy, which should be sufficient since the enemy has withdrawn four of his best divisions. With this I hope to turn and break the Gothic Line, break into the Po valley, and ultimately advance by Trieste and the Ljubljana Gap to Vienna. Even if the war came to an end at an early date I have told Alexander to be ready for a dash with armoured cars."

CHAPTER VII

ROME. THE GREEK PROBLEM

Alexander Prepares to Attack the Gothic Line – Field-Marshal Smuts Surveys the Situation, August 12 – I Visit the Front, August 17 – Two Days at Siena – The Weakening of the Fifteenth Army Group – A Visit to General Mark Clark – Sombre Reflections – I Fly to Rome, August 21 – Preparations to Liberate Greece – My Telegram to the President, August 17 – His Reply – A Meeting with M. Papandreou – The Future of the Greek Monarchy – I Report to Mr. Eden, August 22 – I Meet Some Italian Politicians – Audience with Pope Pius XII – The Crown Prince Umberto, Lieutenant of the Realm.

DURING the early weeks of August Alexander was planning and regrouping his depleted forces to attack the main Gothic Line, with whose outpost positions he was already in close contact. Skilfully sited to make full use of the natural difficulties of the country, the main defences firmly barred all likely lines of approach from the south, leaving enticingly weak only those sectors which were almost inaccessible.

The difficulties of launching the attack directly through the mountains from Florence to Bologna were obvious, and Alexander decided that the Eighth Army should make the first major stroke on the Adriatic side, where the succession of river valleys, difficult though they would be, offered less unfavourable ground provided heavy rain held off. Kesselring could not afford to have his eastern flank turned and Bologna captured behind his main front, so it was certain that if our attack made good progress he would have to reinforce the flank by troops from the centre. Alexander therefore planned to have ready a second attack, to be carried out by Mark Clark's Fifth Army, towards Bologna and Imola, which would be launched when enemy reserves had been committed and their centre weakened.

The preparatory troop and air movements, carried out in great secrecy during the third week of August, were skilfully accomplished. Leaving the XIIIth British Corps east of Florence to come under Fifth Army command, two whole corps of the Eighth Army were transferred eastwards and concentrated near Pergola, on the left of the Polish Corps. When all was completed Alexander held ready for battle the equivalent of twenty-three divisions, of which rather more than half were with the Eighth Army. Opposing him Kesselring had twenty-six German divisions, in good shape, and two reconstituted Italian divisions; nineteen of them were disposed for the defence of his main positions.

★   ★   ★   ★   ★

Smuts fully realised what was at stake, as the following telegram shows.

Field-Marshal Smuts to Prime Minister   12 Aug 44

Knowing how deeply occupied you are, I have refrained from worrying you unduly with correspondence. I myself have also been fully engaged with difficult local problems. I am glad to hear you are once more in Italy, in close contact with that important section of our war front, and wish you a very successful and happy trip, as well as health and strength for the exacting tasks still awaiting you.

2. I assumed that one of the objects of your present visit will be to strengthen Alexander's hands by combing our Mediterranean theatre as drastically as possible. Considerable forces must still be in reserve there for contingencies now no longer important. The end can best be hastened by the concentration of our forces in the few decisive theatres, of which Alexander's is one. With Turkey now lost to the enemy and Bulgaria more and more shaky, we can afford to ignore the theatres for which large forces have been kept in the Middle East and assemble whatever we have to strengthen Alexander's move, which may lead to very great results both in the Balkans and in Hitler's European fortress. I would cut off the frills elsewhere in order to improve the alluring chances before this move. A front extending along North Italy, the Adriatic, and the route through Trieste to Vienna is one worthy of our concentrated effort and of one of the ablest generals this war has produced. I am sure both Wilson and Paget agree that this is the correct strategy for us to follow in order to complete our task and gather the mature fruits of our great Mediterranean campaigns. Any further assistance I could still give from here is in the air, and I have already offered to man a number of additional squadrons from personnel released from the air training schools now being closed down

in the Union. Already I am manning some R.A.F. squadrons with South Africans, and by this means could probably man six more for Alexander's operations. As we are now scraping the bottom in our recruiting campaign, and owing to the dispersal of our man-power in many other directions, I am unable to do more for the infantry than keep up the strength of the 6th South African Division. The additional air help would only be possible if the Air Ministry accept my offer already made to them. They have the details.

A decisive stage has now been reached in the war, and an all-out offensive on all three main fronts against Germany must lead to the grand finale this summer. If the present tremendous successful offensive can only be maintained the end cannot be long deferred, especially in view of what we now know of conditions inside the German Army.

I shall be glad to see the "Dragoon" correspondence, disheartening as it may be. The way things are going now, Southern France has ceased to be a theatre of real military significance, and our large forces and resources detached for it will have small bearing on the tremendous decisions elsewhere. I even doubt whether the enemy will trouble to reinforce it.

* * * * *

On the morning of August 17 I set out by motor to meet General Alexander. I was delighted to see him for the first time since his victory and entry into Rome. He drove me all along the old Cassino front, showing me how the battle had gone and where the main struggles had occurred. The monastery towered up, a dominating ruin. Anyone could see the tactical significance of this stately crag and building which for so many weeks played its part in stopping our advance. When we had finished it was time for lunch, and a picnic table had been prepared in an agreeable grove. Here I met General Clark and eight or ten of the leading British officers of the Fifteenth Army Group. Alexander then took me in his own plane, with which I was familiar, by a short flight to Siena, the beautiful and famous city, which I had visited in bygone peaceful days. Thence we visited our battle-front on the Arno. We had the south bank of the river and the Germans the north. Considerable efforts were made by both sides to destroy as little as possible, and the historic bridge at Florence was at any rate preserved. We were lodged in a beautiful but dismantled château a few miles to the west of Siena, and here I passed a couple of days, mostly working in bed, reading, and dictating telegrams. Of course in all these journeys I kept close to me the nucleus of my Private Office and the necessary ciphering staff, which enabled me to receive and to answer all messages from hour to hour.

Alexander brought his chief officers to dinner, and explained to me fully his difficulties and plans. The Fifteenth Group of Armies had indeed been skinned and starved. The far-reaching projects we had cherished must now be abandoned. It was still our duty to hold the Germans in the largest numbers on our front. If this purpose was to be achieved an offensive was imperative; but the well-integrated German armies were almost as strong as ours, composed of so many different contingents and races. It was proposed to attack along the whole front early on the 26th. Our right hand would be upon the Adriatic, and our immediate objective Rimini. To the westward, under Alexander's command, lay the Fifth American Army. This had been stripped and mutilated for the sake of "Anvil", but would nevertheless advance with vigour.

On August 19 I set off to visit General Mark Clark at Leghorn. This was a long drive, and everywhere we stopped to visit brigades and divisions. Mark Clark received me at his headquarters. We lunched in the open air by the sea. In our friendly and confidential talks I realised how painful the tearing to pieces of this fine army had been to those who controlled it. I toured the harbour, which had often played a part in our naval affairs, in a motor torpedo-boat. Then we went to the American batteries. A pair of new 9-inch guns had just been mounted, and I was asked to fire the first shot. Everyone stood clear—I tugged a lanyard—there was a loud bang and a great recoil, and the observation post reported that the shell had hit its mark. I claim no credit for the aim. Later I was asked to inspect and address a parade of the Brazilian Brigade, the forerunners of the Brazilian Division, which had just arrived and made an imposing spectacle, together with Negro and Japanese-American units.

All the time amid these diversions I talked with Mark Clark. The General seemed embittered that his army had been robbed of what he thought—and I could not disagree—was a great opportunity. Still, he would drive forward to his utmost on the British left and keep the whole front blazing. It was late and I was thoroughly tired out when I got back to the château at Siena, where Alexander came again to dine.

When one writes things on paper to decide or explain large questions affecting action there is mental stress. But all this bites much deeper when you see and feel it on the spot. Here was this splendid army, equivalent to twenty-five divisions, of which a quarter were American, reduced till it was just not strong enough to produce decisive results against the immense power of the defensive. A very little more, half what had been taken from us, and we could have broken into the valley of the Po, with all the gleaming possibilities and prizes which lay open towards Vienna. As it was our·forces, about a million strong, could play a mere secondary part in any commanding strategic conception. They could keep the enemy on their front busy at the cost and risk of a hard offensive. They could at least do their duty. Alexander maintained his soldierly cheerfulness, but it was in a sombre mood that I went to bed. In these great matters failing to gain one's way is no escape from the responsibility for an inferior solution.

*   *   *   *   *

As Alexander's offensive could not start till the 26th I flew to Rome on the morning of the 21st. Here another set of problems and a portentous array of new personages to meet awaited me. Brooke had arrived, and also "Peter" Portal. Walter Moyne, so soon to die from an assassin's bullet, had come from Cairo, and Mr. Leeper was also present.* Again the issue was in most cases not what ought to be done—that would have been too easy—but what was likely to be agreed to not only at home but between Allies.

First I had to deal with the impending Greek crisis, which had been one of the chief reasons for my Italian visit. On July 7 the King of Greece had telegraphed from Cairo that after two months of "cunning and futile arguments" the E.A.M. extremists had repudiated the Lebanon agreement which their leaders had signed in May.† He begged us to declare once again that we would support the Government of M. Papandreou, because it represented most of the Greek nation except the extremists and was the only body which could stop civil war and unite the country against the Germans. He also asked us to denounce E.L.A.S. and withdraw the military missions which we had sent to help them

---

* The Chief of the Imperial General Staff, the Chief of the Air Staff, and our Ambassadors to Egypt and Greece.
† See Volume V (Cassell's edition), p. 487.

fight against Hitler. The British Government agreed to support M. Papandreou, but after a long talk on July 15 with Colonel Woodhouse, a British officer who was serving with the missions in Greece, I consented to let them remain for the time being. He argued that they were a valuable restraint on E.A.M. and that it might be difficult and dangerous to get them out, but I feared that one day they would be taken as hostages and I asked for them to be reduced.

Rumours of the German evacuation of Greece raised intense excitement and discord in M. Papandreou's Cabinet, and revealed the frail and false foundation upon which common action stood. This made it all the more necessary for me to see Papandreou and those he trusted.

Before I left London the following telegrams had passed:

Prime Minister to Foreign Secretary   6 Aug 44

Surely we should tell M. Papandreou he should continue as Prime Minister and defy them all. The behaviour of E.A.M. is absolutely intolerable. Obviously they are seeking nothing but the Communisation of Greece during the confusion of the war, without allowing the people to decide in any manner understood by democracy.

2. We cannot take a man up as we have done Papandreou and let him be thrown to the wolves at the first snarlings of the miserable Greek [Communist] banditti. Difficult as the world is now, we shall not make our course easier by abandoning people whom we have encouraged to take on serious jobs by promises of support. . . .

4. Should matters go downhill and E.A.M. become master we should have to reconsider keeping any of our mission there and put the Greek people bluntly up against Bolshevism. The case seems to me to have reached the following point: either we support Papandreou, if necessary with force as we have agreed, or we disinterest ourselves utterly in Greece.

I also warned our Chiefs of Staff.

Prime Minister to C.I.G.S.   6 Aug 44

It may be that within a month or so we shall have to put 10,000 or 12,000 men into Athens, with a few tanks, guns, and armoured cars. You have a division in England which has above 13,000 troops. Such a force could be embarked now, and would probably be in time for the political crisis, which is of major consequence to the policy of His Majesty's Government. Such a force could be supported by troops from the airfields of the Delta, and by scrapings and combings from the 200,000 tail we have in Egypt.

2. I repeat there is no question of trying to dominate Greece or going outside the immediate curtilage of Athens, but this is the centre of government, and, with the approaches to it, must be made secure. Bren gun carriers would be very useful. If you have a better plan let me know it.

3. It is to be presumed that the Germans have gone or are streaking away to the north, and that the force landed at the Piræus would be welcomed by the great majority of the population of Athens, including all notables. The utmost secrecy must enwrap this project. The whole matter will be debated in a Staff conference, with Ministers present, on Tuesday or Wednesday.

4. You should note that time is more important than numbers, and that 5,000 men in five days is better than 7,000 men in seven days. The force is not of course expected to be mobile. Pray speak to me at the first opportunity.

Matters were arranged accordingly.

*　*　*　*　*

After reaching Naples I began to make the necessary arrangements.

Prime Minister (Italy) to Foreign Secretary　　　　　16 Aug 44

I am not aware of, and certainly never consciously agreed to, any British Cabinet decision that the King of Greece should be advised not to return to Greece until after the plebiscite has taken place, but to come to London. It would be much better to see how events develop, particularly as it may not be possible for a plebiscite to be held under orderly conditions for many months. Perhaps Papandreou's new Government, once safely installed in Athens, may be prepared to invite the King, who would not of course start for Greece at once, but would stay behind in Cairo awaiting developments. I can meet Papandreou in Rome on 21st, when Mr. Leeper should be present.

Regarding our expedition to Greece, General Wilson and his staff are already taking action on the Chiefs of Staff telegram, which I have read. . . . I have strongly emphasised that the operation must be regarded as one of reinforced diplomacy and policy rather than an actual campaign, and that it is to be confined to Athens, with possibly a detachment at Salonika.

As soon as the landing-ground has been secured by the 1,500 British parachutists the Greek Government would follow almost immediately, and within a very few hours should be functioning in Athens, where the people would probably receive the British parachutists with rapture. The arrival of the parachutists in the neighbourhood of

Athens could be effected with complete surprise, and might well be effected before E.A.M. had taken any steps to seize the capital. It might be possible to rely on the two Greek aviation squadrons as part of the air force mentioned, but this can be settled at a later date.

Our small expedition, not exceeding 10,000 men, should start from Alexandria or from the heel of Italy, at about the same time that the parachute landing took place, and, after entering the Piræus when the mines were out of the way, relieve the parachutists, who will be needed elsewhere. Most careful consideration will need to be given to the date. We should however be there first, and another unopposed landing might thus be ensured.

Providing the minesweepers were available, and especially if there was a friendly Government installed in Athens, the considerable process of sweeping the mined approaches to the Piræus, on which the C.-in-C. Mediterranean dwelt a great deal, could no doubt be effected in a few days. The C.-in-C. would like about a month's notice to carry out all the necessary preparations.

It is of course necessary in an integrated Anglo-American Staff that the Americans should share in planning such a movement. They have up to now shared fully in post-war planning for Greece in common with the rest of the Mediterranean. American carrying aircraft will be needed for the operation, and we shall have to detach a portion of the minesweepers from "Dragoon". With the large naval resources available this should not be difficult.

I had also telegraphed to the President.

Prime Minister to President Roosevelt   17 Aug 44
We have always marched together in complete agreement about Greek policy, and I refer to you on every important point. The War Cabinet and Foreign Secretary are much concerned about what will happen in Athens, and indeed in Greece, when the Germans crack or when their divisions try to evacuate the country. If there is a long hiatus after German authorities have gone from the city before organised government can be set up it seems very likely that E.A.M. and Communist extremists will attempt to seize the city and crush all other forms of Greek expression but their own.

2. You and I have always agreed that the destinies of Greece are in the hands of the Greek people, and that they will have the fullest opportunity of deciding between monarchy or republic as soon as tranquillity has been restored, but I do not expect you will relish more than I do the prospect either of chaos and street-fighting or of a tyrannical Communist Government being set up. This could only serve to delay and hamper all the plans which are being made by

U.N.R.R.A. for the distribution of relief to the sorely tried Greek people. I therefore think that we should make preparations through the Allied Staff in the Mediterranean to have in readiness a British force, not exceeding 10,000 men, which could be sent by the most expeditious means into the capital when the time is ripe. The force would include parachute troops, for which the help of your Air Force would be needed. I do not myself expect that anything will happen for a month, and it may be longer, but it is always well to be prepared. As far as I can see there will be no insuperable difficulty. I hope therefore you will agree that we may make these preparations by the Staffs out here in the usual way. If so, the British Chiefs of Staff will submit to the Combined Chiefs of Staff draft instructions to General Wilson.

His reply, which arrived more than a week later, was decisive.

President Roosevelt to Prime Minister   26 Aug 44
I have no objection to your making preparations to have in readiness a sufficient British force to preserve order in Greece when the German forces evacuate that country. There is also no objection to the use by General Wilson of American transport aeroplanes that are available to him at that time and that can be spared from his other operations.

*   *   *   *   *

I met M. Papandreou in Rome on the evening of August 21. He said that E.A.M. had joined his Government because the British had been firm towards them, but the Greek State itself still had no arms and no police. He asked for our help to unite Greek resistance against the Germans. At present only the wrong people had arms, and they were a minority. I told him we could make no promise and enter into no obligations about sending British forces into Greece, and that even the possibility should not be talked about in public; but I advised him to transfer his Government at once from Cairo, with its atmosphere of intrigue, to somewhere in Italy near the headquarters of the Supreme Allied Commander. This he agreed to do.

At this point Lord Moyne joined us, and conversation turned to the position of the Greek King. I said there was no need for him to make any new declarations, because he had already said he would follow his Government's advice about going back to his country. The British nation felt friendly and chivalrous towards him for his conduct at a difficult moment in both our histories. We had no intention of interfering with the solemn right of the Greek people to choose between monarchy and a republic. But it must be for the Greek people as a whole, and not a handful of doctrinaires, to decide so grave an issue. Although I personally gave my loyalty to the constitutional monarchy which had taken shape in England, His Majesty's Government were quite indifferent as to which way the matter was settled for Greece provided there was a fair plebiscite.

I observed that now that E.A.M. had stopped demanding his withdrawal and were asking to join him M. Papandreou was head of a truly national Government. But I warned him against subversive influences. We agreed that the Greek mutineers ought not to be released from custody at this climax of the war, and that we should wait and see how they and their representatives behaved before sending any more arms to E.L.A.S. We should try instead to form a National Army for Greece.

M. Papandreou also complained that the Bulgarians were still occupying Greek soil. I said we would order them back to their own frontiers as soon as we were able to make sure they would obey us, but that Greek claims against them here and in the Dodecanese must wait till after the war. In the meantime we would do all we could for the relief and reconstruction of his country, which had suffered much and deserved the best possible treatment. They too must pull their weight, and the best thing he could do was to establish a Greek Government in Greece. Frontier questions must wait for the peace settlement.

* * * * *

I told Mr. Eden about all this.

Prime Minister (Rome) to Foreign Secretary   22 Aug 44

For reasons which will presently become apparent, I shall be returning to Alexander's army on the night of the 22nd–23rd, and hope to be at Chequers in time for Matins next Sunday.

2. We hope to effect some simplifications in the military commands here, and the C.I.G.S. is working in collaboration with Alexander and later with Wilson to secure to the maximum extent the unique position held by Alexander in Italy.

3. As to the King of Greece, they none of them want him to make a fresh declaration now. Regarding his proposal to return to London, I have advised him to wait till he sees M. Papandreou when he gets back, and then come on home. At a later date a visit to Italy might be considered, and he could then revisit the purged and penitent Greek Brigade here, preferably when they are in the line.

4. I like Papandreou, and there are great advantages in the removal of the Greek Government from the Cairo atmosphere. I think it will do good to save an alert in Greece both of foes and friends such as will be produced by its movement. But while the military affairs are being planned and sorted out here under my direction in accordance with the wishes you expressed a date cannot be fixed; it must be fitted in with other needs, unless the situation itself takes charge. I cannot be ready to act for a month, but thereafter we may be able to pounce when the going is good. Moyne is working with General Wilson this morning, subdividing the departments which remain and those which came forward to Italy. Of course the heavy international organisations and dumps will remain where they are.

I am very glad you had a tour in France in these thrilling and decisive days.

* * * * *

I stayed while in Rome at the Embassy, and our Ambassador, Sir Noel Charles, and his wife devoted themselves to my business and comfort. Guided by his advice, I met most of the principal figures in the *débris* of Italian politics produced by twenty years of dictatorship, a disastrous war, revolution, invasion, occupation, Allied control, and other evils. I had talks with, among others, Signor Bonomi and General Badoglio, also with Comrade Togliatti, who had returned to Italy at the beginning of the year after a long sojourn in Russia. The leaders of all the Italian parties were invited to meet me. None had any electoral mandate, and their party names, revived from the past, had been chosen with an eye to the future. "What is your party?" I asked one group. "We are the Christian Communists," their chief replied. I could not help saying, "It must be very inspiring to your party, having the Catacombs so handy." They did not seem to see the point, and, looking back, I am afraid their minds must have turned to the cruel mass executions which the Germans had so recently perpetrated in these ancient sepulchres. One may however be pardoned for making historical references in Rome. The Eternal City, rising on every side, majestic and apparently invulnerable, with its monuments and palaces, and with its splendour of ruins not produced by bombing, seemed to contrast markedly with the tiny and transient beings who flitted within its bounds.

On August 23 I was received in audience by the Pope. I had visited his predecessor when I came to Rome as Chancellor of the Exchequer with Randolph, then very young, in 1926, and I preserved most agreeable memories of the kindness with which we had been received. Those were the days of Mussolini. Now I was received by Pope Pius XII with the highest ceremony. Not only did the Papal Guard in all their stately array line the long series of ante-rooms and galleries through which we passed, but the Noble Guards, formed of representatives of the highest and most ancient families of Rome, with a magnificent mediæval uniform I had never seen before, were present. The Pope received me in his study with the dignity and informality which he can so happily combine. We had no lack of topics for conversation. The one that bulked the largest at this audience, as it had done with his predecessor eighteen years before, was the danger of Communism. I have always had the greatest dislike of it; and should I ever have the honour of another audience with the Supreme Pontiff I should not hesitate to recur to the subject.

Our Minister to the Vatican, Sir D'Arcy Osborne, drove me back to the Embassy. Here I met for the first time the Crown Prince Umberto, who, as Lieutenant of the Realm, was commanding the Italian forces on our front. His powerful and engaging personality, his grasp of the whole situation, military and political, were refreshing, and gave one a more lively feeling of confidence than I had experienced in my talks with the politicians. I certainly hoped he would play his part in building up a constitutional monarchy in a free, strong, united Italy. However, this was none of my business. I had enough on hand as it was. The Warsaw rising had now been in progress for nearly a month. The insurgents were in desperate straits, and I was engaged in a tense correspondence with Stalin and the President which will be set forth in another chapter.

CHAPTER VIII

ALEXANDER'S SUMMER OFFENSIVE

I Return to Alexander's Headquarters at Siena, August 24 – I Visit the New Zealand Division – A Meeting with General Devers – A Visit to General Leese – My Telegram to Smuts, August 26 – The Attack Begins, August 26 – A Magnificent View – An Adventurous Motor Drive – My Telegram to the President, August 28 – His Reply – Further Correspondence – The President's Vain Hopes – I Fly Home to England: August 28 – My Message to the Italian People.

EARLY on August 24, after my short visit to Rome, I returned by air to Alexander's headquarters at Siena, living in the château a few miles away. The offensive was now fixed for the 26th. I took the opportunity of visiting the New Zealand Division. The last time I had inspected them was at Tripoli in February 1943. I did not wish to have another formal parade, and instead the soldiers gathered along the route and gave me an informal, enthusiastic welcome. I was delighted to see Freyberg and his officers again. To Mr. Fraser I telegraphed:

Prime Minister to Prime Minister of New Zealand   25 Aug 44

It was with great pleasure that I saw about 15,000 men of your really magnificent division in the best of spirits. The division is sorely needed in the forthcoming operation. I had lunch with General Freyberg and his officers yesterday. I told them a good many things they had not heard, and would not hear in the ordinary course. Freyberg sends his respects and good wishes, and so do I.

We were to fly to General Leese's battle headquarters of the Eighth Army, on the Adriatic side, on the afternoon of the 25th. Before starting I spent some hours with Alexander in his headquarters camp. While I was there General Devers and another high American general arrived unexpectedly. The much-disputed operation "Anvil", now called "Dragoon", was at present under

General Patch, but Devers as Deputy to General Wilson had for many weeks been drawing units and key men ruthlessly from the Fifteenth Army Group, and particularly from the Fifth Army under Mark Clark. It was known that the troops of "Dragoon" were likely to be raised to an Army Group Command, and that Devers would be designated as its chief. Naturally he sought to gather all the forces he could for the great enterprise to be entrusted to him, and to magnify it in every way. I saw very soon, although no serious topic was broached, that there was a coolness between him and Alexander. Gay, smiling, debonair, Alexander excused himself after the first few minutes, and left me in the mess tent with our two American visitors. As General Devers did not seem to have anything particular to say to me, and I did not wish to enter upon thorny ground, I also confined myself to civilities and generalities. I expected Alexander to return, but he did not, and after about twenty minutes Devers took his leave. There was of course no public business to be done. I wished him all good luck in his operation, and his courtesy call came to an end. I was conscious however of the tension between these high officers beneath an impeccable surface of politeness.

Presently Alexander came to tell me that we should now drive to the airfield. We took off in his plane and flew north-eastwards for half an hour to Loreto, whence we drove to Leese's camp behind Monte Maggiore. Here we had tents overlooking a magnificent panorama to the northward. The Adriatic, though but twenty miles away, was hidden by the mass of Monte Maggiore. General Leese told us that the barrage to cover the advance of his troops would begin at midnight. We were well placed to watch the long line of distant gun-flashes. The rapid, ceaseless thudding of the cannonade reminded me of the First World War. Artillery was certainly being used on a great scale. After an hour of this I was glad to go to bed, for Alexander had planned an early start and a long day on the front. He had also promised to take me wherever I wanted to go.

* * * * *

Before going to sleep I dictated the following message to Smuts, with whom my correspondence was continuous:

Prime Minister to Field-Marshal Smuts 26 Aug 44
A very considerable battle starts this morning and afternoon, and will

reach its full power to-morrow. Hence my presence here for a couple of days. I must then return to England, visit France, and thereafter go on to Canada, for conferences starting about the middle of September. I tried to see the South Africans yesterday, but they were on the march.

So far "Anvil" has had the opposite effects for which its designers intended it. Firstly, it has attracted no troops away from General Eisenhower at all. On the contrary, two and a half to three divisions of German rearguard troops will certainly reach the main battle-front before the Allied landed troops. Secondly, a stage of stagnation has been enforced here by the breaking in full career of these two great armies, the Fifth and the Eighth, and by the milking out of the key personnel in them. The consequence of this has been the withdrawal from the Italian front of three German divisions, including one very strong Panzer having an active strength of 12,500. These have proceeded direct to the Chalons area. Thus about five divisions have been deployed against Eisenhower, which would not have happened had we continued our advance here in the direction of the Po and ultimately on the great city [Vienna]. I still hope that we may achieve this. Even if the war comes to an end suddenly I can see no reason why our armour should not slip through and reach it, as we can.

* * * * *

Alexander and I started together at about nine o'clock. His aide-de-camp and Tommy came in a second car. We were thus a conveniently small party. The advance had now been in progress for six hours, and was said to be making headway. But no definite impressions could yet be formed. We first climbed by motor up a high outstanding rock pinnacle, upon the top of which a church and village were perched. The inhabitants, men and women, came out to greet us from the cellars in which they had been sheltering. It was at once plain that the place had just been bombarded. Masonry and wreckage littered the single street. "When did this stop?" Alexander asked the small crowd who gathered round us, grinning rather wryly. "About a quarter of an hour ago," they said. There was certainly a magnificent view from the ramparts of bygone centuries. The whole front of the Eighth Army offensive was visible. But apart from the smoke puffs of shells bursting seven or eight thousand yards away in a scattered fashion there was nothing to see. Presently Alexander said that we had better not stay any longer, as the enemy would naturally be firing at observation posts like this and might begin again. So we motored two or three miles to the westward, and had a picnic lunch on the broad slope of a hillside, which gave almost as good a view as the peak and was not likely to attract attention.

News was now received that our troops had pushed on a mile or two beyond the river Metauro. Here Hasdrubal's defeat had sealed the fate of Carthage, so I suggested that we should go across too. We got into our cars accordingly, and in half an hour were across the river, where the road ran into undulating groves of olives, brightly patched with sunshine. Having got an officer guide from one of the battalions engaged, we pushed on through these glades till the sounds of rifle and machine-gun fire showed we were getting near to the front line. Presently warning hands brought us to a standstill. It appeared there was a minefield, and it was only safe to go where other vehicles had already gone without mishap. Alexander and his aide-de-camp now went off to reconnoitre towards a grey stone building which our troops were holding, which was said to give a good close-up view. It was evident to me that only very loose fighting was in progress. In a few minutes the aide-de-camp came back and brought me to his chief, who had found a very good place in the stone building, which was in fact an old château overlooking a rather sharp declivity. Here one certainly could see all that was possible. The Germans were firing with rifles and machine-guns from thick scrub on the farther side of the valley, about five hundred yards away. Our front line was beneath us. The firing was desultory and intermittent. But this was the nearest I got to the enemy and the time I heard most bullets in the Second World War. After about half an hour we went back to our motor-cars and made our way to the river, keeping very carefully to our own wheel tracks or those of other vehicles. At the river we met the supporting columns of infantry, marching up to lend weight to our thin skirmish line, and by five o'clock we were home again at General Leese's headquarters, where the news from the whole of the Army front was marked punctually on the maps. On the whole the Eighth Army had advanced since daybreak about seven thousand yards on a ten- or twelve-mile front, and the losses had not been at all heavy. This was an encouraging beginning.

* * * * *

The next morning plenty of work arrived, both by telegram

and pouch. It appeared that General Eisenhower was worried by the approach of the German divisions I had mentioned to Smuts as having been withdrawn from Italy. I was glad that our offensive, prepared under depressing conditions, had begun. I drafted a telegram to the President explaining the position as I had learned it from the generals on the spot and from my own knowledge. I wished to convey in an uncontroversial form our sense of frustration, and at the same time to indicate my hopes and ideas for the future. If only I could revive the President's interest in this sphere we might still keep alive our design of an ultimate advance to Vienna.

Prime Minister to President Roosevelt   28 Aug 44

General Alexander received a telegram from S.H.A.E.F. asking for efforts to be made to prevent the withdrawal of more [German] divisions from the Italian front. This of course was the consequence of the great weakening of our armies in Italy, and has taken place entirely since the attack on the Riviera. In all, four divisions have left, including a very strong Panzer *en route* for Chalons. However, in spite of the weakening process Alexander began about three weeks ago to plan with Clark to turn or pierce the Apennines. For this purpose the British XIIIth Corps of four divisions has been placed under General Clark's orders, and we have been able to supply him with the necessary artillery, of which his army had been deprived. This army of eight divisions—four American and four British—is now grouped around Florence on a northerly axis.

2. By skinning the whole front and holding long stretches with nothing but anti-aircraft gunners converted to a kind of artillery-infantry and supported by a few armoured brigades, Alexander has also been able to concentrate ten British or British-controlled divisions representative of the whole British Empire on the Adriatic flank. The leading elements of these attacked before midnight on the 25th, and a general barrage opened and an advance began at dawn on the 26th. An advance of about nine miles was made over a large area, but the main position, the Gothic Line, has still to be encountered. I had the good fortune to go forward with this advance, and was consequently able to form a much clearer impression of the modern battlefield than is possible from the kinds of pinnacles and perches to which I have hitherto been confined.

3. The plan is that the Eighth Army of ten divisions, very heavily weighted in depth, will endeavour to pierce the Gothic Line and turn the whole enemy's position, entering the Po valley on the level of Rimini; but at the right moment, depending on the reactions of the

enemy, Mark Clark will strike with his eight divisions and elements of both armies should converge to Bologna. If all goes well I hope that the advance will be much more rapid after that and that the continued heavy fighting will prevent further harm being done to Eisenhower by the withdrawal of divisions from Italy.

4. I have never forgotten your talks to me at Teheran about Istria, and I am sure that the arrival of a powerful army in Trieste and Istria in four or five weeks would have an effect far outside purely military values. Tito's people will be awaiting us in Istria. What the condition of Hungary will be then I cannot imagine, but we shall at any rate be in a position to take full advantage of any great new situation.

I did not send this message off till I reached Naples, whither I flew on the 28th, nor did I receive the answer till three days after I got home.

President Roosevelt to Prime Minister   31 Aug 44

I was very glad to receive your account of the way in which General Wilson has concentrated his forces in Italy and has now renewed the offensive. My Chiefs of Staff feel that a vigorous attack, using all the forces available, should force the enemy into the Po valley. The enemy may then choose to withdraw entirely from Northern Italy. Since such action on his part might enable the enemy to release divisions for other fronts, we must do our best to destroy his forces while we have them in our grasp. I am confident that General Wilson has this as his objective. With an offensive under way and being pressed full strength in Italy, I am sure that General Eisenhower will be satisfied that everything possible is being done in the Mediterranean to assist him by mauling German divisions which might otherwise be moved against his forces in the near future. I understand all available British resources in the Mediterranean are being put into Italy. We are pressing into France all reinforcements and resources we can in order to guarantee that General Eisenhower will be able to maintain the impetus of the joint victories our forces have already won. With the smashing success of our invasion of Southern France and the Russians now crumbling the enemy flank in the Balkans, I have great hopes that complete and final victory will not be long delayed.

It is my thought that we should press the German Army in Italy vigorously with every facility we have available, and suspend decision of the future use of General Wilson's armies until the results of his campaign are better known and we have better information as to what the Germans may do.

We can renew our Teheran talk about Trieste and Istria at "Octagon" [Quebec].

I was struck by the emphasis which this message laid upon General Wilson.

Prime Minister to President Roosevelt   31 Aug 44

All operations in Italy are conceived and executed by General Alexander in accordance with his general directives from the Supreme Commander. You will see that he is now in contact for twenty miles on the Adriatic flank with the Gothic Line, and a severe battle will be fought by the Eighth Army. Also General Clark with the Fifth Army has made an advance from the direction of Florence. I have impressed most strongly upon General Alexander the importance of pressing with his utmost strength to destroy the enemy's armed forces as well as turn his line. It will not be easy for the Germans to effect a general retreat from the Gothic Line over the Alps, especially if we can arrive in the neighbourhood of Bologna. The western passes and tunnels into France are already blocked by your advance into the Rhone valley. Only the direct route to Germany is open. We shall do our utmost to engage, harry, and destroy the enemy. The decisive battle has yet however to be fought.

2. In view of the fact that the enemy on the Italian front has been weakened by four of his best divisions, we no longer ask for further American reinforcements beyond the 92nd Division, which I understand will shortly reach us. On the other hand, I take it for granted that no more will be withdrawn from Italy—*i.e.*, that the four divisions of Clark's army and the elements remaining with them will continue there, and that General Alexander should make his plans on that basis. So much for the present.

3. As to the future, continuous employment against the enemy will have to be found for the Eighth and Fifth Armies once the German armies in Italy have been destroyed or unluckily have made their escape. This employment can only take the form of a movement first to Istria and Trieste and ultimately upon Vienna. Should the war come to an end in a few months, as may well be possible, none of these questions will arise. Anyhow, we can talk this over fully at Quebec.

4. I congratulate you upon the brilliant success of the landings in Southern France. I earnestly hope the retreating Germans may be nipped at Valence or Lyons and rounded up. Another mob of about 90,000 is apparently streaming back from the south via Poitiers.

Roosevelt sent me another telegram on September 4.

President Roosevelt to Prime Minister   4 Sept 44

I share your confidence that the Allied divisions we have in Italy are sufficient to do the task before them and that the battle commander will press the battle unrelentingly with the objective of shattering the enemy forces. After breaking the German forces on the Gothic Line we must go on to use our divisions in the way which best aids General Eisenhower's decisive drive into the enemy homeland.

As to the exact employment of our forces in Italy in the future, this is a matter we can discuss at Quebec. It seems to me that American forces should be used to the westward, but I am completely open-minded on this, and in any event this depends on the progress of the present battle in Italy, and also in France, where I strongly feel that we must not stint in any way the forces needed to break quickly through the western defences of Germany.

The credit for the great Allied success in Southern France must go impartially to the combined Allied force, and the perfection of execution of the operation from its beginning to the present belong to General Wilson and his Allied staff and to Patch and his subordinate commanders. With the present chaotic conditions of the Germans in Southern France, I hope that a junction of the north and south forces may be obtained at a much earlier date than was first anticipated.

We shall see that both these hopes proved vain. The army which we had landed on the Riviera at such painful cost to our operations in Italy arrived too late to help Eisenhower's first main struggle in the north, while Alexander's offensive failed, by the barest of margins, to achieve the success it deserved and we so badly needed. Italy was not to be wholly free for another eight months; the right-handed drive to Vienna was denied to us; and, except in Greece, our military power to influence the liberation of South-Eastern Europe was gone.

On August 28 I flew home from Naples. Before leaving Italy I set myself to composing a short message of encouragement and hope for the Italian people, for whom I have always had, except when we were actually fighting, a great regard. I had been deeply touched by the kindness with which I was welcomed in all the villages and small towns through which I had driven in traversing the entire front. In return I offered a few words of counsel.

28 Aug 44

. . . It has been said that the price of freedom is eternal vigilance. The question arises, What is freedom? There are one or two quite simple, practical tests by which it can be known in the modern world in peace conditions, namely:

Is there the right to free expression of opinion and of opposition and criticism of the Government of the day?

Have the people the right to turn out a Government of which

they disapprove, and are constitutional means provided by which they can make their will apparent?

Are their courts of justice free from violence by the Executive and from threats of mob violence, and free of all association with particular political parties?

Will these courts administer open and well-established laws which are associated in the human mind with the broad principles of decency and justice?

Will there be fair play for poor as well as for rich, for private persons as well as Government officials?

Will the rights of the individual, subject to his duties to the State, be maintained and asserted and exalted?

Is the ordinary peasant or workman who is earning a living by daily toil and striving to bring up a family free from the fear that some grim police organisation under the control of a single party, like the Gestapo, started by the Nazi and Fascist parties, will tap him on the shoulder and pack him off without fair or open trial to bondage or ill-treatment?

These simple, practical tests are some of the title-deeds on which a new Italy could be founded. . . .

This does not seem to require any alteration to-day.

CHAPTER IX

THE MARTYRDOM OF WARSAW

The Russians Cross the Vistula – Germany's Collapse on the Eastern Front – The Broadcast from Moscow, July 29, Calling for a General Rising in Warsaw – The Insurrection Begins, August 1 – My Telegram to Stalin of August 4 – A Grim Reply – The German Counter-Attack – A Distressing Message from Warsaw – My Telegram to Eden, August 14 – Vyshinsky's Astonishing Statement, and Stalin's Telegram of August 16 – The President and I Send a Joint Appeal, August 20 – Stalin's Answer – The Agony of Warsaw Reaches its Height – Mr. Roosevelt's Message to Me of August 24 – Our Need of Soviet Airfields – The President is Adverse – Anger of the British War Cabinet – Their Telegram to Moscow, September 4 – Mr. Roosevelt's Message of September 5 – Apparent Change in Soviet Tactics – Our Heavy Bombers Drop Supplies on Warsaw, September 18 – End of the Tragedy.

THE Russian summer offensive brought their armies in late July to the river Vistula. All reports indicated that in the very near future Poland would be in Russian hands. The leaders of the Polish Underground Army, which owed allegiance to the London Government, had now to decide when to raise a general insurrection against the Germans, in order to speed the liberation of their country and prevent them fighting a series of bitter defensive actions on Polish territory, and particularly in Warsaw itself. The Polish commander, General Bor-Komorowski, and his civilian adviser were authorised by the Polish Government in London to proclaim a general insurrection whenever they deemed fit. The moment indeed seemed opportune. On July 20 came the news of the plot against Hitler, followed swiftly by the Allied break-out from the Normandy beach-head. About July 22 the Poles intercepted wireless messages from the German Fourth

Panzer Army ordering a general withdrawal to the west of the Vistula. The Russians crossed the river on the same day, and their patrols pushed forward in the direction of Warsaw. There seemed little doubt that a general collapse was at hand. In the Nuremberg trials General Guderian described the situation in these terms:

On July 21, 1944, I received a new appointment as Chief of Staff of the German forces on the Eastern Front. After my appointment the whole front—if it can be called a front—was hardly more than an agglomeration of the remains of our armies which were endeavouring to withdraw to the line of the Vistula; twenty-five divisions were completely annihilated.

General Bor therefore decided to stage a major rising and liberate the city. He had about forty thousand men, with reserves of food and ammunition for seven to ten days' fighting. The sound of Russian guns across the Vistula could now be heard. The Soviet Air Force began bombing the Germans in Warsaw from recently captured airfields near the capital, of which the closest was only twenty minutes' flight away. At the same time a Communist Committee of National Liberation had been formed in Eastern Poland, and the Russians announced that liberated territory would be placed under their control. Soviet broadcasting stations had for a considerable time been urging the Polish population to drop all caution and start a general revolt against the Germans. On July 29, three days before the rising began, the Moscow radio station broadcast an appeal from the Polish Communists to the people of Warsaw, saying that the guns of liberation were now within hearing, and calling upon them as in 1939 to join battle with the Germans, this time for decisive action. "For Warsaw, which did not yield but fought on, the hour of action has already arrived." After pointing out that the German plan to set up defence points would result in the gradual destruction of the city, the broadcast ended by reminding the inhabitants that "all is lost that is not saved by active effort", and that "by direct active struggle in the streets, houses, etc., of Warsaw the moment of final liberation will be hastened and the lives of our brethren saved".

On the evening of July 31 the Underground command in Warsaw got news that Soviet tanks had broken into the German defences east of the city. The German military wireless announced, "To-day the Russians started a general attack on Warsaw from the south-east." Russian troops were now at points less than ten miles away. In the capital itself the Polish Underground command ordered a general insurrection at 5 p.m. on the following day. General Bor has himself described what happened:

At exactly five o'clock thousands of windows flashed as they were flung open. From all sides a hail of bullets struck passing Germans, riddling their buildings and their marching formations. In the twinkling of an eye the remaining civilians disappeared from the streets. From the entrances of houses our men streamed out and rushed to the attack. In fifteen minutes an entire city of a million inhabitants was engulfed in the fight. Every kind of traffic ceased. As a big communications centre where roads from north, south, east, and west converged, in the immediate rear of the German front, Warsaw ceased to exist. The battle for the city was on.

The news reached London next day, and we anxiously waited for more. The Soviet radio was silent and Russian air activity ceased. On August 4 the Germans started to attack from strongpoints which they held throughout the city and suburbs. The Polish Government in London told us of the agonising urgency of sending in supplies by air. The insurgents were now opposed by five hastily concentrated German divisions. The Hermann Goering Division had also been brought from Italy, and two more S.S. divisions arrived soon afterwards.

I accordingly telegraphed to Stalin:

Prime Minister to Marshal Stalin   4 Aug 44
At urgent request of Polish Underground Army we are dropping, subject to weather, about sixty tons of equipment and ammunition into the south-west quarter of Warsaw, where it is said a Polish revolt against the Germans is in fierce struggle. They also say that they appeal for Russian aid, which seems to be very near. They are being attacked by one and a half German divisions. This may be of help to your operation.

The reply was prompt and grim.

Marshal Stalin to Prime Minister   5 Aug 44
I have received your message about Warsaw.
I think that the information which has been communicated to you by the Poles is greatly exaggerated and does not inspire confidence.

One could reach that conclusion even from the fact that the Polish emigrants have already claimed for themselves that they all but captured Vilna with a few stray units of the Home Army, and even announced that on the radio. But that of course does not in any way correspond with the facts. The Home Army of the Poles consists of a few detachments, which they incorrectly call divisions. They have neither artillery nor aircraft nor tanks. I cannot imagine how such detachments can capture Warsaw, for the defence of which the Germans have produced four tank divisions, among them the Hermann Goering Division.

Meanwhile the battle went on street by street against the German "Tiger" tanks, and by August 9 the Germans had driven a wedge right across the city through to the Vistula, breaking up the Polish-held districts into isolated sectors. The gallant attempts of the R.A.F., with Polish, British, and Dominion crews, to fly to the aid of Warsaw from Italian bases were both forlorn and inadequate. Two planes appeared on the night of August 4, and three four nights later.

*   *   *   *   *

The Polish Prime Minister, Mikolajczyk, had been in Moscow since July 30 trying to establish some kind of terms with the Soviet Government, which had recognised the Polish Communist Committee of National Liberation as the future administrators of the country. These negotiations were carried on throughout the early days of the Warsaw rising. Messages from General Bor were reaching Mikolajczyk daily, begging for ammunition and anti-tank weapons and for help from the Red Army. Meanwhile the Russians pressed for agreement upon the post-war frontiers of Poland and the setting up of a joint Government. A last fruitless talk took place with Stalin on August 9.

On August 12 I telegraphed to him:

Prime Minister to Marshal Stalin   12 Aug 44

I have had the following distressing message from the Poles in Warsaw, who after ten days are still fighting against considerable German forces, who have cut the city into three:

[*Begins.*] "To the President of the Republic, the Government, and the Commander-in-Chief, from the Vice-Premier:

"Tenth day. We are conducting a bloody fight. The town is cut by three routes. . . . All these routes are strongly held by German tanks and their crossing is extremely difficult (all the buildings along them are burnt out). Two armoured trains on the railway line from Gdansk Station to West Station and artillery from Praga fire continuously on the town, and are supported by air forces.

"In these conditions the fight continues. We receive from you only once a small drop. On the German-Russian front silence since the 3rd. We are therefore without any material or moral support, as, with the exception of a short speech by the [Polish] Vice-Prime Minister (from London), which took place on the eighth day, we have not had from you even an acknowledgment of our action. The soldiers and the population of the capital look hopelessly at the skies, expecting help from the Allies. On the background of smoke they see only German aircraft. They are surprised, feel deeply depressed, and begin to revile.

"We have practically no news from you, no information with regard to the political situation, no advice and no instructions. Have you discussed in Moscow help for Warsaw? I repeat emphatically that without immediate support, consisting of drops of arms and ammunition, bombing of objectives held by the enemy, and air landing, our fight will collapse in a few days.

"With the above-mentioned help the fight will continue.

"I expect from you the greatest effort in this respect." [Ends.]

They implore machine-guns and ammunition. Can you not give them some further help, as the distance from Italy is so very great?

* * * * *

On the 14th I telegraphed to Mr. Eden from Italy, whither I had gone to see General Alexander's army:

It will cause the Russians much annoyance if the suggestion that the Polish patriots in Warsaw were deserted gets afoot, but they can easily prevent it by operations well within their power. It certainly is very curious that at the moment when the Underground Army has revolted the Russian armies should have halted their offensive against Warsaw and withdrawn some distance. For them to send in all the quantities of machine-guns and ammunition required by the Poles for their heroic fight would involve only a flight of 100 miles. I have been talking to [Air Marshal] Slessor, trying to send all possible assistance from here. But what have the Russians done? I think it would be better if you sent a message to Stalin through Molotov referring to the implications that are afoot in many quarters and requesting that the Russians should send all the help they can. This course would be more impersonal than that I should do it through Stalin. Last night twenty-eight aircraft did the 700 miles flight from Italy. Three were lost. This was the fourth flight made from here under these quite exceptional conditions.

On the night of August 16 Vyshinsky asked the United States

Ambassador in Moscow to call, and, explaining that he wished to avoid the possibility of misunderstanding, read out the following astonishing statement:

The Soviet Government cannot of course object to English or American aircraft dropping arms in the region of Warsaw, since this is an American and British affair. But they decidedly object to American or British aircraft, after dropping arms in the region of Warsaw, landing on Soviet territory, since the Soviet Government do not wish to associate themselves either directly or indirectly with the adventure in Warsaw.

On the same day I received the following message couched in softer terms from Stalin:

Marshal Stalin to Prime Minister   16 Aug 44
After the conversation with M. Mikolajczyk I gave orders that the command of the Red Army should drop arms intensively in the Warsaw sector. A parachutist liaison officer was also dropped, who, according to the report of the command, did not reach his objective as he was killed by the Germans.

Further, having familiarised myself more closely with the Warsaw affair, I am convinced that the Warsaw action represents a reckless and terrible adventure which is costing the population large sacrifices. This would not have been if the Soviet command had been informed before the beginning of the Warsaw action and if the Poles had maintained contact with it.

In the situation which has arisen the Soviet command has come to the conclusion that it must dissociate itself from the Warsaw adventure, as it cannot take either direct or indirect responsibility for the Warsaw action.

According to Mikolajczyk's account, the first paragraph of this telegram is quite untrue. Two officers arrived safely in Warsaw and were received by the Polish command. A Soviet colonel had also been there for some days, and sent messages to Moscow via London urging support for the insurgents.

*   *   *   *   *

On the 18th I telegraphed again to Mr. Eden.

Prime Minister to Foreign Secretary   18 Aug 44
I have seen the extremely lukewarm telegram of August 15 from the American Joint Chiefs of Staff to General Eisenhower, which was received after I had sent you my last message.

The air authorities out here assured me that the Americans wished help sent from England to Warsaw, and that the operation was quite practicable, providing of course the Russians gave their consent. It seems hardly credible to me that the request for landing facilities would have been submitted to the Russians unless the practicability of the operation had been examined by General Doolittle. It is most important that you should find out whether it is practicable or not.

Before the President or I, or both, make any personal or joint appeals to Stalin it is of course necessary that the military difficulties should be resolved.

At the same time I appealed to the President.

Prime Minister (Italy) to President Roosevelt   18 Aug 44
An episode of profound and far-reaching gravity is created by the Russian refusal to permit American aircraft to bring succour to the heroic insurgents in Warsaw, aggravated by their own complete neglect to provide supplies by air when only a few score of miles away. If, as is almost certain, a wholesale massacre follows the German triumph in that capital no measure can be put upon the full consequences that will arise.

2. I am prepared to send a personal message to Stalin if you think this wise, and if you will yourself send a separate similar message. Better far than two separate messages would be a joint message signed by us both.

3. The glorious and gigantic victories being achieved in France by the United States and British forces are vastly changing the situation in Europe, and it may well be that the victory gained by our armies in Normandy will eclipse in magnitude anything that the Russians have achieved on any particular occasion. I feel therefore that they will have some respect for what we say so long as it is expressed plainly and simply. We are nations serving high causes, and must give true counsels towards world peace even at the risk of Stalin resenting it. Quite possibly he wouldn't.

Two days later we sent the following joint appeal, which the President had drafted:

Prime Minister (Italy) and President Roosevelt to   20 Aug 44
Marshal Stalin
We are thinking of world opinion if the anti-Nazis in Warsaw are in effect abandoned. We believe that all three of us should do the utmost to save as many of the patriots there as possible. We hope that you will drop immediate supplies and munitions to the patriot Poles in Warsaw, or will you agree to help our planes in doing it very

quickly? We hope you will approve. The time element is of extreme importance.

This was the reply we got:

Marshal Stalin to Prime Minister and President Roosevelt   22 Aug 44

I have received the message from you and Mr. Roosevelt about Warsaw. I wish to express my opinions.

2. Sooner or later the truth about the group of criminals who have embarked on the Warsaw adventure in order to seize power will become known to everybody. These people have exploited the good faith of the inhabitants of Warsaw, throwing many almost unarmed people against the German guns, tanks, and aircraft. A situation has arisen in which each new day serves, not the Poles for the liberation of Warsaw, but the Hitlerites who are inhumanly shooting down the inhabitants of Warsaw.

3. From the military point of view, the situation which has arisen, by increasingly directing the attention of the Germans to Warsaw, is just as unprofitable for the Red Army as for the Poles. Meanwhile the Soviet troops, which have recently encountered new and notable efforts by the Germans to go over to the counter-attack, are doing everything possible to smash these counter-attacks of the Hitlerites and to go over to a new wide-scale attack in the region of Warsaw. There can be no doubt that the Red Army is not sparing its efforts to break the Germans round Warsaw and to free Warsaw for the Poles. That will be the best and most effective help for the Poles who are anti-Nazis.

$$* \quad * \quad * \quad * \quad *$$

Meanwhile the agony of Warsaw reached its height.

Prime Minister to President Roosevelt   24 Aug 44

The following is an eye-witness account of the Warsaw rising. A copy has already been given to the Soviet Ambassador in London.

1. August 11.

The Germans are continuing, despite all efforts of A.K.,* their ruthless terror methods. In many cases they have burnt whole streets of houses and shot all the men belonging to them and turned the women and children out on to the street, where battles are taking place, to find their way to safety. On Krolewska Street many private houses have been bombed out. One house was hit by four separate bombs. In one house, where lived old retired professors of Polish universities, the S.S. troops forced an entrance and killed many of them. Some succeeded in escaping through the cellars to the other

---

* The Polish Underground Army.

houses. The morale of A.K. and the civilian population is of the highest standard. The watchword is "Death to the Germans."

2. August 11.

The German tank forces during last night made determined efforts to relieve some of their strong-points in the city. This is no light task however, as on the corner of every street are built huge barricades, mostly constructed of concrete pavement slabs torn up from the streets especially for this purpose. In most cases the attempts failed, so the tank crews vented their disappointment by setting fire to several houses and shelling others from a distance. In many cases they also set fire to the dead, who litter the streets in many places. . . . The German Tank Corps have begun to have a great respect for the Polish barricade, for they know that behind each one wait determined troops of A.K. with petrol bottles. These petrol bottles have caused great destruction to many of their comrades.

3. August 13.

The German forces have brutally murdered wounded and sick people, both men and women, who were lying in the SS. Lazarus and Karol and Marsa hospitals.

When the Germans were bringing supplies by tanks to one of their outposts they drove before them 500 women and children to prevent the troops of A.K. from taking action against them. Many of them were killed and wounded. The same kind of action has been reported from many other parts of the city.

Despite lack of weapons, the Polish forces continue to hold the initiative in the battle for Warsaw. In some places they have broken into German strongholds and captured much-needed arms and ammunition. On August 12 11,600 rounds of rifle ammunition, five machine-guns, 8,500 [rounds of] small arms ammunition, twenty pistols, thirty anti-tank mines, and transports were captured. The German forces are fighting desperately. When A.K. set fire to a building which the Germans were holding as a fortress two German soldiers tried to escape to the Polish lines with a white flag, but an S.S. officer saw them and shot them dead. During the night of August 12-13 A.K. received some weapons from Allied aircraft.

4. August 15.

The dead are buried in backyards and squares. The food situation is continually deteriorating, but as yet there is no starvation. To-day there is no water at all in the pipes. It is being drawn from the infrequent wells and house supplies. All quarters of the town are under shell fire, and there are many fires. The dropping of supplies has intensified the morale. Everyone wants to fight and will fight, but the uncertainty of a speedy conclusion is depressing.

5. August 16.

Fighting continues to be very bitter in Warsaw. The Germans fight for every inch of ground. It is reported that in some places whole districts have been burnt and the inhabitants either shot or taken to Germany. The inhabitants continue to repeat, "When we get weapons we will pay them back."

Fighting for the electric power station began on August 1 at 5.10 p.m. Twenty-three soldiers of the Polish Home Army were stationed in the works before that hour, because they were employed in the normal course of things, expecting the outbreak of the rising. The Germans had on the day before raised the strength of the garrison to 150 militarised police, stationed in concrete pill-boxes and block-houses, also in all the works buildings. The signal for action was the explosion of a mine under one of the buildings. After nineteen hours of fighting the electric power station was fully in Polish hands. The Polish losses were seventeen killed and twenty-seven wounded. The German losses were twenty killed and twenty-two wounded, with fifty-six taken as prisoners of war. The detachment which captured the station consisted solely of manual and metal workers of the works. In spite of the fact that the buildings of the station are daily bombarded with 75-mm. shells by the Germans, the personnel has succeeded in maintaining the supply of current to the civil population without the slightest interruption.

The battle also raged literally underground. The only means of communication between the different sectors held by the Poles lay through the sewers. The Germans threw hand grenades and gas bombs down the manholes. Battles developed in pitch-darkness between men waist-deep in excrement, fighting hand to hand at times with knives or drowning their opponents in the slime. Above ground German artillery and fighters set alight large areas of the city.

I thought that some of this tale of villainy and horror should reach the world.

Prime Minister (Italy) to Minister of Information　　　　23 Aug 44

Is there any stop on the publicity for the facts about the agony of Warsaw, which seem, from the papers, to have been practically suppressed? It is not for us to cast reproaches on the Soviet Government, but surely the facts should be allowed to speak for themselves? There is no need to mention the strange and sinister behaviour of the Russians, but is there any reason why the consequences of such behaviour should not be made public?

★　　★　　★　　★　　★

The President now replied to my telegram.

President Roosevelt to Prime Minister   24 Aug 44

Thanks for your telegram describing the inhuman behaviour of the Nazis and the dreadful situation of the Poles in Warsaw.

Stalin's reply to our joint proposal for assisting the Warsaw Poles is far from encouraging.

The supply by us of the Warsaw Poles is, I am informed, impossible unless we are permitted to land and take off from Soviet airfields. Their use for the relief of Warsaw is at present prohibited by the Russian authorities.

I do not see what further steps we can take at the present time that promise results.

I replied next day:

Prime Minister to President Roosevelt   25 Aug 44

As Stalin's reply evades the definite questions asked and adds nothing to our knowledge, I propose a reply on the following lines:

[*Begins.*] "We earnestly desire to send U.S. aircraft from England. Is there any reason why the refuelling ground assigned to us behind the Russian lines should not be used by them to land on without inquiry as to their activities on the way? In this way your Government could preserve the principle of dissociation from this particular episode. We feel confident that if disabled British or American aircraft come down behind the lines of your armies your usual consideration will ensure their being succoured. Our sympathies are aroused for these 'almost unarmed people' whose special faith has led them to attack German tanks, guns, and planes, but we are not concerned to form a judgment about the persons who instigated the rising, which was certainly called for repeatedly by Moscow Radio. We cannot think that Hitler's atrocities will end with their resistance, but rather that that is the moment when they will probably begin with full ferocity. The Warsaw massacre will certainly be a matter fraught with trouble for us when we all meet at the end of the war. We therefore propose to send the aircraft unless you directly forbid it." [*Ends.*]

In the event of his failing to reply to this my feeling is that we ought to send the planes and see what happens. I cannot believe that they would be ill-treated or detained. Since this was signed I have seen that the Russians are even endeavouring to take away your airfields which are located at Poltava and elsewhere behind their lines.

The reply was adverse.

President Roosevelt to Prime Minister   26 Aug 44

I do not consider it would prove advantageous to the long-range general war prospect for me to join with you in the proposed message to Stalin, but I have no objection to your sending such a message if you consider it advisable to do so. In arriving at this conclusion I have taken into consideration Uncle J.'s present attitude towards the relief of the Underground forces in Warsaw, as indicated in his message to you and to me, his definite refusal to allow the use by us of Russian airfields for that purpose, and the current American conversations on the subject of the subsequent use of other Russian bases.

* * * * *

I had hoped that the Americans would support us in drastic action. On September 1 I received the Polish Premier, Miko-lajczyk, on his return from Moscow. I had little comfort to offer. He told me that he was prepared to propose a political settlement with the Lublin Committee, offering them fourteen seats in a combined Government. These proposals were debated under fire by the representatives of the Polish Underground in Warsaw itself. The suggestion was accepted unanimously. Most of those who took part in these decisions were tried a year later for "treason" before a Soviet court in Moscow.

When the Cabinet met on the night of September 4 I thought the issue so important that though I had a touch of fever I went from my bed to our underground room. We had met together on many unpleasant affairs. I do not remember any occasion when such deep anger was shown by all our members, Tory, Labour, Liberal alike. I should have liked to say, "We are sending our aeroplanes to land in your territory, after delivering supplies to Warsaw. If you do not treat them properly all convoys will be stopped from this moment by us." But the reader of these pages in after-years must realise that everyone always has to keep in mind the fortunes of millions of men fighting in a world-wide struggle, and that terrible and even humbling submissions must at times be made to the general aim. I did not therefore propose this drastic step. It might have been effective, because we were dealing with men in the Kremlin who were governed by calcula-tion and not by emotion. They did not mean to let the spirit of Poland rise again at Warsaw. Their plans were based on the Lublin Committee. That was the only Poland they cared about. The cutting off of the convoys at this critical moment in their great advance would perhaps have bulked in their minds as much as considerations of honour, humanity, decent commonplace good faith, usually count with ordinary people. The telegrams which follow show the best that we thought it wise to do.

Prime Minister (London) to President Roosevelt　　　　　　4 Sept 44

The War Cabinet are deeply disturbed at the position in Warsaw and at the far-reaching effect on future relations with Russia of Stalin's refusal of airfield facilities.

2. Moreover, as you know, Mikolajczyk has sent his proposals to the Polish Committee of Liberation for a political settlement. I am afraid that the fall of Warsaw will not only destroy any hope of progress, but will fatally undermine the position of Mikolajczyk himself.

3. My immediately following telegrams contain the text of a telegram which the War Cabinet in their collective capacity have sent to our Ambassador in Moscow, and also of a message which the women of Warsaw have communicated to the Pope and which has been handed by the Vatican to our Minister.

4. The only way of bringing material help quickly to the Poles fighting in Warsaw would be for United States aircraft to drop supplies, using Russian airfields for the purpose. Seeing how much is in jeopardy, we beg that you will again consider the big stakes involved. Could you not authorise your air forces to carry out this operation, landing, if necessary, on Russian airfields without their formal consent? In view of our great successes in the West, I cannot think that the Russians could reject this *fait accompli*. They might even welcome it as getting them out of an awkward situation. We would of course share full responsibility with you for any action taken by your Air Force.

Prime Minister to President Roosevelt　　　　　　　　　　4 Sept 44

Following is text of telegram sent to Moscow this evening, mentioned in my immediately preceding telegram:

"The War Cabinet at their meeting to-day considered the latest reports of the situation in Warsaw, which show that the Poles fighting against the Germans there are in desperate straits.

"The War Cabinet wish the Soviet Government to know that public opinion in this country is deeply moved by the events in Warsaw and by the terrible sufferings of the Poles there. Whatever the rights and wrongs about the beginnings of the Warsaw rising, the people of Warsaw themselves cannot be held responsible for the decision taken. Our people cannot understand why no material help has been sent from outside to the Poles in Warsaw. The fact that such help could not be sent on account of your Government's refusal to allow United States aircraft to land on aerodromes in Russian hands is now becoming publicly known. If on top of all this the Poles in Warsaw should now be overwhelmed by the Germans, as we are told they must be within two or three days, the shock to public opinion here will be incalculable. The War Cabinet themselves find it hard to understand your Government's refusal to take account of the obligations of the British and American Governments to help the Poles in Warsaw. Your Government's action in preventing this help being sent seems to us at variance with the spirit of Allied co-operation to which you and we attach so much importance both for the present and the future.

"Out of regard for Marshal Stalin and for the Soviet peoples, with whom it is our earnest desire to work in future years, the War Cabinet have asked me to make this further appeal to the Soviet Government to give whatever help may be in their power, and above all to provide facilities for United States aircraft to land on your airfields for this purpose."

Prime Minister to President Roosevelt   4 Sept 44

Following is text of message from women of Warsaw referred to in my earlier telegram:

"Most Holy Father, we Polish women in Warsaw are inspired with sentiments of profound patriotism and devotion for our country. For three weeks, while defending our fortress, we have lacked food and medicine. Warsaw is in ruins. The Germans are killing the wounded in hospitals. They are making women and children march in front of them in order to protect their tanks. There is no exaggeration in reports of children who are fighting and destroying tanks with bottles of petrol. We mothers see our sons dying for freedom and the Fatherland. Our husbands, our sons, and our brothers are not considered by the enemy to be combatants. Holy Father, no one is helping us. The Russian armies which have been for three weeks at the gates of Warsaw have not advanced a step. The aid coming to us from Great Britain is insufficient. The world is ignorant of our fight. God alone is with us. Holy Father, Vicar of Christ, if you can hear us, bless us Polish women who are fighting for the Church and for freedom."

President Roosevelt to Prime Minister   5 Sept 44

Replying to your telegrams, I am informed by my Office of Military Intelligence that the fighting Poles have departed from Warsaw and that the Germans are now in full control.

The problem of relief for the Poles in Warsaw has therefore unfortunately been solved by delay and by German action, and there now appears to be nothing we can do to assist them.

I have long been deeply distressed by our inability to give adequate assistance to the heroic defenders of Warsaw, and I hope that we may together still be able to help Poland to be among the victors in this war with the Nazis.

* * * * *

On September 10, after six weeks of Polish torment, the Kremlin appeared to change their tactics. That afternoon shells from the Soviet artillery began to fall upon the eastern outskirts of Warsaw, and Soviet planes appeared again over the city. Polish Communist forces, under Soviet orders, fought their way into the fringe of the capital. From September 14 onwards the Soviet Air Force dropped supplies; but few of the parachutes opened and many of the containers were smashed and useless. The following day the Russians occupied the Praga suburb, but went no farther. They wished to have the non-Communist Poles destroyed to the full, but also to keep alive the idea that they were going to their rescue. Meanwhile, house by house, the Germans proceeded with their liquidation of Polish centres of resistance throughout the city. A fearful fate befell the population. Many were deported by the Germans. General Bor's appeals to the Soviet commander, Marshal Rokossovsky, were unanswered. Famine reigned.

My efforts to get American aid led to one isolated but large-scale operation. On September 18 a hundred and four heavy bombers flew over the capital, dropping supplies. It was too late. On the evening of October 2 Premier Mikolajczyk came to tell me that the Polish forces in Warsaw were about to surrender to the Germans. One of the last broadcasts from the heroic city was picked up in London.

This is the stark truth. We were treated worse than Hitler's satellites, worse than Italy, Roumania, Finland. May God, Who is just, pass judgment on the terrible injustice suffered by the Polish nation, and may He punish accordingly all those who are guilty.

Your heroes are the soldiers whose only weapons against tanks, planes, and guns were their revolvers and bottles filled with petrol. Your heroes are the women who tended the wounded and carried messages under fire, who cooked in bombed and ruined cellars to feed children and adults, and who soothed and comforted the dying. Your heroes are the children who went on quietly playing among the smouldering ruins. These are the people of Warsaw.

Immortal is the nation that can muster such universal heroism. For those who have died have conquered, and those who live on will fight on, will conquer and again bear witness that Poland lives when the Poles live.

These words are indelible. The struggle in Warsaw had lasted more than sixty days. Of the 40,000 men and women of the Polish Underground Army about 15,000 fell. Out of a population of a million nearly 200,000 had been stricken. The suppression of the revolt cost the German Army 10,000 killed, 7,000 missing, and 9,000 wounded. The proportions attest the hand-to-hand character of the fighting.

When the Russians entered the city three months later they found little but shattered streets and the unburied dead. Such was their liberation of Poland, where they now rule. But this cannot be the end of the story.

CHAPTER X

THE SECOND QUEBEC CONFERENCE

I Sail from the Clyde, September 5 – British Plans for the War with Japan – Would Germany be Defeated in 1944? – Our Need to Forestall the Russians in Central Europe – We Land at Halifax, September 10 – Our First Plenary Meeting in the Citadel of Quebec, September 13 – My Review of the Progress of the War – The Campaign in Burma – I Offer to Send the British Fleet to the Pacific and Place it Under the United States Supreme Command – American Operations in the Pacific – Lord Portal's Plans for the Royal Air Force – My Telegram Home of September 13 – "A Blaze of Friendship" – The Morgenthau Plan – Report of the Combined Chiefs of Staff, September 16 – Advantages of a Northern Thrust Through Germany – The Battle in Italy – Balkan Plans – The Defeat of Japan – A Farewell Visit to Hyde Park – The Voyage Home.

ON Tuesday, September 5, we sailed once again from the Clyde in the *Queen Mary*. All the Chiefs of Staff came with me, and met daily, and sometimes twice a day, during our six days' voyage. I wanted, before meeting our American friends, to harmonise and grip the many plans and projects which were now before us. In Europe, "Overlord" was not only launched, but triumphant. How, when, and where could we strike at Japan, and assure for Britain an honourable share in the final victory there? We had lost as much, if not more, than the United States. Over 160,000 British prisoners and civilian internees were in Japanese hands. Singapore must be redeemed and Malaya freed. For nearly three years we had persisted in the strategy of "Germany First". The time had now come for the liberation of Asia, and I was determined that we should play our full and equal part in it. What I feared most at this stage of the war was that the United States would say in after-years, "We came to your help in Europe and you left us alone to finish off Japan." We had to regain on the field of battle our rightful possessions in the Far East, and not have them handed back to us at the peace table.

Our main contribution must obviously be on the sea and in the air. Most of our Fleet was now free to move eastwards, and I resolved that our first demand on our American Allies should be for its full participation in the main assault on Japan. The Royal Air Force should follow as soon as possible after Germany was defeated.

The military operations were much more complicated. Things were going badly in China, and Admiral Mountbatten was being pressed both to hurry on the opening of the Burma Road by advancing into Central Burma—an operation called "Capital"—and to increase the supplies which were being flown over the Himalayas. Another project which promised more immediate results was an amphibious expedition to sail across the Bay of Bengal, capture Rangoon, move a few miles inland, and cut off the Japanese troops from their bases and lines of communication in Siam. This was known as Operation "Dracula". At the same time our troops in Central Burma would sweep down and join the force which was to land at Rangoon. It was hoped that this would clear the country and enable us to make an amphibious assault on Sumatra.

But all these tasks called for men and material, and there were not enough in South-East Asia. The only place they could come from was Europe. The landing-craft would have to be taken either from the Mediterranean or from "Overlord", and the troops from Italy and elsewhere, and they would have to leave soon. It was now September. Rangoon lies forty miles up a winding estuary, complicated with backwaters and mud-banks. The monsoon starts in early May, and we should therefore have to attack by April 1945 at the latest. Was it yet safe to start weakening our effort in Europe?

I was by no means certain that Germany would be defeated in 1944. It was true that we had had nearly seven weeks of unbroken military success. Paris was liberated and large areas of France had been cleared of the enemy. Our advance in Italy continued. The Soviet offensive, though temporarily halted, might at any moment surge forward again. Greece would soon be free.

Hitler's "secret weapons" were almost mastered, and there was no evidence that he had learned how to make the atomic bomb. All these and many other factors induced a belief in our military circle that the Nazis would soon collapse. But I was not convinced. I remembered the German onslaught in March 1918. At a meeting of the Chiefs of Staff over which I presided on September 8 I accordingly warned them against basing their plans on an imminent German collapse. I pointed out that resistance in the West had stiffened and the Americans had been sharply checked at Nancy. German garrisons, I said, were offering stout resistance at most of the ports; the Americans had not taken St. Nazaire; and the enemy showed every intention of fighting hard on the shores of the estuary of the Scheldt leading to Antwerp, which we so badly needed.

Another matter lay heavy on my mind. I was very anxious to forestall the Russians in certain areas of Central Europe. The Hungarians, for instance, had expressed their intention of resisting the Soviet advance, but would surrender to a British force if it could arrive in time. If the Germans either evacuated Italy or retired to the Alps I much desired that Alexander should be enabled to make his amphibious thrust across the Adriatic, seize and occupy the Istrian peninsula, and try to reach Vienna before the Russians. It seemed much too early to start sending his troops to South-East Asia. The C.I.G.S. agreed that there should be no question of withdrawing any of Alexander's troops until Kesselring had been driven across the Piave. Our front would then be considerably less than half its present width. For the time being only the first of the Indian divisions needed for the assault on Rangoon would be taken from Alexander. I was discontented even at this prospect. As for landing in the Istrian peninsula, I was told that we should either have to borrow American landing-craft which were due to leave for the Pacific or weaken the campaign in France. The rest of our craft were needed for taking Rangoon. This must be done before the May monsoon, and if we used these vessels in the Adriatic they would not get there in time.

As a result of our lengthy talks on the voyage we reached agreement about what we should say to our great Ally.

*   *   *   *   *

We landed at Halifax on September 10, and reached Quebec the following morning. The President and Mrs. Roosevelt, who were our guests, had arrived just before us, and the President waited at the station to greet me. Once again the Citadel was our home, while the staffs again monopolised the Château Frontenac.

On the morning of Wednesday, September 13, we held our first plenary meeting. With me were Brooke, Portal, Cunningham, Dill, Ismay, and Major-General Laycock, who had succeeded Mountbatten as Chief of Combined Operations. The President had with him Leahy, Marshall, King, and Arnold. But this time, alas, there was no Harry Hopkins. He had sent me a telegram just before I left England: "Although I am now feeling much better, I still must take things easy, and I therefore feel that I should not run the risk of a setback in health by attempting to fight the Battle of Quebec on the Plains of Abraham, where better men than I have been killed." I was not then aware of the change in the character of his relationship with the President, but I was sure that he would be sorely missed.

Mr. Roosevelt asked me to open the discussion. I thereupon made a general survey of the war which I had prepared on the voyage. Since our meeting in Cairo the affairs of the United Nations had taken a revolutionary turn for the better. Everything we had touched had turned to gold, and during the last seven weeks there had been an unbroken run of military success. The manner in which the situation had developed since the Teheran Conference gave the impression of remarkable design and precision of execution. First there had been the Anzio landing, and then on the day before the launching of "Overlord" we had captured Rome. This seemed the most perfect timing. I congratulated the United States Chiefs of Staff on the gratifying results of "Dragoon". It seemed that eighty or ninety thousand prisoners had already been captured, and the south and western parts of France were being systematically cleared of the enemy. Future historians would surely say that since Teheran our Allied war machine had worked with extraordinary efficiency.

I was also glad to record that although the British Empire had now entered the sixth year of the war it was still keeping its position, with a total population, including the Dominions and Colonies, of only seventy million white people. Our effort in

Europe, measured by divisions in the field, was about equal to that of the United States. This was as it should be, and I was proud that we could claim equal partnership with our great Ally. Our strength had now reached its peak, whereas our Ally's was ever increasing. There was complete confidence in General Eisenhower, and his relations with General Montgomery were of the best, as were those between General Montgomery and General Bradley. The part played by General Bedell Smith in directing and cementing the staffs was of the highest order. An efficient integrated American-British Staff machine had been built up, and the battle was being brilliantly exploited.

In Italy General Alexander had resumed the offensive at the end of August. Since then the Eighth Army had suffered about 8,000 casualties and the Fifth Army about 1,000. The Fifth Army had hitherto not been so heavily engaged, but they were expected to attack that very day. In this theatre there was the most representative British Empire army there had ever been. There were in all sixteen British Empire divisions, namely, eight British, two Canadian, one New Zealand, one South African, and four British-Indian. I explained that I had been anxious lest General Alexander might be short of certain essentials for the vigorous prosecution of his campaign, but I now understood that the Combined Chiefs of Staff had agreed to withdraw nothing from his army until either Kesselring's troops had been destroyed or were on the run out of Italy.

General Marshall confirmed this undertaking, and I accordingly emphasised that in that case we should have to look for fresh woods and pastures new. It would never do for our armies to remain idle. I said I had always been attracted by a right-handed movement to give Germany a stab in the Adriatic armpit. Our objective should be Vienna. If German resistance collapsed we should of course be able to reach the city more quickly and more easily. If not I had given considerable thought to aiding this movement by capturing Istria and occupying Trieste and Fiume. I had been relieved to learn that the United States Chiefs of Staff were willing to leave in the Mediterranean certain landing-craft now engaged in the attack on the south of France to provide an amphibious lift for such an operation, if this was found desirable and necessary. Another reason for this right-handed movement was the rapid encroachment of the Russians into the

Balkan peninsula and the dangerous spread of Soviet influence there.

*   *   *   *   *

I then reviewed the campaign in Burma. This had been on a considerable scale. 250,000 men had been engaged, and the fighting for Imphal and Kohima had been extremely bitter. General Stilwell was to be congratulated on his brilliant capture of Myitkyina. We had suffered 40,000 battle casualties and 288,000 sick. Of the sick, happily, the greater proportion recovered and returned to duty. As a result of this campaign the air line to China had been kept open and India made safe from attack. It was estimated that the Japanese had lost 100,000 men. The Burma campaign was the largest land engagement of Japanese forces so far attained.

In spite of these successes it was, I continued, most undesirable that the fighting in the Burmese jungles should go on indefinitely. For this reason the British Chiefs of Staff had proposed Operation "Dracula", the capture of Rangoon. Difficulties were being experienced in gathering the necessary forces and transporting them to South-East Asia in time to take Rangoon before the monsoon of 1945. The present situation in Europe, favourable as it was, did not permit a decision being taken now to withdraw forces. What we wanted was to keep an option open for as long as possible, and every effort was being directed to this end.

Certain trouble-makers had said that we would take no share in the war against Japan once Germany had been defeated. Far from shirking this task, the British Empire was eager to play the greatest possible part in it. We had every reason for doing so. Japan was as much the bitter enemy of the British Empire as of the United States. British territory had been captured in battle and grievous losses had been suffered. The offer I now made was for the British main fleet to take part in the major operations against Japan under United States Supreme Command. We should have available a powerful and well-balanced force. We hoped that by the end of 1945 this would include our newest battleships. A fleet train of adequate proportions would be built up, making the warships independent of shore-based resources for considerable periods.

The President intervened to say that the British Fleet was no sooner offered than accepted. In this, though the fact was not mentioned, he overruled Admiral King's opinion.

I continued that placing a British fleet in the Central Pacific would not prevent us sending a detachment to General MacArthur in the South-West Pacific if this was desired. We had of course no intention of interfering in any way with his command. As a further contribution to the defeat of the enemy, the Royal Air Force would like to take part in the heavy bombardment of Japan. A bomber force of no mean size could be made available, and would feel honoured to share with their American colleagues the dangers of striking at the heart of the enemy. As for land forces, when Germany had been beaten we should probably be able to send six divisions from Europe to the East, and perhaps six more later on. In South-East Asia we had sixteen divisions, which might ultimately be drawn upon. I had always advocated an advance across the Bay of Bengal and operations to recover Singapore, the loss of which had been a grievous and shameful blow to British prestige and must be avenged. There was nothing cast-iron in these ideas. First we should capture Rangoon and then survey the situation. If a better plan could be evolved it should certainly not be ruled out in advance. The key-note should be to engage the largest number of our own forces against the largest number of the enemy at the earliest possible moment.

*   *   *   *   *

The President thanked me for this review, and said it was a matter of profound satisfaction that at each succeeding Conference between the Americans and the British there had been ever-increasing solidarity of outlook and identity of basic thought. Added to this there had always been an atmosphere of cordiality and friendship. Our fortunes had prospered, but it was still not quite possible to forecast when the war with Germany would end. It was clear that the Germans were withdrawing from the Balkans, and it seemed likely that in Italy they would retire to the Alps. The Russians were on the edge of Hungary. The Germans had shown themselves good at staging withdrawals, and had been able to save many men, although they had lost much material. If Alexander's battle went well we should reach the Piave reasonably soon. All forces in Italy should be engaged to the maximum intensity. In the West it seemed probable that the Germans would retire behind the Rhine. Its right bank would be the western rampart of their defence and would present a formidable obstacle. We should have to attack them either from the east or from the west, and our plans must therefore be flexible. The Germans could not yet be counted out. One more big battle would have to be fought, and our operations against Japan would to some extent depend on what happened in Europe.

The President agreed that we should not remain in Burma any longer than was necessary to clean up the Japanese in that theatre. The American plan was to regain the Philippines and to dominate the mainland of Japan from there or Formosa, and from bridge-heads which would be seized in China. If forces could be established on the mainland of China, China could be saved. American experience had been that the "end run" method paid a handsome dividend. Rabaul was an example of this, by-passing technique which had brought considerable success at small cost of life. Would it not be possible, he asked, to by-pass Singapore by seizing an area to the north or east of it, such as Bangkok? He said that he had not hitherto been greatly attracted to the Sumatra plan, but now the operation had acquired greater merit.

I said that all these projects were being examined and would be put in order. No decision could be reached until after we had taken Rangoon. It should not be overlooked that Stalin had volunteered a solemn undertaking at Teheran that Russia would enter the war against Japan the day that Hitler was beaten. There was no reason to doubt that Stalin would be as good as his word. The Russians undoubtedly had great ambitions in the East. If Hitler was beaten by, say, January, and Japan was confronted with the three most powerful nations in the world, she would undoubtedly think twice about continuing the fight.

I then cast back to make sure where we stood, and asked for a definite undertaking about employing the British Fleet in the main operations against Japan.

"I should like," said the President, "to see the British Fleet wherever and whenever possible."

Admiral King said that a paper had been prepared for the Combined Chiefs of Staff, and the question was being actively studied.

"The offer of the British Fleet has been made," I repeated. "Is it accepted?"

"Yes," said Mr. Roosevelt.

"Will you also let the British Air Force take part in the main operations?"

Here it was much more difficult to get a direct answer. Marshall said that General Arnold and he were trying to see how to use the greatest number of aircraft they could. "Not so long ago," he explained, "we were crying out for planes. Now we have a glut. If you are going to be heavily engaged in South-East Asia and Malaya won't you need most of your Air Force? Or is Portal's plan to bomb Japan something quite separate?"

"Quite separate," answered Portal. "If our Lancaster bombers are refuelled in the air they can go nearly as far as your B.29s."

I said that for the sake of good relations, on which so much depended in the future, it was of vital importance that the British should be given their fair share in the main operations against Japan. The United States had given us the most handsome assistance in the fight against Germany. It was only to be expected that the British Empire in return should wish to give the United States all the help in their power towards defeating Japan.

* * * * *

After the meeting I telegraphed home.

Prime Minister to Deputy Prime Minister and   13 Sept 44
War Cabinet

The Conference has opened in a blaze of friendship. The Staffs are in almost complete agreement already. There is to be no weakening of Alexander's army till Kesselring has bolted beyond the Alps or been destroyed. We are to have all the landing-craft in the Mediterranean to work up in the Northern Adriatic in any amphibious plan which can be made for Istria, Trieste, etc.

2. The idea of our going to Vienna, if the war lasts long enough and if other people do not get there first, is fully accepted here.

3. After their work in the Adriatic the landing-craft will of course be free to go on to the Bay of Bengal, or farther, as circumstances may require.

I was also able to reassure our commanders in the Mediterranean.

Prime Minister to General Wilson and   13 Sept 44
General Alexander

Everything has opened here very well so far as your affairs are concerned. There is to be no weakening of Alexander's army till after Kesselring is disposed of, which our Intelligence indicates as probable.

2. Moreover, Admiral King is not making any claims on the landing-craft in the Mediterranean, and the Americans are quite ready to agree that as many as necessary of these should be used for any amphibious work in the Northern Adriatic that may be found practicable.

3. Pray therefore address yourselves to this greatly improved situation in a spirit of audacious enterprise. The Americans talk without any hesitation of our pushing on to Vienna, if the war lasts long enough. I am greatly relieved at the reception all our ideas have met here. We must turn these advantages to the best account.

* * * * *

During the days which followed I had a number of conversations with the President and his advisers. I had been surprised to find when I arrived at Quebec that the President was accompanied by Mr. Morgenthau, the Secretary of the United States Treasury, though neither the Secretary of State nor Harry Hopkins was present. But I was glad to see Morgenthau, as we were anxious to discuss financial arrangements between our two countries for the period between the conquest of Germany and the defeat of the Japanese. The President and his Secretary of the Treasury were however much more concerned about the treatment of Germany after the war. They felt very strongly that military strength rested on industrial strength. We had seen during the nineteen-thirties how easy it was for a highly industrialised Germany to arm herself and threaten her neighbours, and they asserted that there was no need for so much manufacturing in a country as large as Germany, who could to all intents and purposes feed herself. The United Kingdom had lost so many overseas investments that she could only pay her way when peace came by greatly increasing her exports, so that for economic as well as military reasons we ought to restrict German industry and encourage German agriculture. At first I violently opposed this idea. But the President, with Mr. Morgenthau—from whom we had much to ask—was so insistent that in the end we agreed to consider it.

The so-called Morgenthau Plan, which I had not time to examine in detail, seems to have carried these ideas to an ultra-logical conclusion. Even if it had been practicable I do not think it would have been right to depress Germany's standard of life in such a way; but at that time, when German militarism based on

German industry had done such appalling damage to Europe, it did not seem unfair to agree that her manufacturing capacity need not be revived beyond what was required to give her the same standards of life as her neighbours. All this was of course subject to the full consideration of the War Cabinet, and in the event, with my full accord, the idea of "pastoralising" Germany did not survive.

*　　*　　*　　*　　*

We held our last meeting at midday on Saturday, September 16. The Combined Chiefs of Staff had now completed their final report to the President and myself, and at Mr. Roosevelt's request Admiral Leahy read it out to us paragraph by paragraph. The principal passages were as follows:

9. The Supreme Commander's broad intention is to press on with all speed to destroy the German armed forces and occupy the heart of Germany. He considers his best opportunity of defeating the enemy in the West lies in striking at the Ruhr and Saar, since he is convinced that the enemy will concentrate the remainder of his available forces in the defence of these essential areas. The first operation will be to break the Siegfried Line and seize crossings over the Rhine. In doing this his main effort will be on the left. He will then prepare logistically and otherwise for a deep thrust into Germany.

10. We have approved General Eisenhower's proposals and drawn his attention

(a) to the advantages of the northern line of approach into Germany, as opposed to the southern, and
(b) to the necessity for the opening up of the north-west ports, particularly Antwerp and Rotterdam, before bad weather sets in.

I had no quarrel with these broad intentions, but the reader will remember the doubts which I had voiced to the British Chiefs of Staff during our voyage across the Atlantic about the imminence of Germany's defeat. I had also written a paper in this sense which will be printed in a later chapter. Rundstedt's counter-stroke was still to come, and the crossing of the Rhine was not to be achieved for more than another six months.

*　　*　　*　　*　　*

The military recommendations about Italy were as follows:

11. We have examined a report by General Wilson on operations

within his theatre. In so far as the battle in Italy is concerned, he considers that operations will develop in one of two ways:

(a) Either Kesselring's forces will be routed, in which case it should be possible to undertake a rapid regrouping and a pursuit towards the Ljubljana Gap (and across the Alps through the Brenner Pass), leaving a small force to clear up North-West Italy; or

(b) Kesselring's army will succeed in effecting an orderly withdrawal, in which event it does not seem possible that we can do more than clear the Lombardy plains this year. Difficult terrain and severe weather in the Alps during winter would prevent another major offensive until the spring of 1945.

The report continued as follows:

12. We have agreed:

(a) That no major units should be withdrawn from Italy until the outcome of General Alexander's present offensive is known.

(b) That the desirability of withdrawing formations of the United States Fifth Army should be reconsidered in the light of the results of General Alexander's present offensive and of a German withdrawal in Northern Italy, and in the light of the views of General Eisenhower.

(c) To inform General Wilson that if he wishes to retain for use in the Istrian peninsula the amphibious lift at present in the Mediterranean he should submit his plan to the Combined Chiefs of Staff as soon as possible, and not later than October 10. We have instructed the Supreme Allied Commander accordingly.

Here I had to beware of bargains. No major units to be withdrawn until we knew the result of Alexander's offensive; so far, so good. But how far was the offensive to be pushed? If he was only to be allowed to go to the Rimini line, for instance, then the proposal was quite unacceptable. I accordingly said I presumed he would be allowed to invade and dominate the valley of the Po, and I was much relieved when Marshall and Leahy agreed that this was what they meant.

I then thanked Admiral King for promising to lend us his landing-craft for an attack on the Istrian peninsula. The Admiral stressed that they would also be wanted for the assault on Rangoon, and we must therefore make up our minds about invading Istria by October 15.

* * * * *

The next paragraph of the report set forth our joint proposals about operations in the Balkan peninsula. It read as follows:

13. General Wilson considers that a situation can be anticipated in which the bulk of the German forces south of a line Trieste–Ljubljana–Zagreb and the Danube will be immobilised, and will so remain until their supplies are exhausted, in which case they would be ready to surrender to us or will be liquidated by Partisans or the Russian forces. We have noted that as long as the battle in Italy continues there will be no forces available in the Mediterranean to employ in the Balkans except:

(a) the small force of two British brigades from Egypt which is being held ready to occupy the Athens area and so pave the way for commencement of relief and establishment of law and order and the Greek Government;

(b) the small land forces in the Adriatic which are being actively used primarily for Commando type operations.

This was accepted by all of us without amendment or discussion.

* * * * *

The proposals for war in the Pacific dwelt on the importance of flexibility and short-cuts. The Allied superiority of naval and air power should enable us to avoid, wherever possible, costly land campaigns. In South-East Asia it was agreed that the land advance into Burma from the north should be combined with the amphibian capture of Rangoon. I said that while I accepted the British obligation to secure the air route and attain overland communication with China, any tendency to overdo it would rule out our assault on Rangoon, which both the Chiefs of Staff and I wanted to capture before the monsoon of 1945.

The rest of the report was approved with little or no discussion. The planning date for the end of the war against Japan was set for the time being at eighteen months after the defeat of Germany.

The following passage requires verbatim statement.

33. Upon the collapse of organised resistance by the German Army the following subdivision of that part of Germany not allocated to the Soviet Government for disarmament, policing, and the preservation of order is acceptable from a military point of view by the Combined Chiefs of Staff.

34. For disarmament, policing, and preservation of order:

(a) The British forces, under a British commander, will occupy

Germany west of the Rhine and east of the Rhine north of the line from Koblenz, following the northern border of Hesse and Nassau to the border of the area allocated to the Soviet Government.

(b) The forces of the United States, under a United States commander, will occupy Germany east of the Rhine, south of the line Koblenz–northern border of Hesse-Nassau and west of the area allocated to the Soviet Government.

(c) Control of the ports of Bremen and Bremerhaven, and the necessary staging areas in that immediate vicinity, will be vested in the commander of the American Zone.

(d) American area to have in addition access through the western and north-western seaports and passage through the British-controlled area.

(e) Accurate delineation of the above outlined British and American areas of control can be made at a later date.

★   ★   ★   ★   ★

On Sunday, September 17, I left Quebec by train with my wife and daughter Mary to pay a farewell visit to the President at Hyde Park.

I lunched there on September 19. Harry Hopkins was present. He was obviously invited to please me. He explained to me his altered position. He had declined in the favour of the President. There was a curious incident at luncheon, when he arrived a few minutes late and the President did not even greet him. It was remarkable how definitely my contacts with the President improved and our affairs moved quicker as Hopkins appeared to regain his influence. In two days it seemed to be like old times. He said to me, "You must know I am not what I was." He had tried too much at once. Even his fullness of spirit broke under his variegated activities.

After dinner I left for New York, and boarded the *Queen Mary* the following morning. The voyage home was without incident. We arrived in the Clyde on September 25, and left immediately by train for London.

CHAPTER XI

ADVANCE IN BURMA

The Relief of Imphal, June 1944 – Ruinous Japanese Losses – The Advance of the Fourteenth Army – The Fight Against the Monsoon – General Stilwell Captures Myitkyina, August 3 – His "Mars Brigade" – Mountbatten Visits London to Explain His Plans – My Minute of September 12 About the Operations – German Resistance Compels Us to Postpone the Assault on Rangoon – Hard Tidings for Mountbatten, October 5 – The Advance Continues – Changes in the American High Command – Crisis in China – The President's Telegram of December 1 – Withdrawal of Two Chinese Divisions and of the Transport Squadrons – The Advance on Mandalay – Reopening of the Burma Road, January 1945 – My Telegram to Mountbatten of January 23 – Winter Fighting in the Arakan – The Capture of Akyab.

*T*HE swaying battle in Burma has already been described up to the point when the initiative was about to pass into our hands.* Japan's invasion of India collapsed on the mountain plateau of Imphal at the end of June 1944, when relieving troops from the north met the outward thrust of General Scoones' garrison. The road to Dimapur was open and the convoys flowed in. But the three Japanese divisions had still to be thrust back to and beyond the river Chindwin, whence they had come. Their losses had been ruinous. Over thirteen thousand dead were counted on the battlefields, and, allowing for those who died of wounds, disease, or hunger, the total amounted on a Japanese estimate to 65,000 men. The monsoon, now at its height, had in previous years brought active operations to a standstill, and the Japanese doubtless counted on a pause during which they could extricate and rebuild their shattered Fifteenth Army. They were given no such respite.

* Volume V, Chapter XXXI, "Burma and Beyond."

The British-Indian Fourteenth Army, under the able and forceful leadership of General Slim, took the offensive. Their XXXIIIrd Corps first cleared up around Ukhrul, while the IVth recaptured the southern part of the Imphal plain. By the end of July Japanese resistance was broken and the XXXIIIrd took up a general pursuit to the Chindwin. All along the mountain tracks they found evidence of disaster—quantities of abandoned guns, transport, and equipment; thousands of the enemy lay dead or dying. The 5th Indian Division, thrusting south towards Tiddim, had at first a harder task. The Japanese 33rd Division, which opposed them, had not been handled so roughly as the others and had been reinforced. The road twisted its narrow way through mountainous country and was easy to defend. One by one the Japanese positions were overcome, the 221 Group R.A.F., under Air Marshal Vincent, providing violent bombardments immediately before the infantry assaults. Here, as everywhere in Burma at this time, progress, measured in miles a day, was very slow. But our men were fighting in tropical rainfall, soaked to the skin by day and night. The so-called roads were mostly fair-weather dust-tracks, which were now churned into deep mud, through which guns and vehicles had often to be manhandled. It is not the slowness of advance but the fact that any advance was made at all that should cause surprise.

In the Arakan our troops were held on an active defensive. In that tangle of jungle-covered hills, with its narrow coastal strip of ricefields and mangrove swamps, the monsoon rainfall, which sometimes reached twenty inches a week,* stopped serious operations. On the northern front General Stilwell's forces made steady progress. The capture of Myitkyina on August 3 gave him a forward base for future land operations, and, even more important, provided a staging post for the American air-lift to China. The famous "Hump" traffic no longer had to make the direct and often dangerous flight from Northern Assam over the great mountains to Kunming. Work was progressing on the long road from Northern Assam, destined later to link up with the former road from Burma to China. The strain on rearward communications in Assam was relieved by a new oil pipe-line 750 miles long laid from Calcutta, a greater span than the famous desert pipe-line from Iraq to Haifa.

* The average *yearly* rainfall in London is about twenty-four inches.

For his southward advance Stilwell reorganised his five Chinese divisions into two "Armies", one directed from Myitkyina on Bhamo and Namkhan, the other on Shwegu and Katha. The latter advance was led by the British 36th Division, which had been placed under Stilwell's orders. It replaced the Chindit brigades,* under General Lentaigne, when, after nearly six months of arduous and exacting operations, in which they had fought and overcome at least eleven enemy battalions, they were withdrawn for long-needed rest and recoupment. As a reserve in his own hands Stilwell retained his "Mars Brigade", a mobile, lightly equipped force of about ten thousand men, whose principal component was an American regiment. With these forces he began his advance in early August to cross the river Irrawaddy, and, on his eastern flank, to get touch with the Chinese "Yunnan" armies, about 100,000 strong, which were advancing from the river Salween towards Namkhan.

* * * * *

The policy for future operations in South-East Asia was again under review at this time, and, after consultation with his Commanders-in-Chief, Admiral Somerville, General Giffard, and Air Chief Marshal Peirse, Mountbatten came to London to explain his plans. He was already committed to a land advance into Central Burma, which was to continue until the Fourteenth Army was across the Chindwin and had joined hands with Stilwell's forces coming from the north. But with his ever-lengthening line of communications and the limited number of supply aircraft on which he so greatly depended it was doubtful whether he could advance from Mandalay as far as Rangoon. He therefore proposed to carry out the large-scale amphibious attack on Rangoon mentioned in the previous chapter and given the codename "Dracula". Once firmly established there, his troops could thrust northwards and meet the Fourteenth Army. This was an excellent idea, but it demanded many more troops and much more shipping than Mountbatten possessed. They could only be found from North-West Europe.

My view on this and its variants are given in a minute written at Quebec.

---

* The late General Wingate's Long-Range Penetration Force.

Prime Minister to General Ismay, for C.O.S. Committee   12 Sept 44

THE WAR AGAINST JAPAN

The British share in this war may take the form either of direct participation in particular United States enterprises in the Far East or of British diversionary enterprises on a major scale calculated to wear down the enemy forces by land and air, and also to regain British possessions conquered by the Japanese. Of the two I favour the latter, because:

(a) It is nearly always a sound war policy to engage the largest number of the enemy as closely and continuously as possible, at the earliest moment and for the longest time.

(b) This can best be achieved by a direct thrust across the short-haul of the Bay of Bengal, aimed at "Dracula" [Rangoon], "Culverin" [Sumatra], or other attainable preliminary objectives.

(c) A great diminution of the forces engaged with the enemy results from lengthening the communications. A gush has to be poured into the pipe-line at one end to produce only a trickle at the other, so great is the leakage as the route lengthens.

2. It follows, for the above reasons, that I am opposed to sending any British troops to join the Australians and New Zealanders under General MacArthur. The contribution would be both petty and tardy. On the other hand, I do not object to supporting General MacArthur with a British Naval Task Force, including carriers, or with R.A.F. squadrons, provided that the detachment of these does not weaken our major operations across the Bay of Bengal.

3. Admiral Leahy informed me yesterday that it had been decided to accept the British offer to send our Fleet to participate in the main operations against Japan. It would not be inconsistent with this policy to make a detachment from it for the purpose of sustaining General MacArthur's operations.

4. To sum up, our policy should be to give naval assistance on the largest scale to the main American operations, but to keep our own thrust for Rangoon as a preliminary operation, or one of the preliminary operations, to a major attack upon Singapore. Here is the supreme British objective in the whole of the Indian and Far Eastern theatres. It is the only prize that will restore British prestige in this region, and in pursuing it we render the maximum aid to the United States operations by engaging the largest numbers of the enemy in the most intense degree possible and at the earliest moment.

In our discussions at Quebec we had carried the Americans

with us on the Rangoon plan. This promised many advantages. Six months' fighting in the hills and jungles of Burma and on the frontier of India was estimated to have cost the British and Imperial forces 288,000 losses from sickness alone, but a seaborne stroke against Rangoon and a northward advance would cut the enemy's communications and divide his forces. The destruction of the Japanese in Burma would liberate a considerable army, which could immediately attack such targets across the Bay of Bengal as might be considered to be most beneficial to the common cause, the wearing down of Japanese troops, and above all air forces. For this purpose we had resolved to strain every nerve to attack Rangoon by March 15, 1945. It was thought that five or six divisions would be needed for such an operation, but Mountbatten could only supply two or three and not more than one could be spared from the United Kingdom. Failure meant not only needless sacrifices through prolonging in Burma operations ravaged by disease, but the setback of our whole further deployment against the Malay peninsula and beyond until 1946.

Thus the solution, I had suggested, was to send one or two United States divisions to Burma instead of to Europe. This was better than taking two divisions from Montgomery's army which were actually fighting, and would bring more troops rapidly into action against Japan without withdrawing any of those who were engaged against Germany. I explained at Quebec that I did not want a decision there and then, but only that the United States Chiefs of Staff should examine my suggestions. This General Marshall agreed to do, but for various reasons my proposals were not adopted. The sanguine hopes, which I had not shared, that Germany would collapse before the end of the year failed. At the end of September it was obvious that German resistance would continue into and beyond the winter, and Mountbatten was instructed, not for the first time, that he must do what he could with what he had got. I telegraphed accordingly:

Prime Minister to Admiral Mountbatten   5 Oct 44

The Defence Committee have been forced to the conclusion that March "Dracula" is off, and Chiefs of Staff have made this proposal to United States Chiefs of Staff. You will receive official instructions in due course. Meanwhile you should know that the postponement of the operation is due to the working of far larger forces in the Western theatre rather than to any attitude which you or S.E.A.C.

have adopted. You have now to address yourself to the problem of bringing "Dracula" on in November [1945]. I am very sorry indeed that we have not been able to carry out this operation, on which I had set my heart, but the German resistance both in France and Italy has turned out to be far more formidable than we had hoped. We must clean them out first.

* * * * *

All this time our Fourteenth and Stilwell's Armies had been forging slowly ahead. The 5th Indian Division captured Tiddim on October 18, and with the help of concentrated pin-point bombing soon cleared the enemy from the dominating 8,000-feet Kennedy Peak. Thence they fought on towards Kalemyo. The XXXIIIrd Corps, after taking Tamu, sent an East African brigade eastwards. It established a valuable bridgehead across the river Chindwin at Sittaung. The rest of the 11th East African Division went south along the Kabaw valley towards Kalemyo, which they entered hand-in-hand with the 5th Indian Division on November 14. This was a remarkable march against great physical difficulties through an area notorious for malaria and scrub-typhus. The good hygiene discipline now practised by all our units in Burma, the use of the new drug mepacrine, and constant spraying with D.D.T. insecticide kept the sick rate admirably low. But the Japanese were not versed in these precautions and died in hundreds. From Kalemyo the East Africans pushed on to Kalewa and crossed the Chindwin. Here the engineers built a bridge nearly four hundred yards long in twenty-eight working hours, not the least of their many achievements throughout the campaign. Thus on the central front in early December General Slim's Fourteenth Army, with two bridgeheads across the Chindwin, was poised for his main advance into the central plain of Burma.

Changes among the senior United States officers took place in November. General Stilwell was recalled by Washington. His widespread, multifarious duties were taken over by three others. General Wedemeyer succeeded him as military adviser to Chiang Kai-shek, General Wheeler became Deputy to Mountbatten, and General Sultan took over the northern front. Here the Allied forces slowly threw back the two divisions of the Japanese Thirty-third Army. By mid-November Bhamo was closely invested, but held out stubbornly for a further month. The 36th British

Division took Indaw on December 10. Six days later they made contact there with the 19th Indian Division, which had crossed the Chindwin at the Sittaung bridgehead and thrust eastwards. So at last, after more than a year of hard endeavour, marked by many ups and downs, the two Allied armies joined hands.

* * * * *

But formidable administrative problems lay ahead. Far away in South-East China, some months before, the Japanese had begun an advance on Chungking, the Generalissimo's capital, and Kunming, the delivery point of the American supply air-lift. In November General Wedemeyer took a serious view of this situation. Already the forward bases of the U.S. Air Force in China, which had been operating against enemy coastwise shipping, were being overrun. The Chinese troops gave little promise, and Wedemeyer appealed for two of the Chinese divisions in North Burma, and also for more American air squadrons, in particular for three transport squadrons.

The President addressed me.

President Roosevelt to Prime Minister   1 Dec 44

A telegram has been received from General Wedemeyer outlining the gravity of the situation in China and stating that he concurs in the decision of the Generalissimo to transfer the two best trained divisions of Chinese troops from Burma to the Kunming area. You have undoubtedly seen this message, which went to Mountbatten and has been furnished to your mission here in Washington, so I shall not repeat it.

We have General Wedemeyer's view on the ground as to the gravity of the situation, along with his knowledge of the situation and the plans for operations in Burma. I feel that he is better informed as to the general situation and requirements than any other individual at this moment. Furthermore, we are faced by the fact that the Generalissimo, in a grave crisis which threatens the existence of China, has decided that he must recall these two divisions in order to check the Japanese drive on Kunming. It would avail us nothing to open a land-line to China if the Japanese seized the Kunming terminal for air and ground. Under the circumstances I therefore am of the opinion that we are not in a position to bring pressure on the Generalissimo to alter his decision.

These were hard tidings, but we had no choice but to accept.

Prime Minister to General Hollis, for C.O.S. Committee   2 Dec 44

There can be no dispute about the right of the Generalissimo to withdraw any division he requires to defend himself against the Japanese attack upon his vitals. I have little doubt he will first wish to bring home the two divisions [trained by the Americans]. We cannot make a fight about this. If he claims them he must have them. What happens [afterwards] in Burma demands urgent but subsequent study. Pray let me have a telegram drafted agreeing with the Americans about the withdrawal of the divisions.

The loss of two good Chinese divisions was not so grave an inconvenience to the Burma operations as that of the transport squadrons. The Army was four hundred miles beyond its railhead and General Slim relied on air supply to help the tenuous road link. Mountbatten's general plans depended on his transport aircraft. The squadrons needed for China had to go, and although they were later replaced, mostly from British sources, their absence at a critical time caused severe delay to the campaign.

In spite of all this the Fourteenth Army broke out of the hills into the plain north-west of Mandalay. While the leading division of General Messervy's IVth Corps drove south in secret to establish a bridgehead over the Irrawaddy south of its junction with the Chindwin, General Stopford's XXXIIIrd Corps, supported by the 221 Group R.A.F., occupied the north bank of the Irrawaddy upstream from that junction. The 19th Indian Division was already across the river in two places forty miles north of Mandalay. By the end of January General Sultan's forces had reached Namkhan, on the old Burma–China road, and made contact with the Yunnan force farther east. The land route to China, closed by the Japanese invasion of Burma in the spring of 1942, was open again. The first road convoy from Assam reached the Chinese frontier on January 28.

Prime Minister to Admiral Mountbatten (South-East Asia)   23 Jan 45

On behalf of His Majesty's Government I send you our warmest congratulations on having reopened the land route to China in fulfilment of the first part of the directive given to you at Quebec. It reflects the greatest credit on yourself, all your field commanders, and above all upon the well-tried troops of the Fourteenth Army, that this should have been achieved despite your many disappointments in the delay of promised reinforcements.

His Majesty's Government warmly and gratefully recognise, as you have done throughout, the ready assistance given in all possible ways by the forces of the United States and also by those of China.

★   ★   ★   ★   ★

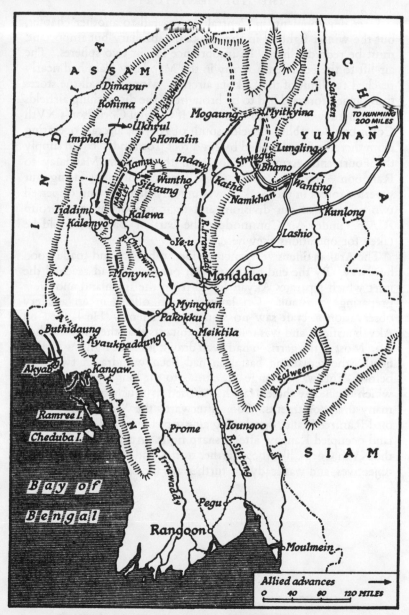

BURMA, JULY 1944 – JANUARY 1945

Later developments in Central Burma fall to another chapter, but the winter fighting in the Arakan, subsidiary but important, must be recorded here. Its importance lay in two spheres. The air-lift to the Fourteenth Army in the Mandalay plain had nearly reached the limit of the Dakota aircraft. Moreover, all the stores thus carried forward had to be brought to the dispatching airfields by the hard-worked Assam railway. If General Christison's XVth Corps could establish airfields south of Akyab, aircraft operating from there, and replenished by sea direct from India, could supply the Fourteenth Army in a southern thrust from Mandalay to Rangoon. Secondly, if the single Japanese division facing our superior forces in the Arakan was quickly defeated and dispersed two or three of our divisions and their supporting 224 Group R.A.F., under Air Commodore the Earl of Bandon, could be taken for operations elsewhere.

The Arakan offensive opened on December 12, and made good headway. By the end of the month our troops had reached the inlet which separates Akyab Island from the mainland and were preparing to assault. On January 2 an officer in an artillery observation aircraft saw no sign of the enemy. He landed on Akyab airfield, and was told by inhabitants that the Japanese had left. Most of the garrison had been drawn into the fighting farther north; the remaining battalion had been withdrawn two days before. This was a strange anticlimax to the long story of Akyab, which for nearly three years had caused us much tribulation and many disappointments. Soon afterwards the XVth Corps occupied Ramree Island, developing air-strips there, and on the mainland occupied Kangaw after a sharp fight. At the end of January the XVth Corps, like those farther north, had reached its primary objectives, and was ready for further advances.

CHAPTER XII

THE BATTLE OF LEYTE GULF

Ocean War Against Japan – Creation of the British Pacific Fleet – The Growing Maritime Strength of the United States – American Tactics and Japanese Defences – The Landing at Saipan, June 15 – Admiral Spruance Gains a Decisive Victory, June 20 – Conquest of the Marianas – Dismay in Tokyo – The Advance to the Philippines – Air Battles over Formosa – The Americans Land at Leyte Gulf, October 20 – The Japanese Commander-in-Chief Resolves to Intervene – Admiral Halsey and the Enemy Trap – Night Action in the Surigao Strait – The American Landings in Peril – Arrival of the Suicide Bomber – Admiral Kurita Turns Back – Twenty-seven Japanese Warships Destroyed – The Landing in Lingayen Gulf, January 9, 1945 – The Fall of Manila – The United States Gain Command of the South China Seas.

OCEAN war against Japan now reached its climax. From the Bay of Bengal to the Central Pacific Allied maritime power was in the ascendant. By April 1944 three British capital ships, two carriers, and some light forces were assembled in Ceylon. These were augmented by the American carrier *Saratoga*, the French battleship *Richelieu*, and a Dutch contingent. A strong flotilla of British submarines also arrived in February, and at once began to take toll of enemy shipping in the Malacca Strait. As the year advanced two more British carriers arrived, and the *Saratoga* returned to the Pacific. With these forces Admiral Somerville could do much more. In April his carriers struck at Sabang, at the northern end of Sumatra, and in May at the oil refinery and engineering works at Sourabaya, in Java. This operation lasted twenty-two days, and the fleet steamed seven thousand miles. In the following months the Japanese sea route to Rangoon was severed by British submarines and aircraft.

In August Admiral Somerville, who had commanded the Eastern Fleet through all the troubled times since March 1942, was relieved by Admiral Sir Bruce Fraser, and soon afterwards succeeded Admiral Noble at the head of our Naval Delegation at Washington. A month later the progress of the war in Europe enabled us to reduce the Home Fleet to no more than a single battleship, with supporting forces. The move to the Far East was hastened, and two modern battleships, the *Howe* and *King George V*, joined Admiral Fraser. On November 22, 1944, the British Pacific Fleet came officially into existence, and subsequently took part in a series of operations which fall to a later chapter.

* * * * *

In the Pacific the organisation and production of the United States were in full stride, and had attained astonishing proportions. A single example may suffice to illustrate the size and success of the American effort. In the autumn of 1942, at the peak of the struggle for Guadalcanal, only three American aircraft-carriers were afloat; a year later there were fifty; by the end of the war there were more than a hundred. This achievement had been matched by an increase in aircraft production which was no less remarkable. The advance of these great forces was animated by an aggressive strategy and an elaborate, novel, and effective tactic. The task which confronted them was formidable.

A chain of island-groups, nearly two thousand miles long, stretches southward across the Pacific from Japan to the Marianas and the Carolines. Many of these islands had been fortified by the enemy and equipped with good airfields, and at the southernmost end of the chain was the Japanese naval base of Truk. Behind this shield of archipelagos lay Formosa, the Philippines, and China, and in its shelter ran the supply routes for the more advanced enemy positions. It was thus impossible to invade or bomb Japan itself. The chain must be broken first. It would take too long to conquer and subdue every fortified island, and the Americans had accordingly advanced leap-frog fashion. They seized only the more important islands and by-passed the rest; but their maritime strength was now so great and was growing so fast that they were able to establish their own lines of communication and break the enemy's, leaving the defenders of the by-passed islands immobile and powerless. Their method of assault was equally successful. First came softening attacks by planes from the aircraft-carriers, then heavy and sometimes prolonged bombardment from the sea, and finally amphibious landing and the struggle ashore. When an island had been won and garrisoned land-based planes moved in and beat off counter-attacks. At the same time they helped in the next onward surge. The fleets worked in echelons. While one group waged battle another prepared for a new leap. This needed very large resources, not only for the fighting, but also for developing bases along the line of advance. The Americans took it all in their stride.

$$\star \quad \star \quad \star \quad \star \quad \star$$

Earlier volumes have described the two-pronged American attack across the Pacific, and when this narrative opens in June 1944 it was well advanced. In the south-west General MacArthur had nearly completed his conquest of New Guinea, and in the centre Admiral Nimitz was pressing deep into the chain of fortified islands. Both were converging on the Philippines, and the struggle for this region was soon to bring about the destruction of the Japanese Fleet. The Fleet had already been much weakened and was very short of carriers, but Japan's only hope of survival lay in victory at sea. To conserve her strength for this perilous but vital hazard her main fleet had withdrawn from Truk and was now divided between the East Indies and her home waters; but events soon brought it to battle. At the beginning of June Admiral Spruance struck with his carriers at the Marianas, and on the 15th he landed on the fortified island of Saipan. If he captured Saipan and the adjacent islands of Tinian and Guam the enemy's defence perimeter would be broken. The threat was formidable, and the Japanese Fleet resolved to intervene. That day five of their battleships and nine carriers were sighted near the Philippines, heading east. Spruance had ample time to make his dispositions. His main purpose was to protect the landing at Saipan. This he did. He then gathered his ships, fifteen of which were carriers, and waited for the enemy to the west of the island. On June 19 Japanese carrier-borne aircraft attacked the American carrier fleet from all directions, and air-fighting continued throughout the day. The Americans suffered little damage, and so shattered the Japanese air squadrons that their carriers had to withdraw.

That night Spruance searched in vain for the vanished enemy. Late in the afternoon of the 20th he found them about 250 miles away. Attacking just before sunset, the American airmen sank one carrier and damaged four others, besides a battleship and a heavy cruiser. The previous day American submarines had sunk two other large carriers. No further attack was possible, and remnants of the enemy fleet managed to escape, but its departure sealed the fate of Saipan. Though the garrison fought hard the landings continued, the build-up progressed, and by July 9 all organised resistance came to an end. The neighbouring islands of Guam and Tinian were overcome, and by the first days of August the American grip on the Marianas was complete.

The fall of Saipan was a great shock to the Japanese High Command, and led indirectly to the dismissal of General Tojo's Government. The enemy's concern was well founded. The fortress was little more than 1,300 miles from Tokyo. They had believed it was impregnable; now it was gone. Their southern defence regions were cut off and the American heavy bombers had gained a first-class base for attacking the very homeland of Japan. For a long time United States submarines had been sinking Japanese merchantmen along the China coast, and now the way was open for other warships to join in the onslaught. Japan's oil and raw materials would be cut off if the Americans advanced any farther. The Japanese Fleet was still powerful, but unbalanced, and so weak in destroyers, carriers, and air-crews that it could no longer fight effectively without land-based planes. Fuel was scarce, and not only hampered training, but made it impossible to keep the ships concentrated in one place, so that in the late summer most of the heavy vessels and cruisers lay near Singapore and the oil supplies of the Dutch East Indies, while the few surviving carriers remained in home waters, where their new air groups were completing their training.

The plight of the Japanese Army was little better. Though still strong in numbers, it sprawled over China and South-East Asia or languished in remote islands beyond reach of support. The more sober-minded of the enemy leaders began to look for some way of ending the war; but their military machine was too strong for them. The High Command brought reinforcements from Manchuria and ordered a fight to the finish both in Formosa and the Philippines. Here and in the homeland the troops would die where they stood. The Japanese Admiralty were no less resolute. If they lost the impending battle for the islands the oil from the East Indies would be cut off. There was no purpose, they argued, in preserving ships without fuel. Steeled for sacrifice but hopeful of victory, they decided in August to send the entire Fleet into battle.

* * * * *

On September 15 the Americans made another advance. General MacArthur seized Morotai Island, midway between the western tip of New Guinea and the Philippines, and Admiral Halsey, who had now assumed command of the United States naval forces, captured an advanced base for his fleet in the Palau group. These simultaneous moves were of high importance. At the same time Halsey continually probed the enemy's defences with his whole force. Thus he hoped to provoke a general action at sea which would enable him to destroy the Japanese Fleet, particularly its surviving carriers. The next leap would be at the Philippines themselves, and there now occurred a dramatic change in the American plan. Till then our Allies had purposed to invade the southernmost portion of the Philippines, the island of Mindanao, and planes from Halsey's carriers had already attacked the Japanese airfields both there and in the large northern island of Luzon. They destroyed large numbers of enemy aircraft, and discovered in the clash of combat that the Japanese garrison at Leyte was unexpectedly weak. This small but now famous island, lying between the two larger but strategically less important land masses of Mindanao and Luzon, became the obvious point for the American descent. On September 13, while the Allies were still in conference at Quebec, Admiral Nimitz, at Halsey's suggestion, urged its immediate invasion. MacArthur agreed, and within two days the American Chiefs of Staff resolved to attack on October 20, two months earlier than had been planned. Such was the genesis of the Battle of Leyte Gulf.

The Americans opened the campaign on October 10 with raids on airfields between Japan and the Philippines. Devastating and repeated attacks on Formosa provoked the most violent resistance, and from the 12th to the 16th there followed a heavy and sustained air battle between ship-borne and land-based aircraft. The Americans inflicted grievous losses both in the air and on the ground, but suffered little themselves, and their carrier fleet withstood powerful land-based air attack. The result was decisive. The enemy's Air Force was broken before the battle for Leyte was joined. Many Japanese naval aircraft destined for the fleet carriers were improvidently sent to Formosa as reinforcements and there destroyed. Thus in the supreme naval battle which now impended the Japanese carriers were manned by little more than a hundred partially trained pilots.

*   *   *   *   *

To comprehend the engagements which followed a study of the accompanying maps is necessary. The two large islands of the Philippines, Luzon in the north and Mindanao in the south, are separated by a group of smaller islands, of which Leyte is the key and centre. This central group is pierced by two navigable straits, both destined to dominate this famous battle. The northerly strait is San Bernardino, and about 200 miles south of it, leading directly to Leyte, is the strait of Surigao. The Americans, as we have seen, intended to seize Leyte, and the Japanese were resolved to stop them and to destroy their fleet. The plan was simple and desperate. Four divisions under General MacArthur would land on Leyte, protected by the guns and planes of the American fleet—so much they knew or guessed. Draw off this fleet, entice it far to the north, and engage it in a secondary battle—such was the first step. But this would be only a preliminary. As soon as the main fleet was lured away two strong columns of warships would sail through the straits, one through San Bernardino and the other through Surigao, and converge on the landings. All eyes would be on the shores of Leyte, all guns trained on the beaches and the heavy ships and the big aircraft-carriers which alone could withstand the assault would be chasing the decoy force in the far north. The plan very nearly succeeded.

On October 17 the Japanese Commander-in-Chief ordered his fleet to set sail. The decoy force, under Admiral Ozawa, the Supreme Commander, sailing direct from Japan, steered for Luzon. It was a composite force, including carriers, battleships, cruisers, and destroyers. Ozawa's task was to appear on the eastern coast of Luzon, engage the American fleet, and draw it away from the landings in Leyte Gulf. The carriers were short of both planes and pilots, but no matter. They were only bait, and bait is made to be eaten. Meanwhile the main Japanese striking forces made for

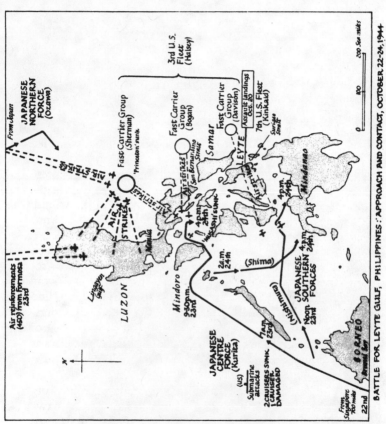

BATTLE FOR LEYTE GULF, PHILIPPINES ': APPROACH AND CONTACT, OCTOBER 22-24, 1944

the straits. The larger, or what may be termed the Centre Force, coming from Singapore, and consisting of five battleships, twelve cruisers, and fifteen destroyers, under Admiral Kurita, headed for San Bernardino to curl round Samar Island to Leyte; the smaller, or Southern Force, in two independent groups, comprising in all two battleships, four cruisers, and eight destroyers, sailed through Surigao.

On October 20 the Americans landed on Leyte. At first all went well. Resistance on shore was weak, a bridgehead was quickly formed, and General MacArthur's troops began their advance. They were supported by Admiral Kinkaid's Seventh United States Fleet, which was under MacArthur's command, and whose older battleships and small aircraft-carriers were well suited to amphibious operations. Farther away to the northward lay Admiral Halsey's main fleet, shielding them from attack by sea.

I was on my way home from Moscow at the time, but Field-Marshal Brooke and I recognised the importance of what had happened and. we sent the following telegram:

Prime Minister and C.I.G.S. to General MacArthur   22 Oct 44
Hearty congratulations on your brilliant stroke in the Philippines. All good wishes.

The crisis however was still to come. On October 23 American submarines sighted the Japanese Centre Force (Admiral Kurita) off the coast of Borneo and sank two of its heavy cruisers, one of which was Kurita's flagship, and damaged a third. Next day, October 24, planes from Admiral Halsey's carriers joined in the attack. The giant battleship *Musashi*, mounting nine eighteen-inch guns, was sunk, other vessels were damaged, and Kurita turned back. The American airmen brought back optimistic and perhaps misleading reports, and Halsey concluded, not without reason, that the battle was won, or at any rate this part of it. He knew that the second or Southern enemy force was approaching the Surigao Strait, but he judged, and rightly, that it could be repelled by Kinkaid's Seventh Fleet.

But one thing disturbed him. During the day he had been attacked by Japanese naval planes. Many of them were shot down, but the carrier *Princeton* was damaged and had later to be abandoned. The planes, he reasoned, probably came from carriers. It was most unlikely that the enemy had sailed without them, yet none had been found. The main Japanese fleet, under Kurita, had been located, and was apparently in retreat, but Kurita had no carriers, neither were there any in the Southern Force. Surely there must be a carrier force, and it was imperative to find it. He accordingly ordered a search to the north, and late in the afternoon of October 24 his flyers came upon Admiral Ozawa's decoy force, far to the north-east of Luzon and steering south. Four carriers, two battleships equipped with flying decks, three cruisers, and ten destroyers! Here, he concluded, was the source of the trouble and the real target. If he could now destroy these carriers, he and his Chief of Staff, Admiral Carney, rightly considered that the power of the Japanese fleet to intervene in future operations would be broken irretrievably. This was a dominating factor in his mind, and would be of particular advantage when MacArthur came later to attack Luzon. Halsey could not know how frail was their power, nor that most of the attacks he had endured came not from carriers at all but from airfields in Luzon itself. Kurita's Centre Force was in retreat. Kinkaid could cope with the Southern Force and protect the landings at Leyte, the way was clear for a final blow, and Halsey ordered his whole fleet to steam northward and destroy Admiral Ozawa next day. Thus he fell into the trap. That same afternoon, October 24, Kurita again turned east, and sailed once more for the San Bernardino Strait. This time there was nothing to stop him.

* * * * *

Meanwhile the Southern Japanese Force was nearing Surigao Strait, and that night they entered it in two groups. A fierce battle followed, in which all types of vessel, from battleships to light coastal craft, were closely engaged.* The first group was annihilated by Kinkaid's fleet, concentrated at the northern exit under the distinguished command of Admiral Oldendorf; the second tried to break through but was driven back. All seemed to be going well, but the Americans had still to reckon with Admiral Kurita. While Kinkaid was fighting in the Surigao Strait and Halsey was in hard pursuit of the decoy force far to the north Kurita had passed unchallenged in the darkness through the Strait of San Bernardino, and in the early morning of October 25 he fell upon a group of escort carriers who were supporting General Mac-

---

* Among them were two Australian warships, the cruiser *Shropshire* and the destroyer *Arunta.*

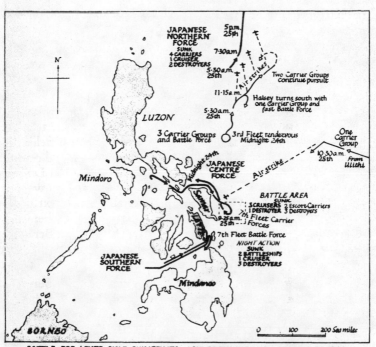

BATTLE FOR LEYTE GULF, PHILIPPINES: THE DECISIVE PHASE, OCTOBER 25, 1944

Arthur's landings. Taken by surprise and too slow-moving to escape, the carriers could not at once rearm their planes to repel the onslaught from the sea. For about two and a half hours the small American ships fought a valiant retreat under cover of smoke. Two of their carriers, three destroyers, and over a hundred planes were lost, one of the carriers by suicide bomber attack; but they succeeded in sinking three enemy cruisers and damaging others.* Help was far away. Kinkaid's heavy ships were well south of Leyte, having routed the Southern Force, and were short of ammunition and fuel. Halsey, with ten carriers and all his fast battleships, was yet more distant, and although another of his carrier groups had been detached to refuel and was now recalled it could not arrive for some hours. Victory seemed to be in Kurita's hands. There was nothing to stop him steaming into Leyte Gulf and destroying MacArthur's amphibious fleet.

But once again Kurita turned back. His reasons are obscure. Many of his ships had been bombed and scattered by Kinkaid's light escort carriers, and he now knew that the Southern Force had met with disaster. He had no information about the fortunes of the decoys in the north and was uncertain of the whereabouts of the American fleets. Intercepted signals made him think that Kinkaid and Halsey were converging on him in overwhelming strength and that MacArthur's transports had already managed to escape. Alone and unsupported, he now abandoned the desperate venture for which so much had been sacrificed and which was about to gain its prize, and, without attempting to enter Leyte Gulf, he turned about and steered once more for the San Bernardino Strait. He hoped to fight a last battle on the way with Halsey's fleet, but even this was denied him. In response to Kinkaid's repeated calls for support Halsey had indeed at last turned back with his battleships, leaving two carrier groups to continue the pursuit to the north. During the day these destroyed all four of Ozawa's carriers. But Halsey himself got back to San Bernardino too late. The fleets did not meet. Kurita escaped. Next day Halsey's and MacArthur's planes pursued the Japanese admiral and sank another cruiser and two more destroyers. This was the end of the battle. It may well be that Kurita's mind had

---

* Suicide bombers made their first appearance in the Leyte operations. The Australian cruiser *Australia*, operating with Kinkaid's fleet, had been hit by one a few days before and had suffered casualties but no serious damage.

become confused by the pressure of events. He had been under constant attack for three days, he had suffered heavy losses, and his flagship had been sunk soon after starting from Borneo. Those who have endured a similar ordeal may judge him.

*   *   *   *   *

The Battle of Leyte Gulf was decisive. At a cost to themselves of three carriers, three destroyers, and a submarine the Americans had conquered the Japanese Fleet. The struggle had lasted from October 22 to October 27. Three battleships, four carriers, and twenty other enemy warships had been sunk, and the suicide bomber was henceforward the only effective naval weapon left to the foe. As an instrument of despair it was still deadly, but it carried no hope of victory.

This time there was no doubt about the result, and we hastened to send our congratulations.

Prime Minister to President Roosevelt   27 Oct 44

Pray accept my most sincere congratulations, which I tender on behalf of His Majesty's Government, on the brilliant and massive victory gained by the sea and air forces of the United States over the Japanese in the recent heavy battles.

We are very glad to know that one of His Majesty's Australian cruiser squadrons had the honour of sharing in this memorable event.

The scale of the battle may be judged from the following table:

TOTAL LOSSES

Japanese	United States
Three battleships.	One light fleet carrier.
One fleet carrier.	Two escort carriers.
Three light carriers.	Three destroyers.
Six heavy cruisers.	One submarine.
Four light cruisers.	
Nine destroyers.	
One submarine.	

Long should this victory be treasured in American history. Apart from valour, skill, and daring, it shed a light on the future more vivid and far-reaching than any we had seen. It shows a battle fought less with guns than by predominance in the air. I have told the tale fully because at the time it was almost unknown to the harassed European world. Perhaps the most important

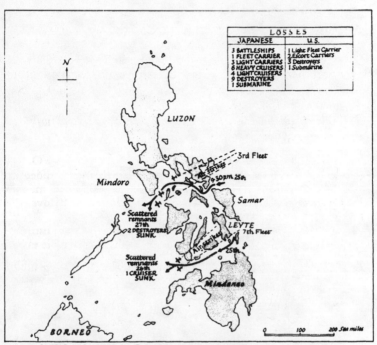

BATTLE FOR LEYTE GULF, PHILIPPINES: THE PURSUIT, OCTOBER 26-27, 1944

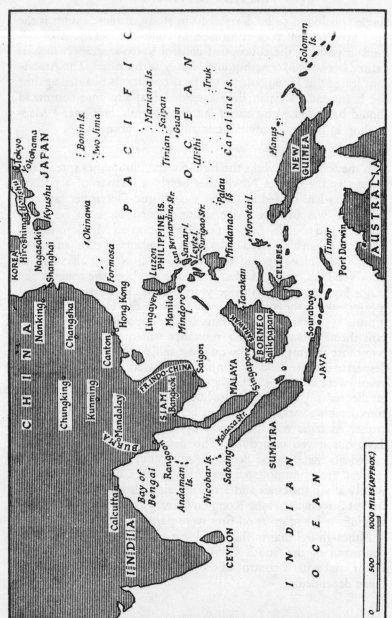

THE SOUTH-WEST PACIFIC

single conclusion to be derived from study of these events is the vital need for unity of command in conjoint operations of this kind in place of the concept of control by co-operation such as existed between MacArthur and Halsey at this time. The Americans learnt this lesson, and in the final operations planned against the homeland of Japan they intended that supreme command should be exercised by either Admiral Nimitz or General Mac-Arthur, as might be advisable at any given moment.

*　　*　　*　　*　　*

In the following weeks the fight for the Philippines spread and grew. By the end of November nearly a quarter of a million Americans had landed in Leyte, and by mid-December Japanese resistance was broken. MacArthur pressed on with his main advance, and soon landed without opposition on Mindoro Island, little more than a hundred miles from Manila itself. On January 9, 1945, a new phase opened with the landing of four divisions in Lingayen Gulf, north of Manila, which had been the scene of the major Japanese invasion three years before. Elaborate deception measures kept the enemy guessing where the blow would fall. It came as a surprise and was only lightly opposed. As the Americans thrust towards Manila resistance stiffened, but they made two more landings on the west coast and surrounded the city. A desperate defence held out until early March, when the last survivors were killed. Sixteen thousand Japanese dead were counted in the ruins. Attacks by suicide aircraft were now inflicting considerable losses, sixteen ships being hit in a single day. The cruiser *Australia* was again unlucky, being hit five times in four days, but stayed in action. This desperate expedient however caused no check to the fleets. In mid-January Admiral Halsey's carriers broke unmolested into the South China Sea, ranging widely along the coast and attacking airfields and shipping as far west as Saigon. At Hong Kong on January 16 widespread damage was inflicted, and great oil fires were started at Canton.

Although fighting in the islands continued for several months, command of the South China seas had already passed to the victor, and with it control of the oil and other supplies on which Japan depended.

CHAPTER XIII

THE LIBERATION OF WESTERN EUROPE

ENERAL EISENHOWER, in accordance with pre-vious and agreed arrangements, assumed direct command of the land forces in Northern France on September 1. These comprised the British Twenty-first Army Group, under Field-Marshal Montgomery, and the American Twelfth Army Group, under General Omar Bradley, whose operations Mont-gomery had hitherto controlled. Eisenhower disposed of five armies in all. In Montgomery's Twenty-first Army Group were the First Canadian Army, under General Crerar, and the Second British Army, under General Dempsey; a total of fourteen divisions and seven armoured brigades. On their right, under the Twelfth U.S. Army Group, were the First Army, under General Hodges, the Third Army, under General Patton, and the Ninth Army, not yet operational, under General Simpson. Eisenhower thus wielded more than thirty-seven divisions, or over half a million fighting men. Each Army Group had its own tactical air force, the whole being under the control of Air Chief Marshal Leigh-Mallory.

This great array was driving before it the remnants of the German armies in the West, who were harassed day and night by our dominating air forces. The enemy were still about seventeen divisions strong, but until they could re-form and were reinforced from the homeland there was little fight left in most of them. General Speidel, Rommel's former Chief of Staff, has described their plight:

An orderly retreat became impossible. The Allied motorised armies surrounded the slow and exhausted German foot divisions in separate groups and smashed them up. . . . There were no German ground forces of any importance that could be thrown in, and next to nothing in the air.*

Eisenhower planned to thrust north-eastwards in the greatest possible strength and to the utmost limit of his supplies. The main effort was to be made by the British Twenty-first Army Group, whose drive along the Channel coast would not only overrun the launching sites of the flying bomb, but also take Antwerp. Without the vast harbour of this city no advance across the lower Rhine and into the plains of Northern Germany was possible. The Twelfth U.S. Army Group was also to pursue the enemy, its First Army keeping abreast of the British, while the remainder, bearing eastwards towards Verdun and the upper Meuse, would prepare to strike towards the Saar.

Montgomery made two counter-proposals, one in late August that his Army Group and the Twelfth U.S. Army Group should strike north together with a solid mass of nearly forty divisions, and the second on September 4, that only one thrust should be made, either towards the Ruhr or the Saar. Whichever was chosen the forces should be given all the resources and maintenance they needed. He urged that the rest of the front should be restrained for the benefit of the major thrust, which should be placed under one commander, himself or Bradley as the case might be. He believed it would probably reach Berlin, and considered that the Ruhr was better than the Saar.

But Eisenhower held to his plan. Germany still had reserves in

* *We Defended Normandy*, Speidel, pp. 152–153

the homeland, and he believed that if a relatively small force were thrust far ahead across the Rhine it would play into the enemy's hands. He thought it was better for the Twenty-first Army Group to make every effort to get a bridgehead over the Rhine, while the Twelfth advanced as far as they could against the Siegfried Line.

Strategists may long debate these issues.

Their discussion caused no check in the pursuit. The number of divisions that could be sustained, and the speed and range of their advance, depended however entirely on harbours, transport, and supplies. Relatively little ammunition was being used, but food, and above all petrol, governed every movement. Cherbourg and the "Mulberry" harbour at Arromanches were the only ports we had, and these were daily being left farther behind. The front line was still sustained from Normandy, and each day about 20,000 tons of supplies had to be carried over ever-increasing distances, together with much material for mending roads and bridges and building airfields. The Brittany ports, when captured, would be even more remote, but the Channel ports from Havre northwards, and especially Antwerp, if we could capture it before it was too seriously damaged, were prizes of vital consequence.

Antwerp was thus the immediate aim of Montgomery's Army Group, which now had its first chance to show its mobility. The Second Army led the advance north of the Seine towards Belgium, grounding one corps and using its transport to sustain the others. The XXXth Corps was in the van. Its 11th Armoured Division captured the commander of the Seventh German Army at his breakfast in Amiens on August 31. The frontier towns so well known to the British Expeditionary Force of 1940, and, at least by name, to their predecessors a quarter of a century before —Arras, Douai, Lille, and many others—were soon reached. Brussels, hastily evacuated by the Germans, was entered by the Guards Armoured Division on September 3, and, as everywhere in Belgium, our troops had a splendid welcome and were much helped by the well-organised Resistance. Thence the Guards turned east for Louvain, and the 11th Armoured entered Antwerp on September 4, where, to our surprise and joy, they found the harbour almost intact. So swift had been the advance—over 200 miles in under four days—that the enemy had been run off their legs and given no time for their usual and thorough demolition.

Farther to the west the XIIth Corps met more resistance, but reached their principal objective, Ghent, on September 5.

Of course this pace could not last. The forward leap was over and the check was evident before we sailed to Quebec. The enemy managed to destroy the crossings over the Albert Canal between Antwerp and Hasselt, and the XXXth Corps found it defended by about ten battalions, some of them quite fresh. The Guards forced a crossing west of Hasselt on September 6, but they had a hard fight and it was not till four days later that they reached the Meuse-Scheldt canal and took a bridge which was still intact.

*　　*　　*　　*　　*

Meanwhile the First Canadian Army had the heavy and responsible task of clearing the western flank. Their commander, General Crerar, had under him the Ist British Corps and the IInd Canadian Corps, which included the Polish Armoured Division. Their main task was to clear the Channel ports from Havre to the north, occupy the flying bomb sites, and establish themselves on the south shore of the Scheldt. Although Antwerp was in our hands our ships could only reach it through the winding, difficult estuary of the Scheldt, and the Germans held both banks. These hard and costly operations were to fall principally on this Canadian Army, and much depended on their success.

The Ist British Corps crossed the Seine near Rouen, swung left-handed, and on September 2 the 51st Highland Division occupied St. Valery, the scene of the tragedy to its parent unit in June of 1940. The left of the corps turned and advanced on Havre, where a garrison of over 11,000 resisted fiercely. In spite of bombardment from the sea by 15-inch guns, and more than 10,000 tons of bombs from the air, the Germans did not surrender Havre till September 12. Meanwhile the Canadian Corps, on their right, had moved swiftly. Dieppe, where they repaid their old scores of 1942, fell to them on September 1. Boulogne and Calais were invested by September 6, then Dunkirk. By the 9th the Canadian Army had cleared all the Pas de Calais, with its flying-bomb launching sites, and reached Bruges. Ghent was taken by the Polish Armoured Division. Boulogne, with nearly 10,000 prisoners, fell on September 22, and Calais on the 30th. Dunkirk, with its garrison of 12,000, was only masked, as the advance to the Scheldt was far more urgent. Here for the moment

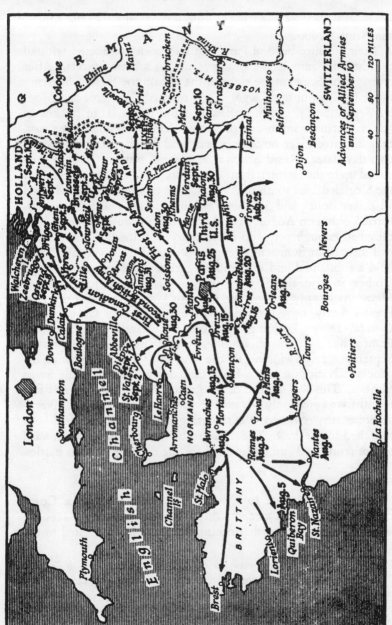

THE PURSUIT

Advances of Allied Armies
until September 10

0   40   80   120 MILES

we must leave the Canadians to follow the fortunes of the American Army Group.

Their advance beyond Paris had also been conducted with all the thrustful impulse of Bradley and his ardent officers. After crossing the Seine on the right of the British the First U.S. Army made for Namur and Liége. They reached Charleroi and Mons by September 3, cutting off and capturing a large pocket of 30,000 Germans south-east of Mons; then, wheeling to the east, they liberated Liége on September 8 and the city of Luxembourg two days later. Resistance was increasing, but on the 12th they closed up to the German frontier on a sixty-mile front and pierced the Siegfried Line south of Aachen. In a fortnight they had freed all Luxembourg and Southern Belgium. The Third Army captured Verdun on August 31 and crossed the Meuse. A week later they had enough petrol to advance to the Moselle. The enemy had scraped up sufficient strength to defend the river, and Metz held a substantial and determined garrison. However, by September 16 bridgeheads were won at Nancy and just south of Metz. As already related, the Seventh U.S. Army and the First French Army, now formed into the Sixth Army Group under General Devers, coming up from their landing in Southern France, had met patrols from Patton's army west of Dijon on September 11. Swinging to the east, they drew level with the general advance on a line from Épinal southwards to the Swiss frontier. This was the end of the great pursuit. For the next few months we could only advance after very hard fighting. Everywhere enemy resistance was stiffening, and our supplies had been stretched to the limit. These had to be restored, and the forward troops reinforced and replenished for the coming autumn battles.

$$* \quad * \quad * \quad * \quad *$$

During our voyage to Quebec our Joint Intelligence Committee had furnished a report on "German Capacity to Resist", which I deemed somewhat optimistic, and I had minuted to the Chiefs of Staff as follows:

Prime Minister to General Ismay, for C.O.S. Committee   8 Sept 44

I have now read this report, and have not noticed any facts in it of which I was not already aware. Generally speaking, I consider it errs on the side of optimism. At the present time we are at a virtual standstill and progress will be very slow. I trust the assumption of a decisive

Russian offensive on the Eastern Front will be realised; but it is at present only an assumption.

2. On the other side there are factors to be noted. Apart from Cherbourg and Arromanches, we have not yet obtained any large harbours. The Germans intend to defend the mouth of the Scheldt, and are still resisting in the northern suburbs of Antwerp. Brest has not been taken in spite of very heavy fighting, and at least six weeks will be needed after it is taken before it is available. Lorient still holds out. No attempt has been made to take and clear the port of St. Nazaire, which is about twice as good as Brest and twice as easy to take. No attempt has been made to get hold of Bordeaux. Unless the situation changes remarkably the Allies will still be short of port accommodation when the equinoctial gales are due.

3. One can already foresee the probability of a lull in the magnificent advances we have made. General Patton's army is heavily engaged on the line Metz–Nancy. Field-Marshal Montgomery has explained his misgivings as to General Eisenhower's future plan. It is difficult to see how the Twenty-first Army Group can advance in force to the German frontier until it has cleared up the stubborn resistance at the Channel ports and dealt with the Germans at Walcheren and to the north of Antwerp. . . .

6. No one can tell what the future may bring forth. Will the Allies be able to advance in strength through the Siegfried Line into Germany during September, or will their forces be so limited by supply conditions and the lack of ports as to enable the Germans to consolidate on the Siegfried Line? Will the Germans withdraw from Italy?—in which case they will greatly strengthen their internal position. Will they be able to draw on their forces, at one time estimated at between twenty-five and thirty-five divisions, in the Baltic States? The fortifying and consolidating effect of a stand on the frontier of the native soil should not be underrated. *It is at least as likely that Hitler will be fighting on January 1 as that he will collapse before then.** If he does collapse before then, the reason will be political rather than purely military.

My view was unhappily to be justified.

* * * * *

But there was still the chance of crossing the lower Rhine. Eisenhower thought this prize so valuable that he gave it priority over clearing the shores of the Scheldt estuary and opening the port of Antwerp. To renew Montgomery's effort Eisenhower gave him additional American transport and air supply. The First Airborne Army, under the American General Brereton,

* Author's subsequent italics.

composed of the 1st and 6th British Airborne Divisions, three
U.S. divisions, and a Polish brigade, with a great complement of
British and American aircraft, stood ready to strike from England.
Montgomery resolved to seize a bridgehead at Arnhem by the
combined action of airborne troops and the XXXth Corps, who
were fighting in a bridgehead across the Meuse–Scheldt canal on
the Dutch border. He planned to drop the 1st British Airborne
Division, supported later by the Polish Brigade, on the north bank
of the lower Rhine to seize the Arnhem bridge. The 82nd U.S.
Division was to capture the bridges at Nijmegen and Grave,
while the 101st U.S. Division secured the road from Grave to
Eindhoven. The XXXth Corps, led by the Guards Armoured
Division, would force their way up the road to Eindhoven and
thence to Arnhem along the "carpet" of airborne troops, hoping
to find the bridges over the three major water obstacles already
safely in their hands.

The preparations for this daring stroke, by far the greatest
operation of its kind yet attempted, were complicated and urgent,
because the enemy were growing stronger every day. It is remarkable that they were completed by the set date, September 17.
There were not sufficient aircraft to carry the whole airborne
force simultaneously, and the movement had to be spread over
three days. However, on the 17th the leading elements of the
three divisions were well and truly taken to their destinations by
the fine work of the Allied air forces. The 101st U.S. Division
accomplished most of their task, but a canal bridge on the road to
Eindhoven was blown and they did not capture the town till the
18th. The 82nd U.S. Division also did well, but could not seize
the main bridge at Nijmegen.

From Arnhem the news was scarce, but it seemed that some
of our Parachute Regiment had established themselves at the
north end of the bridge. The Guards Armoured Division of the
XXXth Corps began to advance in the afternoon up the Eindhoven road, preceded by an artillery barrage and rocket-firing
planes. The VIIIth Corps on the right and the XIIth on the left
protected the flanks of the XXXth. The road was obstinately
defended, and the Guards did not reach the Americans till the
afternoon of the 18th. German attacks against the narrow Eindhoven–Nijmegen salient began next day and grew in strength.
The 101st Division had great difficulty in keeping the road open.

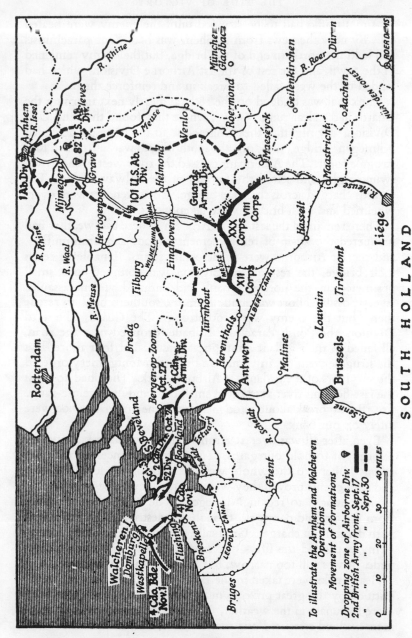

To illustrate the Arnhem and Walcheren
Operations

Movement of formations
" " " Sept.17
" " " Sept.30

Dropping zone of Airborne Div.
2nd British Army front, Sept.17

0 10 20 30 40 MILES

SOUTH HOLLAND

173

At times traffic had to be stopped until the enemy were beaten off. By now the news from Arnhem was bad. Our parachutists still held the northern end of the bridge, but the enemy remained in the town, and the rest of the 1st Airborne Division, which had landed to the west, failed to break in and reinforce them.

The canal was bridged on the 18th, and early next morning the Guards had a clear run to Grave, where they found the 82nd U.S. Division. By nightfall they were close to the strongly defended Nijmegen bridge, and on the 20th there was a tremendous struggle for it. The Americans crossed the river west of the town, swung right, and seized the far end of the railway bridge. The Guards charged across the road bridge. The defenders were overwhelmed and both bridges were taken intact.

There remained the last lap to Arnhem, where bad weather had hampered the fly-in of reinforcements, food, and ammunition, and the 1st Airborne were in desperate straits. Unable to reach their bridge, the rest of the division was confined to a small perimeter on the northern bank and endured violent assaults. Every possible effort was made from the southern bank to rescue them, but the enemy were too strong. The Guards, the 43rd Division, the Polish Parachute Brigade, dropped near the road, all failed in their gallant attempts at rescue. For four more days the struggle went on, in vain. On the 25th Montgomery ordered the survivors of the gallant 1st Airborne back. They had to cross the fast-flowing river at night in small craft and under close-range fire. By daybreak about 2,400 men out of the original 10,000 were safely on our bank.

Even after all was over at Arnhem there was hard fighting for a fortnight to hold our gains. The Germans conceived that our salient imperilled the whole western bank of the lower Rhine, and later events proved they were right. They made many heavy counter-attacks to regain Nijmegen. The bridge was bombed from the air, and damaged, though not destroyed, by swimmers with demolition charges. Gradually the three corps of the Second Army expanded the fifty-mile salient until it was twenty miles wide. It was still too narrow, but for the moment it sufficed.

Heavy risks were taken in the Battle of Arnhem, but they were justified by the great prize so nearly in our grasp. Had we been more fortunate in the weather, which turned against us at critical moments and restricted our mastery in the air, it is probable that we should have succeeded. No risks daunted the brave men, including the Dutch Resistance, who fought for Arnhem.

* * * * *

It was not till I returned from Canada, where the glorious reports had flowed in, that I was able to understand all that had happened. General Smuts was grieved at what seemed to be a failure, and I telegraphed:

Prime Minister to Field-Marshal Smuts 9 Oct 44

I like the situation on the Western Front, especially as enormous American reinforcements are pouring in and we hope to take Antwerp before long. As regards Arnhem, I think you have got the position a little out of focus. The battle was a decided victory, but the leading division, asking, quite rightly, for more, was given a chop. I have not been afflicted by any feeling of disappointment over this and am glad our commanders are capable of running this kind of risk.

* * * * *

Clearing the Scheldt estuary and opening the port of Antwerp had been delayed for the sake of the Arnhem thrust. Thereafter it was given first priority. During the last fortnight of September a number of preliminary actions had set the stage. The IInd Canadian Corps had forced the enemy back from the line Antwerp–Ghent–Bruges into the restricted Breskens "island", bounded on the south by the Leopold Canal. East of Antwerp the Ist Corps, also under the Canadian Army command, had reached and crossed the Antwerp–Turnhout canal.

The problem was threefold: the capture of the Breskens "island"; the occupation of the peninsula of South Beveland; finally, the capture of Walcheren Island by attacks from east, south, and west. The first two proceeded simultaneously. Breskens "island", defended by an experienced German division, proved tough, and there was hard fighting to cross the Leopold Canal. The scales were turned by a Canadian brigade, which embarked upstream, landed at the eastern extremity of the "island", and forced a way along the shore towards Breskens, which fell on October 22. Meanwhile the 1st Corps had steadily advanced north-west from the Antwerp–Turnhout canal, meeting increased opposition as they went. The South Beveland isthmus was sealed off, and plans could be made for continuing the operations westwards towards Walcheren.

This hard task was undertaken by the 2nd Canadian Division, which forced its way westwards through large areas of flooding, their men often waist-deep in water. They were helped by the greater part of the 52nd Division, who were ferried across the Scheldt and landed on the south shore at Baarland. By the end of the month, after great exertions, the whole isthmus was captured. Meantime the last pockets of enemy on Breskens "island" were being eliminated and all was set for the Walcheren attack. The Canadian Army's success was an essential preliminary to more spectacular operations. In four weeks of hard fighting, during which the 2nd Tactical Air Force, under Air Marshal Coningham, gave them conspicuous support, they took no fewer than 12,500 German prisoners, who were anything but ready to surrender.

★　★　★　★　★

The island of Walcheren is shaped like a saucer and rimmed by sand-dunes which stop the sea from flooding the central plain. At the western edge, near Westkapelle, is a gap in the dunes where the sea is held by a great dyke, thirty feet high and over a hundred yards wide at the base. The garrison of nearly 10,000 men was installed in strong artificial defences, and supported by about thirty batteries of artillery, some of large calibre in concrete emplacements. Anti-tank obstacles, mines, and wire abounded, for the enemy had had four years in which to fortify the gateway to Antwerp.

Early in October the Royal Air Force struck the first blow. In a series of brilliant attacks they blew a great gap, nearly four hundred yards across, in the Westkapelle dyke. Through it poured the sea, flooding all the centre of the saucer and drowning such defences and batteries as lay within. But the most formidable emplacements and obstacles were on the saucer's rim, and their capture, already admirably described,★ can be told here only in outline. The attack was concentric. In the east the 2nd Canadian Division tried to advance from South Beveland over the connecting causeway, and finally seized a bridgehead with the help of a brigade of the 52nd Division. In the centre, on November 1, No. 4 Commando was ferried across from Breskens, and boldly landed on the sea-front of Flushing. This first move was followed rapidly by troops of the 52nd Division, who battled their way into

★ *The Green Beret*, by H. St. G. Saunders.

the town. The main attack was from the west, launched by three Marine Commandos under Brigadier Leicester. Embarking at Ostend, they sailed for Westkapelle, and at 7 a.m. on November 1 they sighted the lighthouse tower. As they approached the naval bombarding squadron opened fire. Here were H.M.S. *Warspite* and the two 15-inch-gun monitors *Erebus* and *Roberts*, with a squadron of armed landing-craft. These latter came close inshore, and, despite harsh casualties, kept up their fire until the two leading Commandos were safely ashore. No. 41, landing at the northern end of the gap in the sea-wall, captured the village of Westkapelle and drove on towards Domburg. No. 48, landing south of the gap, soon met fierce resistance. Invaluable though the naval covering fire had been, a principal adjunct was lacking. A heavy bombardment had been planned for the previous day, but mist prevented our aircraft from taking off. Very effective fighter bomber attacks helped the landing at a critical moment, but the Marines met much stronger opposition, from much less damaged defences, than we had hoped.

That evening No. 48 Commando had advanced two miles along the fringe towards Flushing, but was held up by a powerful battery embedded in concrete. The whole of the artillery of the IInd Canadian Corps, firing across the water from the Breskens shore, was brought to bear, and rocket-firing aircraft attacked the embrasures. In the gathering darkness the Commando killed or captured the defenders. Next morning it pressed on and took Zouteland by midday. There No. 47 took up the attack, and, with a weakening defence, reached the outskirts of Flushing. On November 3 they joined hands with No. 4 Commando after its stiff house-to-house fighting in the town. In a few days the whole island was in our hands, with 8,000 prisoners.

Many other notable feats were performed by Commandos during the war, and though other troops and other Services played their full part in this remarkable operation the extreme gallantry of the Royal Marines stands forth. The Commando idea was once again triumphant.

* * * * *

Minesweeping began as soon as Flushing was secure, and in the next three weeks a hundred craft were used to clear the seventy-mile channel. On November 28 the first convoy arrived, and

Antwerp was opened for the British and American Armies. Flying bombs and rockets plagued the city for some time, and caused many casualties, but interfered with the furtherance of the war no more than in London.

Antwerp's ordeal was not the only reason for trying to thrust the Germans farther away. When the 2nd Canadian Division swung west into South Beveland there were still four German divisions in a pocket south of the river Meuse and west of the Nijmegen corridor. It was an awkward salient, which by November 8 was eliminated by the Ist and the XIIth Corps.* On the other flank of the Nijmegen corridor there was still an obstinate enemy, west of the Meuse, in a pocket centred on Venlo. Farther south the First U.S. Army breached the Siegfried Line north of Aachen in the first week of October. The town was attacked from three sides and surrendered on October 21. On their flank the Third Army were twenty miles east of the Moselle. The Seventh Army and the First French Army had drawn level and were probing towards the High Vosges and the Belfort Gap. The Americans had all but outrun their supplies in their lightning advances of September, and a pause was essential to build up stocks and prepare for large-scale operations in November.

* * * * *

The Strategic Air Forces played a big part in the Allied advance to the frontiers of France and Belgium. In the autumn they reverted to their primary rôle of bombing Germany, with oil installations and the transportation systems as specific targets. The enemy's Radar screen and early warning system had been thrust back behind his frontier, and our own navigation and bombing aids were correspondingly advanced. Our casualty rate decreased; the weight and accuracy of our attacks grew. The long-continued onslaught had forced the Germans to disperse their factories very widely. For this they now paid a heavy penalty, since they depended all the more on good communications. Urgently needed coal piled up at pitheads for lack of wagons to move it. Every day a thousand or more freight trains were halted for lack of fuel. Industry, electricity, and gas plant were beginning to close down. Oil production and reserves

* The Ist Corps at this time was a remarkable example of Allied integration. It consisted of four divisions, English, Canadian, American, and Polish.

dropped drastically affecting not only the mobility of the troops, but also the activities and even the training of their air forces.

In August Speer had warned Hitler that the entire chemical industry was being crippled through lack of by-products from the synthetic oil plants, and the position grew worse as time went on. In November he reported that if the decline in railway traffic continued it would result in "a production catastrophe of decisive significance", and in December he paid a tribute to our "far-reaching and clever planning".* At long last our great bombing offensive was reaping its reward.

* Tedder, *Air-Power in War*, pp. 118, 119.

CHAPTER XIV

PRELUDE TO A MOSCOW VISIT

Progress of the Russian Offensive – The Red Army Reaches the Baltic – The Liberation of Belgrade, October 20 – My Desire for Another Meeting with Stalin – Our Concern with the Future of Poland and Greece – The World Organisation and the Deadlock at Dumbarton Oaks – Telegrams from General Smuts – I Plan a Visit to Moscow – The President Approves – Stalin Sends a Cordial Invitation – Russia and the Far East – I Start for Moscow, October 5 – The Campaign in Italy.

THE story of the immense Russian offensive of the summer of 1944 has been told in these pages only to the end of September, when, aided by the Roumanian revolution, the Soviet armies drove up the valley of the Danube to the frontier of Hungary and paused to consolidate their supply. We must now carry it forward to the end of the autumn.

We followed with deep interest and growing hopes the fortunes of this tremendous campaign. The German garrisons in the northern Baltic States had been practically cut off by the Russian advances far to the south, and were extracted with difficulty. The first attacks fell upon them in mid-September from either end of Lake Peipus. These spread swiftly outwards, and in three weeks reached all the Baltic shore from Riga to the north.

On September 24 the southern front again flared into activity. The offensive began with an advance south of the Danube into Yugoslav territory. The Russians were supported on their left flank by the Bulgarian Army, which had readily changed sides. Jointly they made contact with Tito's irregular forces and helped to harry the Germans in their hard but skilful withdrawal from Greece. Hitler, despite the obvious dangers impending in Poland, set great store on the campaign in Hungary, and reinforced it obstinately. The main Russian attack, supported by the Roumanian Army, began on October 6, and was directed on Budapest from the south-east, with a subsidiary thrust from the Carpathians in the north. Belgrade, by-passed on both banks of the Danube, was liberated on October 20 and its German garrison annihilated.

* * * * *

The arrangements which I had made with the President in the summer to divide our responsibilities for looking after particular countries affected by the movements of the armies had tided us over the three months for which our agreement ran. But as the autumn drew on everything in Eastern Europe became more intense. I felt the need of another personal meeting with Stalin, whom I had not seen since Teheran, and with whom, in spite of the Warsaw tragedy, I felt new links since the successful opening of "Overlord". The Russian armies were now pressing heavily upon the Balkan scene, and Roumania and Bulgaria were in their power. As the victory of the Grand Alliance became only a matter of time it was natural that Russian ambitions should grow. Communism raised its head behind the thundering Russian battle-front. Russia was the Deliverer, and Communism the gospel she brought.

I had never felt that our relations with Roumania and Bulgaria in the past called for any special sacrifices from us. But the fate of Poland and Greece struck us keenly. For Poland we had entered the war; for Greece we had made painful efforts. Both their Governments had taken refuge in London, and we considered ourselves responsible for their restoration to their own country, if that was what their peoples really wished. In the main these feelings were shared by the United States, but they were very slow in realising the upsurge of Communist influence, which slid on before, as well as followed, the onward march of the mighty armies directed from the Kremlin. I hoped to take advantage of the better relations with the Soviets to reach satisfactory solutions of these new problems opening between East and West.

Besides these grave issues which affected the whole of Central Europe, the questions of World Organisation were now thrusting themselves upon all our minds. A lengthy conference had been held at Dumbarton Oaks, near Washington, between August and October, at which the United States, Great Britain, the U.S.S.R., and China had produced the now familiar scheme for keeping the peace of the world. They proposed that all peace-loving States should join a new organisation called the United Nations. This would consist of a General Assembly and a Security Council. The Assembly would discuss and consider how to promote and preserve world peace and advise the Security Council what to do about it. Each State would belong to the Assembly and have a vote, but the Assembly could only recommend and pass declarations; it could take no executive action. The Security Council would investigate any disputes between the United Nations, and in effect settle them by force if they could not be settled in peace. This was very different from the League of Nations. Under the new scheme the Assembly could discuss and recommend; the Council alone could act. The discretion of the Council was unfettered by definitions of "aggression" and rules about when force could be used and when sanctions could be applied.

There had been much discussion about who should belong to the Council and how they should use its great authority. Eventually it had been settled that the "Big Three" and China should be permanent members, joined by France in due course, and that the Assembly should elect six more States to sit on it for two years at a time. There remained the question of votes. Every member of the Assembly should have a vote, but they could only deliberate and make recommendations, and in this there was little substance. It had been much more difficult to settle the method of voting in the Security Council. The discussions had revealed many differences between the three great Allies, which will appear as this account proceeds. The Kremlin had no intention of joining an international body on which they would be out-voted by a host of small Powers, who, though they could not influence the course of the war, would certainly claim equal status in the victory. I felt sure we could only reach good decisions with Russia while we had the comradeship of a common foe as a bond. Hitler and Hitlerism were doomed; but after Hitler what?

* * * * *

General Smuts' meditations at his farm in the veldt led him along the same paths of thought, and during the conference he had telegraphed to me:

Field-Marshal Smuts to Prime Minister 20 Sept 11

The crisis arising from the deadlock with Russia in World Organisation talks fills me with deep concern, and in any case comes at a most unfortunate moment before the final end of the war. In this, as in other cases, we are, I fear, being hurried at breakneck pace into momentous decisions. Telecommunications, international aviation, etc., all tell the same tale. As the consequences of this *impasse* may however be particularly calamitous here, I may be excused for sending a warning note.

The Soviet attitude struck me at first as absurd and their contention as one not to be conceded by other Great Powers, and likely to be turned down by smaller Powers also. But second thoughts have inclined the other way. I assume that Molotov sincerely states the Soviet attitude, and that it is correctly interpreted by Cadogan and Clark Kerr as one which involves the honour and standing of Russia among her Allies. She questions whether she is trusted and treated as an equal or whether she is still the pariah and outcast. A misunderstanding here amounts to more than a mere difference. It may, by touching Russian *amour propre* and inducing an inferiority complex, poison European relations, with far-reaching results. Knowing her power, Russia may become more grasping than ever. Her reaction and sense of power are shown by the lack of any attempt on her part to discover a solution. What will be her future relations with countries such as Germany and Japan, even France, not to mention lesser countries? Should a World Organisation be formed which does not include Russia she will become the power centre of another group. We shall then be heading towards a third World War. If the United Nations do not set up such an organisation they will stand stultified before history. This creates a very grave dilemma, and we must at all costs avoid the position into which we may be drifting.

Mindful of these dangers, the small Powers should be prepared to make a concession to Russia's *amour propre*, and on this issue should not demand theoretical equality of status. Such a demand, if pressed, may carry with it for smaller Powers the most devastating results. It would be most unwise in dealing with questions which involve power and security to raise theoretical issues of sovereign equality, and it is for the United States of America and the United Kingdom to use their influence in favour of common sense and safety first rather than the small countries' status.

On the merits the principle of unanimity among the Great Powers has much to recommend it, at least for the years immediately following on this war. If this principle proves unworkable in practice the situation could subsequently be reviewed when mutual confidence has been established and a more workable basis laid down. A clash at the present juncture should be avoided at all costs. In the event of unanimity for the Powers being adopted, even including their voting on questions directly concerning their interests, the result will require that the U.S.A. and the United Kingdom should exert all their influence to get Russia to act moderately and sensibly and not to flout world opinion. And in this it is likely that they will be largely successful. Should Russia prove intransigent it may be necessary for the Organisation to act, but the blame will attach to her. The principle of unanimity will at the worst only have the effect of a veto, of stopping action where it may be wise, or even necessary. Its effect will be negative; it will retard action. But it will also render it impossible for Russia to embark on courses not approved of by the U.S.A. and the United Kingdom.

A brake like unanimity may not be so bad a thing to have where people are drunk with new-won power. I do not defend it; I dislike it; but I do not consider it at present so bad an instrument that on this issue the future of world peace and security should be sacrificed.

The talks have so far been conducted on an official advisory level, although there may no doubt have been intervention on a higher level. Before definite decisions are reached on the highest level the whole situation should, I think, be most carefully reviewed in all its far-reaching implications, and the Great Powers should endeavour to find some *modus vivendi*, even if only of a temporary character, which would prevent a catastrophe of the first magnitude. Where so much is at stake for the future we simply must agree, and cannot afford to differ.

And further:

Field-Marshal Smuts to Prime Minister   26 Sept 44
May the results of your arduous labours in Canada justify your efforts. . . . My warm congratulations on your and Mrs. Churchill's safe return.

While the campaign in Italy again has gone much slower than was anticipated, and the approaching rainy season may even further upset your hopes in that area, Alexander could still press on and maintain our prestige in the Balkan area. In spite of all the help Tito has obtained only from us, he has not behaved loyally to us. I fear that our interests will suffer by his supremacy in Yugoslavia.

What is happening in Greece seems still worse. There the E.A.M. is obtaining control, unfortunately largely with our help. I hope this may still be prevented from the point of view of our great Mediterranean interests, as well as those of the suffering people of Greece, and that our loyal Greek friends may be heartened by positive action on our part. Papandreou is rapidly coming under the power of the E.A.M. elements, who no doubt rely on Soviet backing. You will, I hope, find time to discuss with the King of Greece the best means of safeguarding our and Greek vital interests. The shape of the future Mediterranean set-up is rapidly developing, and in a way not favourable to us.

I do not say this in any spirit of hostility to Russia. It is upon close co-operation between the Big Three that our best hope rests for the near future, and my advice given in the *impasse* at Dumbarton Oaks is a proof of that. But the more firmly Russia can establish herself in the saddle now the farther she will ride in the future and the more precarious our holdfast will become. Our position in the Mediterranean and in Western Europe must be strengthened rather than weakened. In neither of these areas we may have the support of Russia, or even of de Gaullist France. From this standpoint the future dispositions as regards Germany assume an importance for us which may be far greater than and very different from that which they appear to have at present. A new situation will be created for us in Europe and the world by the elimination of Germany through this war. This calls for a searching reconsideration of our entire foreign policy for the future. While a World Organisation is necessary, it is equally essential that our Commonwealth and Empire should emerge from this ordeal as strong and influential as possible, making us an equal partner in every sense for the other Big Two.

From this viewpoint also I much regret the increasing tendency towards the breaking up of your Inter-Party Pact, which has had such magnificent war-time success. Your great influence will, I hope, prevent a break-up prematurely and before the new settlement of Europe and the world has been achieved. The end is not yet. So please take care of your health.

Field-Marshal Smuts to Prime Minister 27 Sept 44

Thank you very much for all your four messages, which I received after sending you my message on your return. The *impasse* at Dumbarton Oaks is dealt with in the first two, and I note your view and the course proposed for later meeting, which from many points of view appears admirable. As regards war plans in Europe and Asia, dealt with in the third message, I am pleased with your arrangements for the campaigns in Italy and the Balkans. Since the enemy appears to be retiring from Greece, it would seem advisable for us to appear there soon and prevent Greece coming under the heel of E.A.M. and charging us with having deserted them. This matter was mentioned in my previous message as one of particular concern for the British Empire.

As regards the Pacific war, it has to be feared that after the war against Germany is won war fever in the United States will cool off and American ardour be transferred to trade and industry. So they should be grateful for your full participation. I am also glad to know that Mountbatten, who has had a raw deal, will get his chance in Burma and Malaya. As events are now developing in China, it is possible that Japan, while being driven from the occupied islands, will become entrenched on the Chinese mainland, from which it will be no easy task to eject her. Stalin will come in very useful there. Unless he does so the Japanese war may last longer than is at present expected.

Division of zones in Germany between United Kingdom and United States appears fair. The destiny of Prussia under occupation by the Russians seems likely to be a Bolshevised Soviet province or protectorate. So much for Hitler's dream. But it shows that Europe's two-thousand-year-old problem of Germany will remain as great as ever.

* * * * *

The Dumbarton Oaks conference ended without agreement, but I felt acutely the need to see Stalin, with whom I always considered one could talk as one human being to another.

Prime Minister to Chief of the Air Staff   27 Sept 44
(For your eyes alone)

It may be necessary for me to go to Moscow with Mr. Eden. It would be a comfort to have the new machine, which was promised by October 15, but we might want to go even earlier than that. I imagine we could fly from here to Cairo at one bound in the new machine, and in two in the York, fuelling, I suppose, at Naples or Malta. From Cairo one would have to watch the weather most carefully. In clear weather the President was able to go over the mountains at five or six thousand feet. I can certainly do eight or ten thousand for a short time. When we went in August 1942 at what height did "Commando" cross the Caucasus Mountains? My idea is we flew at 11,000 feet, and then only for a short time. The whole passage of the mountains took, I think, under three hours. After that it is good low-flying over the Caspian and the plains of Russia. There would be no need to make the *détour* that we did on the last occasion. What is essential is to wait at Cairo or Teheran till there is absolutely clear weather.

Please let me have a report and clear plan.

I telegraphed to Stalin on the same day:

Prime Minister to Marshal Stalin   27 Sept 44

I was gratified to hear from Ambassador Clark Kerr the praise which you gave to British and American operations in France. We value very much such expressions from the leader of the heroic Russian armies. I shall take occasion to repeat to-morrow in the House of Commons what I have said before, that it is the Russian Army that tore the guts out of the German military machine, and is at the present moment holding by far the larger portion of the enemy on its front.

2. I have just returned from long talks with the President, and I can assure you of our intense conviction that on the agreement of our three nations, Britain, United States of America, and U.S.S.R., stand the hopes of the world. I was very sorry to learn that you had not been feeling well lately, and that your doctors did not like your taking long journeys by air. The President had the idea that The Hague would be a good place for us to meet. We have not got it yet, but it may be the course of the war, even before Christmas, may alter the picture along the Baltic shore to such an extent that your journey would not be tiring or difficult. However, we shall have much hard fighting to do before any such plan can be made.

3. *Most Private.* The President intends to visit England, and thereafter France and the Low Countries, immediately after the election, win or lose. My information leads me to believe that he will win.

4. I most earnestly desire, and so, I know, does the President, the intervention of Soviet Russia in the Japanese war, as promised by you at Teheran as soon as the German Army was beaten and destroyed. The opening of a Russian military front against Japan would force them to burn and bleed, especially in the air, in a manner which would vastly accelerate their defeat. From all I have learnt about the internal state of Japan and the sense of hopelessness weighing on their people, I believe it might well be that once the Nazis are shattered a triple summons to Japan to surrender, coming from our three Great Powers, might be decisive. Of course, we must go into all these plans together. I will gladly come to Moscow in October if I can get away from here. If I cannot, Eden will be very ready to take my place. Meanwhile I send you and Molotov my most sincere good wishes.

* * * * *

Roosevelt was impressed with Smuts' views.

President Roosevelt to Prime Minister 28 Sept 44

I have read with great interest your telegram [of September 20] from Field-Marshal Smuts, and I think we are all in agreement with him as to the necessity of having the U.S.S.R. as a fully accepted and equal member of any association of the Great Powers formed for the purpose of preventing international war.

It should be possible to accomplish this by adjusting our differences through compromise by all the parties concerned, and this ought to tide things over for a few years until the child learns how to toddle.

The "child" in this case was the World Instrument.

I replied:

Prime Minister to President Roosevelt 29 Sept 44
... In a conversation with Clark Kerr and Harriman the other night U.J. [Stalin] was most expansive and friendly. He however "grumbled about his own health". He said he never kept well except in Moscow, and even his visits to the front did him harm. His doctors were averse to his flying, and it took him a fortnight to recover from Teheran, etc.

In these circumstances Anthony and I are seriously thinking of flying there very soon. The route is shorter now. Stalin has not yet replied to our suggestion. The two great objects we have in mind would be, firstly, to clinch his coming in against Japan, and, secondly, to try to effect an amicable settlement with Poland. There are other points too concerning Yugoslavia and Greece which we would also discuss. We should keep you informed of every point. Averell's assistance would of course be welcomed by us, or perhaps you could send Stettinius or Marshall. I feel certain that personal contact is essential.

That Germany is not going to be conquered this year seems to me pretty clear. In a telegram I have seen Omar Bradley is already thinking in terms of an operation across the Rhine in the middle of November, and I have noted other signs of stiffening German resistance.

Off the record: It was with much gusto that I read your speech, and I was delighted to see you in such vigorous form.

Every good wish.

On the same day came the following:

Marshal Stalin to Prime Minister 29 Sept 44
I have received the message from you and Mr. Roosevelt about the Conference at Quebec, with information about your further military plans. It is clear from your communication what important tasks have to be settled by American and British armed strength. Permit me to wish you and your armies every kind of success.

At the present time the Soviet armies are busy with the liquidation of the group of German armies along the Baltic which threaten our right flank. Without the liquidation of this group it is not possible for us to penetrate deeply into Eastern Germany. Besides that, our armies have two nearer tasks: to remove Hungary from the war, and to probe the defence of the Germans on the Eastern Front by means of an attack by our troops, and, given favourable conditions, to overcome it.

The President liked our plan of going to Moscow. On September 30 he cabled:

Please let me know after you hear from Uncle J. the date when you and Anthony will arrive in Moscow. It is my opinion that Stalin is at the present time sensitive about any doubt as to his intention to help us in the Orient.

At your request I will direct Harriman to give you any assistance that you may desire. It does not appear practicable or advantageous for me to be represented by Stettinius or Marshall.

Stalin now sent me a cordial invitation.

Marshal Stalin to Prime Minister   30 Sept 44
I have received your message of September 27.

I share your conviction that firm agreement between the three leading Powers constitutes a true guarantee of future peace and answers to the best hopes of all peace-loving peoples. The continuation of our Government in such a policy in the post-war period as we have achieved it during this great war will, it seems to me, have a decisive influence. Of course I have a great desire to meet with you and the President. I attach great importance to it from the point of view of the interests in our common business. But, as far as I am concerned, I must make one reservation. The doctors advise me not to undertake long journeys. For a certain period I must take account of this.

I warmly welcome your wish to come to Moscow in October. We shall have to consider military and other questions, which are of great importance. If anything prevents you from coming to Moscow we should of course be very ready to meet Mr. Eden. Your information about the President's plans for a journey to Europe is of great interest to me. I also am sure of his success in the election. As regards Japan, our position is the same as it was at Teheran.

I and Molotov send you our best wishes.

On this I put matters in train.

Prime Minister to Sir A. Clark Kerr (Moscow)   1 Oct 44
You will have seen Marshal Stalin's telegram of September 30, which is most friendly. Anthony and I propose to start, weather permitting, on Saturday night. The journey is much shorter now, as we can avoid the *détour* by the Atlantic and Spain, and also that by the mountains and Teheran. It should not take more than three days, or perhaps two. The Air Ministry will make the arrangements with Moscow.

2. I think it would be better to go as the guests of the Russian Government, because their high sense of hospitality helps business. But of course we must have festivities at the Embassy. Will you sound them on this?

3. I am thinking that Mrs. Churchill might come with me. She has her Red Cross there now, and people in England would be glad to know she was on the spot to look after me. I wonder how this would fit in. Of course she would not expect to go to the Kremlin banquet,

which would be men only. But I presume there are things she could see apart from her own Red Cross. Would it be an embarrassment to the Russians, as there is no Mrs. Stalin? Pray advise me quite freely on this.

Reply most urgent.

The Ambassador replied next day that he and the Russians were delighted that I had decided to come, and that Anthony would be with me. "The iron stands hot for the striking. The Russians expect you to be their guest. The idea that Mrs. Churchill should come with you has been warmly welcomed."

My wife however decided not to go at this moment. I asked Roosevelt to tell Stalin that he approved of our mission, and that Mr. Harriman would take part in the discussions. I inquired what I could say about the United States' Far Eastern war plans.

Prime Minister to President Roosevelt   4 Oct 44
... We want to elicit the time it will take after the German downfall for a superior Russian Army to be gathered opposite the Japanese on the frontiers of Manchukuo, and to hear from them the problems of this campaign, which are peculiar owing to the lines of communication being vulnerable in the later stages.

Of course the bulk of our business will be about the Poles; but you and I think so much alike about this that I do not need any special guidance as to your views.

The point of Dumbarton Oaks will certainly come up, and I must tell you that we are pretty clear that the only hope is that the Great Powers are agreed [*i.e.*, unanimous]. It is with regret that I have come to this conclusion, contrary to my first thought. Please let me know if you have any wishes about this matter, and also instruct Averell accordingly.

The President then sent his fullest assurances of approval and goodwill.

President Roosevelt to Prime Minister   4 Oct 44
I can well understand the reasons why you feel that an immediate meeting between yourself and Uncle Joe is necessary before the three of us can get together. The questions which you will discuss there are ones which are of course of real interest to the United States, as I know you will agree. I have therefore instructed Harriman to stand by and to participate as my observer, if agreeable to you and Uncle Joe, and I have so informed Stalin. While naturally Averell will not be in a position to commit the United States—I could not permit anyone to commit me in advance—he will be able to keep me fully informed, and

I have told him to return and report to me as soon as the conference is over.

I am only sorry that I cannot be with you myself, but I am prepared for a meeting of the three of us any time after the elections here, for which your meeting with Uncle Joe should be a useful prelude, and I have so informed Uncle Joe.

Like you, I attach the greatest importance to the continued unity of our three countries. I am sorry that I cannot agree with you however that the voting question should be raised at this time. That is a matter which the three of us can, I am sure, work out together, and I hope you will postpone discussion of it until our meeting. There is, after all, no immediate urgency about this question, which is so directly related to public opinion in the United States and Great Britain and in all the United Nations.

I am asking our military people in Moscow to make available to you our joint Chiefs' statement to Stalin.

You carry my best wishes with you, and I will eagerly await word of how it goes.

Prime Minister to President Roosevelt   5 Oct 44

Thank you very much for what you say, and for your good wishes. I am very glad that Averell should sit in at all principal conferences, but you would not, I am sure, wish this to preclude private *tête-à-tête* between me and U.J. or Anthony and Molotov, as it is often under such conditions that the best progress is made. You can rely on me to keep you constantly informed of everything that affects our joint interests, apart from the reports which Averell will send.

2. I gather from your last sentence but one that you have sent some general account of your Pacific plans to your people in Moscow, which will be imparted to U.J., and which I shall see on arrival. This will be most convenient.

3. Should U.J. raise the question of voting, as he very likely will do, I will tell him that there is no hurry about this and that I am sure we can get it settled when we are all three together.

*   *   *   *   *

All the major issues had thus been settled. It only remained to plan the journey.

Prime Minister to Marshal Stalin   4 Oct 44

Your people are anxious about the route I have been advised to take. It is not good for me to go much above 8,000 feet, though I can, if necessary, do so for an hour or so. We think it less of a risk to fly across the Ægean and Black Sea. I have satisfied myself that on the whole this is the best and involves no inappropriate risk.

2. So long as we can get down safely to refuel if necessary at Simferopol or at any other operational landing-ground on the coast which you may prefer, I shall be quite content with the facilities available. I have everything I want in my plane. The only vital thing is that we may send an aeroplane on ahead to establish with you a joint signal station regulating our homing and landing. Please have the necessary orders given.

3. I am looking forward to returning to Moscow under the much happier conditions created since August 1942.

Marshal Stalin to Prime Minister   5 Oct 44

The landing on the aerodrome Sarabuz, near Simferopol, has been arranged. Send your signal aircraft to that aerodrome.

★   ★   ★   ★   ★

Eden and I with Brooke and Ismay started in two planes on the night of the 5th. At Naples we had four hours' discussion with Generals Wilson and Alexander. I was much distressed by their tale. Five weeks had passed since I left Italy at the outset of Alexander's offensive in the last days of August. It will be convenient to carry the story forward to its end in the autumn.

The Eighth Army attack had prospered and augured well. It surprised the Germans, and by September 1 had penetrated the Gothic Line on a twenty-mile front. Kesselring, as ever, was quick to recover, and began to send reinforcements from the central sector. Just in time they manned the Coriano Ridge, barring the way to Rimini, and for a week they resisted all attacks. Then we took it.

Prime Minister to General Alexander   15 Sept 44

Many congratulations upon the storming of the Coriano Ridge and the passage of the Marano. I can see that this has been a grand feat of arms upon the part of the troops involved. Pray give them my compliments. I hope this success may cast a brighter light on your immediate prospects.

From his centre and right Kesselring sent across seven divisions, and there were three days of heavy fighting at San Fortunato. When it was taken, by skilful combination of ground and air attack, the enemy withdrew, and Rimini fell on September 20.

By weakening his centre Kesselring gave Alexander the occasion he had awaited for the Fifth Army. The enemy had withdrawn from their forward positions to economise troops, and we were able to close up to the main position without having

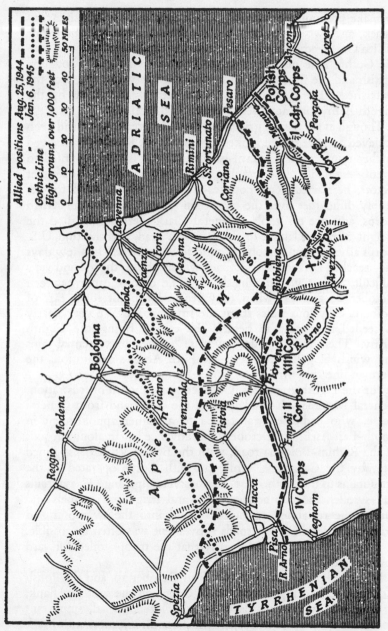

THE GOTHIC LINE

to make preliminary attacks. On September 13 the Fifth Army struck, and two days later the 8th Indian Division, leading our XIIIth Corps, advanced across trackless mountains and broke into the Gothic Line on the road to Faenza. By the 18th both the British and the IInd U.S. Corps on their left were on top of the watershed. The Gothic Line, turned at its eastern end by the Eighth Army, had now been pierced in the centre.

Though at the cost of grievous casualties, great success had been achieved and the future looked hopeful. But Kesselring received further reinforcements, until his German divisions amounted to twenty-eight in all. Scraping up two divisions from quiet sectors, he started fierce counter-attacks, which, added to our supply difficulties over the mountain passes, checked the XIIIth Corps' advance on Imola. General Clark thereupon shifted the weight of his onslaught to the Bologna road, and the IInd U.S. Corps advanced on October 1 with four divisions. In a few days they reached Loiano. The defence was stubborn, the ground very difficult, and it was raining hard. The climax came between October 20 and 24, when they reached a point south-east of Bologna only four miles from the Imola road. We very nearly succeeded in cutting in behind the enemy facing the Eighth Army. Then, in Alexander's words, "assisted by torrential rains and winds of gale force, and the Fifth Army's exhaustion, the German line held firm."

For the Eighth Army too October was a month of frustration. General McCreery had taken over the command from General Leese, who had been transferred to a higher appointment in South-East Asia. On October 7 his advance began along the axis of the Rimini–Bologna road, with the British Vth Corps, joined later by the Canadians, while the Xth Corps operated in the mountains to the south. The weather was appalling. Heavy rains had swollen the numberless rivers and irrigation channels and turned the reclaimed agricultural land into the swamp it had originally been. Off the roads movement was often impossible. It was with the greatest difficulty that the troops toiled forward towards Bologna.

Despite all, Cesena was reached on October 19, and the Polish Corps, who had replaced the Xth Corps on the southern flank, struggled forward towards the Forli–Florence road, important because it would give a shorter lateral communication with Mark

Clark's army. That army, as we have seen, was drawing very close to Bologna, and in the critical situation the German commander took the courageous decision to transfer three good divisions from the eastern to his central front. These doubtless just saved the day in the centre of his line. The Eighth Army also had its subtractions. The 4th Indian Division and the Greek Brigade had to be detached to deal with the crisis in Greece, the story of which falls to a later chapter.

* * * * *

I reported to the President about all this from Moscow, so far as it had developed, on October 10, adding:

The pressure in Dutch salient seems to me to be growing very severe, and our advances are slow and costly. In these circumstances we have with much sorrow had to recommend that we should put off "Dracula" [the amphibious attack on Rangoon] from March to November and leave the British 3rd Division in France, as well as sending there the 52nd Division, one of our best, about 22,000 strong in fighting troops, and the 6th Airborne Division to the Netherlands. Eisenhower is counting on these for the impending operation on the Rhine, and of course this was much the quickest way to bring additional troops into France.

3. Could you not deflect two, or better still three, American divisions to the Italian front, which would enable them to join Mark Clark's Fifth Army and add the necessary strength to Alexander? They would have to be there in three or four weeks. I consider the fact that we shall be sending Eisenhower these extra two divisions gives me a case for your generous consideration.

4. With regard to Istria, Trieste, etc., General Wilson is forwarding his plan to Combined Chiefs of Staff. This plan will be in accord with overall strategic objective, namely, the expulsion from or destruction in Italy of Kesselring's army.

He replied some days later:

President Roosevelt to Prime Minister (Moscow)   16 Oct 44

I appreciate your report on the Italian campaign, where, up to the present, our combined effort has cost us nearly 200,000 battle casualties, 90,000 of them American. My Chiefs of Staff accept Wilson's estimate that we cannot now expect to destroy Kesselring's army this winter and that the terrain and weather conditions in the Po valley will prevent any decisive advance this year. They further consider that the Germans are free to transfer five or six divisions from Italy to the

Western Front whenever they consider such action more profitable than using these divisions in containing our forces south of the Po. Provision of additional U.S. divisions will not affect the campaign in Italy this year. All of us are now faced with an unanticipated shortage of man-power, and overshadowing all other military problems is the need for quick provision of fresh troops to reinforce Eisenhower in his battle to break into Germany and end the European war. While the divisions in Italy are undoubtedly tiring as the result of fighting in the present battle since August 25, Eisenhower is now fighting the decisive battle of Germany with divisions which have been in continuous combat since they landed on the Normandy beaches in the first part of June. The need for building up additional divisions on the long front from Switzerland to the North Sea is urgent. Even more urgent is the need for fresh troops to enable Eisenhower to give some rest to our front-line soldiers, who have been the spear-point of the battle since the first days in Normandy. On the basis of General Marshall's reports on the present situation we are now taking the very drastic step of sending the infantry regiments of the divisions ahead of the other units in order that General Eisenhower may be able to rotate some of our exhausted front-line soldiers.

Diversion of any forces to Italy would withhold from France vitally needed fresh troops, while committing such forces to the high attrition of an indecisive winter campaign in Northern Italy. I appreciate the hard and difficult task which our armies in Italy have faced and will face, but we cannot withhold from the main effort forces which are needed in the Battle of Germany.

From General Marshall's reports on the problem now facing General Eisenhower, I am sure that both of them agree with my conviction that no divisions should be diverted from their destination in France.

The rest of the story is soon told. Although hopes of decisive victory had faded, it remained the first duty of the armies in Italy to keep up the pressure and deter the enemy from sending help to the hard-pressed German armies on the Rhine. And so the Eighth Army fought forward whenever there was a spell of reasonably fine weather, taking Forli on November 9, and soon afterwards clearing all the road to Florence. Thereafter no major offensive was possible. Small advances were made as opportunity offered, but not until the spring were the armies rewarded with the victory they had so well earned, and so nearly won, in the autumn.

CHAPTER XV

OCTOBER IN MOSCOW

I Arrive at Moscow, October 9 – Our First Meeting in the Kremlin – A Half-Sheet of Paper – My Telegram to the President of October 11 – Stalin Dines at the British Embassy – I Send Harry Hopkins the News – Balkan Tangles – My Note of October 11 on Eastern Europe – A Report to the Cabinet, October 12 – Russia and Roumania – Great Britain and Greece – A Meeting with the Poles, October 13 – Command Performance at the Bolshoi Theatre – Russian Plans for War with Japan – The Second Military Meeting in the Kremlin, October 15 – The Curzon Line – My Telegram to the King of October 16 – "All Poles' Day" – I Report to the President, October 22 – The Future of Germany – Closer Contact with the Soviets than Ever Before.

WE alighted at Moscow on the afternoon of October 9, and were received very heartily and with full ceremonial by Molotov and many high Russian personages. This time we were lodged in Moscow itself, with every care and comfort. I had one small, perfectly appointed house, and Anthony another near by. We were glad to dine alone together and rest. At ten o'clock that night we held our first important meeting in the Kremlin. There were only Stalin, Molotov, Eden, and I, with Major Birse and Pavlov as interpreters. It was agreed to invite the Polish Prime Minister, M. Romer, the Foreign Minister, and M. Grabski, a grey-bearded and aged academician of much charm and quality, to Moscow at once. I telegraphed accordingly to M. Mikolajczyk that we were expecting him and his friends for discussions with the Soviet Government and ourselves, as well as with the Lublin Polish Committee. I made it clear that refusal to come to take part in the conversations would amount to a definite rejection of our advice and would relieve us from further responsibility towards the London Polish Government.

The moment was apt for business, so I said, "Let us settle about our affairs in the Balkans. Your armies are in Roumania and Bulgaria. We have interests, missions, and agents there. Don't let us get at cross-purposes in small ways. So far as Britain and Russia are concerned, how would it do for you to have ninety per cent. predominance in Roumania, for us to have ninety per cent. of the say in Greece, and go fifty-fifty about Yugoslavia?" While this was being translated I wrote out on a half-sheet of paper:

Roumania	
Russia	90%
The others	10%
Greece	
Great Britain	90%
(in accord with U.S.A.)	
Russia	10%
Yugoslavia	50–50%
Hungary	50–50%
Bulgaria	
Russia	75%
The others	25%

I pushed this across to Stalin, who had by then heard the translation. There was a slight pause. Then he took his blue pencil and made a large tick upon it, and passed it back to us. It was all settled in no more time than it takes to set down.

Of course we had long and anxiously considered our point, and were only dealing with immediate war-time arrangements. All larger questions were reserved on both sides for what we then hoped would be a peace table when the war was won.

After this there was a long silence. The pencilled paper lay in the centre of the table. At length I said, "Might it not be thought rather cynical if it seemed we had disposed of these issues, so fateful to millions of people, in such an offhand manner? Let us burn the paper." "No, you keep it," said Stalin.

I also raised the question of Germany, and it was agreed that our two Foreign Ministers, together with Mr. Harriman, should go into it. I told Stalin that the Americans would be outlining to him during the course of our future discussions their plan of operations in the Pacific for 1945.

* * * * *

We then sent a joint message to Roosevelt on our first talk.

10 Oct 44

We have agreed not to refer in our discussions to Dumbarton Oaks issues, and that these shall be taken up when we three can meet together. We have to consider the best way of reaching an agreed policy about the Balkan countries, including Hungary and Turkey. We have arranged for Mr. Harriman to sit in as an observer at all meetings where business of importance is to be transacted, and for General Deane to be present whenever military topics are raised. We have arranged for technical contacts between our high officers and General Deane on military aspects, and for any meetings which may be necessary later in our presence and that of the two Foreign Secretaries, together with Mr. Harriman. We shall keep you fully informed ourselves about the progress we make.

We take this occasion to send you our heartiest good wishes, and to offer our congratulations on the prowess of United States forces and upon the conduct of the war in the West by General Eisenhower.

I now reported privately to the President.

11 Oct 44

We have found an extraordinary atmosphere of goodwill here, and we have sent you a joint message. You may be sure we shall handle everything so as not to commit you. The arrangements we have made for Averell are, I think, satisfactory to him, and do not preclude necessary intimate contacts, which we must have to do any good. Of all these I shall give you a faithful report.

2. It is absolutely necessary we should try to get a common mind about the Balkans, so that we may prevent civil war breaking out in several countries, when probably you and I would be in sympathy with one side and U.J. with the other. I shall keep you informed of all this, and nothing will be settled except preliminary agreements between Britain and Russia, subject to further discussion and melting down with you. On this basis I am sure you will not mind our trying to have a full meeting of minds with the Russians.

3. I have not yet received your account of what part of the Pacific operations we may mention to Stalin and his officers. I should like to have this, because otherwise in conversation with him I might go beyond what you wish to be said. Meanwhile I will be very careful. We have not touched upon Dumbarton Oaks, except to say it is barred, at your desire. However, Stalin at lunch to-day spoke in praise of the

meeting and of the very great measure of agreement that has been arrived at there. Stalin also in his speech at this same luncheon animadverted harshly upon Japan as being an aggressor nation. I have little doubt from our talks that he will declare war upon them as soon as Germany is beaten. But surely Averell and Deane should be in a position not merely to ask him to do certain things, but also to tell him, in outline at any rate, the kind of things you are going to do yourself, and we are going to help you to do.

* * * * *

In the evening of October 11 Stalin came to dine at the British Embassy. This was the first time that the British Ambassador had succeeded in making such an arrangement. Every precaution was taken by the police. One of my guests, M. Vyshinsky, on passing some of the N.K.V.D. armed guards on our staircase, remarked, "Apparently the Red Army has had another victory. It has occupied the British Embassy." Till the small hours of the morning we ranged over the whole field of discussion in an informal atmosphere. Among other topics we discussed the next General Election in England. Stalin said that he had no doubt about the result: the Conservatives would win. It is even harder to understand the politics of other countries than those of your own.

* * * * *

I also cabled to Hopkins on various matters.

Prime Minister to Mr. Harry Hopkins   12 Oct 44

Everything is most friendly here, but the Balkans are in a sad tangle. Tito, having lived under our protection for three or four months at Vis, suddenly levanted, leaving no address, but keeping sentries over his cave to make out that he was still there. He then proceeded to Moscow, where he conferred, and yesterday M. Molotov confessed this fact to Mr. Eden. The Russians attribute this graceless behaviour to Tito's suspicious peasant upbringing, and say that they did not tell us out of respect for his wish for secrecy. The Bulgarians are treating our people ill, having arrested some of our officers still remaining both in Greece and Yugoslavia. I saw a tale of their having treated very cruelly American officers when prisoners of theirs. Russian attitude is that they are of course willing to indict Bulgaria for her many offences, but only in spirit of a loving parent—"This hurts me more than it does you." They are taking great interest in Hungary, which, they mentioned erroneously, was their neighbour. They claim fullest responsibility in Roumania, but are prepared largely to disinterest themselves in Greece. All these matters are being flogged out by Mr. Eden and Molotov.

2. Under dire threats from us we persuaded Mikolajczyk and the Poles to accept the invitation we had extracted from the Russians. We hope they will be here to-morrow.

3. We are seeing a great deal of Averell, and he is giving a dinner to-morrow night on Teheran lines—*i.e.*, only the secret ones there. He is sitting in on the military discussions and on the future of Germany talks, as well of course as the Polish conversations when they begin. We have so many bones to pick about the Balkans at the present time that we would rather carry the matter a little further *à deux* in order to be able to talk more bluntly than at a larger gathering. I will cable fully to the President about this in a day or two. Will you very kindly show this to him? I shall be very glad to hear from him.

The President now sent us an encouraging message.

President Roosevelt to Prime Minister and Marshal Stalin　　　　12 Oct 44
　　Thanks for your joint message of October 10.

I am most pleased to know that you are reaching a meeting of your two minds as to international policies in which, because of our present and future common efforts to prevent international wars, we are all interested.

*　　*　　*　　*　　*

After our first meeting I reflected on our relations with Russia throughout Eastern Europe, and in order to clarify my ideas drafted a letter to Stalin on the subject, enclosing a memorandum stating our interpretation of the percentages which we had accepted across the table. In the end I did not send this letter, deeming it wiser to let well alone. I print it only as an authentic account of my thought.

MOSCOW
October 11, 1944

I deem it profoundly important that Britain and Russia should have a common policy in the Balkans which is also acceptable to the United States. The fact that Britain and Russia have a twenty-year alliance makes it especially important for us to be in broad accord and to work together easily and trustfully and for a long time. I realise that nothing we can do here can be more than preliminary to the final decisions we shall have to take when all three of us are gathered together at the table of victory. Nevertheless I hope that we may reach understandings, and in some cases agreements, which will help us through immediate emergencies, and will afford a solid foundation for long-enduring world peace.

201

These percentages which I have put down are no more than a method by which in our thoughts we can see how near we are together, and then decide upon the necessary steps to bring us into full agreement. As I said, they would be considered crude, and even callous, if they were exposed to the scrutiny of the Foreign Offices and diplomats all over the world. Therefore they could not be the basis of any public document, certainly not at the present time. They might however be a good guide for the conduct of our affairs. If we manage these affairs well we shall perhaps prevent several civil wars and much bloodshed and strife in the small countries concerned. Our broad principle should be to let every country have the form of government which its people desire. We certainly do not wish to force on any Balkan State monarchic or republican institutions. We have however established certain relations of faithfulness with the Kings of Greece and Yugoslavia. They have sought our shelter from the Nazi foe, and we think that when normal tranquillity is re-established and the enemy has been driven out the peoples of these countries should have a free and fair chance of choosing. It might even be that Commissioners of the three Great Powers should be stationed there at the time of the elections so as to see that the people have a genuine free choice. There are good precedents for this.

However, besides the institutional question there exists in all these countries the ideological issue between totalitarian forms of government and those we call free enterprise controlled by universal suffrage. We are very glad that you have declared yourselves against trying to change by force or by Communist propaganda the established systems in the various Balkan countries. Let them work out their own fortunes during the years that lie ahead. One thing however we cannot allow —Fascism or Nazism in any of their forms, which give to the toiling masses neither the securities offered by your system nor those offered by ours, but, on the contrary, lead to the build-up of tyrannies at home and aggression abroad. In principle I feel that Great Britain and Russia should feel easy about the internal government of these countries, and not worry about them or interfere with them once conditions of tranquillity have been restored after this terrible blood-bath which they, and indeed we, have all been through.

It is from this point of view that I have sought to adumbrate the degrees of interest which each of us takes in these countries with the full assent of the other, and subject to the approval of the United States, which may go far away for a long time and then come back again unexpectedly with gigantic strength.

In writing to you, with your experience and wisdom, I do not need to go through a lot of arguments. Hitler has tried to exploit the fear of an aggressive, proselytising Communism which exists throughout Western Europe, and he is being decisively beaten to the ground. But, as you know well, this fear exists in every country, because, whatever the merits of our different systems, no country wishes to go through the bloody revolution which will certainly be necessary in nearly every case before so drastic a change could be made in the life, habits, and outlook of their society. We feel we were right in interpreting your dissolution of the Comintern as a decision by the Soviet Government not to interfere in the internal political affairs of other countries. The more this can be established in people's minds the smoother everything will go. We, on the other hand, and I am sure the United States as well, have Governments which stand on very broad bases, where privilege and class are under continual scrutiny and correction. We have the feeling that, viewed from afar and on a grand scale, the differences between our systems will tend to get smaller, and the great common ground which we share of making life richer and happier for the mass of the people is growing every year. Probably if there were peace for fifty years the differences which now might cause such grave troubles to the world would become matters for academic discussion.

At this point, Mr. Stalin, I want to impress upon you the great desire there is in the heart of Britain for a long, stable friendship and cooperation between our two countries, and that with the United States we shall be able to keep the world engine on the rails.

To my colleagues at home I sent the following :

12 Oct 44

The system of percentage is not intended to prescribe the numbers sitting on commissions for the different Balkan countries, but rather to express the interest and sentiment with which the British and Soviet Governments approach the problems of these countries, and so that they might reveal their minds to each other in some way that could be comprehended. It is not intended to be more than a guide, and of course in no way commits the United States, nor does it attempt to set up a rigid system of spheres of interest. It may however help the United States to see how their two principal Allies feel about these regions when the picture is presented as a whole.

2. Thus it is seen that quite naturally Soviet Russia has vital interests in the countries bordering on the Black Sea, by one of whom, Roumania, she has been most wantonly attacked with twenty-six divisions, and with the other of whom, Bulgaria, she has ancient ties. Great Britain feels it right to show particular respect to Russian views about these two countries, and to the Soviet desire to take the lead in a practical way in guiding them in the name of the common cause.

3. Similarly, Great Britain has a long tradition of friendship with Greece, and a direct interest as a Mediterranean Power in her future. In this war Great Britain lost 30,000 men in trying to resist the German-Italian invasion of Greece, and wishes to play a leading part in guiding Greece out of her present troubles, maintaining that close agreement with the United States which has hitherto characterised Anglo-American policy in this quarter. Here it is understood that Great Britain will take the lead in a military sense and try to help the existing Royal Greek Government to establish itself in Athens upon as broad and united a basis as possible. Soviet Russia would be ready to concede this position and function to Great Britain in the same sort of way as Britain would recognise the intimate relationship between Russia and Roumania. This would prevent in Greece the growth of hostile factions waging civil war upon each other and involving the British and Russian Governments in vexatious arguments and conflict of policy.

4. Coming to the case of Yugoslavia, the numerical symbol 50–50 is intended to be the foundation of joint action and an agreed policy between the two Powers now closely involved, so as to favour the creation of a united Yugoslavia after all elements there have been joined together to the utmost in driving out the Nazi invaders. It is intended to prevent, for instance, armed strife between the Croats and Slovenes on the one side and powerful and numerous elements in Serbia on the other, and also to produce a joint and friendly policy towards Marshal Tito, while ensuring that weapons furnished to him are used against the common Nazi foe rather than for internal purposes. Such a policy, pursued in common by Britain and Soviet Russia, without any thought of special advantages to themselves, would be of real benefit.

5. As it is the Soviet armies which are obtaining control of Hungary, it would be natural that a major share of influence should rest with them, subject of course to agreement with Great Britain and probably the United States, who, though not actually operating in Hungary, must view it as a Central European and not a Balkan State.

6. It must be emphasised that this broad disclosure of Soviet and British feelings in the countries mentioned above is only an interim guide for the immediate war-time future, and will be surveyed by the Great Powers when they meet at the armistice or peace table to make a general settlement of Europe.

*　　*　　*　　*　　*

At five o'clock on the evening of October 13 we had our meeting at the Soviet Government Hospitality House, known as Spiridonovka, to hear Mikolajczyk and his colleagues **put their** case. These talks were held as a preparation for a further meeting at which the British and American delegations would meet the Lublin Poles. I pressed Mikolajczyk hard to consider two things, namely, *de facto* acceptance of the Curzon Line, with interchange of population, and a friendly discussion with the Lublin Polish Committee so that a united Poland might be established. Changes, I said, would take place, but it would be best if unity were established now, at this closing period of the war, and I asked the Poles to consider the matter carefully that night. Mr. Eden and I would be at their disposal. It was essential for them to make contact with the Polish Committee and to accept the Curzon Line as a working arrangement, subject to discussion at the Peace Conference.

At ten o'clock the same evening we met the so-called Polish National Committee. It was soon plain that the Lublin Poles were mere pawns of Russia. They had learned and rehearsed their part so carefully that even their masters evidently felt they were overdoing it. For instance, M. Bierut, the leader, spoke in these terms: "We are here to demand on behalf of Poland that Lvov shall belong to Russia. This is the will of the Polish people." When this had been translated from Polish into English and Russian I looked at Stalin and saw an understanding twinkle in his expressive eyes, as much as to say, "What about that for our Soviet teaching!" The lengthy contribution of another Lublin leader, Osóbka-Morawski, was equally depressing. Mr. Eden formed the worst opinion of the three Lublin Poles.

The whole conference lasted over six hours, but the achievement was small.

* * * * *

On the 14th there was a Command Performance at the Bolshoi Theatre—first a ballet, then opera, and finally some splendid dancing and singing by the Red Army choir. Stalin and I occupied the Royal Box, and we had a rapturous ovation from the entire audience. After the theatre we had a most interesting and successful military discussion at the Kremlin. Stalin had with him Molotov and General Antonov. Harriman brought General Deane. I had Brooke, Ismay, and General Burrows, head of our Military Mission in Moscow.

We began by telling them our future intentions in North-

West Europe, Italy, and Burma. Deane followed with a statement about the campaign in the Pacific, and gave an outline of the sort of help which would be particularly valuable from the Soviets, once they were at war with Japan. General Antonov then made a very frank statement about the situation on the Eastern Front, the difficulties which confronted Russian armies, and their plans for the future. Stalin intervened from time to time to emphasise points of special significance, and concluded by assuring us that the Russian armies would press vigorously and continuously into Germany and that we need not have the slightest anxiety that the Germans would be able to withdraw any troops from their Eastern Front.

There was no doubt whatever that the Soviets intended to enter the war against Japan as soon after the defeat of Germany as they could collect the necessary forces and supplies in the Far East. Stalin would not commit himself definitely to a date. He spoke of a period of "several months" after the German defeat. We got the impression that this might be interpreted as about three or four. The Russians agreed to an immediate start in building up stocks of food and fuel in their Far Eastern oilfields, and to let the Americans use the airfields and other facilities in the maritime provinces which they needed for their Strategical Air Force. Stalin did not seem anxious about the effect of these preparations on the Japanese. In fact, he hoped they would make a "premature attack", as this would encourage the Russians to fight their best. "The Russians," he remarked, "would have to know what they were fighting for."

On the 15th I had a high temperature and could not attend the second military meeting, which was held in the Kremlin that evening. Eden took my place, and had with him Brooke, Ismay, and Burrows; while Stalin, in addition to Molotov and Antonov, had Lieut.-General Shevchenko, Chief of Staff of the Soviet Army in the Far East. Harriman was again there, with General Deane. The only subject discussed was Soviet participation against Japan. Substantial conclusions were reached.

Stalin first of all agreed that we should concert our war plans. He asked for United States help in building up a two to three months' reserve of fuel, food, and transport in the Far East, and said that if this could be done and the political issues could be clarified the U.S.S.R. would be ready to attack Japan about three months after Germany had been defeated. He also promised to prepare airfields in the maritime provinces for the United States and Soviet Strategic Air Forces, and to receive American four-engined planes and instructors without delay. Meetings between the Soviet and American military staffs in Moscow would begin at once, and he promised to take part in the first one himself.

* * * * *

As the days passed only slight improvement was made with the festering sore of Soviet-Polish affairs. The Poles were willing to accept the Curzon Line "as a line of demarcation between Russia and Poland". The Russians insisted on the words "as a basis of frontier between Russia and Poland". Neither side would give way. Mikolajczyk declared that he would be repudiated by his own people, and Stalin at the end of a talk of two hours and a quarter which I had with him alone remarked that he and Molotov were the only two of those he worked with who were favourable to dealing "softly" with Mikolajczyk. I was sure there were strong pressures in the background, both party and military.

Stalin did not think it desirable to proceed with an attempt to form a united Polish Government without the frontier question being agreed. Had this been settled he would have been quite willing that Mikolajczyk should head the new Government. I myself thought that difficulties not less obstinate would arise in a discussion for a merger of the Polish Government with the Lublin Poles, whose representatives continued to make the worst possible impression on us, and who, I told Stalin, were "only an expression of the Soviet will". They had no doubt also the ambition to rule Poland, and were thus a kind of Quislings. In all the circumstances the best course was for the two Polish delegations to return whence they had come. I felt very deeply the responsibility which lay on me and the Foreign Secretary in trying to frame proposals for a Russo-Polish settlement. Even forcing the Curzon Line upon Poland would excite criticism.

In other directions considerable advantages had been gained. The resolve of the Soviet Government to attack Japan on the overthrow of Hitler was obvious. This would have supreme value in shortening the whole struggle. The arrangements made about the Balkans were, I was sure, the best possible. Coupled with successful military action, they should now be effective in saving Greece, and I had no doubt that our agreement to pursue a fifty-fifty joint policy in Yugoslavia was the best solution for our difficulties in view of Tito's behaviour and of the arrival of Russian and Bulgarian forces under Russian command to help his eastern flank.

There is no doubt that in our narrow circle we talked with an ease, freedom, and cordiality never before attained between our two countries. Stalin made several expressions of personal regard which I feel sure were sincere. But I became even more convinced that he was by no means alone. As I said to my colleagues at home, "Behind the horseman sits black care."

* * * * *

Prime Minister to the King 16 Oct 44

With humble duty, the Prime Minister hopes His Majesty has had a successful and interesting visit to the armies in the Netherlands and is now safely back home. He trusts His Majesty is well after these exertions.

2. Here in Moscow the weather is brilliant but crisp, and the political atmosphere is extremely cordial. Nothing like it has been seen before. The Prime Minister and Mr. Eden in their various talks with Marshal Stalin and M. Molotov have been able to deal with the most delicate problems in a frank, outspoken manner without the slightest sign of giving offence. The Prime Minister attended a special performance of the ballet, which was very fine, and received a prolonged ovation from an enormous audience. Presently when Marshal Stalin came into the box for the first time in this war and stood beside him there was an almost passionate demonstration. At or after the very lengthy feasts, with very numerous cordial toasts, it has been possible to touch on many grave matters in an easy fashion. The nights are very late, lasting till three or even four o'clock; but the Prime Minister also keeps late hours, and much work is done from about noon onwards, with conferences of various kinds.

3. We had three hours on the entire military scene. After Field-Marshal Brooke and the Prime Minister had explained the situation and plans in the West, in Italy, and in Burma, Mr. Harriman and General Deane, U.S.A., gave a full account of the Pacific, past, present, and future, which appeared to interest Marshal Stalin greatly. Later the Russian Deputy Chief of Staff told us much about Russian plans against Germany which we had never heard before, the gist of which was extremely satisfactory. On account of secrecy I will defer all further reference to what he said until I get home. To-night at six

o'clock we are to have a Russian statement on the Far Eastern theatre, which is likely to be satisfactory and of the greatest interest.

4. The day before yesterday was "All Poles' Day". Our lot from London are, as your Majesty knows, decent but feeble, but the delegates from Lublin could hardly have been under any illusions as to our opinion of them. They appeared to me to be purely tools, and recited their parts with well-drilled accuracy. I cross-examined them fairly sharply, and on several points Marshal Stalin backed me up. We shall be wrestling with our [London] Poles all to-day, and there are some hopes that we may get a settlement. If not we shall have to hush the matter up and spin it out until after the [American] Presidential election.*

5. There are still many subjects to be discussed, like the future treatment of Germany.

Mr. Churchill, with his humble duty, remains Your Majesty's faithful servant.

* * * * *

On the evening of October 17 we held our last meeting. The news had just arrived that Admiral Horthy had been arrested by the Germans as a precaution now that the whole German front in Hungary was disintegrating. I remarked that I hoped the Ljubljana Gap could be reached as fast as possible, and added that I did not think the war would be over before the spring. We then had our first talk on the question of Germany. We discussed the merits and drawbacks of the Morgenthau Plan. It was decided that the European Advisory Commission should study the problem in detail.

* * * * *

While flying home I gave the President further details of our talks.

Prime Minister to President Roosevelt   22 Oct 44

On our last day at Moscow Mikolajczyk saw Bierut, who admitted his difficulties. Fifty of his men had been shot in the last month. Many Poles took to the woods rather than join his forces. Approaching winter conditions behind the front would be very hard as the Russian Army moved forward, using all transport. He insisted however that if Mikolajczyk were Premier he (Bierut) must have 75 per cent. of the Cabinet. Mikolajczyk proposed that each of the five Polish parties should be represented, he having four out of the five of their best men, whom he would pick from personalities not obnoxious to Stalin.

---

* This took place on November 7. Mr. Roosevelt was re-elected President of the United States for the fourth time by a majority of over three and a half million votes.

2. Later, at my request, Stalin saw Mikolajczyk and had an hour and a half's very friendly talk. Stalin promised to help him, and Mikolajczyk promised to form and conduct a Government thoroughly friendly to the Russians. He explained his plan, but Stalin made it clear that the Lublin Poles must have the majority.

3. After the Kremlin dinner we put it bluntly to Stalin that unless Mikolajczyk had fifty-fifty plus himself the Western World would not be convinced that the transaction was *bona fide* and would not believe that an independent Polish Government had been set up. Stalin at first replied he would be content with fifty-fifty, but rapidly corrected himself to a worse figure. Meanwhile Eden took the same line with Molotov, who seemed more comprehending. I do not think the composition of the Government will prove an insuperable obstacle if all else is settled. Mikolajczyk had previously explained to me that there might be one announcement to save the prestige of the Lublin Government and a different arrangement among the Poles behind the scenes.

4. Apart from the above, Mikolajczyk is going to urge upon his London colleagues the Curzon Line, including Lvov, for the Russians. I am hopeful that even in the next fortnight we may get a settlement. If so I will cable you the exact form so that you can say whether you want it published or delayed.

5. On major war criminals U.J. took an unexpectedly ultra-respectable line. There must be no executions without trial; otherwise the world would say we were afraid to try them. I pointed out the difficulties in international law, but he replied if there were no trials there must be no death sentences, but only lifelong confinements.

6. We also discussed informally the future partition of Germany. U.J. wants Poland, Czecho, and Hungary to form a realm of independent, anti-Nazi, pro-Russian States, the first two of which might join together. Contrary to his previously expressed view, he would be glad to see Vienna the capital of a federation of South German States, including Austria, Bavaria, Württemberg, and Baden. As you know, the idea of Vienna becoming the capital of a large Danubian federation has always been attractive to me, though I should prefer to add Hungary, to which U.J. is strongly opposed.

7. As to Prussia, U.J. wished the Ruhr and the Saar detached and put out of action and probably under international control, and a separate State formed in the Rhineland. He would also like the internationalisation of the Kiel Canal. I am not opposed to this line of thought. However, you may be sure that we came to no fixed conclusions pending the triple meeting.

8. I was delighted to hear from U.J. that you had suggested a triple meeting towards the end of November at a Black Sea port. I think this

a very fine idea, and hope you will let me know about it in due course. I will come anywhere you two desire.

9. U.J. also raised formally the Montreux Convention, wishing for modification for the free passage of Russian warships. We did not contest this in principle. Revision is clearly necessary, as Japan is a signatory and Inönü missed his market last December. We left it that detailed proposals should be made from the Russian side. He said they would be moderate.

10. About recognising the present French Administration as the Provisional Government of France, I will consult the Cabinet on my return. Opinion of United Kingdom is very strongly for immediate recognition. De Gaulle is no longer sole master, but is better harnessed than ever before. I still think that when Eisenhower proclaims a large zone of the interior for France it would not be possible to delay this limited form of recognition. Undoubtedly de Gaulle has the majority of the French nation behind him, and the French Government need support against potential anarchy in large areas. I will however cable you again from London.

I am now in the air above Alamein of Blessed Memory. Kindest regards.

He replied:

President Roosevelt to Prime Minister   22 Oct 44

I am delighted to learn of your success at Moscow in making progress toward a compromise solution of the Polish problem.

When and if a solution is arrived at I should like to be consulted as to the advisability from this point of view of delaying its publication for about two weeks. You will understand.

Everything is going well here at the present time.

Your statement of the present attitude of Uncle J. towards war criminals, the future of Germany, and the Montreux Convention is most interesting. We should discuss these matters, together with our Pacific war effort, at the forthcoming three-party meeting.

*   *   *   *   *

On leaving after this profoundly interesting fortnight, in which we got closer to our Soviet Allies than ever before—or since—I had written to Stalin :

20 Oct 44

Eden and I have come away from the Soviet Union refreshed and fortified by the discussions which we had with you, Marshal Stalin, and with your colleagues. This memorable meeting in Moscow has shown that there are no matters that cannot be adjusted between us when we meet together in frank and intimate discussion. Russian hospitality, which is renowned, excelled itself on the occasion of our visit. Both in Moscow and in the Crimea, where we spent some enjoyable hours, there was the highest consideration for the comfort of myself and our mission. I am most grateful to you and to all those who were responsible for these arrangements. May we soon meet again.

CHAPTER XVI

PARIS

Need for a Representative Administration in France – General de Gaulle's National Committee – My Speech to the House of Commons of September 28 – My Telegram to the President of October 14 – The Committee of National Liberation Becomes the Provisional Government of France, October 20 – I Fly to Paris, November 10 – The Procession in the Champs Élysées, November 11 – A Trip to the Vosges – My Telegram to the President and to Stalin, November 15 – The French Army and the Occupation of Germany – Exchange of Congratulatory Messages with de Gaulle – Stalin's Telegram of November 20 – Rumours of a Western Bloc – My Telegram to Stalin of November 25 – The Winter Battle in the West – General de Gaulle's Visit to Moscow – Proposed Franco-Soviet Pact – France's Post-War Frontier on the Rhine – Correspondence with the President – Signature of the Franco-Soviet Pact, December 10 – Question of an Anglo-French Treaty.

AS our armies moved eastwards and southwards it became increasingly urgent to set up a unified and broadly representative Administration in France. We were anxious not to impose a ready-made Committee from abroad, and we tried first to gauge the feelings of the people themselves as the liberation progressed. I had long pondered this problem, and as early as July 10 had minuted to Mr. Eden:

Prime Minister to Foreign Secretary   10 July 44
Surely it would be most unwise of us to make up our minds on this [proposal to ask the U.S.A. and the U.S.S.R. to join with us in recognising the French Committee of National Liberation as the Provisional Government of France] until the result of the President's honeymoon with de Gaulle is made known. Clearly we shall have to go as far as the United States go, and after their decision is declared

we may press them to go farther. Should the President make a *volte-face* and come to terms with de Gaulle we shall have a very good case to present to Parliament, showing how foolish it would have been to have had a premature debate which might have spoiled all this happy kissing.

Five weeks later the break-out from Normandy had been accomplished and Patton was at the gates of Paris, but I was still averse to taking any decisive steps and minuted again:

Prime Minister to Foreign Secretary   18 Aug 44

I should deprecate taking any decisions about France till we can see more clearly what emerges from the smoke of battle. Should the great success of our operations secure the liberation of the West and South of France, including Paris, as may easily be the case, there will be a large area from which a real Provisional Government might be drawn instead of one being composed entirely of the French National Liberation Committee, whose interest in seizing the title-deeds of France is obvious.

I therefore strongly deprecate commitments of any kind at this stage to the French National Committee beyond those which have already been agreed to. One does not know at all what may happen, and it is as well to keep our hands free. I think a broader basis should be established before we commit ourselves

Throughout the following weeks we watched the rallying of the Maquis and of public opinion to General de Gaulle's National Committee. Hitherto, by force of circumstances, it could not be a body representative of France as a whole, but by the end of September progress had been made, and on the 28th I said in my review of the war to the House of Commons:

Naturally that body has new elements, especially among those who formed the Maquis and Resistance movements, and among those who raised the glorious revolt in Paris, which reminded us of the famous days of the Revolution, when France and Paris struck a blow that opened the path broadly for all the nations of the world. Naturally, we, and, I believe, the United States and the Soviet Union, are most anxious to see emerge an entity which can truly be said to speak in the name of the people of France—the whole people of France. It would now seem possible to put into force the decree of the Algiers Committee whereby, as an interim stage, the Consultative Assembly would be transformed into an elected body, reinforced by the addition of new elements drawn from inside France. To this body the French Committee of National Liberation would be responsible. Such a step, once taken, when seen to have the approval of the French people, would greatly strengthen the position of France, and would render possible that recognition of the Provisional Government of France, and all those consequences thereof, which we all desire to bring about at the earliest moment. I close no doors upon a situation which is in constant flux and development.

The welcome which the Maquis gave to the Committee seemed to me to be a decisive point in favour of its more formal recognition. I therefore telegraphed to the President:

Prime Minister (Moscow) to President Roosevelt   14 Oct 44

I have been reflecting about the question of the recognition of the French Provisional Government. I think events have now moved to a point where we could take a decision on the matter consistently with your own policy and my latest statement in the House of Commons.

2. In your telegram you said that you thought that we should wait until France was cleared of the enemy, and you implied that in any case de Gaulle must first show himself ready to take over from Eisenhower full responsibility for the administration of part of France as an interior zone. I for my part took the line in Parliament that the reorganisation of the Consultative Assembly on a more representative basis ought to precede recognition.

3. I understand that Eisenhower is anxious to comply with the request he has already had from the French to constitute a large part of France into an interior zone. Negotiations between Supreme Headquarters and the French are making good progress, and it appears that we may expect about three-quarters of France to become an interior zone very shortly.

4. The enlargement of the Consultative Assembly is also making good progress. Duff Cooper reports that owing to very real difficulties of communications in France the French have found it impracticable to proceed with the original Algiers plan of getting members of an enlarged Assembly confirmed in their mandates by elections in liberated departments. They propose instead to add selected delegates from the Resistance Movement and Parliamentary groups. I understand it is hoped to settle the matter shortly and publish a new decree defining attributions of the reformed Assembly and giving it increased powers over the Executive. It is thought that the enlarged Assembly should be able to meet at the end of this month.

5. There is no doubt that the French have been co-operating with Supreme Headquarters and that their Provisional Government has the support of the majority of French people. I suggest therefore that we can now safely recognise General de Gaulle's Administration as the Provisional Government of France.

6. One procedure might be to tell the French now that we will recognise [it] as soon as the enlarged Assembly has met and has given de Gaulle's Administration a vote of confidence.

7. An alternative procedure would be to recognise as soon as the interior zone has been formally established. I am inclined to think that this alternative is preferable, as it would connect recognition with what will be a mark of satisfactory co-operation between the French authorities and the Allied Armies in the common cause against Germany.

8. Please tell me what you think. If you agree that we should settle the matter by one or other of the procedures suggested above, the Foreign Office and State Department might at once compare their ideas upon the actual terms in which we should give recognition. It is important that we should take the same line, although we need not necessarily adopt exactly the same wording. We should of course have also to inform the Soviet Government of what we intend.

9. Recognition would not of course commit us on the separate question of French membership of the European Advisory Commission or similar bodies.

The President replied:

President Roosevelt to Prime Minister   20 Oct 44

I think until the French set up a real zone of interior that we should make no move towards recognising them as a Provisional Government. The enlargement of the Consultative Assembly, which has already been extended and made more representative, is almost as important, and I should be inclined to hang recognition on the effective completion of both these acts. I would not be satisfied with de Gaulle merely saying that he was going to do it.

I agree with you that there must be no implication, if and when we do recognise a Provisional Government, that this means a seat on the European Advisory Council, etc. These matters can be taken up later on their merits.

I am anxious to handle this matter, for the present, directly between you and me, and would prefer, for the moment, that the *modus operandi* should not become a matter of discussion between the State Department and your Foreign Office.

I do hope you are free of the temperature and really feeling all right again.

★   ★   ★   ★   ★

Our discussions proceeded on these lines. The French Assembly was strengthened and enlarged by members of the Resistance organisations and the old Parliamentary group. Already in August we had concluded a Civil Affairs Agreement with the

French Provisional Administration, dividing France into a forward zone, under the Supreme Allied Commander, and an interior zone, where the administration would be in the hands of the French authorities. . On October 20 it was announced that, with the agreement of the Allied High Command, an interior zone comprising the larger part of France, including Paris, had been set up. The Committee of National Liberation was thus finally transformed into a Provisional Government of France.

I was now prepared to recommend, in concert with our Allies, the official acceptance of this body as the Government of Liberated France. After last-minute hesitations by the State Department the public announcement was made during my visit to Moscow, where I discussed the final stages of formal recognition with the Russians. This came sooner than I expected, and I telegraphed to the President:

Prime Minister (Moscow) to President Roosevelt   23 Oct 44

I was naturally surprised at the very sharp turn taken by the State Department, and on arrival here I find the announcement is to be made to-morrow. We shall of course take similar and simultaneous action. I think it likely that the Russians will be offended. Molotov in conversation said that he expected they would be made to appear the ones who were obstructing, whereas they [the Russians] would have recognised long ago, but had deferred to American and British wishes. I hope therefore it has been possible to bring them in too.

* * * * *

I said in my speech in the House of Commons on October 27:

I have been myself for some weeks past satisfied not only that the present French Government, under General de Gaulle, commands the full assent of the vast majority of the French people, but that it is the only Government which can possibly discharge the very heavy burdens which are being cast upon it, and the only Government which can enable France to gather its strength in the interval which must elapse before the constitutional and Parliamentary processes, which it has declared its purpose to reinstate, can again resume their normal functions.

Thus we completed the processes begun in the dark and far-off days of 1940.

* * * * *

It was thought fitting that my first visit to Paris should be on Armistice Day, and this was publicly announced. There were many reports that collaborators would make attempts on my life and extreme precautions were taken. On the afternoon of November 10 I landed at Orly airfield, where de Gaulle received me with a guard of honour, and we drove together through the outskirts of Paris and into the city itself until we reached the Quai d'Orsay, where my wife and Mary and I were entertained in state. The building had long been occupied by the Germans, and I was assured I should sleep in the same bed and use the same bathroom as had Goering. Everything was mounted and serviced magnificently, and inside the palace it was difficult to believe that my last meeting there, described in a previous volume, with Reynaud's Government and General Gamelin in May 1940 was anything but a bad dream. At eleven o'clock on the morning of November 11 de Gaulle conducted me in an open car across the Seine and through the Place de la Concorde, with a splendid escort of Gardes Républicains in full uniform with all their breastplates. They were several hundred strong, and provided a brilliant spectacle, on which the sun shone brightly. The whole of the famous avenue of the Champs Élysées was crowded with Parisians and lined with troops. Every window was filled with spectators and decorated with flags. We proceeded through wildly cheering multitudes to the Arc de Triomphe, where we both laid wreaths upon the tomb of the Unknown Warrior. After this ceremony was over the General and I walked together, followed by a concourse of the leading figures of French public life, for half a mile down the highway I knew so well. We then took our places on a dais, and there was a splendid march past of French and British troops. Our Guards detachment was magnificent. When this was over I laid a wreath beneath the statue of Clemenceau, who was much in my thoughts on this moving occasion.

De Gaulle entertained me at a large luncheon at the Ministry of War, and made a most flattering speech about my war services. But many problems had still to be settled.

On the night of the 12th after dinner at the Embassy I left with General de Gaulle for Besançon. The General was anxious for me to see the attack on a considerable scale which was planned for the French Army under General de Lattre de Tassigny. All the arrangements for the journey in a luxurious special train were most carefully made, and we arrived in plenty of time for the battle. We were to go to an observation point in the mountains, but

owing to bitter cold and deep snow the roads were impassable
and the whole operation had to be delayed. I passed the day
driving with de Gaulle, and we found plenty to talk about in a
long and severe excursion, inspecting troops at intervals. The
programme continued long after dark. The French soldiers
seemed in the highest spirits. They marched past in great style
and sang famous songs with moving enthusiasm. My personal
party—my daughter Mary and my naval aide Tommy—feared
that I should have another go of pneumonia, since we were out
at least ten hours in terrible weather. But all went well, and in
the train the dinner was pleasant and interesting. I was struck by
the awe, and even apprehension, with which half a dozen high
generals treated de Gaulle in spite of the fact that he had only one
star on his uniform and they had lots.

During the night our train divided. De Gaulle returned to
Paris, and our half went on to Rheims, arriving next morning,
when I went to Ike's headquarters. In the afternoon I flew back
to Northolt.

* * * * *

On my return to London I sent the President a report. A copy
was also passed to Stalin.

Prime Minister to President Roosevelt   15 Nov 44
... Thank you for your kind wishes about the Paris-de Gaulle trip.
I certainly had a wonderful reception from about half a million
French in the Champs Élysées, and also from the partly Opposition
Centre at the Hôtel de Ville. I re-established friendly private relations
with de Gaulle.

I see statements being put out in the French Press and other quarters
that all sorts of things were decided by us in Paris. You may be sure
that our discussions about important things took place solely on an
ad referendum basis to the three Great Powers, and of course especially
to you, who have by far the largest forces in France. Eden and I had
a two hours' talk with de Gaulle and two or three of his people after
luncheon on the 11th. De Gaulle asked a number of questions which
made me feel how very little they were informed about anything that
had been decided or was taking place. He is of course anxious to obtain
full modern equipment for eight more divisions, which can only be
supplied by you. S.H.A.E.F. reasonably contends that these will not
be ready for the defeat of Germany in the field and that shipping must
be devoted to the upkeep of the actual forces that will win the battles
of the winter and spring. I reinforced this argument.

At the same time I sympathise with the French wish to take over more of the line, to have the best share they can in the fighting or what is left of it—and there may be plenty—and not to have to go into Germany as a so-called conqueror who has not fought. I remarked that this was a sentimental point which ought nevertheless to receive consideration. The important thing for France was to have an Army prepared for the task which it would actually have to discharge, namely, their obligation first to maintain a peaceful and orderly country behind the front of our armies, and secondly to assist in the holding down of parts of Germany later on.

On this second point the French pressed very strongly to have a share in the occupation of Germany, not merely as sub-participation under British or American command, but as a French command. I expressed my sympathy with this, knowing well that there will be a time not many years distant when the American armies will go home and when the British will have great difficulty in maintaining large forces overseas, so contrary to our mode of life and disproportionate to our resources, and I urged them to study the type of army fitted for that purpose, which is totally different in form from the organisation by divisions required to break the resistance of a modern war-hardened enemy army. They were impressed by this argument, but nevertheless pressed their view.

I see a Reuter message, emanating no doubt unofficially from Paris, that it was agreed France should be assigned certain areas—the Ruhr, the Rhineland, etc.—for their troops to garrison. There is no truth in this, and it is obvious that nothing of this kind can be settled on such a subject except in agreement with you. All I said to de Gaulle on this was that we had made a division of Germany into Russian, British, and United States spheres; roughly, the Russians had the east, the British the north, and the Americans the south. I further said that, speaking for His Majesty's Government, the less we had of it the better we should be pleased, and that we should certainly favour the French taking over as large a part as their capacity allowed, but that all this must be settled at an inter-Allied table. I could of course issue something which would be a disclaimer of any loose statements made by Reuter, but you may not think this necessary in view of the obvious facts. I am telegraphing to U.J. in the same sense. We did not attempt to settle anything finally or make definite agreements.

It is evident however that there are a number of questions which press for decision at a level higher than that of the High Commands, without which decisions no clear guidance can be given. Here is another reason why we should have a triple meeting if U.J. will not come, or a quadruple meeting if he will. In the latter case the French would be in on some subjects and out on others. One must always realise that before five years are out a French army must be made to take on the main task of holding down Germany. The main question of discussion between Eden and Bidault was Syria, which was troublesome, lengthy, and inconclusive, but primarily our worry.

I thought I would give you this account at once in case of further tendentious statements being put out in the Press.

I thought very well of Bidault. He looks like a younger Reynaud, especially in speech and smiling. He made a very favourable impression on all of us, and there is no doubt that he has a strong share in the power. Giraud was at the banquet, and apparently quite content. What a change in fortune since Casablanca! Generally I felt in the presence of an organised Government, broadly based and of rapidly growing strength, and I am certain that we should be most unwise to do anything to weaken it in the eyes of France at this difficult, critical time. I had a considerable feeling of stability in spite of Communist threats, and that we could safely take them [the French] more into our confidence. I hope you will not consider that I am putting on French clothes when I say this. Let me know your thoughts. I will cable you later about the meeting. . . .

I also exchanged warm messages with General de Gaulle.

Prime Minister to General de Gaulle   16 Nov 44
Now that I am back home, let me express to Your Excellency and to your colleagues of the French Government my profound appreciation of the splendid hospitality and innumerable kindnesses and courtesies shown me and my friends during the memorable days which I have just spent in France. I shall always recall as one of the proudest and most moving occasions of my life the wonderful reception which the people of Paris gave to their British guests on this our first visit to your capital after its liberation. I was also most grateful for the opportunity of seeing for myself something of the ardour and high quality of French troops, which are completing the liberation of their native soil under the skilful leadership of General de Lattre de Tassigny. The welcome extended to us was indeed a happy augury for that continued friendship between our two countries essential to the safety and to the future peace of Europe.

General de Gaulle to Prime Minister   20 Nov 44
Je vous remercie au nom du Gouvernement de votre message. La France, sa capitale, et son Armée, ont été heureuses d'acclamer dans votre personne non seulement le Premier Ministre d'un grand pays qui leur est cher, mais encore le glorieux combattant qui a maintenu la coalition dans la guerre aux jours les plus sombres et lui vaut ainsi

la victoire. Laissez-moi vous dire combien j'ai été personnellement heureux de vous revoir.

Prime Minister to General de Gaulle (Paris)   25 Nov 44

If you think well, please give the following message from me to de Lattre:

I send all my congratulations on the brilliant exploits of your young Army. It must be wonderful to be a Frenchman twenty years old with good weapons in his hands and France to avenge and save.

*   *   *   *   *

On November 20 Stalin sent a friendly reply to my telegram of November 15.

Marshal Stalin to Prime Minister   20 Nov 44

Thank you for your information about your conversations with de Gaulle. I have acquainted myself with your communication with interest. I have nothing against your proposal about a possible meeting between us three and the French, provided the President also agrees with this, but it is necessary first to settle definitely about the time and place of the meeting between us three.

General de Gaulle expressed recently his wish to come to Moscow to establish contact with the leaders of the Soviet Government. We replied agreeing to this. The French are expected to arrive in Moscow towards the end of this month. The French have not yet specified the questions which they would wish to discuss. In any case, after our conversations with General de Gaulle I will let you know about it.

This raised the whole issue of the future organisation of Europe. There had been many rumours in the Press and elsewhere about forming a Western *bloc* when the war was over. Such a plan seemed to be particularly popular in Foreign Office circles, although it would burden us with heavy military commitments. I felt that the Cabinet should be consulted very soon, particularly as Franco-Soviet talks were approaching.

In consultation with Mr. Eden I now sent the following reply to Stalin:

Prime Minister to Marshal Stalin   25 Nov 44

Your message of November 20. I am glad de Gaulle is coming to see you, and I hope you will talk over the whole field together. There has been some talk in the Press about a Western *bloc*. I have not yet considered this. I trust first of all to our Treaty of Alliance and close collaboration with the United States to form the mainstays of a World Organisation to ensure and compel peace upon the tortured world

It is only after and subordinate to any such world structure that European arrangements for better comradeship should be set on foot, and in these matters we shall have no secrets from you, being well assured that you will keep us equally informed of what you feel and need.

2. The battle in the West is severe and the mud frightful. The main collision is on the axis Aix-la-Chapelle–Cologne. This is by no means decided in our favour yet, though Eisenhower still has substantial reserves to throw in. To the north-west Montgomery's armies are facing north, holding back the Germans on the line of the Dutch Maas. This river permits us an economy of force on this front. To the east we are making slow but steady progress and keeping the enemy in continual battle. One must acclaim the capture of Metz and the driving of the enemy back towards the Rhine as a fine victory for the Americans. In the south the French have had brilliant success, particularly in reaching the Rhine on a broad front and in taking Strasbourg, and these young French soldiers, from eighteen to twenty-one years old, are showing themselves worthy of their glorious chance to cleanse the soil of France. I think highly of General de Lattre de Tassigny. De Gaulle and I travelled there in order to see the opening of this battle from a good viewpoint. However, a foot of snow fell in the night and everything was put off for three days.

3. In a week or ten days it should be possible to estimate whether the German armies will be beaten decisively west of the Rhine. If they are we can go on in spite of the weather. Otherwise there may be some lull during the severity of the winter, after which one more major onslaught should break the organised German resistance in the West.

4. Do you think it is going to be a hard winter, and will this suit your strategy? We all greatly liked your last speech. Please do not fail to let me know privately if anything troublesome occurs, so that we can smooth it away and keep the closing grip on Nazidom at its most tense degree.

General de Gaulle had meanwhile arrived in Moscow and conversations with the Russians had begun. Stalin lost no time in informing me of the general points.

Marshal Stalin to Prime Minister   2 Dec 44

There is every evidence that de Gaulle and his French friends, having arrived in the Soviet Union, will raise two questions:

1. The conclusion of a French-Soviet Pact of Mutual Assistance similar to the Anglo-Soviet Pact.

We can hardly object. But I should like to know your view on this subject. Please give your advice.

2. De Gaulle will probably raise the question of changing France's eastern frontier and extending the French frontier to the left bank of the Rhine. It is also common knowledge that there is a scheme for forming a Rhenish-Westphalian province under international control. Possibly French participation in this control is also contemplated. Thus the proposal of the French to transfer the boundary line to the Rhine will compete with the scheme for creating a Rhine province under international control.

Please give your advice on this question also.

I have sent a similar message to the President.

And again the following day:

Marshal Stalin to Prime Minister   3 Dec 44

The meeting with General de Gaulle has provided an opportunity for a friendly exchange of views on questions of Franco-Soviet relations. During the conversations General de Gaulle persisted, as I had expected, with two main questions: the frontier of France on the Rhine and the conclusion of a Franco-Soviet Pact of Mutual Assistance of the type of the Anglo-Soviet Treaty.

As regards the frontier of France on the Rhine, I expressed myself to the effect that it was impossible to decide this question without the knowledge and agreement of our chief Allies, whose armies are waging a battle of liberation against the enemy on the territory of France. I emphasised the complexity of a solution of this question.

With regard to the proposal of a Franco-Soviet Pact of Mutual Assistance, I pointed out the necessity of a study of this question from all sides, and the necessity for clarification of the juridical aspect of such a pact, in particular of the question who would ratify such a pact in France in present conditions.

Consequently the French still have to furnish a number of explanations, which we have up till now not received from them.

In sending you this information I shall be grateful for a reply from you and for your comments on these questions.

I have conveyed the same message to the President.

I send you my best wishes.

On December 4 the Cabinet met to survey the possibilities of a Western *bloc* and de Gaulle's talks in Moscow. I read out to my colleagues the latest exchange of correspondence with Stalin, and the results of our deliberations were embodied in a message which I sent him in the early hours of December 5.

Prime Minister to Marshal Stalin   5 Dec 44

Your telegram about de Gaulle's visit and the two questions he will raise. We have no objection whatever to a Franco-Soviet Pact of Mutual Assistance similar to the Anglo-Soviet Pact. On the contrary, His Majesty's Government consider it desirable and an additional link between us all. Indeed, it also occurs to us that it might be best of all if we were to conclude a tripartite treaty between the three of us which would embody our existing Anglo-Soviet Treaty, with any improvements. In this way the obligations of each one of us would be identical and linked together. Please let me know if this idea appeals to you, as I hope it may. We should both of course tell the United States.

2. The question of changing the eastern frontier of France to the left bank of the Rhine, or alternatively of forming a Rhenish-Westphalian province under international control, together with other alternatives, ought to await settlement at the peace table. There is however no reason why when three heads of Governments meet we should not come much closer to conclusions about all this than we have done so far. As you have seen, the President does not expect de Gaulle to come to the meeting of the Three. I would hope that this could be modified to his coming in later on, when decisions especially affecting France were under discussion.

3. Meanwhile would it not be a good thing to let the European Advisory Commission sitting in London, of which France is a member, explore the topic for us all without committing in any way the heads of Governments?

4. I am keeping the President informed.

Mr. Roosevelt also kept in close touch with me.

President Roosevelt to Prime Minister   6 Dec 44

I have this date sent the following message to Uncle Joe:

"Thank you for your two informative messages of December 2 and December 3.

"In regard to a proposed Franco-Soviet Pact along the lines of the Anglo-Soviet Pact of Mutual Assistance, this Government would have no objection in principle if you and General de Gaulle considered such a pact in the interests of both your countries and European security in general.

"I am in complete agreement with your replies to General de Gaulle with regard to the post-war frontier of France. It appears to me at the present time that no advantage to our common war effort would result from an attempt to settle this question now and that its settlement subsequent to the collapse of Germany is preferable."

And later:

President Roosevelt to Prime Minister   6 Dec 44

You will have seen from my reply to Stalin on his talks with de

225

Gaulle that our views are identical on the two questions which he raised.

I still adhere to my position that any attempt to include de Gaulle in the meeting of the three of us would merely introduce a complicating and undesirable factor.

In regard to your suggestion to Uncle Joe that the question of France's post-war frontiers be referred to the European Advisory Commission, I feel that since the Commission is fully occupied with questions relating to the surrender of Germany it would be a mistake to attempt to bring up at this stage before it any questions of post-war frontiers. It seems to me preferable to leave this specific topic for further exploration between us.

I fully appreciate the advantages which you see in a possible tripartite Anglo-Franco-Soviet Pact. I am somewhat dubious however as to the effect of such an arrangement on the question of an international Security Organisation, to which, as you know, I attach the very highest importance. I fear that a tripartite pact might be interpreted by public opinion here as a competitor to a future World Organisation, whereas a bilateral arrangement between France and the Soviet Union similar to the Soviet-British Pact would be more understandable. I realise however that this is a subject which is of primary concern to the three countries involved.

Stalin telegraphed next day:

Marshal Stalin to Prime Minister   7 Dec 44
I have received your reply to my message about a Franco-Soviet Pact and about the frontier of France on the Rhine. I thank you for your advice.

At the time of receiving your reply we had already begun discussions with the French about the pact. Your proposal in preference for a tripartite Anglo-Franco-Soviet Pact as an improvement in comparison with the Anglo-Soviet Pact has been approved by myself and my colleagues. We have made a proposal to de Gaulle for the conclusion of such a tripartite pact, but we have not yet had his reply.

I have delayed my reply to your other messages. I hope to reply soon.

But events took a slightly different course. The French were determined for domestic reasons to come away from Moscow with a strictly Franco-Soviet pact. This was signed on December 10, and Stalin telegraphed the same day:

Marshal Stalin to Prime Minister   10 Dec 44
I communicated to General de Gaulle your opinion about your

226

preference for an Anglo-French-Soviet Pact of Mutual Assistance, and spoke in favour of accepting your proposal. However, General de Gaulle insisted on concluding a Franco-Soviet Pact, saying that a three-party pact should be concluded at the next stage, as that question demanded preparation. At the same time a message came from the President, who informed me that he had no objection to a Franco-Soviet Pact. In the result we reached agreement about concluding a pact and it was signed to-day. The pact will be published after General de Gaulle's arrival in Paris.

I think that General de Gaulle's visit has had positive results, and will assist not only in strengthening Franco-Soviet relations, but will also be a contribution to the common work of the Allies.

It was for the French now to make a similar agreement with us, if they felt so inclined. I informed Stalin of this possibility in a jocular way.

Prime Minister to Marshal Stalin   19 Dec 44

I saw last night for the second time the film which you have given me called *Kutuzov*. The first time I greatly admired it, but as it was all in Russian I could not understand the exact meaning of each situation. Last night I saw it with the English captions, which made exactly intelligible the whole thing, and I must tell you that in my view this is one of the most masterly film productions I have ever witnessed. Never has the conflict of two will-powers been more clearly displayed. Never has the importance of fidelity in commanders and men been more effectively inculcated by the film pictures. Never have the Russian soldiers and the Russian nation been presented by this medium so gloriously to the British nation. Never have I seen the art of the camera better used.

2. If you thought it fit privately to communicate my admiration and thanks to those who have laboured in producing this work of art and high morale I should thank you. Meanwhile I congratulate you.

3. I like to think we were together in that deadly struggle, as in this Thirty Years War. I do not suppose that you showed the film to de Gaulle, any more than I shall show him *Lady Hamilton* when he comes over here to make a similar treaty to that which you have made with him, and we have made together.

Salutations.

On December 25 he replied that he would "of course welcome the conclusion of an Anglo-French treaty". I felt that there was no hurry about this and that we should await a move from the French. On December 31 I minuted to Mr. Eden:

You may feel inclined to see how the proposals which have come on the *tapis* meanwhile for a bilateral treaty between Britain and France shape themselves. You said to me that if de Gaulle attempted to say that there could be no Anglo-French Treaty until we had settled everything about Syria you would let him wait. It is for him to make the proposal, not us.

Meantime we are losing nothing from the point of view of security, because the French have practically no Army and all the other nations concerned are prostrate or still enslaved. We must be careful not to involve ourselves in liabilities which we cannot discharge and in engagements to others for which there is no corresponding return. I do not know what our financial position will be after the war, but I am sure we shall not be able to maintain armed forces sufficient to protect all these helpless nations even if they make some show of recreating their armies. Anyhow, the first thing to do is to set up the World Organisation, on which all depends.

CHAPTER XVII

COUNTER-STROKE IN THE ARDENNES

NOVEMBER brought the Allied cause the loss of Field-Marshal Sir John Dill, the head of our Inter-Service Mission in Washington. After forty years of a full Army life, which began in the South African War, he was appointed Chief of the Imperial General Staff in May 1940. In that great position his balanced judgment and steadfast temperament were a great stand-by in our days of peril. After Pearl Harbour he was transferred to Washington, where he interpreted our views to the United States Chiefs of Staff. He soon endeared himself there, and formed a firm personal friendship with General Marshall, which proved invaluable in overcoming the frets and frictions that inevitably arise between Allies. This was the climax of his career. He would surely have lived far beyond his sixty-three years but for his selfless devotion to duty; but even when a very sick man he would not give in. As a last tribute to him and to all he stood for, he was accorded the signal distinction of burial in the National Cemetery at Arlington, where America's great men lie. An equestrian statue was erected to his memory by the American Army.

The President sent me a message saying, "America joins with Great Britain in sorrow at the loss of your distinguished soldier, whose personal admirers here are legion." I thanked him, and to General Marshall I said: "I read with emotion the message which the United States Joint Chiefs of Staff have addressed to their British colleagues about the death of our friend Sir John Dill. Let me express my own thanks for all your kind thoughts. He did all he could to make things go well, and they went well."

To fill the gap required important changes in our commands.

Prime Minister to General Wilson (Italy)   21 Nov 44

It is of great importance that Field-Marshal Dill's position should be filled by someone who would have access to the President from time to time and a status which would enable him to be in very close touch with General Marshall. It goes without saying that the officer selected must be one who works well with Americans and be in full possession of the general outlook upon the war as a whole. I can find only one officer with the necessary credentials and qualities, namely, yourself. I have therefore proposed to the President that you should succeed Dill as Head of the British Military Mission and my official representative in military matters at Washington. The President has cordially agreed, and you are assured of a warm welcome in Washington. I hope therefore that you will feel able to let me know at once that you accept this extremely important appointment.

2. I also proposed to the President that General Alexander should become Allied Supreme Commander in the Mediterranean in succession to you, with General McNarney as his Deputy, and that General Mark Clark should take over the group of armies on the Italian front.

3. The President has replied that these proposals are entirely agreeable to the United States Chiefs of Staff and to himself.

4. I should like you to come home this next week for a day or two for a preliminary discussion. I hope that you can manage this. My "York" will come out at once. I hope you will bring Macmillan with you.

"I appreciate the compliment," the President had cabled, "you pay General Clark in suggesting that he take over the Army Group in Italy as General Alexander's successor."

* * * * *

There had meanwhile been much preparation on the Western Front for the advance to the Rhine. The November rains were the worst for many years, flooding the rivers and streams, and making quagmires through which the infantry had to struggle. In the British sector Dempsey's Second Army drove the enemy from their large salient west of Venlo back across the Meuse. Farther south our XXXth Corps had come into the line between Maeseyck and Geilenkirchen, where they joined hands with the Ninth U.S. Army. Together they took Geilenkirchen on November 19, after intensive artillery preparation, and then toiled over saturated country towards the river Roer. The right of the Ninth Army reached the river near Jülich on December 3, while the First Army on their flank had a bitter struggle in the Hürtgen Forest. Seventeen Allied divisions were engaged. The enemy had almost as many and the fighting was severe.

It would have been rash as yet to cross the river, because its level was controlled by massive dams a score of miles to the south. These were still in enemy hands, and by opening the sluices he could have cut off our troops on the far bank. Heavy bombers tried to burst the dams and release the water, but in spite of several direct hits no gap was made, and on December 13 the First U.S. Army had to renew their advance to capture them.

South of the Ardennes Patton's Third Army had meanwhile crossed the Moselle on each side of Thionville and thrust eastwards to the German frontier. Metz was entered on November 20, though the Germans still clung to the surrounding forts, of which the last held out till December 13. From Metz and Nancy the Army swung up towards the river Saar, which they reached on a broad front, and threw bridgeheads across it near Saarlautern on December 4. Here they confronted the strongest part of the Siegfried defences, consisting of a forward line along the north bank of the river, and behind it a zone over two miles deep of mutually supporting concrete works. Against such formidable and obstinately held fortifications the Third Army came to a halt.

On the right of the line General Devers's Sixth Army Group, from Luneville and Épinal, forced their way through the Vosges

THE FRONTIER REGIONS

and the Belfort Gap. The American Seventh Army had a stiff fight for the mountain crests, but the French First Army, after a week's battle, the opening of which I had hoped to see, captured Belfort on November 22 and reached the Rhine north of Bâle. Thence they swung down the river towards Colmar. This turned the German flank in the Vosges and the enemy withdrew. Strasbourg was entered on November 23, and during the next few weeks the Seventh Army cleared all Northern Alsace, wheeled up on the right of the Third Army, and, crossing the German frontier on a wide front, penetrated the Siegfried Line near Wissembourg. There was still a large pocket of Germans on French soil at Colmar, thirty miles deep and broad, which the French had been unable to clear. This was to prove an embarrassment a few weeks later.

*　*　*　*　*

I sent my comments on the whole situation to Smuts.

Prime Minister to Field-Marshal Smuts   3 Dec 44

. . . 2. In spite of Metz and Strasbourg and other successes, we have of course sustained a strategic reverse on the Western Front. Before this offensive was launched we placed on record our view that it was a mistake to attack against the whole front and that a far greater mass should have been gathered at the point of desired penetration. Montgomery's comments and predictions beforehand have in every way been borne out. I imagine some readjustments will be made giving back to Montgomery some of the scope taken from him after the victory he gained in Normandy. You must remember however that our armies are only about one-half the size of the American and will soon be little more than one-third. All is friendly and loyal in the military sphere in spite of the disappointment sustained. We must now re-group and reinforce the armies for a spring offensive. There is at least one full-scale battle to fight before we get to the Rhine in the north, which is the decisive axis of advance. I am trying meanwhile to have Holland cleaned up behind us. But it is not so easy as it used to be for me to get things done.

3. Our armies in Italy were delayed by "Anvil" and greatly weakened for its sake. Consequently we cleared the Apennines only to find the valley of the Po a bog. Thus both in the mountains and on the plains our immense armour superiority has been unable to make itself felt, and now the bad weather in Italy, as on the Western Front, greatly diminishes the tactical air-power in which we have so great a predominance. Hitherto in Italy we have held twenty-eight German

divisions, and therefore no reproach can be made against our activities. On the contrary, General Marshall is astonished we have done so well. This is only however because the Germans have delayed a withdrawal through the Brenner and Ljubljana, presumably in order to bring their forces home from the Balkans. We cannot look for any very satisfactory events in Northern Italy at present, though we are still attacking. . . .

5. In Burma too we have been compelled to work downwards from the north through the jungles I had hoped to avoid, and Mountbatten was doing pretty well. Now however the disasters in China have overtaken Kunming, and may soon affect Chungking. The Generalissimo is withdrawing his best Chinese troops from the southward advance in Burma to defend his capital, his air terminus, and, I expect, his life and régime. I cannot blame him, but this gravely affects the success of Mountbatten's well-conducted but already unappetising operations. We seem condemned to wallow at half-speed through these jungles, and I cannot so far procure agreement for a far-flung amphibious strategical movement across the Bay of Bengal. Everything has to be chewed up by the Combined Staffs and "Safety first" overloads every plan. The Americans are having hard fighting at Leyte, but their advance in the Pacific has been admirable during the year, and I hope our Fleet will join them in growing strength in 1945. As old Fisher said, "The Royal Navy always travels first class," and you can imagine the enormous demands in man-power, ancillary vessels, and preparations of all kinds which the Admiralty blithely put forward.

6. Meanwhile there approaches the shadow of the General Election, which before many months have passed will certainly break up the most capable Government England has had or is likely to have. Generally we have a jolly year before us. Our financial future fills in any spaces in the horizon not already overcast with clouds. However, I am sure we shall master all these troubles as they come upon us, singly or in company, even though the tonic element of mortal danger is lacking.

7. Of all the messages which reached me on my birthday, none was more movingly phrased or gave me more encouragement than yours, my old and trusted friend.

Three days later I addressed myself to the President.

Prime Minister to President Roosevelt   6 Dec 44

As we are unable to meet, I feel that the time has come for me to place before you the serious and disappointing war situation which faces us at the close of this year. Although many fine tactical victories have been gained on the Western Front and Metz and Strasbourg are trophies, the fact remains that we have definitely failed to achieve the strategic object which we gave to our armies five weeks ago. We have not yet reached the Rhine in the northern part and most important sector of the front, and we shall have to continue the great battle for many weeks before we can hope to reach the Rhine and establish our bridgeheads. After that, again, we have to advance through Germany.

2. In Italy the Germans are still keeping twenty-six divisions—equivalent to perhaps sixteen full strength or more—on our front. They could however at any time retreat through the Brenner and Ljubljana and greatly shorten their line by holding from Lake Garda to, say, the mouth of the Adige. By this they might save half their Italian forces for home defence. Even after that there are the Alps to which they could fall back, thus saving more men. It seems to me that their reason for standing so long in Italy may have been to extricate the twelve divisions in the Balkans, etc., which are now fighting their way back to Hungary and Austria. Apart from the Air and Partisans and small Commando forces, there are no means of preventing this, and my opinion is that the greater part will escape. About half of these might be available for adding to what may be saved from Italy. This would be a powerful reinforcement to the German homeland, available, according to events, either in the East or in the West.

3. We have secured weighty advantages from "Dragoon" [the landing in Southern France] for the battle on the main front, but the reason why the Fifteenth Group of Armies has not been able to inflict a decisive defeat on Kesselring is that, owing to the delay caused by the weakening of our forces for the sake of "Dragoon", we did not get through the Apennines till the valley of the Po had become waterlogged. Thus neither in the mountains nor on the plains have we been able to use our superiority in armour.

4. On account of the obstinacy of the German resistance on all fronts, we did not withdraw the five British and British-Indian divisions from Europe in order to enable Mountbatten to attack Rangoon in March, and for other reasons also this operation became impracticable. Mountbatten therefore began, as we agreed at Quebec, the general advance through Burma downstream from the north and the west, and this has made satisfactory progress. Now, owing to the advance of the Japanese in China, with its deadly threat to Kunming and perhaps Chungking, to the Generalissimo and his régime, two and possibly more Chinese divisions have to be withdrawn for the defence of China. I have little doubt that this was inevitable and right. The consequences however are serious so far as Mountbatten's affairs are concerned, and

no decision has yet been taken on how to meet this new misfortune, which at one stroke endangers China and your air terminal as well as the campaign in Northern Burma. All my ideas about a really weighty blow across the Adriatic or across the Bay of Bengal have equally been set back.

5. The vast-scale operations which you have conducted in the Pacific are at present the only part of the war where we are not in a temporary state of frustration.

6. We have however, happily, to consider what the Russians will do. We have Stalin's promise for a winter campaign, starting, I presume, in January. On most of his immense front he seems to have been resting and preparing, though only about three or four German divisions have come over to face Eisenhower. I am not in a position to measure the latest attacks he has launched to the south-west of Budapest. We may however, I think, look forward to more assistance from this and other Russian action than we have had lately, and the German position is so strained that any heavy penetration might bring about a partial if not a total collapse.

7. I have tried to survey the whole scene in its scope and proportion, and it is clear that we have to face, in varying degrees of probability:

(a) A considerable delay in reaching, still more in forcing, the Rhine on the shortest road to Berlin.
(b) A marked degree of frustration in Italy.
(c) The escape home of a large part of the German forces from the Balkan peninsula.
(d) Frustration in Burma.
(e) Elimination of China as a combatant.

When we contrast these realities with the rosy expectations of our peoples, in spite of our joint efforts to damp them down, the question very definitely arises, "What are we going to do about it?" My anxiety is increased by the destruction of all hopes of an early meeting between the three of us and the indefinite postponement of another meeting of you and me with our Staffs. Our British plans are dependent on yours, our Anglo-American problems at least must be surveyed as a whole, and the telegraph and the telephone more often than not only darken counsel. Therefore I feel that if you are unable to come yourself before February I am bound to ask you whether you could not send your Chiefs of Staff over here as soon as practicable, where they would be close to your main armies and to General Eisenhower and where the whole stormy scene can be calmly and patiently studied with a view to action as closely concerted as that which signalised our campaigns of 1944.

Though sympathetic, Mr. Roosevelt did not appear to share my anxieties.

President Roosevelt to Prime Minister   10 Dec 44

. . . Perhaps I am not close enough to the picture to feel as disappointed about the war situation as you are, and perhaps also because six months ago I was not as optimistic as you were on the time element.

On the European front I always felt that the occupation of Germany up to the left bank of the Rhine would be a very stiff job. Because in the old days I bicycled over most of the Rhine terrain, I have never been as optimistic as to the ease of getting across the Rhine with our joint armies as many of the commanding officers have been.

However, our agreed broad strategy is developing according to plan. You and I are now in the position of Commanders-in-Chief who have prepared their plans, issued their orders, and committed their resources to battle according to those plans and orders. For the time being, even if a little behind schedule, it seems to me the prosecution and outcome of the battles lie with our Field Commanders, in whom I have every confidence. We must remember that the winter season is bringing great difficulties, but our ground and air forces are day by day chewing up the enemy's dwindling man-power and resources, and our supply flow is much improved with the opening of Antwerp. General Eisenhower estimates that on the Western front line he is inflicting losses in excess of the enemy's capability to form new units. I still cannot see clearly just when, but soon a decisive break in our favour is bound to come.

As to the Italian front, Alexander's forces are doing their bit in keeping those German divisions in Italy, and we must remember that the Germans are really free to withdraw to the line of the Alps if they so decide.

The same thing applies to their troops in the Balkans. I have never believed that we had the power to capture any large German forces in the Balkans without assistance by the Russians.

On the Russian front we must also give full allowance to the vile weather, and the Russians seem to be doing their bit at the present time. This of course you know more about than I do.

The Far Eastern situation is of course on a somewhat different footing, and I am not at all happy about it.

From the long-range point of view, other than the measures Wedemeyer is now taking, we can do very little to prepare China to conduct a worth-while defence, but Japan is suffering losses in men and ships and materials in the Pacific area that are many times greater than ours, and they too cannot keep this up. Even the Almighty is helping. This magnificent earthquake and tidal wave is a proof.

The time between now and spring, when the freeze is over, will develop many things. We shall know a lot more than we know now.

My Chiefs of Staff are now devoting all of their abilities and energies in directing their organisations towards carrying out the plans we have made and in supporting our forces throughout the world. Practically all of these forces are, for the time being, committed. That is why I do not feel that my Chiefs should leave their posts at this time, since no requirement exists for broad strategic decisions to guide our Field Commanders. . . .

★   ★   ★   ★   ★

A heavy blow now impended. Within six days a crisis burst upon us. The Allied decision to strike hard from Aachen in the north as well as through Alsace in the south had left our centre very weak. In the Ardennes sector a single corps, the VIIIth American, of four divisions, held a front of seventy-five miles. The risk was foreseen and deliberately accepted, but the consequences were grave and might have been graver. By a remarkable feat the enemy gathered about seventy divisions on their Western Front, of which fifteen were armoured. Many were under strength and needed rest and re-equipment, but one formation, the Sixth Panzer Army, was known to be strong and in good fettle. This potential spear-head had been carefully watched while it lay in reserve east of Aachen. When the fighting on that front died down in early December it vanished for a while from the ken of our Intelligence, and bad flying weather hindered our efforts to trace it. Eisenhower suspected that something was afoot, though its scope and violence came as a surprise.

The Germans had indeed a major plan. Rundstedt assembled two Panzer armies, the Fifth and Sixth, and the Seventh Army, a total of ten Panzer and fourteen infantry divisions. This great force, led by its armour, was intended to break through our weak centre in the Ardennes to the river Meuse, swing north and northwest, cut the Allied line in two, seize the port of Antwerp, and sever the life-line of our northern armies. This bold bid was planned by Hitler, who would brook no changes in it on the part of his doubting generals. In its support the remnants of the German Air Force were assembled for a final effort, while paratroops, saboteurs, and agents in Allied uniforms were all given parts to play.

The attack began on December 16 under a heavy artillery

barrage. At its northern flank the Sixth Panzer Army ran into the right of the First U.S. Army in the act of advancing towards the Roer dams. After a swaying battle the enemy were held. Farther south the Germans broke through on a narrow front, but the determined defence of St. Vith, where the 7th U.S. Armoured Division specially distinguished itself, hindered them for several critical days. The Sixth Panzer Army launched a new spear-head to strike west and then northwards at the Meuse above Liége. The Fifth Panzer Army meanwhile drove through the centre of the VIIIth U.S. Corps, by-passed St. Vith and Bastogne, and penetrated deeply to Marche and towards the Meuse at Dinant.

Although the time and weight of the attack surprised the Allied High Command its importance and purpose were quickly recognised. They resolved to strengthen the "shoulders" of the break-through, hold the Meuse crossings both east and south of Namur, and mass mobile troops to crush the salient from north and south. Eisenhower acted speedily. He stopped all Allied attacks in progress and brought up four American divisions from reserve, and six more from the south. Two airborne divisions, one of them the 6th British, came from England. North of the salient the British XXXth Corps, of four divisions, which had just come out of the line on the river Roer, was concentrated between Liége and Louvain behind the American First and Ninth Armies. These latter threw in all their reserves to extend a defensive flank westwards from Malmedy.

By severing the front of General Bradley's Twelfth Army Group the Germans had made it impossible for him to exercise effective command from his headquarters in Luxembourg over his two armies north of the bulge. General Eisenhower therefore very wisely placed Montgomery in temporary command of all Allied troops in the north, while Bradley retained the Third U.S. Army and was charged with holding and counter-attacking the enemy from the south. Corresponding arrangements were made for the tactical air forces.

I telegraphed to Smuts.

Prime Minister to Field-Marshal Smuts   22 Dec 44

Montgomery and also we here in England have, as you are aware, pressed for several months for the emphasis of the advance to the north of the Ruhr, and have on repeated occasions urged that our strength did not enable us to undertake two major offensives such as the one against Cologne and that across the Saar. In spite of appalling weather conditions our friends however pushed on confidently, and were very much spread from north to south when the enemy began his counterstroke. I spoke to Eisenhower on the telephone during the afternoon of the 20th and suggested that he give to Montgomery the whole command north of the break-through, and to Omar Bradley everything south of the break-through, keeping control himself of the concerted operation. He replied that he had issued orders exactly on these lines in the morning. Montgomery now in fact has under his command eighteen American divisions plus his Twenty-first Army Group, comprising about sixteen divisions. He is forming substantial reserves, and is assuming entire charge of the battle in the area of his command. He should be able to intervene heavily. There is nothing so far to suggest that the Germans have the power to mount a full-scale offensive against the Twenty-first Army Group's main front.

2. Matters are not by any means so clear south of the gap. The Americans are putting up stubborn resistance, but there is a good deal of disorganisation. Naturally an army has been gathered from the Metz region to march north under Patton. The position of the enemy does not strike me as good. As usual I am optimistic; the tortoise has thrust his head out very far.

* * * * *

Three of our reinforcing divisions lined the Meuse south of Namur. Bradley concentrated a corps at Arlon and sent the American 101st Airborne Division to secure the important road junctions at Bastogne. The German armour swung north of Bastogne and sought to break their way north-westwards, leaving their infantry to capture the town. The 101st, with some armoured units, were isolated, and for a week beat off all attacks.

The wheel of the Fifth and Sixth Panzer Armies produced bitter fighting around Marche, which lasted till December 26. By then the Germans were exhausted, although at one time they were only four miles from the Meuse and had penetrated over sixty miles. Bad weather and low ground fogs had kept our air forces out of the first week of the battle, but on December 23 flying conditions got better and they intervened with tremendous effect. Heavy bombers attacked railways and centres of movement behind the enemy lines, and tactical air forces played havoc in his forward areas, starving him of reinforcements, fuel, food, and ammunition. Strategic raids on German refineries helped to deny him petrol and slacken the advance.

Baulked of their foremost objective, the Meuse, the Panzers turned savagely on Bastogne. The American 101st Division had been reinforced on December 26 by part of the 4th U.S. Armoured Division, and though vastly outnumbered held the town grimly for another week. Before the end of December the German High Command must have realised, however unwillingly, that the battle was lost, for Patton's counter-offensive from Arlon, which started on the 22nd, was steadily if slowly progressing over the snow-choked countryside towards Houffalize. The enemy made one last bid, this time in the air. On January 1 they made a violent low-level surprise attack on all our forward airfields. Our losses were heavy, though promptly replaced, but the Luftwaffe lost more than they could afford in their final massed attack of the war.

* * * * *

On January 3 Montgomery also launched his northern counter-offensive against Houffalize to join Patton's attack from the south. I visited the front at this time, and telegraphed to the President:

Prime Minister to President Roosevelt   6 Jan 45

C.I.G.S. and I have passed the last two days with Eisenhower and Montgomery, and they both feel the battle very heavy, but are confident of success. I hope you understand that, in case any troubles should arise in the Press, His Majesty's Government have complete confidence in General Eisenhower and feel acutely any attacks made on him.

2. He and Montgomery are very closely knit, and also Bradley and Patton, and it would be disaster which broke up this combination, which has in 1944 yielded us results beyond the dreams of military avarice. Montgomery said to me to-day that the break-through would have been most serious to the whole front but for the solidarity of the Anglo-American Army.

3. Although I regret our divisions only amount to seventeen and two-thirds, all units are absolutely up to strength, and we have seven or eight thousand reinforcements all ready in addition in France awaiting transfer to their units. The measures we have taken to bring another 250,000 into or nearer the front line enable me to say with confidence that at least our present strength will be maintained throughout the impending severe campaign.

4. I am deeply impressed with the need of sustaining the Foot, who bear two-thirds of the losses but are very often the last to receive

reinforcements. More important even than the sending over of large new units is the keeping up of the infantry strength of divisions already engaged. We are therefore preparing a number of infantry brigades, including several from the Marines, of whom the Navy has 80,000. These brigades will liberate mobile divisions from quasi-static sectors, and at the same time do the particular work which is needed in them. Montgomery welcomed this idea most cordially as regards the Twenty-first Army Group. I gathered from General Eisenhower that he takes the same view, and that he is longing for more infantry drafts— *i.e.*, rifle and bayonet—to maintain the United States divisions at their proper establishment.

5. I most cordially congratulate you on the extraordinary gallantry which your troops have shown in all this battle, particularly at Bastogne and two other places which Montgomery mentioned to me on his own front, one at the peak of the salient, where the 1st and 9th American Divisions fought on and won after extremely heavy losses, and the other in connection with the 7th United States Armoured Division, which seems to have performed the highest acts of soldierly devotion. Also many troops of the First Army have fought to the end, holding cross-roads in the area of incursion, which averted serious perils to the whole armies of the north at heavy personal sacrifice.

6. As I see there have been criticisms in the American papers of our troops having been kept out of the battle, I take this occasion to assure you that they stand absolutely ready at all times to obey General Eisenhower's commands. I believe that the dispositions which he and Field-Marshal Montgomery under him have made are entirely in accordance with strict military requirements, both as regards the employment of troops in counter-attack and their lateral movement, having regard to criss-cross communications. I have not found a trace of discord at the British and American headquarters; but, Mr. President, there is this brute fact: we need more fighting troops to make things move.

7. I have a feeling this is a time for an intense new impulse, both of friendship and exertion, to be drawn from our bosoms and to the last scrap of our resources. Do not hesitate to tell me of anything you think we can do.

* * * * *

At this time Eisenhower and his staff were of course acutely anxious to know whether the Russians could do anything from their side to take off some of the pressure against us in the West. All efforts through the liaison officers in Moscow had failed to obtain any reply from their opposite numbers. In order to put the case to the Soviet Chiefs of Staff in the most effective manner

Eisenhower had sent his Deputy, Air Marshal Tedder, with a special mission. They were considerably delayed by the weather. As soon as I heard of this I said to Eisenhower, "You may find many delays on the staff level, but I expect Stalin would tell me if I asked him. Shall I try?" He asked me to do so, and I therefore sent the following message:

Prime Minister to Marshal Stalin   6 Jan 45
The battle in the West is very heavy, and at any time large decisions may be called for from the Supreme Command. You know yourself from your own experience how very anxious the position is when a very broad front has to be defended after the temporary loss of the initiative. It is Eisenhower's great desire and need to know in outline what you plan to do, as this obviously affects all his and our major decisions. Our envoy, Air Chief Marshal Tedder, was last night reported weather-bound in Cairo. His journey has been much delayed through no fault of yours. In case he has not reached you yet, I shall be grateful if you can tell me whether we can count on a major Russian offensive on the Vistula front, or elsewhere, during January, with any other points you may care to mention. I shall not pass this most secret information to anyone except Field-Marshal Brooke and General Eisenhower, and only under conditions of the utmost secrecy. I regard the matter as urgent.

When one considers how serious was the decision asked for and how many people were involved, it is remarkable that the answer should have been sent me the very next day.

Marshal Stalin to Prime Minister   7 Jan 45
I received your message of January 6, 1945, on the evening of January 7.
Unfortunately Air Marshal Tedder has not yet arrived in Moscow.
It is most important that we should be able to take advantage of our supremacy over the Germans in artillery and in the air. This demands clear flying weather and an absence of low mists, which hinder aimed artillery fire. We are preparing an offensive, but the weather is at present unfavourable. Nevertheless, taking into account the position of our Allies on the Western Front, G.H.Q. of the Supreme Command has decided to accelerate the completion of our preparation, and, regardless of the weather, to commence large-scale offensive operations against the Germans along the whole Central Front not later than the second half of January. You may rest assured that we shall do everything possible to render assistance to the glorious forces of our Allies.

Prime Minister to Marshal Stalin 9 Jan 45

I am most grateful to you for your thrilling message. I have sent it over to General Eisenhower for his eyes only. May all good fortune rest upon your noble venture.

2. The battle in the West goes not too badly. There is a good chance of the Huns being crushed out of their salient with very heavy losses. It is preponderantly an American battle, and their troops have fought splendidly, with heavy losses. We are both shoving everything in we can. The news you give me will be a great encouragement to General Eisenhower, because it gives him the assurance that the German reinforcements will have to be split between both our flaming fronts. The battle in the West will be continuous, according to the generals responsible for fighting it.

I quote this interchange as a good example of the speed at which business could be done at the summit of the Alliance, and also because it was a fine deed of the Russians and their chief to hasten their vast offensive, no doubt at a heavy cost in life. Eisenhower was very pleased indeed at the news I was able to send him. He asked however for any reinforcements that could be sent. Nearly three weeks beforehand the country had been told that another 250,000 men would be found to nourish and sustain troops in contact with the enemy, and that for the first time in our long struggle the British Government proposed to use its powers to compel the women of our fighting services to serve abroad. Not much compulsion was needed. The keenest zeal prevailed. But these drastic measures took time to mature, and although we could make good our autumn losses in the field and keep up a full supply of material, we had little left in hand. The Americans for their part, in addition to 60,000 infantry reinforcements, prepared to send nine fresh divisions from the United States.

* * * * *

From the north two American corps, with the XXXth British on their western flank, pressed down upon the enemy. On January 7 they crossed the Laroche-Vielsalm road, an important escape route for the Germans. Struggling through snowstorms, the two wings of the Allied attack slowly drew closer, until they met at Houffalize on January 16. The Germans were forced steadily eastwards and harassed continually from the air, until by the end of the month they were back behind their frontier, with nothing to show for their supreme effort except ruinous losses of material and casualties amounting to a hundred and twenty thousand men.

One awkward situation during the battle must be recorded, although happily it did not affect the issue. In order to release divisions from the Third Army Eisenhower had ordered Devers's Sixth Army Group to take over part of Patton's front, and authorised, if necessary, a withdrawal from the Rhine to the Vosges. This meant leaving Strasbourg open to the enemy. There was understandable consternation in French political and military circles. What vengeance would fall upon the citizens of Strasbourg, who had rallied so passionately to their deliverers! I chanced to be at Eisenhower's headquarters at St. Germains at this juncture, and he and Bedell Smith listened attentively to my appeal. The enemy did indeed spring into action on the Army Group's front, especially in the Colmar pocket, but were repulsed. Eisenhower cancelled his instructions, and the military necessity which might have made the evacuation of Strasbourg imperative never arose. De Gaulle expressed his gratitude.

This was the enemy's final offensive of the war. At the time it caused us no little anxiety. Our own advance had to be postponed, but we benefited in the end. The Germans could not replace their losses, and our subsequent battles on the Rhine, though severe, were undoubtedly eased. The German High Command, and even Hitler, must have been disillusioned. Taken by surprise, Eisenhower and his commanders acted swiftly, but they will agree that the major credit lies elsewhere. In Montgomery's words, "The Battle of the Ardennes was won primarily by the staunch fighting qualities of the American soldier." *

For my own part, I will quote from a speech I made to the House of Commons on January 18:

"I have seen it suggested that the terrific battle which has been proceeding since December 16 on the American front is an Anglo-American battle. In fact however the United States troops have done almost all the fighting, and have suffered almost all the losses. . . . I never hesitate . . . to stand up for our own soldiers when their achievements have been cold-shouldered or neglected or overshadowed, as they sometimes are; but we must not forget that it is to American homes that the telegrams of personal losses and anxiety have been going during the past month. . . . According

---

* *Normandy to the Baltic*, Field-Marshal Montgomery, p. 181.

to the professional advice which I have at my disposal, what was done to meet von Rundstedt's counter-stroke was resolute, wise, and militarily correct. A gap was torn open, as a gap can always be in a line hundreds of miles long. General Eisenhower at once gave the command to the north of the gap to Field-Marshal Montgomery, and to the south of it to General Omar Bradley. . . . In the result both these highly skilled commanders handled the very large forces at their disposal in a manner which I think I may say without exaggeration may become the model for military students in the future. . . ."

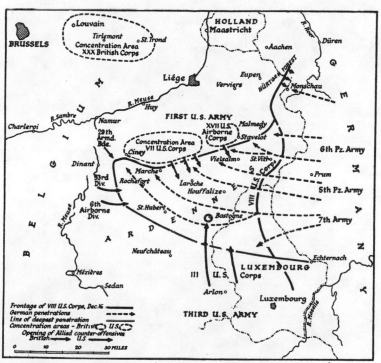

RUNDSTEDT'S COUNTER-OFFENSIVE

CHAPTER XVIII

BRITISH INTERVENTION IN GREECE

Operation "Manna" – German Delay in Quitting Athens – The Caserta Agreement – The Liberation of Athens, October 14 – Greece in Ruins – Mr. Eden's Visit to Athens – Disorder Grows and Spreads – General Scobie is Directed to Make Counter-Preparations Against E.A.M. – Demobilising the Guerrillas – Resignation of the E.A.M. Ministers – Civil War Begins, December 3 – I Order General Scobie to Put Down the Rebellion – Hard Fighting in Athens, and Hard Attacks at Home and in the States – My Speech in the House of Commons, December 8 – American Feeling – Proposals to Set Up a Regency in Greece – Mr. Harold Macmillan and Field-Marshal Alexander Arrive in Athens – Alexander is Given a Free Hand – An Astonishing Leakage – Loyalty of the British Trades Union Congress – A Telegram from the President, December 13 – And from Field-Marshal Smuts, December 14 – My Reply to Mr. Roosevelt, December 17 – Telegrams to the Prime Minister of Canada – Our Policy Vindicated by the English-speaking World.

BEFORE leaving Italy at the end of August I had asked the Chief of the Imperial General Staff to work out the details of a British expedition to Greece in case the Germans there collapsed.* We gave it the code-name "Manna". Its planning was complicated by our strained resources and the uncertainty of Germany's strategic position in the Balkans, but I directed that our forces should be ready to act by September 11, and that the Greek Prime Minister and representatives of the Greek Government in Italy should be prepared to enter Athens without delay. By the end of the first week of September they were installed in a villa near Caserta. Here Papandreou set to work with his new E.A.M. colleagues.† It was essential that there should be no political

* See Chapter VII.
† E.A.M., the Greek "National Liberation Front".

vacuum in Greece. As I minuted on August 29, "It is most desirable to strike out of the blue without any preliminary crisis. It is the best way to forestall the E.A.M." The essence of the plan was to occupy Athens and its airfield with a parachute brigade, bring in four squadrons of fighter aircraft, clear Piræus harbour for further reinforcements from Egypt, and ensure the early arrival of the Greek Ministers. We would then hasten in relief supplies and bring over the Greek Brigade from Italy.

German delay in quitting Athens forced us to modify our project. The garrison of ten thousand men showed no sign of moving, and on September 13 I telegraphed to General Wilson instructing him to prepare a preliminary descent in the Peloponnese, where the Germans were withdrawing northwards to the Corinth area. As from midnight September 13–14 the troops for "Manna" were placed at forty-eight hours' notice. They were commanded by General Scobie, and consisted initially of the 2nd Parachute Brigade from Italy, the 23rd Armoured Brigade acting as infantry, administrative troops from Egypt, and whatever Greek forces were at the disposal of their recognised Government. The 15th Cruiser Squadron, with minesweeping flotillas, and four British and three Greek air squadrons, together with United States transport aircraft, were to sustain this expedition.

The tardy German withdrawal from Athens enabled us however to consolidate the direction of Greek affairs on the eve of the decisive stroke. I was glad that the Greek Government was now at hand in Italy. At the end of September General Wilson summoned Saraphis, the E.L.A.S. general,* and his Nationalist rival Zervas to meet Papandreou at Caserta. Mr. Macmillan, as Minister Resident in the Mediterranean, together with Mr. Leeper, our Ambassador to the Greek Government, were present to advise and direct the political side of this important conference, which had to create a unified command of all Greek forces available in Italy and inside Greece, together with the British forces now poised for the landing.

A comprehensive agreement was signed on September 26. It laid down that all guerrilla forces in the country should place themselves under the orders of the Greek Government, who in their turn put them under the command of General Scobie. The

---

* E.L.A.S., the Greek "People's National Army of Liberation". (Both E.A.M. and E.L.A.S. were Communist-controlled.)

Greek guerrilla leaders declared that none of their men would take the law into their own hands. Any action in Athens would be taken only on the direct orders of the British commander. This document, known as the Caserta Agreement, governed our future action.

The liberation did not begin until October. Commando units were then sent into Southern Greece, and in the early hours of October 4 our troops occupied Patras. This was our first foothold since the tragic exit of 1941. The troops then worked their way along the southern shores of the Gulf of Corinth. On October 12 General Wilson learnt that the Germans were evacuating Athens, and next day British parachutists landed on the Megara airfield, about eight miles west of the capital. On the 14th the rest of the paratroopers arrived, and occupied the city on the heels of the German withdrawal. Our naval forces entered the Piræus, bringing with them General Scobie and the main part of his force, and two days later the Greek Government arrived, together with our Ambassador.

* * * * *

The testing time for our arrangements had now come. At the Moscow conference I had obtained Russian abstention at a heavy price. We were pledged to support Papandreou's Provisional Administration, in which E.A.M. was fully represented. All parties were bound by the Caserta Agreement, and we wished to hand over authority to a stable Greek Government without loss of time. But Greece was in ruins. The Germans destroyed roads and railways as they withdrew northwards. Our Air Force harassed them as they went, but on land we could do little to interfere. E.L.A.S. armed bands filled the gap left by the departing invaders, and their central command made little effort to enforce the solemn promises which had been given. Everywhere was want and dissension. Finances were disordered and food exhausted. Our own military resources were stretched to the limit.

At the end of the month Mr. Eden visited Athens on his way home from Moscow, and received a tumultuous welcome in memory of his efforts for Greece in 1941. With him were Lord Moyne, the Minister Resident in Cairo, and Mr. Macmillan. The whole question of relief was discussed and everything humanly possible was done. Our troops willingly went on half-rations to increase the food supplies, and British sappers started to build emergency communications. By November 1 the Germans had evacuated Salonika and Florina, and ten days later the last of their forces had crossed the northern frontier. Apart from a few isolated island garrisons, Greece was free.

But the Government in Athens had not enough troops to control the country and compel E.L.A.S. to observe the Caserta Agreement. Disorder grew and spread. On November 7 I minuted to the Foreign Secretary:

Prime Minister to Foreign Secretary   7 Nov 44

In my opinion, having paid the price we have to Russia for freedom of action in Greece, we should not hesitate to use British troops to support the Royal Hellenic Government under M. Papandreou.

2. This implies that British troops should certainly intervene to check acts of lawlessness. Surely M. Papandreou can close down E.A.M. newspapers if they call a newspaper strike.

3. I hope the Greek Brigade will soon arrive, and will not hesitate to shoot when necessary. Why is only one Indian brigade of the Indian Division to be sent in? We need another eight or ten thousand foot-soldiers to hold the capital and Salonika for the present Government. Later on we must consider extending the Greek authority. I fully expect a clash with E.A.M., and we must not shrink from it, provided the ground is well chosen.

And the following day:

Prime Minister to General Wilson (Italy) and   8 Nov 44
Mr. Leeper (Athens)

In view of increasing threat of Communist elements in Greece and indications that they plan to seize power by force, I hope that you will consider reinforcing our troops in Athens area by immediate dispatch of the 3rd Brigade of 4th Indian Division or some other formation. . . .

*   *   *   *   *

A revolt by E.A.M. was imminent, and on November 15 General Scobie was directed to make counter-preparations. Athens was to be declared a military area, and authority was given to order all E.L.A.S. troops to leave it. The 4th Indian Division was sent from Italy to Salonika, Athens, and Patras. The Greek Brigade also came from Italy, and became the centre of controversy between Papandreou and his E.A.M. colleagues. The only chance of averting civil war was to disarm the guerrillas and other forces by mutual agreement and establish a new National Army and police force under the direct control of the Government in Athens. Arrangements were made to raise and equip National Guard battalions, each 500 strong. Ultimately there were thirty of these; they proved very useful in rounding up armed hostile civilians and guarding areas cleared by our troops.

A draft decree for the demobilisation of the guerrillas, drawn up at M. Papandreou's request by the E.A.M. Ministers themselves, was presented to the distracted Cabinet. The regular Greek Mountain Brigade and the Sacred Squadron were to remain. E.L.A.S. were to keep a brigade of their own, and E.D.E.S.* were to be given a small force. But at the last moment the E.A.M. Ministers went back on their own proposals, on which they had wasted a precious week, and demanded that the Mountain Brigade should be disbanded. The Communist tactic was now in full swing. On December 1 the six Ministers associated with E.A.M. resigned, and a general strike in Athens was proclaimed for the following day. The rest of the Cabinet passed a decree dissolving the guerrillas, and the Communist Party moved its headquarters from the capital. General Scobie issued a message to the people of Greece stating that he stood firm behind the present constitutional Government "until the Greek State can be established with a legally armed force and free elections can be held". I issued a similar personal statement from London.

On Sunday, December 3, Communist supporters, engaging in a banned demonstration, collided with the police and civil war began. The next day General Scobie ordered E.L.A.S. to evacuate Athens and the Piræus forthwith. Instead their troops and armed civilians tried to seize the capital by force.

At this moment I took a more direct control of the affair. On learning that the Communists had already captured almost all the police stations in Athens, murdering the bulk of their occupants not already pledged to their attack, and were within half a mile of the Government offices, I ordered General Scobie and his 5,000 British troops, who ten days before had been received with rapture as deliverers by the population, to intervene and fire upon the treacherous aggressors. It is no use doing things like this by halves. The mob violence by which the Communists sought to conquer the city and present themselves to the world as the Government demanded by the Greek people could only be met by firearms. There was no time for the Cabinet to be called.

* E.D.E.S., the Greek "National Democratic Army".

Anthony and I were together till about two o'clock, and were entirely agreed that we must open fire. Seeing how tired he was, I said to him, "If you like to go to bed, leave it to me." He did, and at about 3 a.m. I drafted the following telegram:

Prime Minister to General Scobie (Athens). Repeated to   5 Dec 44
General Wilson (Italy)★

I have given instructions to General Wilson to make sure that all forces are left with you and all possible reinforcements are sent to you.

2. You are responsible for maintaining order in Athens and for neutralising or destroying all E.A.M.-E.L.A.S. bands approaching the city. You may make any regulations you like for the strict control of the streets or for the rounding up of any number of truculent persons. Naturally E.L.A.S. will try to put women and children in the van where shooting may occur. You must be clever about this and avoid mistakes. But do not hesitate to fire at any armed male in Athens who assails the British authority or Greek authority with which we are working. It would be well of course if your command were reinforced by the authority of some Greek Government, and Papandreou is being told by Leeper to stop and help. *Do not however hesitate to act as if you were in a conquered city where a local rebellion is in progress.*†

3. With regard to E.L.A.S. bands approaching from the outside, you should surely be able with your armour to give some of these a lesson which will make others unlikely to try. You may count upon my support in all reasonable and sensible action taken on this basis. *We have to hold and dominate Athens. It would be a great thing for you to succeed in this without bloodshed if possible, but also with bloodshed if necessary.*

This telegram was dispatched at 4.50 a.m. on the 5th. I must admit that it was somewhat strident in tone. I felt it so necessary to give a strong lead to the military commander that I intentionally worded it in the sharpest terms. The fact that he had such an order in his possession would not only encourage him to decisive action, but gave him the certain assurance that I should be with him in any well-conceived action he might take, whatever the consequences might be. I felt grave concern about the whole business, but I was sure that there should be no room for doubts or hedging. I had in my mind Arthur Balfour's celebrated telegram in the eighties to the British authorities in Ireland: "Don't hesitate to shoot." This was sent through the open

---

★ The Command had not yet changed hands.
† Author's subsequent italics throughout.

telegraph offices. There was a furious storm about it in the House of Commons of those days, but it certainly prevented loss of life. It was one of the key stepping-stones by which Balfour advanced to power and control. The setting of the scene was now entirely different. Nevertheless "Don't hesitate to shoot" hung in my mind as a prompter from those far-off days.

Later that day I telegraphed to our Ambassador:

Prime Minister to Mr. Leeper (Athens) 5 Dec 44

This is no time to dabble in Greek politics or to imagine that Greek politicians of varying shades can affect the situation. You should not worry about Greek Government compositions. The matter is one of life and death.

2. You must urge Papandreou to stand to his duty, and assure him he will be supported by all our forces if he does so. The day has long gone past when any particular group of Greek politicians can influence this mob rising. His only chance is to come through with us.

3. I have put the whole question of the defence of Athens and the maintenance of law and order in the hands of General Scobie, and have assured him that he will be supported in the use of whatever force is necessary. Henceforward you and Papandreou will conform to his directions in all matters affecting public order and security. You should both support Scobie in every possible way, and you should suggest to him any means which occur to you of making his action more vigorous and decisive.

Every good wish.

*　　*　　*　　*　　*

E.L.A.S. had quickly gained control of most of Athens, except only its very centre, where our troops first held them and then began to counter-attack. Scobie reported:

8 Dec 44

Increased activities on the part of the rebels and widespread sniping limited progress during the fighting, which continued throughout yesterday. By midday the total of rebel prisoners under military guard was 35 officers, 524 other ranks. These figures do not include those held by the police, as it is difficult to obtain accurate figures for them.

Some progress was made by the 23rd Brigade in house-to-house clearing throughout the afternoon. A further sector in the centre of the city was cleared by the Parachute Brigade.

Marine reinforcements had to be landed from H.M.S. *Orion* to deal with serious sniping of Navy House, Piræus, by rebels who infiltrated into the area south of Port Leontos. In face of strong opposition our troops were forced to withdraw in one area.

In the area being cleared by the Greek Mountain Brigade an attack was made by the rebels from the flank. The attack was held, but delayed progress of the brigade.

This showed the scale of the fighting on which we had now embarked.

Prime Minister to General Wilson (Italy) 9 Dec 44

You should send further reinforcements to Athens without the slightest delay. The prolongation of the fight has many dangers. I warned you of the paramount political importance of this conflict. At least two more brigades should hurry to the scene.

2. In addition to the above, why does not the Navy help all the time instead of only landing a small number in a crisis? You guaranteed most strongly that you had already sent enough soldiers.

Prime Minister to General Scobie 8 Dec 44

There is much talk in the Press to-night of a peace offer by E.L.A.S. Naturally we should be glad to have this matter settled, but you should make quite sure, so far as your influence goes, that we do not give away for the sake of kindness what has been won or can still be won by our troops. It would seem to me that anything less satisfactory than the terms agreed upon before the revolt took place should not be accepted. Also it is difficult to see how E.A.M. leaders, with their hands wet with Greek and British blood, should resume their places in the Cabinet. This might however be got over. The great thing is to proceed with caution and to consult us upon the terms when they are made. The clear objective is the defeat of E.A.M. The ending of the fighting is subsidiary to this. I am ordering large reinforcements to come to Athens, and Field-Marshal Alexander will probably be with you in a few days. Firmness and sobriety are what are needed now, and not eager embraces, while the real quarrel is unsettled.

Keep us informed before any compromise is settled in which you or Leeper are concerned.

Rumours were spread by Communists and their like in London that British troops were in sympathy with E.A.M. There was no truth in them.

On the peace offer the answer was:

General Scobie to Prime Minister 10 Dec 44

We would at once inform you should any peace offer be made by E.L.A.S., but neither the Ambassador nor I know of any such approach.

I have clearly before me the main objective you mention. While any one party is able to back its views with a private army Greece can never achieve peace and stability. Fighting may, I hope, be restricted to Athens-Piræus, but I am ready to see it through in the rest of the country if necessary. It is a pity that tear gas may not be used. It would be of great help in this city fighting.

Your assurance that large reinforcements are being sent is most welcome. I have been informed by Allied Force Headquarters that the 4th Division is being dispatched early.

*   *   *   *   *

Now that the free world has learnt so much more than was then understood about the Communist movement in Greece and elsewhere, many readers will be astonished at the vehement attacks to which His Majesty's Government, and I in particular at its head, were subjected. The vast majority of the American Press violently condemned our action, which they declared falsified the cause for which they had gone to war. If the editors of all these well-meaning organs will look back at what they wrote then and compare it with what they think now they will, I am sure, be surprised. The State Department, in the charge of Mr. Stettinius, issued a markedly critical pronouncement, which they in their turn were to regret, or at least reverse, in after-years. In England there was much perturbation. The *Times* and the *Manchester Guardian* pronounced their censures upon what they considered our reactionary policy. Stalin however adhered strictly and faithfully to our agreement of October, and during all the long weeks of fighting the Communists in the streets of Athens not one word of reproach came from *Pravda* or *Isvestia*.

In the House of Commons there was a great stir. I accepted willingly the challenge flung at us in an amendment moved by Sir Richard Acland, the leader and sole member in Parliament of the Commonwealth Party, supported by Mr. Shinwell and Mr. Aneurin Bevan. There was a strong current of vague opinion, and even passion, of which these and other similar figures felt themselves the exponents. Here again any Government which had rested on a less solid foundation than the National Coalition might well have been shaken to pieces. But the War Cabinet stood like a rock against which all the waves and winds might beat in vain.

When we recall what has happened to Poland, to Hungary, and Czechoslovakia in these later years we may be grateful to Fortune for giving us at this critical moment the calm, united strength of determined leaders of all parties. Space does not allow me to quote more than a few extracts from the speech I made on December 8 against the amendment to the Vote of Confidence which we had demanded.

Let me present to the House the charge which is made against us. It is that we are using His Majesty's forces to disarm the friends of democracy in Greece and in other parts of Europe and to suppress those popular movements which have valorously assisted in the defeat of the enemy. Here is a pretty direct issue, and one on which the House will have to pronounce before we separate this evening. Certainly His Majesty's Government would be unworthy of confidence if His Majesty's forces were being used by them to disarm the friends of democracy.

The question however arises, and one may be permitted to dwell on it for a moment, who are the friends of democracy, and also how is the word "democracy" to be interpreted? My idea of it is that the plain, humble, common man, just the ordinary man who keeps a wife and family, who goes off to fight for his country when it is in trouble, goes to the poll at the appropriate time, and puts his cross on the ballot-paper showing the candidate he wishes to be elected to Parliament—that he is the foundation of democracy. And it is also essential to this foundation that this man or woman should do this without fear, and without any form of intimidation or victimisation. He marks his ballot-paper in strict secrecy, and then elected representatives meet and together decide what Government, or even, in times of stress, what form of government, they wish to have in their country. If that is democracy I salute it. I espouse it. I would work for it. . . . I stand upon the foundation of free elections based on universal suffrage, and that is what we consider the foundation for democracy. But I feel quite differently about a swindle democracy, a democracy which calls itself democracy because it is Left Wing. It takes all sorts to make democracy, not only Left Wing, or even Communist. I do not allow a party or a body to call themselves democrats because they are stretching farther and farther into the most extreme forms of revolution. I do not accept a party as necessarily representing democracy because it becomes more violent as it becomes less numerous.

One must have some respect for democracy and not use the word too lightly. The last thing which resembles democracy is mob law, with bands of gangsters, armed with deadly weapons, forcing their way into great cities, seizing the police stations and key points of government, endeavouring to introduce a totalitarian régime with an iron hand, and clamouring, as they can nowadays if they get the power—— [*Interruption.*]

I am sorry to be causing so much distress. I have plenty of time, and if any outcries are wrung from hon. Members opposite I can always take a little longer over what I have to say, though I should regret to do so. I say that the last thing that represents democracy is mob law and the attempt to introduce a totalitarian régime which clamours to shoot everyone who is politically inconvenient as part of a purge of those who are said to have collaborated with the Germans during the occupation. Do not let us rate democracy so low, do not let us rate democracy as if it were merely grabbing power and shooting those who do not agree with you. That is the antithesis of democracy.

Democracy is not based on violence or terrorism, but on reason, on fair play, on freedom, on respecting the rights of other people. Democracy is no harlot to be picked up in the street by a man with a tommy gun. I trust the people, the mass of the people, in almost any country, but I like to make sure that it is the people and not a gang of bandits who think that by violence they can overturn constituted authority, in some cases ancient Parliaments, Governments, and States. . . .

We march along an onerous and painful path. Poor old England! (Perhaps I ought to say "Poor old Britain!") We have to assume the burden of the most thankless tasks, and in undertaking them to be scoffed at, criticised, and opposed from every quarter; but at least we know where we are making for, know the end of the road, know what is our objective. It is that these countries shall be freed from the German armed power, and under conditions of normal tranquillity shall have a free universal vote to decide the Government of their country—except a Fascist régime—and whether that Government shall be of the Left or of the Right.

There is our aim—and we are told that we seek to disarm the friends of democracy. We are told that because we do not allow gangs of heavily armed guerrillas to descend from the mountains and install themselves, with all the bloody terror and vigour of which they are capable, in power in great capitals, we are traitors to democracy. I repulse that claim too. I shall call upon the House as a matter of confidence in His Majesty's Government, and of confidence in the spirit with which we have marched from one peril to another till victory is in sight, to reject such pretensions with the scorn that they deserve. . . .

If I am blamed for this action I will gladly accept my dismissal at the hands of the House; but if I am not so dismissed—make no mistake about it—we shall persist in this policy of clearing Athens and the Athens region of all who are rebels against the authority of the constitutional Government of Greece—of mutineers against the orders of the Supreme Commander in the Mediterranean under whom all the guerrillas have undertaken to serve. I hope I have made the position clear, both generally as it affects the world and the war and as it affects the Government.

Only thirty members faced us in the division lobby. Nearly three hundred voted confidence. Here again was a moment in which the House of Commons showed its enduring strength and authority.

I telegraphed the following day:

Prime Minister to Mr. Leeper (Athens) 9 Dec 44
Do not be at all disquieted by criticisms made from various quarters in the House of Commons. No one knows better than I the difficulties you have had to contend with. I do not yield to passing clamour, and will always stand with those who execute their instructions with courage and precision. In Athens as everywhere else our maxim is "No peace without victory".

*　*　*　*　*

There is no doubt that the emotional expression of American opinion and the train of thought at that time being followed by the State Department affected President Roosevelt and his immediate circle. The sentiments I had expressed in the House of Commons have now become commonplaces of American doctrine and policy and command the assent of the United Nations. But in those days they had an air of novelty which was startling to those who were governed by impressions of the past and did not feel the onset of the new adverse tide in human affairs. In the main the President was with me, and Hopkins sent me a friendly message about the speech.

Earl of Halifax to Prime Minister 8 Dec 44
Harry and Jim Forrestal have just telephoned enthusiastic approval of your speech on Greece, which they both think will have done immense good. I am sure they are right.

Prime Minister to Mr. Harry Hopkins 9 Dec 44
I am very glad you were pleased by my speech. I was much upset by the last sentence of the Stettinius Press release,* which seemed to

---

* This was dated December 5, and ran as follows:
"The Department of State has received a number of inquiries from correspondents in regard to the position of this Government concerning the recent Cabinet crisis in Italy.
"The position of this Government has been consistently that the composition of the Italian Government is purely an Italian affair, except in the case of appointments where important military factors are concerned. This Government has not in any way intimated to the Italian Government that there would be any opposition on its part to Count

reflect on the whole of our foreign policy in Belgium, where we acted under your orders, and in Greece, where our action was fully approved at Quebec. Naturally the prolongation and severity of the fighting in Athens with E.L.A.S. causes me anxiety.

Every good wish.

And again the same day:

I hope you will tell our great friend that the establishment of law and order in and around Athens is essential to all future measures of magnanimity and consolation towards Greece. After this has been established will be the time for talking. My guiding principle is "No peace without victory". It is a great disappointment to me to have been set upon in this way by E.L.A.S. when we came loaded with good gifts and anxious only to form a united Greece which could establish its own destiny. But we have been set upon, and we intend to defend ourselves. I consider we have a right to the President's support in the policy we are following. If it can be said in the streets of Athens that the United States are against us, then more British blood will be shed and much more Greek. It grieves me very much to see signs of our drifting apart at a time when unity becomes ever more important, as danger recedes and faction arises.

2. For you personally. Do not be misled by our majority yesterday. I could have had another eighty by sending out a three-line whip instead of only two. On Fridays, with the bad communications prevailing here, Members long to get away for the week-end. Who would not?

Every good wish.

British troops were still fighting hard in the centre of Athens, hemmed in and outnumbered. We were engaged in house-to-house combat with an enemy at least four-fifths of whom were in plain clothes. Unlike many of the Allied newspaper correspondents in Athens, our troops had no difficulty in understanding the issues involved.

Papandreou and his remaining Ministers had lost all authority. Previous proposals to set up a Regency under the Archbishop Damaskinos had been rejected by the King, but on December 10 Mr. Leeper revived the idea. King George however was against it, and we were reluctant at the time to press him.

Sforza. Since Italy is an area of combined responsibility, we have reaffirmed to both the British and Italian Governments that we expect the Italians to work out their problems of government along democratic lines without influence from outside. *This policy would apply to an even more pronounced degree with regard to Governments of the United Nations in their liberated territories.*"

Amid these tumults Field-Marshal Alexander and Mr. Macmillan arrived in Athens. We received the first reports of their mission on December 11. Our plight was worse than we had expected. Alexander telegraphed, "The British forces are in fact beleaguered in the heart of the city." The road to the airfield was not secure. We were not in control of Piræus harbour, so no ships could be unloaded there. Only six days' rations and three days' reserve of ammunition were left for the troops fighting in the city. Alexander proposed to clear the port and the road to Athens at once, bring in immediate reinforcements from Italy, and build up supply dumps, and, "having linked up securely both ends of the dumb-bell, to undertake the necessary operations to clear the whole of Athens and Piræus". He also pressed Leeper's proposal to appoint the Archbishop as Regent, and asked for stern measures against the rebels and permission to bomb areas inside Athens.

On December 12 the War Cabinet gave Alexander a free hand in all military measures. The 4th British Division, on passage from Italy to Egypt, was diverted, and their arrival during the latter half of the month turned the scale. I told Alexander that the Greek King would not agree to the plan for a Regency. The suggestion that the Archbishop should be called upon to form a Government satisfied no one. The political reaction at home to these events showed a clearer and calmer view.

★　　★　　★　　★　　★

At this moment an astonishing leakage of official secrets occurred. The reader will remember my telegram to General Scobie dispatched at 4.50 a.m. on December 5. This had been marked "Personal and Top Secret. From Prime Minister to General Scobie. Repeated to General Wilson," and of course was in cipher. A few days later an American columnist was able to publish practically an exact copy of it. All our communications were menaced thereby.

I learned on inquiry that all messages sent through General Wilson's Supreme Headquarters in Italy were communicated to several personages, including the American Ambassador in Rome, unless they bore a special restrictive marking. On reading the text of my message sent before dawn on the 5th to General Scobie the Ambassador repeated its substance to the State Department. He was fully within his rights in doing this. What happened after his paraphrase reached the State Department has never been discovered, or at any rate made known, but on the 11th the American journalist made public what might well have been, at that time, an awkward bombshell. It happened that the next day the Trades Union Congress was to meet in London. There was naturally much anxiety about our policy in Greece and Left Wing forces were astir. It seemed probable that the publication of the drastic terms of my message to General Scobie would produce a bad impression. However, the matter was not mentioned at the Trades Union Congress, nor indeed did it attract any attention in Parliament. Mr. Bevin represented the War Cabinet at the Congress, and with characteristic loyalty and courage he defended and vindicated our policy in Greece. He carried the whole conference with him, and by an overwhelming majority the trade unions gave their support to the Government and proved once again their stable and responsible qualities in great matters.

* * * * *

I had meanwhile received a most kindly worded telegram from the President.

President Roosevelt to Prime Minister   13 Dec 44

I have been as deeply concerned as you have yourself in regard to the tragic difficulties you have encountered in Greece. I appreciate to the full the anxious and difficult alternatives with which you have been faced. I regard my rôle in this matter as that of a loyal friend and ally whose one desire is to be of any help possible in the circumstances. You may be sure that in putting my thoughts before you I am constantly guided by the fact that nothing can in any way shake the unity and association between our two countries in the great tasks to which we have set our hands.

As anxious as I am to be of the greatest help to you in this trying situation, there are limitations, imposed in part by the traditional policies of the United States and in part by the mounting adverse reaction of public opinion in this country. No one will understand better than yourself that I, both personally and as Head of State, am necessarily responsive to the state of public feeling. It is for these reasons that it has not been possible for this Government to take a stand along with you in the present course of events in Greece. Even an attempt to do so would bring only temporary value to you, and would in the long run do injury to our basic relationships. I don't need to tell you how much I dislike this state of affairs as between you and me. My one hope is to see it rectified so that we can go along, in this as in everything, shoulder to shoulder. I know that you, as the one on whom the responsibility rests, desire with all your heart a satisfactory solution of the Greek problem, and particularly one that will bring peace to that ravished country. I will be with you wholeheartedly in any solution which takes into consideration the factors I have mentioned above. With this in mind I am giving you at random some thoughts that have come to me in my anxious desire to be of help.

I know that you have sent Macmillan there with broad powers to find such a solution, and it may be that he will have been successful before you get this. I of course lack full details and am at a great distance from the scene, but it has seemed to me that a basic reason—or excuse perhaps—for the E.A.M. attitude has been distrust regarding the intentions of King George II. I wonder if Macmillan's efforts might not be greatly facilitated if the King himself would approve the establishment of a Regency in Greece and would make a public declaration of his intention not to return unless called for by popular plebiscite. This might be particularly effective if accompanied by an assurance that elections will be held at some fixed date, no matter how far in the future, when the people would have full opportunity to express themselves.

Meanwhile might it not be possible to secure general agreement on the disarmament and dissolution of all the armed groups now in the country, including the Mountain Brigade and the Sacred Squadron, leaving your troops to preserve law and order alone until the Greek national forces can be reconstituted on a non-partisan basis and be adequately equipped?

I shall be turning over in my mind this whole question, and hope you will share your thoughts and worries with me.

This however did not give me any practical help. I replied:

14 Dec 44

I will send you over the week-end a considered answer to your telegram, for the kindly tone of which I thank you. I hope that the British reinforcements now coming steadily into Attica may make a more healthy situation in Athens. You will realise how very serious it would be if we withdrew, as we easily could, and the result was a frightful massacre, and an extreme Left Wing régime under Communist inspiration installed itself, as it would, in Athens. My Cabinet colleagues here of all parties are not prepared to act in a manner so dishonourable to our record and name. Ernest Bevin's speech to the Labour conference won universal respect. Stern fighting lies ahead, and even danger to our troops in the centre of Athens. The fact that you are supposed to be against us, in accordance with the last sentence of Stettinius's Press release, has added, as I feared, to our difficulties and burdens. I think it probable that I shall broadcast to the world on Sunday night and make manifest the purity and disinterestedness of our motives throughout, and also of our resolves.

2. Meanwhile I send you a letter I have received from the King of Greece, to whom we have suggested the policy of making the Archbishop of Athens Regent. The King refuses to allow this. Therefore an act of constitutional violence will be entailed if we finally decide upon this course. I know nothing of the Archbishop, except that our people on the spot think he might stop a gap or bridge a gully.

* * * * *

It was a pleasure to hear at the same time from one on whose judgment and instinct in such matters I relied.

Field–Marshal Smuts to Prime Minister 14 Dec 44

I am very distressed at the anxiety and trouble which the situation in Greece is causing you and the Cabinet. I spoke strongly yesterday at Port Elizabeth in favour of policy pursued by the United Kingdom Government. I hope my comments have been conveyed by cable in condensed form. We may, I fear, find, if private Partisan armies and underground movements are kept alive, the peace degenerating in civil convulsions and anarchy not only in Greece but elsewhere also in Europe. . . . I hope that it is possible for the Archbishop to act with greater decision and authority. At this stage firmness is in any case essential, and weakness in dealing with the Partisans may end in real civil war at a later, more inconvenient, stage.

To be frank, I dislike our Ambassador taking this important rôle in change of Greek Government, as it may later be used as an argument against you for undue interference in the affairs of Greece. My own view, for what it is worth, is that after the suppressiom of the E.A.M. revolt the Greek King should return to discharge his proper constitutional functions, and the onus of practically running Greece should no longer be borne by His Majesty's Government.

I also received from the Greek 3rd Mountain Brigade, which had been fighting loyally with us, a message of thanks for our efforts to protect their country and of grief because British blood was being shed. They asked me to become their Honorary Commander.

But from Harry Hopkins came another warning.

16 Dec 44

Public opinion here is deteriorating rapidly because of Greek situation and your statement in Parliament about the United States and Poland.

With the battle joined as it is in Europe and Asia, with every energy required on everyone's part to defeat the enemy, I confess I find myself greatly disturbed at the diplomatic turn of events, which throw into the public gaze our several difficulties.

I do not know what the President or Stettinius may have to say publicly, but it may well be that one or both of them must state in unequivocal terms our determination to do all that we can to seek a free and secure world.

We were all agreed on this aim, but the question was whether it could be achieved by allowing the Communists to seize all power in Athens. That was the issue at stake.

Prime Minister to Mr. Harry Hopkins   17 Dec 44

I am distressed and puzzled by your message. I hope you will not hesitate to telegraph me on any points on which you think we, or I personally, have been in error, and what you would advise, because I have great trust in your judgment and friendship, even if I may at times look at matters from a different angle. All the President's telegrams to me have been most kind and encouraging, and also his telegram to U.J. may do a world of good.

2. Naturally I should welcome any public statements in America which set forth the aims stated in your last sentence. These are also ours. We seek nothing for ourselves from this struggle.

I also sent my promised reply to the President.

Prime Minister to President Roosevelt   17 Dec 44

About Greece. The present position is that our representatives on the spot, Macmillan and Leeper, have strongly recommended the appointment of the Archbishop as Regent. This is obnoxious to the Papandreou Government, though they might be persuaded to advocate a Regency of three, namely, the Archbishop, General Plastiras, and Dragoumis. There is suspicion that the Archbishop is ambitious of obtaining chief political power, and that, supported by E.A.M., he will use it ruthlessly against existing Ministers. Whether this be true or not I cannot say. The facts are changing from hour to hour. I do not feel at all sure that in setting up a one-man Regency we might not be imposing a dictatorship on Greece.

2. There is also to be considered the fact that the King refuses, I think inflexibly, to appoint a Regency, certainly not a one-man Regency of the Archbishop, whom he distrusts and fears. According to the Greek constitution, the Crown Prince is Regent in the absence of the King. The King also states that all his Ministers under Papan-

dreou advise him against such a step, and that, as a constitutional monarch, he cannot be responsible for it.

3. The War Cabinet decided to await for three or four days the course of military operations. Our reinforcements are arriving rapidly, and the British General Staff Intelligence says that there are not more than twelve thousand E.L.A.S. in Athens and the Piræus. The Greek King's estimate is fifteen to twenty-two thousand. Anyhow, we shall by the middle of next week be far superior in numbers. I am not prepared, as at present informed, to give way to unconstitutional violence in such circumstances.

4. Our immediate task is to secure control of Athens and the Piræus. According to the latest reports, E.L.A.S. may agree to depart. This will give us a firm basis from which to negotiate the best settlement possible between the warring Greek factions. It will certainly have to provide for the disarming of the guerrilla forces. The disarmament of the Greek Mountain Brigade, who took Rimini, and the Sacred Squadron, who have fought so well at the side of British and American troops, would seriously weaken our forces, and in any case we could not abandon them to massacre. They may however be removed elsewhere as part of a general settlement.

5. I am sure you would not wish us to cast down our painful and thankless task at this time. We embarked upon it with your full consent. We desire nothing from Greece but to do our duty by the common cause. In the midst of our task of bringing food and relief and maintaining the rudiments of order for a Government which has no armed forces we have become involved in a furious, though not as yet very bloody, struggle. I have felt it much that you were unable to give a word of explanation for our action, but I understand your difficulties.

6. Meanwhile the Cabinet is united and the Socialist Ministers approve Mr. Bevin's declarations at the Labour conference, which on this matter endorsed the official platform by a majority of 2,455,000 votes to 137,000. I could at any time obtain, I believe, a ten to one majority in the House of Commons.

I am sure you will do whatever you can. I will keep you constantly informed.

*   *   *   *   *

Mr. Mackenzie King in Canada also felt the unfavourable reactions to our Greek policy which had been so volubly expressed in the United States. He revealed his embarrassments in several telegrams.

Prime Minister to Prime Minister of Canada   15 Dec 44
    In the House I have done my best to clarify our position. To my

mind the essential point is that, having obtained the written assent of all parties, including the E.A.M., the Greek Prime Minister invited British troops to enter Greece to keep order and safeguard supplies. We accepted this invitation, and must still do our best to carry it out. The task is ungrateful, but we could not in honour shirk our responsibilities. With tempers inflamed on both sides in Athens, the situation is inevitably difficult. But the visit of Alexander was most valuable, and on the whole latest reports are more encouraging.

I also sent Mr. Mackenzie King the telegrams which I had exchanged with the President in August,* and drew his attention to the Caserta Agreement, which by now had been made public. I told him that I had the verbal approval of Stalin to our entering Greece and liberating Athens. "Although," I concluded, "Communists are at the root of the business, Stalin has not so far made any public reflection on our action."

In deference to these facts, arguments, and appeals, Mr. Mackenzie King refrained from any public act of dissociation.

It is odd, looking back on these events, now that some years have passed, to see how completely the policy for which I and my colleagues fought so stubbornly has been justified by events. Myself, I never had any doubts about it, for I saw quite plainly that Communism would be the peril civilisation would have to face after the defeat of Nazism and Fascism. It did not fall to us to end the task in Greece. I little thought however at the end of 1944 that the State Department, supported by overwhelming American opinion, would in little more than two years not only adopt and carry on the course we had opened, but would make vehement and costly exertions, even of a military character, to bring it to fruition. In his evidence before the House of Representatives Foreign Affairs Committee, Mr. Dean Acheson, the United States Acting Secretary of State, is reported to have testified on March 21, 1947, as follows: "A Communist-dominated Government in Greece would be considered dangerous to United States security."

If Greece has escaped the fate of Czechoslovakia and survives to-day as one of the free nations, it is due not only to British action in 1944, but to the steadfast efforts of what was presently to become the united strength of the English-speaking world.

* Chapter VII, pp. 99–100.

CHAPTER XIX

CHRISTMAS AT ATHENS

Street-Fighting in Athens – Grave Correspondence with Field-Marshal Alexander – My Telegram to Smuts, December 22 – My Doubts About a Regency – I Fly to Athens with Mr. Eden, December 24 – Christmas Night on Board H.M.S. "Ajax" – I Report to Mr. Attlee – And to the President – We Meet the Communists in the Greek Foreign Office, December 26 – A Telegram to Mrs. Churchill – We Agree to Ask the King of Greece to Appoint Archbishop Damaskinos as His Regent – The Flight Home, December 28 – Telegrams to the President – A Painful Task – The Royal Announcement – A Comment by Sir Reginald Leeper – General Plastiras Becomes Prime Minister, January 3 – Wise Advice from Field-Marshal Smuts – A Truce is Signed, January 11 – E.L.A.S. Expelled from Athens – The End of the Struggle.

I N Athens the street-fighting swayed to and fro on an enlarging scale. On December 15 Field-Marshal Alexander warned me that it was most important to get a settlement quickly, and the best chance was through the Archbishop. "Otherwise," he telegraphed, "I fear if rebel resistance continues at the same intensity as at present I shall have to send further large reinforcements from the Italian front to make sure of clearing the whole of Piræus–Athens, which is fifty square miles of houses."

Prime Minister to Field-Marshal Alexander (Italy)   17 Dec 44

The E.L.A.S. advance towards the centre of Athens seems to me a very serious feature, and I should like your appreciation of whether, with the reinforcements now arriving, we are likely to hold our own in the centre of the city and defeat the enemy. Have you any other reinforcements in view besides the 4th Division, the Tank Regiment, and the two remaining brigades of the 46th Division? Is there now any danger of a mass surrender of British troops cooped up in the city of Athens, followed by a massacre of Greeks who sided with us? The War Cabinet desire your report on the military situation in this respect.

2. We have no intention of subduing or occupying Greece. Our object is to afford a foundation upon which a broad-based Greek Government can function and raise a national force to preserve itself in Attica. After this we go, as we have no interests in Greece except those of sentiment and honour.

3. The King of Greece has refused categorically in a long and powerfully reasoned letter to appoint a Regent, and especially to appoint the Archbishop, of whom he has personal distrust. I have heard mixed accounts of the Archbishop, who is said to be very much in touch with E.A.M. and to have keen personal ambitions. We have not yet decided whether or in what way to overcome the King's resistance. If this cannot be overcome there will be no constitutional foundation other than an act of violence, to which we must become parties. The matter would be rendered more complicated if, as it may prove and as the King asserts, he is advised not to appoint a Regency by his Prime Minister and Government. In this case we should be punishing the King for obeying his constitutional oath and be ourselves setting up a dictator. The Cabinet have therefore decided to await further developments of the military situation before taking final and fateful decisions.

4. Personally I feel that our military predominance should be plainly established before we make terms, and in any case I should not like to make terms on grounds of weakness rather than of strength. Of course if you tell me it is impossible for us to be in control of Attica within a reasonable time the situation presents difficulties, but not such as should daunt us after all the others we have overcome.

And two days later:

Prime Minister to Field-Marshal Alexander (Italy)   19 Dec 44
The Cabinet feel it better to let the military operations to clear Athens and Attica run for a while rather than embark all our fortunes on the character of the Archbishop. Have you looked up his full record? It is a hard thing to ask me to throw over a constitutional King acting on the true advice of his Ministers, apart from British pressure, in order to install a dictator who may very likely become the champion of the extreme Left.

We are waiting here till the scene clears a little more, after which we shall give all the necessary directions.

Alexander's reply was grave. He had now succeeded General Wilson in the Supreme Command.

Field-Marshal Alexander to Prime Minister   31 Dec 44

In answer to your signal of December 19, I am most concerned that you should know exactly what true situation is and what we can do and cannot do. This is my duty. You would know the strength of British forces in Greece, and what additions I can send from Italian front if forced by circumstances to do so.

Assuming that E.L.A.S. continue to fight, I estimate that it will be possible to clear the Athens–Piraeus area and thereafter to hold it securely, but this will not defeat E.L.A.S. and force them to surrender. We are not strong enough to go beyond this and undertake operations on the Greek mainland. During the German occupation they maintained between six and seven divisions on the mainland, in addition to the equivalent of four in the Greek islands. Even so they were unable to keep their communications open all the time, and I doubt if we will meet less strength and determination than they encountered.

The German intentions on the Italian front require careful watching. Recent events in the West and the disappearance and silence of 16th S.S. Division opposite Fifth U.S. Army indicates some surprise move which we must guard against. I mention these factors to make the military situation clear to you, and to emphasise that it is my opinion that the Greek problem cannot be solved by military measures. The answer must be found in the political field.

Finally, I think you know that you can always rely on me to do everything in my power to carry out your wishes, but I earnestly hope that you will be able to find a political solution to the Greek problem, as I am convinced that further military action after we have cleared the Athens–Piraeus area is beyond our present strength.

I replied:

Prime Minister to Field-Marshal Alexander (Italy)   22 Dec 44

There is no question of our embarking in any military operations away from the Athens–Piraeus area. We must however have a military foundation there on which a Greek Government of some kind or other can function. I have personally great doubts about the Archbishop, who might quite conceivably make himself into a dictator supported by the Left Wing. However, these doubts may be removed in the next few days, and I am hopeful that in these days we shall achieve the mastery in Attica and cleanse Athens.

2. Thereafter we do not intend to stay in Greece except for such reasonable period as may be necessary to let the new Government, whatever it is, gain for itself a National Army or Militia, in the hopes that these may be able to conduct elections, plebiscites, etc. We can achieve no political solution while negotiating from a basis of weakness

and frustration. The political field in the present circumstances can only be entered by the gate of success.

I sent Field-Marshal Smuts my reactions on Greek affairs.

Prime Minister to Field-Marshal Smuts   22 Dec 44

Greece has proved a source of endless trouble to me, and we have indeed been wounded in the house of our friends. With this new chance, Communist and Left Wing forces throughout the world have stirred in sympathy, and our prestige and authority in Greece has to some extent been undermined by the American Press, reporting back. The return of the Greek King would provide no basis for a British policy. We must at all costs avoid giving the impression of forcing him on them by our bayonets.

I have serious doubts about the Regency, which may well assume the form of a dictatorship. I am unable to say whether it would be a dictatorship of the Left, as I do not know enough about the Archbishop. All Leftist forces and our people on the spot have certainly given their support to it. Alexander of course has his heart in the North and strongly dislikes the whole Greek business. But if the powers of evil prevail in Greece, as is quite likely, we must be prepared for a quasi-Bolshevised Russian-led Balkans peninsula, and this may spread to Italy and Hungary. I therefore foresee great perils to the world in these quarters, but am powerless to do anything effective without subjecting the Government to great stresses and quarrelling with America. I am hoping that the next few days may see an improvement in the progress of military operations in Attica, and thus induce a more healthy atmosphere. In the meantime our reinforcements are coming in, and of course in numbers we are already greatly superior to E.L.A.S. The situation is not however very pleasant.

*   *   *   *   *

Two days later I resolved to go and see for myself.

It was December 24, and we had a family and children's party for Christmas Eve. We had a Christmas tree—one sent from the President—and were all looking forward to a pleasant evening, the brighter perhaps because surrounded by dark shadows. But when I had finished reading my telegrams I felt sure I ought to fly to Athens, see the situation on the spot, and especially make the acquaintance of the Archbishop, around whom so much was turning. I therefore set the telephone working and arranged for an aeroplane to be ready at Northolt that night. I also spoilt Mr. Eden's Christmas by the proposal, which he immediately accepted, that he should come too. After having been much reproached by the family for deserting the party, I motored to meet Eden at Northolt, where the Skymaster which General Arnold had recently sent me waited, attentive and efficient. We slept soundly until about eight o'clock, when we landed at Naples to refuel. Here were several generals, and we all had breakfast together or at adjoining tables. Breakfast is not my best hour of the day, and the news we had both from the Italian front and from Athens was bleak. In an hour we were off again, and in perfect weather flew over the Peloponnese and the Straits of Corinth. Athens and the Piraeus unfolded like a map beneath us on a gigantic scale, and we gazed down upon it wondering who held what.

At about noon we landed at the Kalamaki airfield, which was guarded by about two thousand British airmen, all well armed and active. Here were Field-Marshal Alexander, Mr. Leeper, and Mr. Macmillan. They came on board the plane, and we spent nearly three hours in hard discussion of the whole position, military and political. We were, I think, in complete agreement at the end, and about the immediate steps to be taken.

I and my party were to sleep on board the *Ajax*, anchored off the Piraeus, the famous light cruiser of the Plate River battle,* which now seemed a long time ago. The road was reported clear, and with an escort of several armoured cars we traversed the few miles without incident. We boarded the *Ajax* before darkness fell, and I realised for the first time that it was Christmas Day. All preparations had been made by the ship's company for a jolly evening, and we certainly disturbed them as little as possible.

The sailors had a plan for a dozen of them to be dressed up in every kind of costume and disguise, as Chinese, Negroes, Red Indians, Cockneys, clowns—all to serenade the officers and warrant officers, and generally inaugurate revels suitable to the occasion. The Archbishop and his attendants arrived—an enormous tall figure in the robes and high hat of a dignitary of the Greek Church. The two parties met. The sailors thought he was part of their show of which they had not been told, and danced around him enthusiastically. The Archbishop thought this motley gang was a premeditated insult, and might well have departed to the shore but for the timely arrival of the captain, who, after some embarrassment, explained matters satisfactorily. Mean-

* See Volume I (Cassell's 2nd edition), Chapter XXIX.

while I waited, wondering what had happened. But all ended happily.

*　　*　　*　　*　　*

I sent an account of our various discussions to the War Cabinet.

Prime Minister (Athens) to Deputy Prime Minister　　　　26 Dec 44
and others

On our arrival at air-port at Athens Foreign Secretary and I held a conference with Field-Marshal Alexander, Mr. Macmillan, and Mr. Leeper.

2. Field-Marshal Alexander gave an encouraging account of present military situation, which had been grave a fortnight ago but was now much better. The Field-Marshal however had formed the decided view that behind the E.L.A.S. units there was a stubborn core of resistance, Communist in character, which was stronger than we had thought and would be very difficult to eradicate. If we were successful in pushing the E.L.A.S. force outside the boundaries of Athens we should still be faced with a tremendous task if we tried to eliminate them altogether.

3. Mr. Macmillan and Mr. Leeper informed us they had been considering the summoning of a conference of all the political leaders, which E.L.A.S. would be invited to attend. We felt that the convening of such a conference, with the declared object of putting an end to fratricidal strife in Greece, would, even if E.L.A.S. refused the invitation, ensure that our intentions would have been made clear to the world. We also agreed it would be a good move that the Archbishop should be chairman of the conference. At our meeting [in the aeroplane] we drew up the text of a public statement which Messrs. Macmillan and Leeper were to show to the Greek Prime Minister and the Archbishop, text of which has already been telegraphed to you.

4. We expressed our wishes that the conference should rapidly become a conference among Greeks, though we would stay there as long as it was helpful. When the time came to put this to the Archbishop we had been informed beforehand that he would agree to play his part. When he came to see us [on board the *Ajax*] he spoke with great bitterness against the atrocities of E.L.A.S. and the dark, sinister hand behind E.A.M. Listening to him, it was impossible to doubt that he greatly feared the Communist, or Trotskyite as he called it, combination in Greek affairs. He told us that he had issued an encyclical to-day condemning the E.L.A.S. crowd for taking eight thousand hostages, middle-class people, many of them Egyptians, and shooting a few every day, and that he had said that he would report these matters to the Press of the world if the women were not released. After some wrangling he understood that the women would be released.

Generally he impressed me with a good deal of confidence. He is a magnificent figure, and he immediately accepted the proposal of being chairman of the conference. We are asking the U.S. and Soviet representatives in Athens to be present as observers. The conference is fixed for 4 p.m. on December 26.

5. The Archbishop, at my request, is sending me proposals for the agenda of the conference. I cannot foretell what may come out of it. It may be of course that E.L.A.S. will refuse the invitation. If they do so they will be shown before the world as making an unbridled bid for power. If they do accept I do not rate the chance of forming a united Government high. I was impressed, especially from what the Archbishop said, by the intensity of hatred for Communists in the country. We had no doubt of this before we came here. Present position is confirmed by all we have heard so far. There is no doubt how the people of Athens would vote if they had a chance, and we must keep the possibility of getting them that chance steadily in view. We will send you further reports after we have met E.L.A.S., if they come to-morrow.

I had of course kept the President informed.

Prime Minister to President Roosevelt   26 Dec 44

Anthony and I are going out to see what we can do to square this Greek entanglement. Basis of action: the King does not go back until a plebiscite in his favour has been taken. For the rest, we cannot abandon those who have taken up arms in our cause, and must if necessary fight it out with them. It must always be understood that we seek nothing from Greece, in territory or advantages. We have given much, and will give more if it is in our power. I count on you to help us in this time of unusual difficulty. In particular I should like you to tell your Ambassador in Athens to make contact with us and to help all he can in accordance with the above principles.

He replied next day.

President Roosevelt to Prime Minister   27 Dec 44

I have asked our Ambassador to call upon you as soon as possible, and I am ready to be of all assistance I can in this difficult situation.

I hope that your presence there on the spot will result in achieving an entirely satisfactory solution.

*   *   *   *   *

On the morning of the 26th, "Boxing Day", I set out for the Embassy. I remember that three or four shells from the fighting which was going on a mile away on our left raised spouts of water fairly near the *Ajax* as we were about to go ashore. Here an armoured car and military escort awaited us. I said to my Private Secretary, Jock Colville, "Where is your pistol?" and when he said that he had not got one I scolded him, for I certainly had my own. In a few moments, while we were crowding into our steel box, he said, "I have got a tommy gun." "Where did you get it from?" I asked. "I borrowed it from the driver," he replied. "What is *he* going to do?" I asked. "He will be busy driving." "But there will be no trouble unless we are stopped," I answered, "and what is he going to do then?" Jock had no reply. A black mark! We rumbled along the road to the Embassy without any trouble.

There I again met the Archbishop, on whom we were about to stake so much. He agreed to all that was proposed. We planned the procedure at the conference to be held in the afternoon. I was already convinced that he was the outstanding figure in the Greek turmoil. Among other things, I had learned that he had been a champion wrestler before he entered the Orthodox Church. Mr. Leeper has noted that I said, "It would distress me to think that any new task Your Beatitude assumes as Regent might in any way interfere with your spiritual functions." He gave me all the necessary reassurance.

About six o'clock that evening, December 26, the conference opened in the Greek Foreign Office. We took our seats in a large, bleak room after darkness fell. The winter is cold in Athens. There was no heating, and a few hurricane lamps cast a dim light upon the scene. I sat on the Archbishop's right, with Mr. Eden, and Field-Marshal Alexander was on his left. Mr. MacVeagh, the American Ambassador, M. Baelen, the French Minister, and the Soviet military representative had all accepted our invitation. The three Communist leaders were late. It was not their fault. There had been prolonged bickering at the outposts. After half an hour we began our work, and I was already speaking when they entered the room. They were presentable figures in British battle dress. In my speech I said, among other things:

When we came here yesterday we thought it would be a good thing to have a talk round a table. It is better to let every effort be made to remake Greece as a factor in the victory, and to do it now. Therefore we had a talk with M. Papandreou, the Prime Minister. . . . We proposed to him that there should be a conference like this. Mr. Eden and I have come all this way, although great battles are raging in Belgium and on the German frontier, to make this effort to rescue Greece from a miserable fate and raise her to a point of great fame and repute. M. Papandreou told us immediately that he would welcome such a conference, and we have all met here now, in this city, where the sound of firing can be heard from minute to minute at no great distance. The next British step was to invite the Archbishop to be the chairman of this Greek conference. We do not intend to obstruct your deliberations. We British, and other representatives of the great united victorious Powers, will leave you Greeks to your own discussions under this most eminent and most venerable citizen, and we shall not trouble you unless you send for us again. We may wait a little while, but we have many other tasks to perform in this world of terrible storm. My hope is however that the conference which begins here this afternoon in Athens will restore Greece once again to her fame and power among the Allies and the peace-loving peoples of the world, will secure the Greek frontiers from any danger from the north, and will enable every Greek to make the best of himself and the best of his country before the eyes of the whole world. For all eyes are turned upon this table at this moment, and we British trust that whatever has happened in the heat of fighting, whatever misunderstandings there may have been, we shall preserve that old friendship between Greece and Great Britain which played so notable a part in the establishment of Greek independence.

General Alexander added a sharp touch that Greek troops should be fighting in Italy and not against British troops in Greece.

Once we had broken the ice and got the Greeks who had done such terrible injuries to each other to parley round the table under the presidency of the Archbishop, and the formal speeches had been made, the British members of the conference withdrew.

* * * * *

I was glad to get back to the Embassy, where there were a few oil stoves lent by G.H.Q. for the duration of my visit. While we were awaiting news from the conference and dinner I sent the following telegram to my wife, towards whom I felt penitent because of my desertion on Christmas Eve:

Prime Minister to Mrs. Churchill 26 Dec 44
We have had a fruitful day, and so far there is no need to give up hope of some important results. H.M.S. *Ajax* is very comfortable, and one can get a view of the fighting in North Piræus at quite short range.

We have had to move a mile farther away, as we were getting too many of their trench mortar bombs in our neighbourhood. I went into the Embassy up that long road from Piræus to Athens in an armoured car with strong escort, and I addressed all the plucky women on Embassy staff, who have been in continued danger and discomfort for so many weeks, but are in gayest of moods. Mrs. Leeper is an inspiration to them.

2. You will have read about the plot to blow up H.Q. in the Hôtel Grande-Bretagne. I do not think it was for my benefit. Still, a ton of dynamite was put in sewers by extremely skilled hands and with German mechanism between the time my arrival was known and daylight. I have made friends with the Archbishop, and think it has been very clever to work him in as we have done, leaving the constitutional questions for further treatment later.

3. The conference at Greek Foreign Office was intensely dramatic. All those haggard Greek faces round the table, and the Archbishop with his enormous hat, making him, I should think, seven feet high, whom we got to preside. The American, Russian, and French Ambassadors were all very glad to be invited. You will hear speeches on radio no doubt, or see them printed in Wednesday's papers. E.L.A.S. arrived late, three in all. Thanks were proposed, with many compliments to us for coming, by the Greek Government, and supported by E.L.A.S. representative, who added reference to Great Britain, "our great Ally"—all this with guns firing at each other not so far away.

4. After some consideration I shook E.L.A.S. delegate's hand, and it was clear from their response that they were gratified. They are the very top ones. We have now left them together, as it was a Greek show. It may break up at any moment. We shall wait for a day or two if necessary to see. At least we have done our best.

* * * * *

Bitter and animated discussions between the Greek parties occupied all the following day. At 5.30 that evening I had a final discussion with the Archbishop. As the result of his conversations with the E.L.A.S. delegates it was agreed I should ask the King of Greece to make him Regent. He would set about forming a new Government without any Communist members. We undertook to carry on the fighting in full vigour until either E.L.A.S. accepted a truce or the Athens area was clear of them. I told him that we could not undertake any military task beyond Athens and Attica, but that we would try to keep British forces in Greece until the Greek National Army was formed.

Just before this talk I had received a letter from the Communist delegates asking for a private meeting with me. The Archbishop begged me not to assent to this. I replied that as the conference was fully Greek in character I did not feel justified in agreeing to their request.

On the following morning, December 28, Mr. Eden and I left by air for Naples and London. I had no chance to say good-bye to M. Papandreou before leaving. He was about to resign, and was a serious loser by the whole business. I asked our Ambassador to keep in friendly touch with him.

I sent the following telegram to the Chiefs of Staff:

Prime Minister (Athens) to General Ismay, for C.I.G.S.   28 Dec 44
and C.O.S.

It is clear to me that great evils will follow here in Athens, affecting our position all over the world, if we cannot clear up situation quickly —*i.e.*, in two or three weeks. This would entail, according to Alexander, the moving in of the two brigades of the 46th Division, which are already under orders and standing by. On the other hand, the military situation in Western Apennines is such that any serious weakening of the reserves of Fifteenth Army Group might be attended with danger.

2. In these circumstances I wish you to consider and be ready to discuss with me on my return allowing the leading brigade of 5th Division to proceed from Palestine to Italy on schedule arranged before 4th Division was diverted to Greece. It would be a great convenience if we could have a reply to this to-morrow, Thursday. I do not leave Caserta until after midnight. This of course would mean that no violent action could be taken in Palestine, irritating the Jews, such as the search for arms on a large scale, until the situation is easier all round.

Just before leaving Athens I also sent the following telegram to the President, from whom I had had a kindly inquiry:

Prime Minister (Athens) to President Roosevelt   28 Dec 44

Many thanks for your message, which encouraged me amidst many difficulties. Ambassador MacVeagh called yesterday and we had a resumed talk. Like everyone else here, he is convinced that a Regency under the Archbishop is the only course open at the moment. I have seen the Archbishop several times, and he made a very good impression on me by the sense of power and decision which he conveyed, as well as by his shrewd political judgments. You will not expect me to speak here of his spiritual qualities, for I really have not had sufficient opportunity to measure these.

2. The Greek conference, of which you will have had from other sources full account, was unanimous in recommending a Regency. This was strongly supported by E.A.M. However, I do not consider Archbishop is at all Left Wing in Communist sense. On the contrary, he seems to be an extremely determined man, bent on establishing a small, strong executive in Greece to prevent the continuance of civil war.

3. I am therefore returning with Anthony to England to press upon the King of Greece to appoint the Archbishop Regent. Effect of this, if King agrees, will of course mean that Archbishop will form a Government of ten or less of the "best will". I gathered that he would make Plastiras Prime Minister, and that Papandreou would not be included. Naturally I could not probe too far while all these matters are hypothetical.

4. On our return we shall advise our colleagues, who are already inclined to this course, that we should put the strongest pressure on the Greek King to accept advice of his Prime Minister, M. Papandreou, who changed his mind about three times a day but has now promised to send a telegram in his own words.

5. If Ambassador MacVeagh's report should on these matters correspond with mine I should greatly hope that you would feel yourself able to send a personal telegram to the King of Greece during the next few days supporting the representation we shall make to him, of which we shall keep you informed. My idea is that the Regency should be only for one year, or till a plebiscite can be held under conditions of what is called "normal tranquillity".

The Archbishop has left this matter entirely in my hands, so that I can put the case in most favourable manner to the King. Of course if after these difficulties have been surmounted and Archbishop is Regent you felt able to send him a telegram of support that would make our task easier. Mr. President, we have lost over one thousand men, and though the greater part of Athens is now clear it is a painful sight to see this city with street-fighting raging now here, now there, and the poor people all pinched and only kept alive in many cases by rations we are carrying, often at loss of life, to them at the various depots. Anything that you can say to strengthen this new lay-out as the time comes will be most valuable, and may bring about acceptance by E.L.A.S. of the terms of truce set forth by General Scobie. For the rest we are reinforcing as is necessary and military conflict will go on. The vast majority of the people long for a settlement that will free them from the Communist terror.

6. We have to think of an interim arrangement which can be reviewed when our long-hoped-for meeting takes place. This date

should not now be far distant. It will then be possible to correlate our opinions and actions. In the meanwhile we have no choice but to recommend creation of a new and more competent executive Government under the Regency of the Archbishop, and to press on with our heavy and unsought task of clearing Athens from very dangerous, powerful, well-organised, and well-directed elements which are now pressing into the area. I should value a telegram when I return on Friday morning.

*   *   *   *   *

On December 29 we arrived back in London, and I telegraphed again to President Roosevelt.

Ambassador Winant has sent me a copy of your message to the Greek King. We are all very much obliged to you for acting so promptly. Anthony and I have just returned. The War Cabinet have endorsed all our actions, and have authorised us to urge the King of Greece to-night to appoint the Archbishop as Regent. The Archbishop left it to me to discuss the period of the Regency with the King, so that this gives a little latitude.

2. Failing agreement, His Majesty's Government will advise the Archbishop to assume the office of Regent and assure him that we will recognise him and the Government he forms as the Government of Greece.

Later that same night I sent him more solid news.

Prime Minister to President Roosevelt   30 Dec 44

Anthony and I sat up with the King of Greece till 4.30 this morning, at the end of which time His Majesty agreed to the following announcement. I have sent this to Ambassador Leeper in Athens in order that the Archbishop may go to work at once. The Greek translation is now being made, and I will furnish you with a copy of it at the earliest moment.

This has been a very painful task to me. I had to tell the King that if he did not agree the matter would be settled without him and that we should recognise the new Government instead of him. I hope you will be able to give every support and encouragement to the Archbishop and his Government.

This was the announcement:

We, George II, King of the Hellenes, having deeply considered the terrible situation into which our well-loved people have fallen through circumstances alike unprecedented and uncontrollable, and being ourselves resolved not to return to Greece unless summoned by a free and fair expression of the national will, and having full confidence in your loyalty and devotion, do now by this declaration appoint you, Archbishop Damaskinos, to be our Regent during this period of emergency; and we accordingly authorise and require you to take all steps necessary to restore order and tranquillity throughout our kingdom. We further declare our desire that there should be ascertained, by processes of democratic government, the freely expressed wishes of the Greek people as soon as these storms have passed, and thus abridge the miseries of our beloved country, by which our heart is rent.

I sent the royal announcement at once to Mr. Leeper in Athens, saying that the Archbishop from the moment when he received it should consider himself free to proceed with all the functions of his office and could be assured of the resolute support of His Majesty's Government.

President Roosevelt replied on the same day: "I am happy to know of your safe arrival, and wish you every success in the solution of the Greek problem, which seems very promising as a result of your journey."

I answered:

Prime Minister to President Roosevelt   31 Dec 44
The Greek King behaved like a gentleman and with the utmost dignity, and I am sure a private message from you would give him comfort. I shall send only a civil acknowledgment to E.L.A.S. for the published message they have sent me, and hand the matter over to the Archbishop. It is clearly his job now.

The great battle in the West seems to be turning steadily in our favour, and I remain of the opinion that Rundstedt's sortie is more likely to shorten than to lengthen the war.

Mr. Leeper (now Sir Reginald Leeper, G.B.E., K.C.M.G.) in his account of these events in his book *When Greek Meets Greek* comments:

The King's declaration, which endorsed the unanimous recommendation of the conference, was the direct result of Mr. Churchill's visit. It finally scotched the legend that the British were trying to force the King back on his people. For that reason alone Mr. Churchill's visit to Athens had been abundantly justified. Had his instinct not brought him to the scene of trouble at that moment I doubt very much whether any other influence could have induced all sides to come together in recommending the Regency to the King.*

* Page 127.

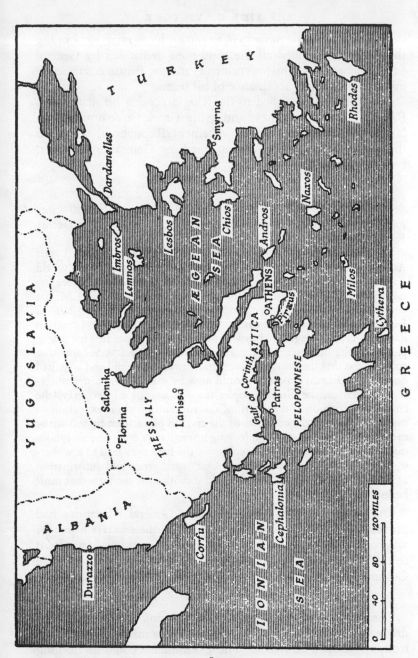

E.L.A.S. addressed a message to me on December 30 claiming that they had fulfilled all the conditions demanded by General Scobie for a truce. This was not true, and the British commander insisted on a formal acceptance of his terms.

The Archbishop replied to the King accepting his mandate as Regent. There was a new and living Greek Government. On January 3 General Plastiras, a vehement Republican, who was the leader of the Army revolt against King Constantine in 1922, became Prime Minister.

*   *   *   *   *

I also received some wise advice from Smuts.

Field-Marshal Smuts to Prime Minister and   30 Dec 44
Foreign Secretary

It is with deep interest and much anxiety that we have followed your Athens mission. It will have a profound and beneficial effect on world opinion. A wholly distorted picture of the true position in Greece has unfortunately been painted by the Press. Hence E.L.A.S.-E.A.M. has come to be viewed as the champion of democracy fighting against British backing of the royal cause. Though this is false, world reaction has been very damaging. Now is the time, I suggest, to give a true picture of the situation, and the Press should paint E.L.A.S. in its true colours. So that the world will see that Britain, as friend and ally, had no choice, a factual exposure should now be made of the bitter suffering inflicted on the Greek people, the dynamiting of property, the ruthless destruction and extortion, the rounding up and execution of innocent hostages, the coercion of the civilian population by terroristic methods in true Nazi style. Following immediately on your courageous mission, a full and accurate statement of the facts may lead to a wholesome reversal of public opinion. Our Intelligence and Information agencies in London and Athens should publish now the facts that must be in their possession.

Our own troops had no illusions. General Alexander had previously sent me a censorship report on their letters home. I was so struck by what I read that I had it printed and circulated to the War Cabinet. It completely disposed of the lie spread in Communist circles that their sympathies were with E.L.A.S.

*   *   *   *   *

The continuous fighting in Athens during December at last drove the insurgents from the capital, and by mid-January British troops controlled all Attica. The Communists could do nothing against our men in open country, and a truce was signed on January 11. All E.L.A.S. forces were to withdraw well clear of Athens, Salonika, and Patras. Those in the Peloponnese were to be given a safe-conduct to return to their homes. British troops would cease fire and stand fast. Prisoners would be released on both sides. These arrangements came into force on the 15th.

Thus ended the six weeks' struggle for Athens, and, as it ultimately proved, for the freedom of Greece from Communist subjugation. When three million men were fighting on either side on the Western Front and vast American forces were deployed against Japan in the Pacific the spasms of Greece may seem petty, but nevertheless they stood at the nerve-centre of power, law, and freedom in the Western world.

BOOK II

THE IRON CURTAIN

CHAPTER XX

PREPARATIONS FOR A NEW
CONFERENCE

*The Advance of the Red Army – The Struggle in Italy – Political
Problems in Eastern Europe – Correspondence Between the President
and Stalin About Poland – The Need for a Three-Power Conference –
I Agree to the Proposal – My Telegram to Mr. Roosevelt of January 1,
and His Reply – My Efforts to Arrange a Meeting of the Combined
Chiefs of Staff – Harry Hopkins Comes to London, January 21 –
Doubts About Yalta – I Fly to Malta, January 29 – The Scene in
Valletta Harbour, February 2 – The Discussions Between the British
and American Chiefs of Staff – A Cold Flight to Saki – The Drive
to Yalta.*

*E*ARLIER chapters have traced the advance of the Soviet
armies to the borders of Poland and Hungary. After
occupying Belgrade on October 20 the Russians had
resumed their thrust up the valley of the Danube, but resistance
stiffened the farther they penetrated across the Hungarian plain.
They forced a bridgehead on November 29 over the Danube
eighty miles below Budapest, and struck north. By the end of
December the capital was completely surrounded, and for six
weeks endured some of the grimmest street-fighting of the war.
On the shores of Lake Balaton solid German resistance and violent
counter-attacks also brought the Russian advance to a halt until
the spring.

In Poland the Russians spent the autumn months in building
up their forces after the extraordinary advances of the summer.
In January they were ready. Striking westwards from their
bridgeheads about Sandomir, at the end of the month they had
crossed the German frontier and pierced deep into the great
industrial basin of Upper Silesia. Farther north, crossing the

Vistula on both sides of Warsaw, they captured the city on January 17, and, investing Posen, fanned out towards the lower reaches of the Oder and towards Stettin and Danzig. Simultaneously they crushed East Prussia from east and south. By the end of January they were in complete possession, except for the heavily defended fortress of Königsberg. Here, as at Danzig, the garrison was to continue a stubborn but hopeless defence until April. The German forces cut off in Courland remained there till the surrender, as Hitler refused to let them come away.

The Soviet High Command, with a superiority of perhaps three to one on land and dominance in the air, used a strategy which recalls Foch's final victory in 1918. A series of battles, now here, now there, along a wide front punched successive holes until the whole line was forced to recoil.

* * * * *

Our own campaign in the West, though on a smaller scale, had equally brought us to the frontiers of Germany, so that at the end of January 1945 Hitler's armies were virtually compressed within their own territory, save for a brittle hold in Hungary and in Northern Italy. There, as has already been recorded, Alexander's skilful but hopelessly mutilated offensive had come to a halt. In November the strategic and tactical air forces had opened a six-months' campaign against the railways from the Reich to Italy. By destroying transformer stations a great part of the Brenner line was forced from electric to steam traction, and elsewhere the movement of enemy reinforcements and supplies was severely impeded. It is not possible to record the strenuous day-by-day operations of the Allied tactical air forces, commanded by General Cannon under the U.S. General Eaker, the Air Commander-in-Chief. Quelling their opponents, in spite of atrocious weather, they had done great things to help the autumn campaign, and they well deserved this tribute from General Alexander's dispatch:

I cannot speak too highly of General Cannon's gifts as a leader or of the encouragement which his assistance and support always gave me. The measure of his achievement can be seen in the complete immunity we enjoyed from enemy air attacks, the close and effective support enjoyed by the ground forces, and the long lines of destroyed enemy vehicles, the smashed bridges, and useless railways found by my armies wherever they advanced into enemy territory.

Nevertheless the liberation of Italy was not to be completed till the spring.

Such was the military position on the eve of the impending Triple Conference.

* * * * *

The political situation, at any rate in Eastern Europe, was by no means so satisfactory. A precarious tranquillity had indeed been achieved in Greece, and it seemed that a free democratic Government, founded on universal suffrage and secret ballot, might be established there within a reasonable time. But Roumania and Bulgaria had passed into the grip of Soviet military occupation, Hungary and Yugoslavia lay in the shadow of the battlefield, and Poland, though liberated from the Germans, had merely exchanged one conqueror for another. The informal and temporary arrangement which I had made with Stalin during my October visit to Moscow could not, and so far as I was concerned was never intended to, govern or affect the future of these wide regions once Germany was defeated.

The whole shape and structure of post-war Europe clamoured for review. When the Nazis were beaten how was Germany to be treated? What aid could we expect from the Soviet Union in the final overthrow of Japan? And once military aims were achieved what measures and what organisation could the three great Allies provide for the future peace and good governance of the world? The discussions at Dumbarton Oaks had ended in partial disagreement. So, in a smaller but no less vital sphere, had the negotiations between the Soviet-sponsored "Lublin Poles" and their compatriots from London which Mr. Eden and I had with much difficulty promoted during our visit to the Kremlin in October 1944. An arid correspondence between the President and Stalin, of which Mr. Roosevelt had kept me informed, had accompanied the secession of M. Mikolajczyk from his colleagues in London, while on January 5, contrary to the wishes of both the United States and Great Britain, the Soviets had recognised the Lublin Committee as the Provisional Government of Poland.

* * * * *

The President had already told me about his exchanges with Stalin. They were as follows:

289

Marshal Stalin to President Roosevelt   27 Dec 44

. . . A number of facts which took place during the time after the last visit of Mikolajczyk to Moscow, and in particular the radio communications with Mikolajczyk's Government intercepted by us from terrorists arrested in Poland—underground agents of the Polish *émigré* Government—with all palpability prove that the negotiations of M. Mikolajczyk with the Polish National Committee served as a screen for those elements who conducted from behind Mikolajczyk's back criminal terrorist work against Soviet officers and soldiers on the territory of Poland. We cannot reconcile with such a situation when terrorists instigated by Polish emigrants kill in Poland soldiers and officers of the Red Army, lead a criminal fight against Soviet troops who are liberating Poland, and directly aid our enemies, whose allies they in fact are. The substitution of Mikolajczyk by Arciszewsky, and in general transpositions of Ministers in the Polish *émigré* Government, have made the situation even worse and have created a precipice between Poland and the *émigré* Government. Meanwhile the Polish National Committee has made serious achievements in the strengthening of the Polish State and the apparatus of Governmental power on the territory of Poland, in the expansion and strengthening of the Polish Army, in carrying into practice a number of important Governmental measures, and, in the first place, the agrarian reform in favour of the peasants. All this has led to consolidation of democratic powers of Poland and to powerful strengthening of authority of the National Committee among wide masses in Poland and among wide social Polish circles abroad.

It seems to me that now we should be interested in the support of the Polish National Committee and all those who want and are capable to work together with it, and that is especially important for the Allies and for the solution of our common task—the speeding of the defeat of Hitlerite Germany. For the Soviet Union, which is bearing the whole burden for the liberation of Poland from German occupationists, the question of relations with Poland under present conditions is the task of daily close and friendly relations with a Power which has been established by the Polish people on its own soil and which has already grown strong and has its own army, which, together with the Red Army, is fighting against the Germans.

I have to say frankly that if the Polish Committee of National Liberation will transform itself into a Provisional Polish Government, then, in view of the above-said, the Soviet Government will not have any serious ground for postponement of the question of its recognition. It is necessary to bear in mind that in the strengthening of a pro-Allied and democratic Poland the Soviet Union is interested more than any other Power, not only because the Soviet Union is bearing the main brunt of the battle for liberation of Poland, but also because Poland is a border State with the Soviet Union and the problem of Poland is inseparable from the problem of security of the Soviet Union. To this I have to add that the successes of the Red Army in Poland in the fight against the Germans are to a great degree dependent on the presence of a peaceful and trustworthy rear in Poland. And the Polish National Committee fully takes into account this circumstance, while the *émigré* Government and its underground agents by their terroristic actions are creating a threat of civil war in the rear of the Red Army and counteract the successes of the latter.

On the other hand, under the conditions which exist in Poland at the present time there are no reasons for the continuation of the policy of support of the *émigré* Government, which has lost all confidence of the Polish population in the country, and besides creates a threat of civil war in the rear of the Red Army, violating thus our common interests of a successful fight against the Germans. I think that it would be natural, just, and profitable for our common cause if the Governments of the Allied countries as the first step have agreed on an immediate exchange of representatives with the Polish National Committee, so that after a certain time it would be recognised as the lawful Government of Poland after the transformation of the National Committee into a Provisional Government of Poland. Otherwise I am afraid that the confidence of the Polish people in the Allied Powers may weaken. I think that we cannot allow the Polish people to say that we are sacrificing the interests of Poland in favour of the interests of a handful of Polish emigrants in London.

Mr. Roosevelt's reply was reported in a message to myself.

President Roosevelt to Prime Minister   30 Dec 44

I have to-day sent the following to Stalin. You will see that we are in step.

"I am disturbed and deeply disappointed over your message of December 27 in regard to Poland, in which you tell me that you cannot see your way clear to hold in abeyance the question of recognising the Lublin Committee as the Provisional Government of Poland until we have had an opportunity at our meeting to discuss the whole question thoroughly. I would have thought no serious inconvenience would have been caused your Government or your armies if you could have delayed the purely juridical act of recognition for the short period of a month remaining before we meet.

"There was no suggestion in my request that you curtail your practical relations with the Lublin Committee, nor any thought that you should deal with or accept the London Government in its present composition. I had urged this delay upon you because I felt you would realise how extremely unfortunate and even serious it would be at this period in the war in its effect on world opinion and enemy morale if your Government should formally recognise one Government of Poland while the majority of the other United Nations, including the United States and Great Britain, continue to recognise and to maintain diplomatic relations with the Polish Government in London.

"I must tell you with a frankness equal to your own that I see no prospect of this Government's following suit and transferring its recognition from the Government in London to the Lublin Committee in its present form. This is in no sense due to any special ties or feelings for the London [Polish] Government. The fact is that neither the Government nor the people of the United States have as yet seen any evidence either arising from the manner of its creation or from subsequent developments to justify the conclusion that the Lublin Committee as at present constituted represents the people of Poland. I cannot ignore the fact that up to the present only a small fraction of Poland proper west of the Curzon Line has been liberated from German tyranny; and it is therefore an unquestioned truth that the people of Poland have had no opportunity to express themselves in regard to the Lublin Committee.

"If at some future date following the liberation of Poland a Provisional Government of Poland with popular support is established the attitude of this Government would of course be governed by the decision of the Polish people.

"I fully share your view that the departure of M. Mikolajczyk from the Government in London has worsened the situation. I have always felt that M. Mikolajczyk, who, I am convinced, is sincerely desirous of settling all points at issue between the Soviet Union and Poland, is the only Polish leader in sight who seems to offer the possibility of a genuine solution of the difficult and dangerous Polish question. I find it most difficult to believe from my personal knowledge of M. Mikolajczyk and my conversations with him when he was here in Washington and his subsequent efforts and policies during his visit at Moscow that he had knowledge of any terrorist instructions.

"I am sending you this message so that you will know the position of this Government in regard to the recognition at the present time of the Lublin Committee as the Provisional Government. I am more than ever convinced that when the three of us get together we can reach a solution of the Polish problem, and I therefore still hope that you can hold in abeyance until then the formal recognition of the Lublin Committee as a Government of Poland. I cannot from a military angle see any great objection to a delay of a month."

Stalin had answered:

Marshal Stalin to President Roosevelt 1 Jan 45
I have received your message of December 30.

I greatly regret that I have not been able to convince you of the correctness of the Soviet Government's attitude towards the Polish question. I nevertheless hope that events will convince you that the Polish National Committee has always rendered and will continue to render to the Allies, and in particular to the Red Army, considerable assistance in the struggle against Hitlerite Germany, whereas the *émigré* Government in London assists the Germans by creating disorganisation in this struggle.

I naturally fully comprehend your suggestion that the Soviet Government's recognition of the Provisional Government of Poland should be postponed for a month. There is however a circumstance here which makes it impossible for me to fulfil your wish. The position is that as early as December 27 the Presidium of the Supreme Soviet of the U.S.S.R. informed the Poles in reply to an inquiry on the subject that it proposed to recognise the Provisional Government of Poland as soon as the latter was formed. This circumstance makes it impossible for me to fulfil your wish.

Permit me to send you my greetings for the New Year and to wish you health and success.

∗ ∗ ∗ ∗ ∗

I now received from Stalin a direct message on Poland.

Marshal Stalin to Prime Minister 4 Jan 45
You are of course already aware of the publication by the Polish National Council in Lublin of the decision to which it has come regarding the transformation of the Polish Committee of National Liberation into the National Provisional Government of the Polish Republic. You are also well aware of our relations with the Polish National Committee, which in our view has already acquired great authority in Poland and is the lawful exponent of the will of the Polish nation. The transformation of the Polish National Committee into a Provisional Government seems to us entirely opportune, especially since Mikolajczyk has ceased to be a member of the *émigré* Polish Government and the latter has thus ceased to possess any semblance of a Government. I consider it impossible to leave Poland without a Government. Accordingly the Soviet Government has consented to recognise the Polish Provisional Government.

I much regret that I was unable completely to convince you of the correctness of the Soviet Government's attitude towards the Polish question. I nevertheless hope that future events will show that our recognition of the Polish Government in Lublin is in the interests of the general Allied cause and will contribute to hasten the defeat of Germany.

I attach for your information two messages from me to the President on the Polish question.

I am aware that the President has your consent to a meeting between us three at the end of this month or the beginning of February. I shall be glad to see you and the President on the territory of our country, and I hope for the success of our joint labours.

I take this opportunity of sending you good wishes for the New Year and wishing you the best of health and success.

Further correspondence did not seem to me to be likely to do much good. Only a personal meeting gave hope.

Prime Minister to Marshal Stalin   5 Jan 45

I thank you for sending me your two messages to the President on the Polish question. Naturally I and my War Cabinet colleagues are distressed at the course events are taking. I am quite clear that much the best thing is for us three to meet together and talk all these matters over, not only as isolated problems but in relation to the whole world situation both of the war and the transition to peace. Meanwhile our attitude as you know it remains unchanged.

*   *   *   *   *

The President was fully convinced of the need for another meeting of "the Three", for which arrangements had been discussed for some time. The usual debate about meeting-places had followed. "If Stalin cannot manage to meet us in the Mediterranean," the President said, "I am prepared to come to the Crimea and have the meeting at Yalta, which appears to be the best place available in the Black Sea, having the best accommodation ashore and the most promising flying conditions. My party will equal that of Teheran—about thirty-five. I still hope the military situation will permit Marshal Stalin to meet us half-way."

I replied:

Prime Minister to President Roosevelt   29 Dec 44

I send you the Admiralty report on Yalta. If this place is chosen it would be well to have a few destroyers on which we can live if necessary. There would be no difficulty in flying from the great air base and weather centre at Caserta. I myself landed in a York at Simferopol. I dare say however Stalin will make good arrangements ashore. Our party will be kept to the smallest dimensions. I think we should aim at the end of January. I must bring Anthony and Leathers.

On the 30th Mr. Roosevelt cabled that he would leave the United States as soon as possible after his Presidential inauguration, and would travel by warship to the Mediterranean, and from there by air to Yalta. I agreed at once to this, and promised to send a passenger ship to Sebastopol to supplement our quarters on shore. I myself proposed to fly direct via Caserta. On December 31 I cabled: "Have you a name for this operation? If not I suggest 'Argonaut', which has a local but not deducible association."

The President's circle however produced an alternative to flying from Caserta. His advisers, medical and otherwise, considered it inadvisable for him to fly at high altitudes over the mountains between Italy and Yalta. Admiral Hewitt recommended going by sea to Malta and flying on thence. I liked this.

Prime Minister to President Roosevelt 1 Jan 45

We shall be delighted if you will come to Malta. I shall be waiting on the quay. You will also see the inscription of your noble message to Malta of a year ago. Everything can be arranged to your convenience. No more let us falter! From Malta to Yalta! Let nobody alter!

I elaborated this for private use:

No more let us alter or falter or palter.
From Malta to Yalta, and Yalta to Malta.

Perhaps it was as well I did not cable it.

President Roosevelt to Prime Minister 2 Jan 45

We plan to arrive by ship at Malta early forenoon February 2, and hope to proceed at once by plane *without faltering*.* It will be grand to meet you on the quay.

Your suggestion of "Argonaut" is welcomed. You and I are direct descendants.

*   *   *   *   *

Lord Halifax reported from Washington that he had seen the President on the day before and did not think he "looked too good". Mr. Roosevelt however told him he was very well and much looking forward to our meeting. He said he thought our

* Author's italics.

295

action in Greece had been of immense value, and he was full of regret at not being able to visit England on the way. He was concerned at Japanese suicide aircraft attacks in the Pacific, which meant constantly losing forty or fifty Americans for one Japanese, and he was not very hopeful about an early end of either war.

This remark and other considerations made me anxious to arrange a meeting of the Combined Chiefs of Staff, at which both of us could either preside or be available before our meeting with Stalin. I therefore sent the following telegram:

Prime Minister to President Roosevelt   5 Jan 45
Would it not be possible for you to spend two or three nights at Malta and let the Staffs have a talk together unostentatiously? Also, Eisenhower and Alexander could both be available there. We think it very important that there should be some conversation on matters which do not affect the Russians—*e.g.*, Japan—and also about future use of the Italian armies. You have but to say the word and we can arrange everything.

2. We are very sorry indeed you will not come to our shores on this journey. We should feel it very much and a very dismal impression would be made if you were to visit France before you came to Britain; in fact, it would be regarded as a slight on your closest ally. I gather however that you will only go to the Mediterranean and Black Sea, in which case it is merely a repetition of Teheran.

3. The C.I.G.S. and I have passed two very interesting days at Eisenhower's headquarters at Versailles. Quite by chance de Gaulle arrived at the same time on the business about which he has sent you and me, as heads of Governments, a telegram concerning the Southern Sector [Strasbourg]. We had an informal conference and the matter has been satisfactorily adjusted so far as he is concerned. Eisenhower has been very generous to him.

4. I am now in Eisenhower's train going to visit Montgomery, the weather having made flying impossible. The whole country is covered with snow. I hope to be back in England Saturday. Every good wish.

The President did not at first think it possible for us to have a preliminary conference at Malta. He said that with favourable weather at sea he could arrive there by February 2, and would have to go on by air the same day in order to keep our date with Stalin. "I regret," he cabled, "that in view of the time available to me for this journey it will not be possible for us to meet your suggestion and have a British-American Staff meeting at Malta before proceeding to 'Argonaut'. I do not think that by not having

a meeting at Malta any time will be lost at Yalta. I am envious
of your visits to the great battle-front, which are denied to me by
distance." Nevertheless I pressed my proposal. The reader will
remember the anxieties which I had expressed about our operations in North-West Europe in my telegram to the President of
December 6.* These still weighed with me. The British and
American Chiefs of Staff had great need for discussion before we
reached Yalta, and I hoped that their principal members might
arrive at Malta two or three days before us and go over the
military ground together, and that the President would invite
Eisenhower, if the battle could spare him. I wanted Alexander to
come too. It seemed consonant with this idea of a preliminary
military conference that there should be a similar meeting of
Foreign Secretaries. I did not know whether the President would
bring Stettinius, recently appointed, with him, nor whether Molotov would come, but I should have liked a conference between
Eden, Stettinius, and Molotov at Alexandria or the Pyramids
about a week before the President and I reached Yalta.

I therefore telegraphed again on January 8, and, after setting
forth this proposal, continued:

I am still thinking it of high importance that our military men should
get together for a few days before we arrive at Yalta. There will no
doubt be opportunities for them to confer together at Sebastopol on days
when we are engaged in politics and do not require technical advice.
All the same, there are a tremendous lot of questions which should be
looked at beforehand, and our agenda ought really to be considered.

What are your ideas of the length of our stay at Yalta? This may
well be a fateful Conference, coming at a moment when the Great
Allies are so divided and the shadow of the war lengthens out before
us. At the present time I think the end of this war may well prove to
be more disappointing than was the last.

The President replied that he had directed Marshall, King,
and Arnold, with their assistants, to arrive in Malta in time for a
conference with the British Staffs in the forenoon of January 30,
but explained that he could not spare Mr. Stettinius for a preliminary meeting of Foreign Secretaries. He himself would be at
sea, and his Secretary of State ought not to be out of the country
so long at the same time. He would join us instead at Malta and
accompany our party to the Triple Conference.

* Chapter XVII, pp. 234–6.

"My idea," he concluded, "of the length of stay at Yalta is that it should not be more than five or six days. I am very desirous of keeping our date with Uncle Joe if it can possibly be done."

This was better than nothing, but it seemed to me that we might run very short of time, and I therefore persisted.

Prime Minister to President Roosevelt   10 Jan 45

Thank you very much about the Combined Chiefs of Staff's preliminary meeting.

2. Eden has particularly asked me to suggest that Stettinius might come on forty-eight hours earlier to Malta with the United States Chiefs of Staff, so that he (Eden) can run over the agenda with him beforehand. Even though Molotov were not invited, I am sure this would be found very useful. I do not see any other way of realising our hopes about World Organisation in five or six days. Even the Almighty took seven. Pray forgive my pertinacity.

3. I have now read very carefully your message to Congress, and I hope you will let me say that it is a most masterly document.

Every good wish.

But the President replied that there was too much business in Washington for Mr. Stettinius to reach Malta before January 31. He promised however to send Harry Hopkins to England to talk things over with Eden and myself. On January 21 Hopkins flew to London to discuss some of the topics which would confront us at the triple meeting, and the differences which had arisen between us in the previous month over Greece, Poland, and Italy. We had a number of very frank conversations during these three days. Hopkins records that I was "volcanic" in my remarks, but that the visit was "very satisfactory". I am said to have told him that from all the reports I had received about conditions at Yalta we could not have found a worse place for a meeting if we had spent ten years in looking for it. At any rate, it seems that the President's advisers were not without misgivings, for on the day of my departure I received the following telegram:

President Roosevelt to Prime Minister   29 Jan 45

The approaches to "Argonaut" appear to be much more difficult than at first reported. I will have my advance party make recommendations as to how I shall travel after Malta.

I agree that we must notify Uncle Joe as soon as we can fix our schedule in the light of present information.

Mr. Roosevelt was by now at sea and there was little one could do about it, but his forebodings were soon to be justified.

★　★　★　★　★

On January 29 I left Northolt in the Skymaster given to me by General Arnold. My daughter Sarah and the official party, together with Mr. Martin and Mr. Rowan, my private secretaries, and Commander Thompson, travelled with me. The rest of my personal staff and some departmental officials travelled in two other planes. We arrived at Malta just before dawn on January 30, and there I learnt that one of these two aircraft had crashed near Pantelleria. Only three of the crew and two passengers survived. Such are the strange ways of fate.

During the journey I developed a high temperature, and under Lord Moran's orders I stayed in bed on the plane until noon. I then took up my quarters in H.M.S. *Orion*, where I rested all day. In the evening I felt better, and both the Governor of Malta and Mr. Harriman dined with me.

On the morning of February 2 the Presidential party, on board the U.S.S. *Quincy*, steamed into Valletta harbour. It was a warm day, and under a cloudless sky I watched the scene from the deck of the *Orion*. As the American cruiser steamed slowly past us towards her berth alongside the quay wall I could see the figure of the President seated on the bridge, and we waved to each other. With the escort of Spitfires overhead, the salutes, and the bands of the ships' companies in the harbour playing "The Star-spangled Banner" it was a splendid scene. I thought I was well enough to lunch on board the *Quincy*, and at six o'clock that evening we had our first formal meeting in the President's cabin. Here we reviewed the report of the Combined Chiefs of Staff and the military discussions which had been taking place in Malta during the previous three days. Our Staffs had done a remarkable piece of work. Their discussions had centred principally round Eisenhower's plans for carrying his forces up to and across the Rhine. There were differences of opinion on the subject, which are related in another chapter.* The opportunity was of course taken to review the whole span of the war, including the war against the U-boats, the future campaigns in South-East Asia and the Pacific, and the Mediterranean situation. We reluctantly agreed to withdraw two divisions from Greece as soon as they

* Chapter XXIV, "Crossing the Rhine".

could be spared, but I made it clear that we should not be obliged to do this until the Greek Government had built up its own military forces. Three divisions were also to be withdrawn from Italy to reinforce North-West Europe, but I stressed that it would be unwise to make any significant withdrawal of amphibious forces. It was very important to follow up any German surrender in Italy, and I told the President that we ought to occupy as much of Austria as possible, as it was "undesirable that more of Western Europe than necessary should be occupied by the Russians". In all these military matters a large measure of agreement was reached, and the discussions had the useful result that the Combined Chiefs of Staff were aware of their respective points of view before engaging in talks with their Russian counterparts.

That evening we all dined together on the *Quincy*, to talk over informally the conversations which had taken place on the previous days between Mr. Eden and Mr. Stettinius on the political issues which should be raised at Yalta. That night the exodus began. The "thirty-five" staff which the President had contemplated had been multiplied by both of us tenfold. Transport planes took off at ten-minute intervals to carry some seven hundred persons, forming the British and American delegations, over fourteen hundred miles to the airfield of Saki, in the Crimea. A Royal Air Force contingent had been stationed there for two months beforehand to deal with the technical preparations. I boarded my plane after dinner, and went to bed. After a long and cold flight we landed on the airfield, which was under deep snow. My plane was ahead of the President's, and we stood for a while awaiting him. When he was carried down the lift from the "Sacred Cow" he looked frail and ill. Together we inspected the guards of honour, the President sitting in an open car, while I walked beside him. Our party then went into a large marquee for refreshment, with Molotov and the Russian delegation, which had come to meet us.

Presently we set off on a long drive from Saki to Yalta. Lord Moran and Mr. Martin came with me in my car. We had taken the precaution of bringing sandwiches with us, but after we had duly eaten them we came to a house where we were told Molotov awaited us, and we were invited to take places at a magnificent luncheon for about ten people. The President's party had apparently slipped past unawares and Molotov was alone with two of his officials. He was in the best of humours, and offered us all the delicacies of the Russian table. We did our best to conceal the fact that we had already blunted our appetites.

The journey took us nearly eight hours, and the road was often lined by Russian soldiers, some of them women, standing shoulder to shoulder in the village streets and on the main bridges and mountain passes, and at other points in separate detachments. As we crossed the mountains and descended towards the Black Sea we suddenly passed into warm and brilliant sunshine and a most genial climate.

CHAPTER XXI

YALTA: PLANS FOR WORLD PEACE

*The Vorontzov Palace – Russian Hospitality – Stalin Calls on Me,
February 4 – The Oder and the Ardennes – The First Plenary Meet-
ing, February 5 – The Future of Germany – Dismemberment and
Reparations – A Momentous Statement by Mr. Roosevelt – The
Second Meeting, February 6 – Need for a French Zone of Occupation
in Germany – Discussions on Dumbarton Oaks – Stalin's Views –
World Organisation and Unanimity Among the Great Powers –
Molotov Accepts the New Plan – The Russian Republics and the
World Organisation – My Telegram to the War Cabinet, February 8 –
Agreement at the Fourth Meeting, February 8 – Dinner with Stalin
at the Yusupov Palace – Grave and Friendly Speeches – Stalin Dis-
cusses the Past.*

THE Soviet headquarters at Yalta were in the Yusupov Palace,
and from this centre Stalin and Molotov and their generals
carried on the government of Russia and the control of their
immense front, now in violent action. President Roosevelt was
given the even more splendid Livadia Palace, close at hand, and
it was here, in order to spare him physical inconvenience, that all
our plenary meetings were held. This exhausted the undamaged
accommodation at Yalta. I and the principal members of the
British delegation were assigned a very large villa about five miles
away which had been built in the early nineteenth century by
an English architect for a Russian Prince Vorontzov, one-time
Imperial Ambassador to the Court of St. James.

My daughter Sarah, Mr. Eden, Sir Alexander Cadogan, Sir
Alan Brooke, Sir Andrew Cunningham, Sir Charles Portal, Field-
Marshal Alexander, Sir Archibald Clark Kerr, General Ismay, and
Lord Moran were among those who stayed with me. The rest
of our delegation were put up in two rest-houses about twenty

minutes away, five or six people sleeping in a room, including high-ranking officers, but no one seemed to mind. The Germans had evacuated the neighbourhood only ten months earlier, and the surrounding buildings had been badly damaged. We were warned that the area had not been completely cleared of mines, except for the grounds of the villa, which were, as usual, heavily patrolled by Russian guards. Over a thousand men had been at work on the scene before our arrival. Windows and doors had been repaired, and furniture and stores brought down from Moscow.

The setting of our abode was impressive. Behind the villa, half Gothic and half Moorish in style, rose the mountains, covered in snow, culminating in the highest peak in the Crimea. Before us lay the dark expanse of the Black Sea, severe, but still agreeable and warm even at this time of the year. Carved white lions guarded the entrance to the house, and beyond the courtyard lay a fine park with sub-tropical plants and cypresses. In the dining-room I recognised the two paintings hanging each side of the fireplace as copies of family portraits of the Herberts at Wilton. It appeared that Prince Vorontzov had married a daughter of the family, and had brought these pictures back with him from England.

Every effort was made by our hosts to ensure our comfort, and every chance remark noted with kindly attention. On one occasion Portal had admired a large glass tank with plants growing in it, and remarked that it contained no fish. Two days later a consignment of goldfish arrived. Another time somebody said casually that there was no lemon-peel in the cocktails. The next day a lemon tree loaded with fruit was growing in the hall. All must have come by air from far away.

* * * * *

At three o'clock on February 4, the day after our arrival, Stalin called on me, and we had an agreeable discussion about the war against Germany. He was optimistic. Germany was short of bread and coal; her transport was seriously damaged. I asked what the Russians would do if Hitler moved south—to Dresden, for example. "We shall follow him," was the reply. He went on to say that the Oder was no longer an obstacle, as the Red Army had several bridgeheads across it and the Germans were using untrained, badly led, and ill-equipped Volkssturm for its defence.

They had hoped to withdraw trained troops from the Vistula and use them to defend the river, but the Russian armour had bypassed them. Now they had only a mobile or strategic reserve of twenty or thirty badly trained divisions. They had some good ones in Denmark, Norway, and Italy, and in the West, but on the whole their front was broken and they were merely trying to patch up the gaps.

When I asked Stalin what he thought of Rundstedt's offensive against the Americans he called it a stupid manœuvre which had harmed Germany and was done for prestige. The German military body was sick and could not be cured by such methods. The best generals had gone and only Guderian was left, and he was an adventurer. If the German divisions cut off in East Prussia had been withdrawn in time they might have been used to defend Berlin, but the Germans were foolish. They still had eleven armoured divisions at Budapest, but they had failed to realise that they were no longer a world-Power and could not have forces wherever they wished. They would understand in due time, but it would be too late.

I then showed him my map-room, already fully mounted by Captain Pim, and after describing our position in the West I asked Field-Marshal Alexander to explain what was happening in Italy. Stalin's comment was interesting. The Germans were unlikely to attack us. Could we not leave a few British divisions on the front and transfer the rest to Yugoslavia and Hungary and direct them against Vienna? Here they could join the Red Army and outflank the Germans who were south of the Alps. He added that we might need a considerable force. It cost him nothing to say this now, but I made no reproaches.

"The Red Army," I answered, "may not give us time to complete the operation."

* * * * *

At five o'clock the President, Stalin, and I met to review the military situation, and in particular the Russian offensive on the Eastern Front. We heard a detailed account of the progress of the Russian Army, and also set the frame for the coming discussions between our respective Chiefs of Staff. I said that one of the questions we should consider was how long it would take the enemy to move eight divisions from Italy to the battle-front against Russia, and what counter-action we should take. Perhaps we ought to transfer divisions from Northern Italy to strengthen our attacking forces elsewhere. Another issue was whether we should try to strike at the head of the Adriatic, through the Ljubljana Gap, and join up with the Russian left flank.

The atmosphere of the meeting was most cordial. General Marshall gave us a brilliantly concise account of Anglo-American operations in the West. Stalin said that the Russian offensive in January had been launched as a moral duty, quite unconnected with the decisions made at Teheran, and he now asked how he could continue to help. I replied that now was the moment, when the three Staffs were met together, to review the whole question of military co-ordination between the Allies.

* * * * *

The first plenary meeting of the Conference started at a quarter-past four on the afternoon of February 5. We met in the Livadia Palace, and took our seats at a round table. With the three interpreters we were twenty-three. With Stalin and Molotov were Vyshinsky, Maisky, Gousev, the Russian Ambassador in London, and Gromyko, the Russian Ambassador in Washington. Pavlov acted as interpreter. The American delegation was headed by President Roosevelt and Mr. Stettinius, and included Admiral Leahy, Byrnes, Harriman, Hopkins, Matthews, Director of European Affairs in the State Department, and Bohlen, special assistant from the State Department, who also interpreted. Eden sat beside me, and my own party included Sir Alexander Cadogan, Sir Edward Bridges, and Sir Archibald Clark Kerr, our Ambassador in Moscow. Major Birse interpreted for us, as he had always done since my first meeting with Stalin at Moscow in 1942.

The discussion opened on the future of Germany. I had of course pondered this problem, and had addressed Mr. Eden a month before.

Prime Minister to Foreign Secretary 4 Jan 45

Treatment of Germany after the war. It is much too soon for us to decide these enormous questions. Obviously, when the German organised resistance has ceased the first stage will be one of severe military control. This may well last for many months, or perhaps for a year or two, if the German underground movement is active.

2. We have yet to settle the practical questions of the partition of Germany, the treatment of the Ruhr and Saar industries, etc. These may be touched upon at our forthcoming meeting, but I doubt

whether any final decision will be reached then. No one can foresee at the present moment what the state of Europe will be or what the relations of the Great Powers will be, or what the temper of their peoples will be. I am sure that the hatreds which Germany has caused in so many countries will find their counterpart here.

3. I have been struck at every point where I have sounded opinion at the depth of the feeling that would be aroused by a policy of "putting poor Germany on her legs again". I am also well aware of the arguments about "not having a poisoned community in the heart of Europe". I do suggest that, with all the work we have on our hands at the present moment, we should not anticipate these very grievous discussions and schisms, as they may become. We have a new Parliament to consider, whose opinions we cannot foretell.

4. I shall myself prefer to concentrate upon the practical issues which will occupy the next two or three years, rather than argue about the long-term relationship of Germany to Europe. I remember so well last time being shocked at the savage views of the House of Commons and of the constituencies, and being indignant with Poincaré when he sent the French into the Ruhr. In a few years however the mood of Parliament and the public changed entirely. Thousands of millions of money were lent to Germany by the United States. I went along with the tolerant policy towards Germany up to the Locarno Treaty and during the rest of Mr. Baldwin's Government on the grounds that Germany had no power to harm us. But thereafter a very swift change occurred. The rise of Hitler began. And thereafter I once again found myself very much out of sympathy with the prevailing mood.

5. It is a mistake to try to write out on little pieces of paper what the vast emotions of an outraged and quivering world will be either immediately after the struggle is over or when the inevitable cold fit follows the hot. These awe-inspiring tides of feeling dominate most people's minds, and independent figures tend to become not only lonely but futile. Guidance in these mundane matters is granted to us only step by step, or at the utmost a step or two ahead. There is therefore wisdom in reserving one's decisions as long as possible and until all the facts and forces that will be potent at the moment are revealed. Perhaps our approaching triple discussions will throw more light upon the problem.

Stalin now asked how Germany was to be dismembered. Were we to have one Government or several, or merely some form of administration? If Hitler surrendered unconditionally should we preserve his Government or refuse to treat with it? At Teheran Mr. Roosevelt had suggested dividing Germany into five parts, and he had agreed with him. I, on the other hand, had hesitated and had only wanted her to be split into two, namely, Prussia and Austria-Bavaria, with the Ruhr and Westphalia under international control. The time had now come, he said, to take a definite decision.

I said that we all agreed that Germany should be dismembered, but the actual method was much too complicated to be settled in five or six days. It would require a very searching examination of the historical, ethnographical, and economic facts, and prolonged review by a special committee, which would go into the different proposals and advise on them. There was so much to consider. What to do with Prussia? What territory should be given to Poland and the U.S.S.R.? Who was to control the Rhine valley and the great industrial zones of the Ruhr and the Saar? These were questions which needed profound study, and His Majesty's Government would want to consider carefully the attitude of their two great Allies. A body should be set up at once to examine these matters, and we ought to have its report before reaching any final decision.

I then speculated on the future. If Hitler or Himmler were to come forward and offer unconditional surrender it was clear that our answer should be that we would not negotiate with any of the war criminals. If they were the only people the Germans could produce we should have to go on with the war. It was more probable that Hitler and his associates would be killed or would disappear, and that another set of people would offer unconditional surrender. If this happened the three Great Powers must immediately consult and decide whether they were worth dealing with or not. If they were, the terms of surrender which had been worked out would be laid before them; if not, the war would be continued and the whole country put under strict military government.

Mr. Roosevelt suggested asking our Foreign Secretaries to produce a plan for studying the question within twenty-four hours and a definite plan for dismemberment within a month. Here, for a time, the matter was left.

Other questions were discussed, but not settled. The President asked whether the French should be given a zone of occupation in Germany. We agreed that this should certainly be done by allocating to them part of the British and American zones, and that the Foreign Secretaries should consider how this area was to be controlled.

At Stalin's request M. Maisky then expounded a Russian scheme for making Germany pay reparations and for dismantling her munitions industries. I said that the experience of the last war had been very disappointing, and I did not believe it would be possible to extract from Germany anything like the amount which M. Maisky had suggested should be paid to Russia alone. Britain too had suffered greatly. Many buildings had been destroyed. We had parted with much of our foreign investments and were faced with the problem of how to raise our exports sufficiently to pay for the imports of food on which we depended. I doubted whether these burdens could be substantially lightened by German reparations. Other countries had also suffered and would have to be considered. What would happen if Germany were reduced to starvation? Did we intend to stand by and do nothing and say it served her right? Or did we propose to feed the Germans, and, if so, who would pay? Stalin said that these questions would arise anyway, and I answered that if you wanted a horse to pull your wagon you had to give him some hay. We eventually agreed that the Russian proposal should be examined by a special commission, which would sit in secret at Moscow.

We then arranged to meet next day and consider two topics which were to dominate our future discussions, namely, the Dumbarton Oaks scheme for world security and Poland.

* * * * *

At this first meeting Mr. Roosevelt had made a momentous statement. He had said that the United States would take all reasonable steps to preserve peace, but not at the expense of keeping a large army in Europe, three thousand miles away from home. The American occupation would therefore be limited to two years. Formidable questions rose in my mind. If the Americans left Europe Britain would have to occupy single-handed the entire western portion of Germany. Such a task would be far beyond our strength.

At the opening of our second meeting on February 6 I accordingly pressed for French help in carrying such a burden. To give France a zone of occupation was by no means the end of the matter. Germany would surely rise again, and while the Americans could always go home the French had to live next door to her. A strong France was vital not only to Europe but to Great Britain. She alone could deny the rocket sites on her Channel coast and build up an army to contain the Germans.

We then turned to the World Instrument for Peace. The President said that in the United States public opinion was decisive. If it was possible to agree on the Dumbarton Oaks proposals or something like them his country would be more likely to take a full part in organising peace throughout the world, because there was a large measure of support in the United States for such a World Organisation. But, as has been recorded in an earlier chapter, the conference at Dumbarton Oaks had ended without reaching complete agreement about the all-important question of voting rights in the Security Council.

On December 5, 1944, the President had made new suggestions to Stalin and myself. They were as follows: Each member of the Council should have one vote. Before any decision could be carried out seven of the eleven members must vote in favour of it. This would suffice for details of procedure. All larger matters, such as admitting or expelling States from the organisation, suppressing and settling disputes, regulating armaments and providing armed forces, would need the concurring votes of all the permanent members. In other words, unless the "Big Four" were unanimous the Security Council was virtually powerless. If the United States, the U.S.S.R., Great Britain, or China disagreed, then it could refuse its assent and stop the Council doing anything. Here was the Veto.

Mr. Roosevelt's proposals had contained one other refinement. The dispute might be settled by peaceful methods. If so, this also would need seven votes, and the permanent members—that is to say, the "Big Four"—would all have to agree. But if any member of the Council (including the "Big Four") were involved in the dispute it could discuss the decision but could not vote on it. Such was the plan which Mr. Stettinius unfolded at this second meeting on February 6.

* * * * *

Stalin said that he would study the proposal and see if he could understand it, but at present it was not altogether clear. He said he feared that, though the three Great Powers were allies to-day, and would none of them commit any act of aggression, in ten years or less the three leaders would disappear and a new generation would come into power which had not experienced the war and would forget what we had gone through. "All of us," he declared, "want to secure peace for at least fifty years. The greatest danger is conflict among ourselves, because if we remain united the German menace is not very important. Therefore we must now think how to secure our unity in the future, and how to guarantee that the three Great Powers (and possibly China and France) will maintain a united front. Some system must be elaborated to prevent conflict between the main Great Powers."

He then expressed his regret that other business had hitherto prevented him from studying the American scheme in detail. As he understood it, the proposal was to divide all conflicts into two categories—first, those which required sanctions, whether economic, political, or military, and, secondly, those which could be settled by peaceful means. Both kinds would be freely discussed. Sanctions could only be applied if the permanent members of the Council were unanimous, and if one of these members was itself a party to the dispute then it could both take part in the discussions and vote. On the other hand, if there was a dispute which could be settled peacefully, then the parties to it could not vote. The Russians, he said, were accused of talking too much about voting. It was true they thought it was very important, because everything would be decided by vote and they would be greatly interested in the results. Suppose, for instance, that China as a permanent member of the Security Council demanded the return of Hong Kong, or that Egypt demanded the return of the Suez Canal, he assumed they would not be alone and would have friends and perhaps protectors in the Assembly or in the Council.

I said that, as I understood it, the powers of the World Organisation could not be used against Britain if she was unconvinced and refused to agree.

Stalin asked if this was really so, and I assured him it was.

Mr. Eden then explained that in such a case China or Egypt could complain, but that no decision involving the use of force could be taken without the concurrence of His Majesty's Government, and Mr. Stettinius confirmed that no sanctions could be imposed unless the permanent members of the Security Council were unanimous. Steps for a peaceful settlement—for instance, by arbitration—might be recommended.

Stalin said he feared that disputes about Hong Kong or Suez might break the unity of the three Great Powers.

I replied that I appreciated the danger, but the World Organisation in no way destroyed normal diplomatic intercourse between States, great or small. The World Organisation was separate and apart, and its members would continue to discuss their affairs among themselves. It would be foolish to raise subjects in the World Organisation if they might break up the unity of the Great Powers.

"My colleagues in Moscow," said Stalin, "cannot forget what happened in December 1939, during the Russo-Finnish War, when the British and the French used the League of Nations against us and succeeded in isolating and expelling the Soviet Union from the League, and when they later mobilised against us and talked of a crusade against Russia. Cannot we have some guarantees that this sort of thing will not happen again?"

Mr. Eden pointed out that the American proposal would make it impossible.

"Can we create even more obstacles?" asked Stalin.

I said that special provision had been made about the unanimity of the Great Powers.

"We have heard of it to-day for the first time," he replied.

I admitted there was a risk of an agitation working up against one of the Great Powers—say, the British—and I could only say that normal diplomacy would be playing its part at the same time. I should not expect the President to start or to back an attack on Great Britain, and I felt certain that everything would be done to stop it. I felt equally certain that Marshal Stalin would not make an attack—verbally, of course—on the British Empire without talking to us first and trying to find some way of coming to a friendly arrangement.

"I agree," he answered.

Mr. Roosevelt said that of course there would be differences between the Great Powers in future. Everybody would know about them, and they would be discussed in the Assembly. But it would not promote disunity to permit their discussion in the Council as well. On the contrary, it would show the confidence which we all had in each other and in our ability to solve such problems. This would strengthen our unity, not weaken it.

Stalin said that this was true, and promised to study the plan and continue the discussion next day.

* * * * *

When we met again on the following afternoon Molotov accepted the new scheme. At Dumbarton Oaks, he explained, the Russians had done all they could to preserve the unity of the three Powers after the war, and they thought that the plans which had emerged from the Conference would secure collaboration between all nations, great and small. They were now satisfied with the new voting procedure, and with the provision that the three Great Powers must be unanimous. There was only one thing to be settled. Should the Soviet Republics be members of the World Organisation, with votes in the Assembly? This had been discussed at Dumbarton Oaks, but now he was going to ask for something different. The Soviet delegation would be content if three, or at any rate two, of their republics became original members, namely, the Ukraine, White Russia, and Lithuania. All were important, all had made great sacrifices in the war; they were the first to be invaded and had suffered greatly. The Dominions of the British Commonwealth had approached independence gradually and patiently. This was an example to Russia, and they had therefore decided on this much narrower proposal. "We fully agree," he ended, "with the President's proposal about voting, and we ask that three, or at any rate two, of our republics should be founder members of the World Organisation."

This was a great relief to us all, and Mr. Roosevelt was quick to congratulate Molotov.

The next thing, said the President, was to invite the nations to come together. When should this be, and whom should we ask? The U.S.S.R. had great masses of people organised in separate republics; the British Empire had large independent groups living a long way from each other; the United States was a single unit, with a single Foreign Minister and no colonies. Then there were other countries, like Brazil, which had less territory than Russia but more than the United States, and at the other end of the scale a number of very small States. Could we stick to one vote for each nation, or should the larger nations have more than one vote in the World Assembly? He suggested referring all this to the three Foreign Ministers.

I also thanked Stalin for his great step in accepting the President's plan for voting, and said that the agreement which we had reached would give relief and satisfaction to people all over the world. Molotov's suggestion was also a great advance. President Roosevelt was quite right in saying that the position of the United States differed from that of the British Empire in this matter of voting. We had four self-governing Dominions, who during the last twenty-five years had played a notable part in the international organisation of peace which had broken down in 1939. All four had worked for peace and democratic progress. When the United Kingdom had declared war against Germany in 1939 all of them had sprung to arms, although they knew how weak we were. We had had no means of compelling them to do this. They had done it freely, of their own accord, on a matter about which it had only been possible to consult them very partially, and we could never agree to any system which excluded them from the position they had held and justified for a quarter of a century. For these reasons I could not but hear the proposals of the Soviet Government with a feeling of profound sympathy. My heart went out to mighty Russia, bleeding from her wounds but beating down the tyrants in her path. I recognised that a nation of a hundred and eighty millions might well look with a questioning eye at the constitutional arrangements of the British Commonwealth, which resulted in our having more than one voice in the Assembly, and I was glad therefore that President Roosevelt had given an answer which could in no way be regarded as a refusal of M. Molotov's request.

I pointed out however that I must not exceed my personal authority. I should like time to discuss M. Molotov's proposal with Mr. Eden, and perhaps send a telegram to the Cabinet, and I asked to be excused from giving a final answer that day. We thereupon agreed to refer the whole matter to our Foreign Ministers. Mr. Roosevelt had also suggested that the nations should meet in March to set up the World Organisation. I said I was doubtful about this, because the battle against Germany would then be at its height, but for the time being I let it pass.

* * * * *

Late that night I telegraphed to Mr. Attlee.

Prime Minister to Deputy Prime Minister   8 Feb 45 (2.49 a.m.)

To-day has been much better. All the American proposals for the

Dumbarton Oaks constitution were accepted by the Russians, who stated that it was largely due to our explanation that they had found themselves in a position to embrace the scheme wholeheartedly. They also cut down their demand for sixteen membership votes of the Assembly to two, making the plea that White Russia and the Ukraine had suffered so much and fought so well that they should be considered for inclusion among the founder members of the new World Organisation. The President by no means rejected this idea, though obviously visualising difficulties from the American standpoint. He suggested that it should be submitted to a conference of the United Nations which he seeks to convene in America during March.

Our position appears to me to be somewhat different. For us to have four or five members, six if India is included, when Russia has only one is asking a great deal of an Assembly of this kind. In view of other important concessions by them which are achieved or pending I should like to be able to make a friendly gesture to Russia in this matter. That they should have two besides their chief is not much to ask, and we will be in a strong position, in my judgment, because we shall not be the only multiple voter in the field.

At the present moment all I ask is that we should be authorised by the Cabinet to give to the Russians the undertaking that when this matter comes up for decision, whether here or at the Conference in America in March, we shall favour their view. I trust the Cabinet will grant us this authority to use or not to use as circumstances may dictate. The undertaking would be in the nature of a gentleman's agreement, and would be no different from that which we gave them long ago about supporting them on the Curzon Line.

If this increased representation is conceded or virtually conceded to Russia the whole business relating to Dumbarton Oaks is settled by unanimous agreement, and anyhow I think it will go through. This must be regarded as a considerable advantage, and one to which the Americans will attach great value for the purposes of politics, predominance, and publicity. It also forms a part of our main scheme for the World Organisation. . . .

In spite of our gloomy warning and forebodings Yalta has turned out very well so far. It is a sheltered strip of austere Riviera, with winding Corniche roads. The villas and palaces, more or less undamaged, are of an extinct imperialism and nobility. In these we squat on furniture carried with extraordinary effort from Moscow. The plumbing and road-making has been done without regard to cost in a few days by our hosts, whose prodigality exceeds belief. All the Chiefs of Staff have taken a holiday to-day to look at the battlefield of Balaclava. This is not being stressed in our conversations with our Russian friends.

As the time is short we shall, unless we hear from you to the contrary, be acting in the sense of this telegram.

* * * * *

The remaining details were settled very quickly. When we met again on the afternoon of February 8 we agreed to support the Russian request to admit two of the Soviet republics to the United Nations and to hold the first conference of the World Organisation on Wednesday, April 25. Only those States which had declared war on the common enemy by March 1 or had already signed the United Nations declaration would be invited. I agreed with Stalin that this would mean asking a certain number of nations who had not played a very good part in the war and had watched until they saw which side was going to win, but it would all help to depress Germany.

* * * * *

That night we all dined together with Stalin at the Yusupov Palace. The speeches were recorded, and may be printed here. Among other things, I said:

It is no exaggeration or compliment of a florid kind when I say that we regard Marshal Stalin's life as most precious to the hopes and hearts of all of us. There have been many conquerors in history, but few of them have been statesmen, and most of them threw away the fruits of victory in the troubles which followed their wars. I earnestly hope that the Marshal may be spared to the people of the Soviet Union and to help us all to move forward to a less unhappy time than that through which we have recently come. I walk through this world with greater courage and hope when I find myself in a relation of friendship and intimacy with this great man, whose fame has gone out not only over all Russia, but the world.

Stalin replied in flattering terms. He said:

I propose a toast for the leader of the British Empire, the most courageous of all Prime Ministers in the world, embodying political experience with military leadership, who when all Europe was ready to fall flat before Hitler said that Britain would stand and fight alone against Germany even without any allies. Even if the existing and possible allies deserted her he said she would continue to fight. To the health of the man who is born once in a hundred years, and who bravely held up the banner of Great Britain. I have said what I feel, what I have at heart, and of what I am conscious.

I then struck a graver note:

I must say that never in this war have I felt the responsibility weigh so heavily on me, even in the darkest hours, as now during this Conference. But now, for the reasons which the Marshal has given, we see that we are on the crest of the hill and there is before us the prospect of open country. Do not let us under-estimate the difficulties. Nations, comrades in arms, have in the past drifted apart within five or ten years of war. Thus toiling millions have followed a vicious circle, falling into the pit, and then by their sacrifices raising themselves up again. We now have a chance of avoiding the errors of previous generations and of making a sure peace. People cry out for peace and joy. Will the families be reunited? Will the warrior come home? Will the shattered dwellings be rebuilt? Will the toiler see his home? To defend one's country is glorious, but there are greater conquests before us. Before us lies the realisation of the dream of the poor—that they shall live in peace, protected by our invincible power from aggression and evil. My hope is in the illustrious President of the United States and in Marshal Stalin, in whom we shall find the champions of peace, who after smiting the foe will lead us to carry on the task against poverty, confusion, chaos, and oppression. That is my hope, and, speaking for England, we shall not be behindhand in our efforts. We shall not weaken in supporting your exertions. The Marshal spoke of the future. This is the most important of all. Otherwise the oceans of bloodshed will have been useless and outrageous. I propose the toast to the broad sunlight of victorious peace.

Stalin answered. I had never suspected that he could be so expansive. "I am talking," he declared, "as an old man; that is why I am talking so much. But I want to drink to our alliance, that it should not lose its character of intimacy, of its free expression of views. In the history of diplomacy I know of no such close alliance of three Great Powers as this, when allies had the opportunity of so frankly expressing their views. I know that some circles will regard this remark as naïve.

"In an alliance the allies should not deceive each other. Perhaps that is naïve? Experienced diplomatists may say, 'Why should I not deceive my ally?' But I as a naïve man think it best not to deceive my ally even if he is a fool. Possibly our alliance is so firm just because we do not deceive each other; or is it because it is not so easy to deceive each other? I propose a toast to the firmness of our Three-Power Alliance. May it be strong and stable; may we be as frank as possible."

And later:

For the group of workers who are recognised only during a war and whose services after a war are quickly forgotten. While there is a war these men are favoured and meet with respect not only of people of their own kind, but also that of the ladies. After a war their prestige goes down and the ladies turn their backs on them.

I raise my glass to the military leaders.

He had no illusions about the difficulties which lay before us.

A change has taken place in European history, a radical change, during these days. It is good to have an alliance of the principal Powers during a war. It would not be possible to win the war without the alliance. But an alliance against the common enemy is something clear and understandable. Far more complicated is an alliance after the war for securing lasting peace and the fruits of victory. That we fought together was a good thing, but it was not so difficult; on the other hand, that in these days the work of Dumbarton Oaks has been consummated and the legal foundations laid for organising security and strengthening peace is a great achievement. It is a turning-point.

I propose a toast for the successful conclusion of Dumbarton Oaks, and that our alliance, born under the stress of battle, be made solid and extended after the war, that our countries should not become engrossed only in their own affairs, but should remember that, apart from their own problems, there is the common cause, and that they should defend the cause of unity with as much enthusiasm in peace as during the war.

Even Molotov was in genial mood. He said:

I propose a toast for the three representatives of the Army, Air Force, and Navy of the country which went to war before we did. They had a hard task and suffered heavily, and we must recognise that they have accomplished their task well. I wish them success and a rapid end of the war in Europe, so that the victorious armies of the Allies may enter Berlin and hoist their banner over that city. I drink to the representatives of the British Army, Air Force, and Navy, Field-Marshal Brooke, Admiral Cunningham, and Air Marshal Portal, and to Field-Marshal Alexander.

* * * * *

As we sat at the dinner table in this cordial atmosphere Stalin began talking with me about the past. Some of his remarks remain on record.

"The Finnish war," he said, "began in the following way. The Finnish frontier was some twenty kilometres from Leningrad [he often called it "Petersburg"]. The Russians asked the Finns to move it back thirty kilometres, in exchange for territorial concessions in the north. The Finns refused. Then some Russian frontier guards were shot at by the Finns and killed. The frontier guards detachment complained to Red Army troops, who opened fire on the Finns. Moscow was asked for instructions. These contained the order to return the fire. One thing led to another and the war was on. The Russians did not want a war against Finland.

"If the British and French had sent a mission to Moscow in 1939 containing men who really wanted an agreement with Russia the Soviet Government would not have signed the pact with Ribbentrop.

"Ribbentrop told the Russians in 1939 that the British and Americans were only merchants and would never fight.

"If we, the three Great Powers, now hold together no other Power can do anything to us."

CHAPTER XXII

RUSSIA AND POLAND: THE SOVIET PROMISE

POLAND was discussed at no fewer than seven out of the eight plenary meetings of the Yalta Conference, and the British record contains an interchange on this topic of nearly eighteen thousand words between Stalin, Roosevelt, and myself. Aided by our Foreign Ministers and their subordinates, who also held tense and detailed debate at separate meetings among themselves, we finally produced a declaration which represented both a promise to the world and agreement between ourselves on our future actions. The painful tale is still unfinished and the true facts are as yet imperfectly known, but what is here set down may perhaps contribute to a just appreciation of our efforts at the last but one of the war-time Conferences. The difficulties and the problems were ancient, multitudinous, and imperative. The

Soviet-sponsored Lublin Government of Poland, or the "Warsaw" Government as the Russians of all names preferred to call it, viewed the London Polish Government with bitter animosity. Feeling between them had got worse, not better, since our October meeting in Moscow. Soviet troops were flooding across Poland, and the Polish Underground Army was freely charged with the murder of Russian soldiers and with sabotage and attacks on their rear areas and their lines of communication. Both access and information were denied to the Western Powers. In Italy and on the Western Front over 150,000 Poles were fighting valiantly for the final destruction of the Nazi armies. They and many others elsewhere in Europe were eagerly looking forward to the liberation of their country and a return to their homeland from voluntary and honourable exile. The large community of Poles in the United States anxiously awaited a settlement between the three Great Powers.

The questions which we discussed may be summarised as follows:

How to form a single Provisional Government for Poland.
How and when to hold free elections.
How to settle the Polish frontiers, both in the east and the west.
How to safeguard the rear areas and lines of communication of the advancing Soviet armies.

* * * * *

The reader should bear in mind the important correspondence between the President and Stalin, and my share in it, about Poland, which is set forth in an earlier chapter. Poland had indeed been the most urgent reason for the Yalta Conference, and was to prove the first of the great causes which led to the breakdown of the Grand Alliance.

When we met on February 6 President Roosevelt opened the discussion by saying that, coming from America, he had a distant view on the Polish question. There were five or six million Poles in the United States, mostly of the second generation, and most of them were generally in favour of the Curzon Line. They knew they would have to give up East Poland. They would like East Prussia and part of Germany, or at any rate something with which to be compensated. As he had said at Teheran, it would make it easier for him if the Soviet Government would grant some concession, such as Lvov, and some of the oil-bearing lands, to counterbalance the loss of Königsberg. But the most important point was a permanent Government for Poland. General opinion in the United States was against recognising the Lublin Government, because it represented only a small section of Poland and of the Polish nation. There was a demand for a Government of national unity, drawn perhaps from the five main political parties.

He knew none of the members of either the London or Lublin Governments. He had been greatly impressed by Mikolajczyk when he had come to Washington, and felt he was an honest man. He therefore hoped to see the creation of a Government of Poland which would be representative, and which the great majority of Poles would support even if it was only an interim one. There were many ways in which it might be formed, such as creating a small Presidential Council to take temporary control and set up a more permanent institution.

I then said it was my duty to state the position of His Majesty's Government. I had repeatedly declared in Parliament and in public my resolution to support the claim of the U.S.S.R. to the Curzon Line as interpreted by the Soviet Government. That meant including Lvov in the U.S.S.R. I had been considerably criticised in Parliament (as had the Foreign Secretary) and by the Conservative Party for this. But I had always thought that, after the agonies Russia had suffered in defending herself against the Germans, and her great deeds in driving them back and liberating Poland, her claim was founded not on force but on right. If however she made a gesture of magnanimity to a much weaker Power, and some territorial concession, such as the President had suggested, we should both admire and acclaim the Soviet action.

But a strong, free, and independent Poland was much more important than particular territorial boundaries. I wanted the Poles to be able to live freely and live their own lives in their own way. That was the object which I had always heard Marshal Stalin proclaim with the utmost firmness, and it was because I trusted his declarations about the sovereignty, independence, and freedom of Poland that I rated the frontier question as less important. This was dear to the hearts of the British nation and the Commonwealth. It was for this that we had gone to war against Germany—that Poland should be free and sovereign.

Everyone knew what a terrible risk we had taken when we had gone to war in 1939 although so ill-armed. It had nearly cost us our life, not only as an Empire but as a nation. Great Britain had no material interest of any kind in Poland. Honour was the sole reason why we had drawn the sword to help Poland against Hitler's brutal onslaught, and we could never accept any settlement which did not leave her free, independent, and sovereign. Poland must be mistress in her own house and captain of her own soul. Such freedom must not cover any hostile design by Poland or by any Polish group, possibly in intrigue with Germany, against Russia; but the World Organisation that was being set up would surely never tolerate such action or leave Soviet Russia to deal with it alone.

At present there were two Governments of Poland, about which we differed. I had not seen any of the present London Government of Poland. We recognised them, but had not sought their company. On the other hand, Mikolajczyk, Romer, and Grabski were men of good sense and honesty, and with them we had remained in informal but friendly and close relations. The three Great Powers would be criticised if they allowed these rival Governments to cause an apparent division between them, when there were such great tasks in hand and they had such hopes in common. Could we not create a Government or governmental instrument for Poland, pending full and free elections, which could be recognised by all? Such a Government could prepare for a free vote of the Polish people on their future constitution and administration. If this could be done we should have taken one great step forward towards the future peace and prosperity of Central Europe. I said I was sure that the communications of the Russian Army, now driving forward in victorious pursuit of the Germans, could be protected and guaranteed.

*　　*　　*　　*　　*

After a brief adjournment Stalin spoke. He said that he understood the British Government's feeling that Poland was a question of honour, but for Russia it was a question both of honour and security; of honour because the Russians had had many conflicts with the Poles and the Soviet Government wished to eliminate the causes of such conflicts; of security, not only because Poland was on the frontiers of Russia, but because throughout history

Poland had been a corridor through which Russia's enemies had passed to attack her. During the last thirty years the Germans had twice passed through Poland. They passed through because Poland had been weak. Russia wanted to see a strong and powerful Poland, so that she would be able to shut this corridor of her own strength. Russia could not keep it shut from the outside. It could only be shut from the inside by Poland herself, and it was for this reason that Poland must be free, independent, and powerful. This was a matter of life and death for the Soviet State. Their policy differed greatly from that of the Czarist Government. The Czars had wanted to suppress and assimilate Poland. Soviet Russia had started a policy of friendship, and friendship moreover with an independent Poland. That was the whole basis of the Soviet attitude, namely, that they wanted to see Poland independent, free, and strong.

He then dealt with some of the points which Mr. Roosevelt and I had put forward. The President, he said, had suggested there should be some modification of the Curzon Line and that Lvov and perhaps certain other districts should be given to Poland, and I had said that this would be a gesture of magnanimity. But the Curzon Line had not been invented by the Russians. It had been drawn up by Curzon and Clemenceau and representatives of the United States at the conference in 1918, to which Russia had not been invited. The Curzon Line had been accepted against the will of Russia on the basis of ethnographical data. Lenin had not agreed with it. He had not wished to see the town and province of Bialystok given to Poland. The Russians had already retired from Lenin's position, and now some people wanted Russia to take less than Curzon and Clemenceau had conceded. That would be shameful. When the Ukrainians came to Moscow they would say that Stalin and Molotov were less trustworthy defenders of Russia than Curzon or Clemenceau. It was better that the war should continue a little longer, although it would cost Russia much blood, so that Poland could be compensated at Germany's expense. When Mikolajczyk had been in Russia during October he had asked what frontier for Poland Russia would recognise in the west, and he had been delighted to hear that Russia thought that the western frontier of Poland should be extended to the Neisse. There were two rivers of that name, said Stalin, one near Breslau, and another farther west. It was the

Western Neisse he had in mind, and he asked the Conference to support his proposal.

* * * * *

Stalin then pointed out that we could not create a Polish Government unless the Poles themselves agreed to it. Mikolajczyk a⁻ᵈ Grabski had come to Moscow during my visit there. They had met the Lublin Government, a measure of agreement had been reached, and Mikolajczyk had gone to London on the understanding that he would come back. Instead of that he had been turned out of office by his colleagues, simply because he favoured an agreement with the Lublin Government. The Polish Government in London were hostile to the very idea of the Lublin Government, and described it as a company of bandits and criminals. The Lublin Government had paid them back in their own coin, and it was now very difficult to do anything about it.

The Lublin or Warsaw Government, as it should now be called, wanted to have nothing to do with the London Government. They had told him they would accept General Zeligowski and Grabski, but they would not hear of Mikolajczyk becoming Prime Minister. "Talk to them if you like," he said in effect. "I will get them to meet you here or in Moscow, but they are just as democratic as de Gaulle, and they can keep the peace in Poland and stop civil war and attacks on the Red Army." The London Government could not do this. Their agents had killed two hundred and twelve Russian soldiers; they were connected with the Polish underground resistance and had raided supply dumps to get arms. Their radio stations were operating without permission and without being registered. The agents of the Lublin Government had been helpful, and the agents of the London Government had done much evil. It was vital for the Red Army to have safe rear areas, and as a military man he would only support the Government which could guarantee to provide them.

* * * * *

It was now late in the evening, and the President suggested adjourning till next day, but I thought it right to state that the United Kingdom and the Soviet Government had different sources of information in Poland and had received different accounts of what had happened. I said that according to our information not more than one-third of the Polish people would support the Lublin Government if they were free to express their opinion. This estimate of course was based on the best information which we could obtain, and we might be mistaken in certain particulars. I assured Stalin that we had greatly feared a collision between the Polish Underground Army and the Lublin Government. We had feared that this would lead to bitterness, bloodshed, arrests, and deportations, and that was why we had been so anxious for a joint arrangement. We feared the effect which all this would have on the Polish question, already difficult enough. We recognised of course that attacks on the Red Army must be punished. But on the facts at my disposal I could not feel that the Lublin Government had a right to say that they represented the Polish nation.

The President was now anxious to end the discussion. "Poland," he remarked, "has been a source of trouble for over five hundred years." "All the more," I answered, "must we do what we can to put an end to these troubles." We then adjourned.

*　*　*　*　*

That evening the President wrote a letter to Stalin, after consultation with and amendment by us, urging that two members of the Lublin Government and two from London or from within Poland should come to the Conference and try to agree in our presence about forming a Provisional Government which we could all recognise to hold free elections as soon as possible. I favoured this course, and supported the President when we met again on February 7. Mr. Roosevelt once more emphasised his concern. Frontiers, he said, were important, but it was quite within our province to help the Poles to set up a united temporary Government, or even to set one up ourselves until they could produce one of their own founded on free elections. "We ought to do something," he said, "that will come like a breath of fresh air in the murk that exists at the moment on the Polish question." He then asked Stalin if he would like to add anything to what he had said the day before.

Stalin replied that he had received the President's letter only about an hour and a half before, and had immediately given instructions for Bierut and Osóbka-Morawski to be found so that he could talk to them on the telephone. He had just learned that they were in Cracow and Lodz respectively, and he promised to ask them how representatives from the opposition camp could be traced, as he did not know their addresses. In case there might not be time to get them to the Conference, Molotov had elaborated some proposals which to some extent met the President's suggestions.

Molotov accordingly took the stage and read out the following summary:

1. It was agreed that the Curzon Line should be the eastern frontier of Poland, with adjustments in some regions of five to eight kilometres in favour of Poland.

2. It was decided that the western frontier of Poland should be drawn from the town of Stettin (which would be Polish) and thence southwards along the river Oder and the Western Neisse.

3. It was considered desirable to add to the Provisional Polish Government some democratic leaders from Polish *émigré* circles.

4. It was considered desirable that the enlarged Provisional Polish Government should be recognised by the Allied Governments.

5. It was considered desirable that the Provisional Polish Government, enlarged as suggested in paragraph 3, should as soon as possible call the population of Poland to the polls for the establishment by general vote of permanent organs of the Polish Government.

6. M. Molotov, Mr. Harriman, and Sir A. Clark Kerr should be entrusted with the discussion of the question of enlarging the Provisional Polish Government and submitting their proposals for the consideration of the three Allied Governments.

Mr. Roosevelt seemed encouraged, and declared that we were making definite progress, but he wanted to talk the matter over with Mr. Stettinius. "I do not like the word *émigré*," he concluded. "I know none of the people concerned except Mikolajczyk, but I do not think we need only contact *émigrés*. We should also find some people in Poland itself." Stalin agreed to postpone the discussion, but I now intervened, and the interchange which followed may be deemed significant in the light of what happened afterwards.

I said that I shared the President's dislike of the word *émigré*. It was a term which originated with the French aristocracy who were driven out after the French Revolution, and was properly applied only to those who had been driven out of their own country by their own people. But the Poles abroad had been driven out of their country by the Germans, and I suggested that the words "Poles abroad" should be substituted for émigrés. Stalin assented. As for the river Neisse, mentioned in the second of Molotov's proposals, I reminded my hearers that in previous talks I had always qualified the moving of the Polish frontier westwards by saying that the Poles should be free to take territory in the west, but not more than they wished or could properly manage. It would be a great pity to stuff the Polish goose so full of German food that it died of indigestion. I was conscious of a large body of opinion in Great Britain which was frankly shocked at the idea of moving millions of people by force. A great success had been achieved in disentangling the Greek and Turkish populations after the last war, and the two countries had enjoyed good relations ever since; but in that case under a couple of millions of people had been moved. If Poland took East Prussia and Silesia as far as the Oder that alone would mean moving six million Germans back to Germany. It might be managed, subject to the moral question, which I would have to settle with my own people.

Stalin observed that there were no Germans in these areas, as they had all run away.

I replied that the question was whether there was room for them in what was left of Germany. Six or seven million Germans had been killed and another million (Stalin suggested two millions) would probably be killed before the end of the war. There should therefore be room for these migrant people up to a certain point. They would be needed to fill the vacancies. I was not afraid of the problem of transferring populations, so long as it was proportionate to what the Poles could manage and to what could be put into Germany. But it was a matter which required study, not as a question of principle, but of the numbers which would have to be handled.

In these general discussions maps were not used, and the distinction between the Eastern and Western Neisse did not emerge as clearly as it should have done. This was however soon to be made clear.

* * * * *

In the early hours of the following morning I telegraphed to Mr. Attlee:

... An impressive letter was sent to Stalin last night by the President, who wrote it after consultation with and amendment by us. In it he proposed that in place of the existing Lublin Government a new All-Polish Government should be formed containing representatives both from the Poles abroad and those within Poland. An answer in five or six heads was put forward by the Russians to-day, Wednesday. It does not challenge in principle any of the broad issues. We have asked for delay till to-morrow. The following counter-proposal drafted by the Foreign Secretary is being telegraphed to you together with the original Soviet proposal.

This matter is by no means settled. It is our plan to fight hard for a Government in Poland which we and United States can recognise and to which we can attract the recognition of all the United Nations. In return for this we require real substantial and effective representation from the Polish element with whom we have at present been associated, especially Mikolajczyk, Grabski, and Romer, as well as from a number of Poles still in Poland, Witos, Sapieha, etc., whom the Americans have listed. If it can be so arranged that eight or ten of these are included in the Lublin Government it would be to our advantage to recognise this Government at once. We could then get ambassadors and missions into Poland, and find out at least to some extent what is happening there and whether the foundations can be laid for the free, fair, and unfettered election which alone can give life and being to a Polish Government. We hope that on this difficult ground you will give us full freedom to act and manœuvre.

After setting forth the Soviet proposal I continued:

Following is the text of our [Anglo-American] revised proposals for Poland:

(i) It was agreed that the Curzon Line should be the eastern frontier of Poland, with adjustments in some regions of five to eight kilometres in favour of Poland.

(ii) It was decided that the territory of Poland in the west should include the free city of Danzig, the region of East Prussia west and south of Königsberg, the administrative district of Oppeln in Silesia, and the lands desired by Poland to the east of the line of the Oder. It was understood that the Germans in the said region should be repatriated to Germany and that all Poles in Germany should at their wish be repatriated to Poland.

(iii) Having regard to the recent liberation of Western Poland by the Soviet Union Army, it was deemed desirable to facilitate the establishment of a fully representative Provisional Polish Government, based upon all the democratic and anti-Fascist forces in Poland, and including democratic leaders from Poles abroad. That Government should be so constituted as to command recognition by the three Allied Governments.

(iv) It was agreed that establishment of such a Provisional Government was primarily the responsibility of the Polish people, and that pending the possibility of free elections representative Polish leaders should consult together on the composition of this Provisional Government. M. Molotov, Mr. Harriman, and Sir Archibald Clark Kerr were entrusted with the task of approaching such leaders and submitting their proposals to the consideration of the Allied Governments.

(v) It was deemed desirable that the Provisional Polish Government thus established should as soon as possible hold free and unfettered elections on the basis of universal suffrage and secret ballot, in which all democratic parties would have the right to participate and to promote candidatures in order to ensure the establishment of a Government truly representative of the will of the Polish people.

* * * * *

When we met again on February 8 Mr. Roosevelt read out his revised proposals based on Molotov's draft. "No objection," he stated, "is perceived to the Soviet proposal that the eastern boundary of Poland should be the Curzon Line, with modifications in favour of Poland in some areas of from five to eight kilometres." Here at least was one matter on which we could all agree, and although I had invited the Russians to make some minor concessions it seemed better not to multiply our difficulties, which were already serious enough. But the President was firm and precise about the frontier in the west. He agreed that Poland should receive compensation at the expense of Germany, "including that portion of East Prussia south of the Königsberg line, Upper Silesia, and up to the line of the Oder; but," he continued, "*there would appear to be little justification for extending it up to the Western Neisse.*"* This had always been my view, and I was to press it very hard when we met again at Potsdam five months later.

There remained the question of forming a Polish Government which we could all recognise and which the Polish nation would accept. Mr. Roosevelt suggested a Presidential Committee of three Polish leaders who would go to Moscow, form a Provisional Government from representatives in Warsaw, London, and inside Poland itself, and hold free elections as soon as possible.

* Author's italics.

After a short adjournment Molotov voiced his disagreement. The Lublin Government, he said, was now at the head of the Polish people. It had been enthusiastically acclaimed by most of them and enjoyed great authority and prestige. The same could not be said of the men from London. If we tried to create a new Government the Poles themselves might never agree, so it was better to try to enlarge the existing one. It would only be a temporary institution, because all our proposals had but one object, namely, to hold free elections in Poland as soon as possible. How to enlarge it could best be discussed in Moscow between the American and British Ambassadors and himself. He said he greatly desired an agreement, and he accepted the President's proposals to invite two out of the five people mentioned in his letter of February 6. There was always the possibility, he said, that the Lublin Government would refuse to talk with some of them, like Mikolajczyk, but if they sent three representatives and two came from those suggested by Mr. Roosevelt conversations could start at once.

"What about the Presidential Committee?" asked Mr. Roosevelt.

"Better avoid it," he answered. "It will mean having two bodies to deal with instead of one."

"This," I said, "is the crucial point of the Conference. The whole world is waiting for a settlement, and if we separate still recognising different Polish Governments the whole world will see that fundamental differences between us still exist. The consequences will be most lamentable, and will stamp our meeting with the seal of failure. On the other hand, of course we take different views about the basic facts in Poland, or at any rate some of them. According to British information, the Lublin Government does not commend itself to the great majority of the Polish people, and we cannot feel that it would be accepted abroad as representing them. If the Conference is to brush aside the existing London Government and lend all its weight to the Lublin Government there will be a world outcry. As far as can be foreseen, the Poles outside of Poland will make a virtually united protest. There is under our command a Polish army of 150,000 men, who have been gathered from all who have been able to come together from outside Poland. This army has fought, and is still fighting, very bravely. I do not believe it will be at all

reconciled to the Lublin Government, and if Great Britain transfers recognition from the Government which it has recognised
since the beginning of the war they will look on it as a betrayal."

"As Marshal Stalin and M. Molotov well know," I proceeded,
"I myself do not agree with the London Government's action,
which has been foolish at every stage. But the formal act of
transferring recognition from those whom we have hitherto
recognised to this new Government would cause the gravest
criticism. It would be said that His Majesty's Government have
given way completely on the eastern frontier (as in fact we have)
and have accepted and championed the Soviet view. It would also
be said that we have broken altogether with the lawful Government of Poland, which we have recognised for these five years of
war, and that we have no knowledge of what is actually going on
in Poland. We cannot enter the country. We cannot see and hear
what opinion is. It would be said we can only accept what the
Lublin Government proclaims about the opinion of the Polish
people, and His Majesty's Government would be charged in
Parliament with having altogether forsaken the cause of Poland.
The debates which would follow would be most painful and em-
barrassing to the unity of the Allies, even supposing that we were
able to agree to the proposals of my friend M. Molotov."

"I do not think," I continued, "that these proposals go nearly
far enough. If we give up the Polish Government in London a
new start should be made from both sides on more or less equal
terms. Before His Majesty's Government ceased to recognise
the London Government and transferred their recognition to
another Government they would have to be satisfied that the
new Government was truly representative of the Polish nation. I
agree that this is only one point of view, as we do not fully know
the facts, and all our differences will of course be removed if a
free and unfettered General Election is held in Poland by ballot
and with universal suffrage and free candidatures. Once this is
done His Majesty's Government will salute the Government that
emerges without regard to the Polish Government in London.
It is the interval before the election that is causing us so much
anxiety."

Molotov said that perhaps the talks in Moscow would have
some useful result. It was very difficult to deal with this question without the participation of the Poles themselves, who would

have to have their say. I agreed, but said that it was so important that the Conference should separate on a note of agreement that we must all struggle patiently to achieve it. The President supported me. He said that it was the great objective of the Americans that there should be an early General Election in Poland. The only problem was how the country was to be governed in the meantime, and he hoped it would be possible to hold elections before the end of the year. The problem was therefore limited in time.

Stalin now took up my complaint that I had no information and no way of getting it.

"I have a certain amount," I replied.

"It doesn't agree with mine," he answered, and proceeded to make a speech, in which he assured us that the Lublin Government was really very popular, particularly Bierut, Osóbka-Morawski, and General Zymierski. They had not left the country during the German occupation, but had lived all the time in Warsaw and came from the Underground movement. That made a deep impression on the Poles, and the peculiar mentality of people who had lived under the German occupation should be borne in mind. They sympathised with all those who had not left the country in difficult times, and they considered the three persons he had named to be people of that kind. He said he did not believe that they were geniuses. The London Government might well contain cleverer people, but they were not liked in Poland because they had not been seen there when the population was suffering under the Hitlerite occupation. It was perhaps a primitive feeling, but it certainly existed.

It was, he said, a great event in Poland that the country had been liberated by Soviet troops, and this had changed everything. It was well known that the Poles had not liked the Russians, because they had three times helped to partition Poland. But the advance of the Soviet troops and the liberation of Poland had completely changed their mood. The old resentment had disappeared, and had given way to goodwill and even enthusiasm for the Russians. That was perfectly natural. The population had been delighted to see the Germans flee and to feel that they were liberated. Stalin said it was his impression that the Polish population considered the driving out of the Germans a great patriotic holiday in Polish life, and they were astonished that the London

Government did not take any part in this festival of the Polish nation. They saw on the streets the members of the Provisional Government, but asked where were the London Poles. This undermined the prestige of the London Government, and was the reason why the Provisional Government, though not great men, enjoyed great popularity.

Stalin thought that these facts could not be ignored if we wanted to understand the feelings of the Polish people. I had said that I feared the Conference separating before agreement was reached. What then was to be done? The various Governments had different information, and drew different conclusions from it. Perhaps the first thing was to call together the Poles from the different camps and hear what they had to say.

There was dissatisfaction, he continued, because the Polish Government was not elected. It would naturally be better to have a Government based on free elections, but the war had so far prevented that. But the day was near when elections could be held. Until then we must deal with the Provisional Government, as we had dealt, for instance, with General de Gaulle's Government in France, which also was not elected. He did not know whether Bierut or General de Gaulle enjoyed greater authority, but it had been possible to make a treaty with General de Gaulle, so why could we not do the same with an enlarged Polish Government, which would be no less democratic? It was not reasonable to demand more from Poland than from France. So far the French Government had carried out no reform which created enthusiasm in France, whereas the Polish Government had enacted a land reform which had aroused great enthusiasm. If we approached the matter without prejudice we should be able to find a common ground. The situation was not as tragic as I thought, and the question could be settled if too much importance was not attached to secondary matters and if we concentrated on essentials.

"How soon," asked the President, "will it be possible to hold elections?"

"Within a month," Stalin replied, "unless there is some catastrophe on the front, which is improbable."

I said that this would of course set our minds at rest, and we could wholeheartedly support a freely elected Government which would supersede everything else, but we must not ask for anything which would in any way hamper the military operations. These were the supreme end. If however the will of the Polish people could be ascertained in so short a time, or even within two months, the situation would be entirely different and no one could oppose it.

We thereupon agreed to let our Foreign Secretaries talk the matter over.

* * * * *

The three Ministers accordingly met at noon on February 9. They were unable to agree. When however the Conference assembled in plenary session at four o'clock in the afternoon Molotov produced some fresh proposals which were much nearer to the American draft. The Lublin Government was to be "reorganised on a wider democratic basis, with the inclusion of democratic leaders from Poland itself, and also from those living abroad". He and the British and American Ambassadors should consult together in Moscow about how this would be done. Once reorganised the Lublin Government would be pledged to hold free elections as soon as possible, and we should then recognise whatever Government emerged. Mr. Stettinius had desired a written pledge that the three Ambassadors in Warsaw should observe and report that the elections were really free and unfettered, but Molotov opposed this, because, he alleged, it would offend the Poles. Subject to this and to a few minor amendments, he accepted the United States plan.

This was a considerable advance, and I said so, but I felt it my duty to sound a general warning. This would be the last but one of our meetings.* There was an atmosphere of agreement, but there was also a desire to put foot in the stirrup and be off. We could not, I declared, afford to allow the settlement of these important matters to be hurried and the fruits of the Conference lost for lack of another twenty-four hours. A great prize was in view and decisions must be unhurried. These might well be among the most important days in our lives.

Mr. Roosevelt declared that the differences between us and the Russians were now largely a matter of words, but both he and I were anxious that the elections should really be fair and free. I told Stalin that we were at a great disadvantage, because we knew

---

* Our meeting on February 11 merely approved the report on the Conference. Serious discussion ended on February 10.

so little of what was going on inside Poland and yet had to take decisions of great responsibility. I knew, for instance, that there was bitter feeling among the Poles and that M. Osóbka-Morawski had used very fierce language, and I had been told that the Lublin Government had openly said it would try as traitors all members of the Polish Home Army and Underground movement. This, I said, caused me anxiety and distress. Of course I put the security of the Red Army first, but I begged Stalin to consider our difficulty. The British Government did not know what was going on inside Poland, except through dropping brave men by parachute and bringing members of the Underground movement out. We had no other means of knowing, and did not like getting our information in this way. How could this be remedied without in any way hampering the movements of the Soviet troops? Could any facilities be granted to the British (and no doubt to the United States) for seeing how these Polish quarrels were being settled? Tito had said that when elections took place in Yugoslavia he would not object to Russian, British, and American observers being present to report impartially to the world that they had been carried out fairly. So far as Greece was concerned, His Majesty's Government would greatly welcome American, Russian, and British observers to make sure the elections were conducted as the people wished. The same question would arise in Italy. When Northern Italy was delivered there would be a vast change in the Italian political situation, and there would have to be an election before it was possible to form a Constituent Assembly or Parliament. The British formula there was the same—Russian, American, and British observers should be present to assure the world that everything had been done in a fair way. It was impossible, I said, to exaggerate the importance of carrying out elections fairly. For instance, would Mikolajczyk be able to go back to Poland and organise his party for the elections?

"That will have to be considered by the Ambassadors and M. Molotov when they meet the Poles," said Stalin.

I replied, "I must be able to tell the House of Commons that the elections will be free and that there will be effective guarantees that they are freely and fairly carried out."

Stalin pointed out that Mikolajczyk belonged to the Peasant Party, which, as it was not a Fascist party, could take part in the elections and put up its candidates. I said that this would be still more certain if the Peasant Party were already represented in the Polish Government, and Stalin agreed that the Government should include one of their representatives.

I said that we should have to leave it at that, and added that I hoped that nothing I had said had given offence, since nothing had been further from my heart.

"We shall have to hear," he answered, "what the Poles have to say." I explained that I wanted to be able to carry the eastern frontier question through Parliament, and I thought this might be done if Parliament was satisfied that the Poles had been able to decide for themselves what they wanted.

"There are some very good people among them," he replied. "They are good fighters, and they have had some good scientists and musicians, but they are very quarrelsome."

"All I want," I answered, "is for all sides to get a fair hearing."

"The elections," said the President, "must be above criticism, like Cæsar's wife. I want some kind of assurance to give to the world, and I don't want anybody to be able to question their purity. It is a matter of good politics rather than principle."

"I am afraid," said Molotov, "that if we insert the American draft the Poles will feel they are not trusted. We had better discuss it with them."

I was not content with this, and resolved to raise it with Stalin later on. The opportunity presented itself next day.

* * * * *

Just before our last effective meeting, on February 10, Mr. Eden and I had a private conversation with Stalin and Molotov at the Yusupov Villa, at which I once more explained how difficult it was for us to have no representatives in Poland who could report what was going on. The alternatives were either an Ambassador with an embassy staff or newspaper correspondents. The latter was less desirable, but I pointed out that I should be asked in Parliament about the Lublin Government and the elections and I must be able to say that I knew what was happening.

"After the new Polish Government is recognised it would be open to you to send an Ambassador to Warsaw," Stalin answered.

"Would he be free to move about the country?"

"As far as the Red Army is concerned, there will be no interference with his movements, and I promise to give the necessary instructions, but you will have to make your own arrangements with the Polish Government."

Stalin also pointed out that de Gaulle had a representative in Poland.

We then agreed to add the following to our declaration:

As a consequence of the above, recognition would entail an exchange of Ambassadors, by whose reports the respective Governments would be informed about the situation in Poland.

This was the best I could get.

* * * * *

When the Conference reassembled at a quarter to five Mr. Eden read out a statement which the three Foreign Secretaries had agreed. I was concerned to note that it said nothing about frontiers, and I said that the whole world would want to know why. We were all agreed in principle about the western frontier, and the only question was where exactly the line should be drawn and how much we should say about it. The Poles should have part of East Prussia and be free to go up to the line of the Oder if they wished, but we were very doubtful about going any farther or saying anything on the question at this stage, and I told the Conference that we had had a telegram from the War Cabinet which strongly deprecated any reference to a frontier as far west as the Western Neisse because the problem of moving the population was too big to manage.

Mr. Roosevelt said he would prefer to hear what the new Polish Government of National Unity said about it, and suggested omitting all reference to the line in the west.

"We should certainly mention the eastern frontier," said Stalin.

I supported him in this, although I knew there would be much criticism.

As for the western frontier, I said that the wishes of the new Polish Government should first be ascertained, and that the frontier itself should be determined as part of the peace settlements. After some further discussion, which was complicated by the President's inability under the United States Constitution to settle matters of this kind without the approval of the Senate, we eventually agreed what to do. The communiqué issued at the end of the Conference accordingly included a joint declaration about Poland, and ran as follows:

11 Feb 45

We came to the Crimea Conference resolved to settle our differences about Poland. We discussed fully all aspects of the question. We reaffirm our common desire to see established a strong, free, independent, and democratic Poland. As a result of our discussions we have agreed on the conditions in which a new Polish Provisional Government of National Unity may be formed in such a manner as to command recognition by the three major Powers.

The agreement reached is as follows:

A new situation has been created in Poland as a result of her complete liberation by the Red Army. This calls for the establishment of a Polish Provisional Government which can be more broadly based than was possible before the recent liberation of Western Poland. The Provisional Government which is now functioning in Poland should therefore be reorganised on a broader democratic basis, with the inclusion of democratic leaders from Poland itself and from Poles abroad. This new Government should then be called the Polish Provisional Government of National Unity.

M. Molotov, Mr. Harriman, and Sir A. Clark Kerr are authorised as a commission to consult in the first instance in Moscow with members of the present Provisional Government and with other Polish democratic leaders from within Poland and from abroad with a view to the reorganisation of the present Government along the above lines. This Polish Provisional Government of National Unity shall be pledged to the holding of free and unfettered elections as soon as possible on the basis of universal suffrage and secret ballot. In these elections all democratic and anti-Nazi parties shall have the right to take part and to put forward candidates.

When a Polish Provisional Government of National Unity has been properly formed in conformity with the above the Government of the Union of Soviet Socialist Republics, which now maintains diplomatic relations with the present Provisional Government of Poland, and the Government of the United Kingdom and the Government of the United States will establish diplomatic relations with the new Polish Government of National Unity, and will exchange Ambassadors, by whose reports the respective Governments will be kept informed about the situation in Poland.

The three heads of Governments consider that the eastern frontier of Poland should follow the Curzon Line, with digressions from it in some regions of five to eight kilometres in favour of Poland. They recognise that Poland must receive substantial accessions of territory in the north and west. They feel that the opinion of the new Polish Provisional Government of National Unity should be sought in due course on the extent of these accessions, and that the final delimitation of the western frontier of Poland should thereafter await the Peace Conference.

CHAPTER XXIII

YALTA FINALE

America; Russia, and the Far East – A Secret Agreement – My Private Talk with Stalin, February 8 – My Telegram to the Dominion Prime Ministers of July 5 – Our Final Dinner at the Vorontzov Palace, February 10 – Stalin and the Toast to the King – My Toast to Stalin – Stalin's Views on the British General Election – "Uncle Joe" – End of the Conference – We Drive to Sebastopol – A Visit to Balaclava – Return to Saki – A Flying Visit to Athens – My Speech in Constitution Square – We Leave for Egypt, February 15 – Family Lunch with the President and an Affectionate Farewell – I Meet King Ibn Saud – An Interchange of Gifts – I Stay at the Casey Villa – We Fly to England, February 19 – The Yalta Debate – Deep Anxieties About Poland.

THE Far East played no part in our formal discussions at Yalta. I was aware that the Americans intended to raise with the Russians the question of Soviet participation in the Pacific war. We had touched on this in general terms at Teheran, and in December 1944 Stalin had made certain specific proposals about Russia's post-war claims in these regions to Mr. Harriman in Moscow. The American military authorities estimated that it would take eighteen months after the surrender of Germany to defeat Japan. Russian help would reduce heavy American casualties. The invasion of the Japanese home islands was at this time still in the planning stage, and General MacArthur had entered Manila only on the second day of the Yalta Conference. The first experimental explosion of the atomic bomb was not to take place for another five months. The large Japanese army in Manchuria could, if Russia still remained neutral, be thrown into the battle for Japan itself.

With all this in mind President Roosevelt and Mr. Harriman

discussed the Russian territorial demands in the Far East with Stalin on February 8. The only other person present apart from the Russian interpreter was Mr. Charles E. Bohlen, of the State Department, who also interpreted. Two days later the conversation was continued, and the Russian terms were accepted, with certain modifications, which Mr. Harriman mentioned in his testimony before the United States Senate in 1951. In return Russia agreed to enter the war against Japan within two or three months after the surrender of Germany.

During the same afternoon, in a private talk with Stalin, I asked him about Russian wishes in the Far East. He said that they wanted a naval base, such as Port Arthur. The Americans would prefer the ports to be under international control, but the Russians wanted to have their interests safeguarded. I replied that we would welcome the appearance of Russian ships in the Pacific, and were in favour of Russia's losses in the Russo-Japanese War being made good. The following day, February 11, I was shown the agreement which had been drafted the previous afternoon by the President and Stalin, and I signed it on behalf of the British Government. This document was kept secret until negotiations were completed between the Soviet Union and the Nationalist Chinese Government, which Stalin categorically agreed to support. There the matter rested until just before we met again at Potsdam.

My record of these negotiations is preserved in the following extract from a telegram which I sent to the Dominion Prime Ministers on July 5.

In the most rigid secrecy Stalin informed Roosevelt and myself at the Crimea Conference of the Soviet Government's willingness to enter the war against Japan two or three months after Germany's surrender, on the conditions stated below:

(a) Preservation of the *status quo* in Outer Mongolia.
(b) Restoration of the Russian rights lost in the year 1904, viz.:

    (i) Recovery of Southern Sakhalin and the islands adjacent to it.
    (ii) Internationalisation of the commercial port of Dairen, with safeguards for the pre-eminent interests of the U.S.S.R. and restoration of the lease of Port Arthur as a Soviet naval base.
    (iii) Joint operation by a Soviet-Chinese company of the

Chinese Eastern Railway and the South Manchuria Railway, providing an outlet to Dairen, on the understanding that the pre-eminent interests of the U.S.S.R. will be safeguarded and that China will retain full sovereignty in Manchuria.

(c) Acquisition by the U.S.S.R. of the Kurile Islands.

2. These conditions were embodied in a personal agreement between Roosevelt, Stalin, and myself. The agreement recognised that Chiang Kai-shek's concurrence to the conditions would be required, and Roosevelt undertook to obtain this concurrence on advice from Stalin. We all three agreed to see that the Soviet claims were fulfilled without question following the defeat of Japan. The agreement contained nothing else, except an expression of Russian readiness to enter into a treaty of alliance with China with the object of helping the latter to throw off the Japanese yoke.

I must make it clear that though on behalf of Great Britain I joined in the agreement neither I nor Eden took any part in making it. It was regarded as an American affair, and was certainly of prime interest to their military operations. It was not for us to claim to shape it. Anyhow, we were not consulted, but only asked to approve. This we did. In the United States there have been many reproaches about the concessions made to Soviet Russia. The responsibility rests with their own representatives. To us the problem was remote and secondary. It would have been wrong for us to get in their way unless we had some very solid reason.

* * * * *

It was my turn to preside at our final dinner on February 10. Several hours before Stalin was due to arrive a squad of Russian soldiers came to the Vorontzov villa. They locked the doors on either side of the reception rooms which were to be used for dinner. Guards were posted and no one was allowed to enter. They then searched everywhere—under the tables and behind the walls. My staff had to go outside the building in order to get from their offices to their own quarters. All being in order, the Marshal arrived in a most cordial mood, and the President a little later.

At the Yusupov dinner Stalin had proposed the King's health in a manner which, though meant to be friendly and respectful, was not to my liking. He had said that in general he had always been against kings, and that he was on the side of the people and not that of any king, but in this war he had learnt to honour and esteem the British people, who honoured and respected their King, so he would propose the health of the King of England. I was not satisfied with this treatment of the toast, and I asked Molotov to explain that Stalin's scruples might be avoided by proposing on future occasions the health of "the three heads of States". This having been agreed, I now put into practice the new procedure:

I propose the health of His Majesty the King, the President of the United States, and President Kalinin of the U.S.S.R., the three heads of the three States.

To this the President, who seemed very tired, replied. "The Prime Minister's toast," he said, "recalls many memories. In 1933 my wife visited a school in our country. In one of the classrooms she saw a map with a large blank space on it. She asked what was the blank space, and was told they were not allowed to mention the place—it was the Soviet Union. That incident was one of the reasons why I wrote to President Kalinin asking him to send a representative to Washington to discuss the opening of diplomatic relations. That is the history of our recognition of Russia."

It was now my task to propose the health of Marshal Stalin. I said:

I have drunk this toast on several occasions. This time I drink it with a warmer feeling than at previous meetings, not because he is more triumphant, but because the great victories and the glory of the Russian arms have made him kindlier than he was in the hard times through which we have passed. I feel that, whatever differences there may be on certain questions, he has a good friend in Britain. I hope to see the future of Russia bright, prosperous, and happy. I will do anything to help, and I am sure so will the President. There was a time when the Marshal was not so kindly towards us, and I remember that I said a few rude things about him, but our common dangers and common loyalties have wiped all that out. The fire of war has burnt up the misunderstandings of the past. We feel we have a friend whom we can trust, and I hope he will continue to feel the same about us. I pray he may live to see his beloved Russia not only glorious in war, but also happy in peace.

Stalin replied in the best of tempers, and I had the feeling that he thought the "heads of States" procedure was well adapted to our triple meetings. I have no record of his actual words. Including interpreters, we were less than a dozen, and after the formalities we talked together in twos and threes. I had mentioned that there would be a General Election in the United Kingdom after the defeat of Hitler. Stalin thought my position was assured, "since the people would understand that they needed a leader, and who could be a better leader than he who had won the victory?" I explained that we had two parties in Britain, and I only belonged to one of them. "One party is much better," said Stalin, with deep conviction. I then thanked him for his hospitality to the British Parliamentary delegation which had recently visited Russia. Stalin said that it had been his duty to show hospitality, and he liked young military fighters like Lord Lovat. Of late he had acquired a new interest in life, an interest in military affairs; in fact, it had become almost his sole interest.

This led the President to speak of the British Constitution. He said that I was always talking about what the Constitution allowed and what it did not allow, but actually there was no Constitution. However, an unwritten Constitution was better than a written one. It was like the Atlantic Charter; the document did not exist, yet all the world knew about it. Among his papers he had found one copy signed by himself and me, but strange to say both signatures were in his own handwriting. I replied that the Atlantic Charter was not a law, but a star.

As our talk continued Stalin spoke of what he called "the unreasonable sense of discipline in the Kaiser's Germany", and recounted an incident which occurred when he was in Leipzig as a young man. He had come with two hundred German Communists to attend an International Conference. Their train arrived punctually at the station, but there was no official to collect their tickets. All the German Communists therefore waited docilely for two hours to get off the platform. So none of them were able to attend the meeting for which they had travelled far.

In this easy manner the evening passed away agreeably. When the Marshal left many of our British party had assembled in the hall of the villa, and I called for "Three cheers for Marshal Stalin", which were warmly given.

* * * * *

There was another occasion during our stay at Yalta when things had not gone so smoothly. Mr. Roosevelt, who was host at a luncheon, said that he and I always referred to Stalin in our secret telegrams as "Uncle Joe". I had suggested that he should tell him this privately, but instead the President made it into a jocular statement to the company. This led to a difficult moment. Stalin took offence. "When can I leave this table?" he asked in anger. Mr. Byrnes saved the situation with an apt remark. "After all," he said, "you do not mind talking about Uncle Sam, so why should Uncle Joe be so bad?" At this the Marshal subsided, and Molotov later assured me that he understood the joke. He already knew that he was called Uncle Joe by many people abroad, and he realised that the name had been given in a friendly way and as a term of affection.

* * * * *

The next day, Sunday, February 11, was the last of our Crimean visit. As usual at these meetings many grave issues were left unsettled. The Polish communiqué laid down in general terms a policy which if carried out with loyalty and good faith might indeed have served its purpose pending the general Peace Treaty. The agreement about the Far East which the President and his advisers had made with the Russians to induce them to enter the war against Japan was not one which concerned us directly. It has since become a matter of fierce controversy in the United States. The President was anxious to go home, and on his way to pay a visit to Egypt, where he could discuss the affairs of the Middle East with various potentates. Stalin and I lunched with him in the Czar's former billiard-room at the Livadia Palace. During the meal we signed the final documents and official communiqués. All now depended upon the spirit in which they were carried out.

* * * * *

The same afternoon Sarah and I drove to Sebastopol, where the liner *Franconia* was berthed. She had come through the Dardanelles and acted as the headquarters ship, which could also be used in case accommodation on shore at Yalta broke down. We went on board, where I was joined by Sir Alan Brooke and the other Chiefs of Staff. From the deck we looked out over the port, which the Germans had practically destroyed, though now it was full of activity again and in the night-time its ruins blazed with lights.

I was anxious to see the field of Balaclava, and I asked Brigadier Peake of the War Office Intelligence Staff to look up all the details of the action and prepare himself to show us round. On the afternoon of February 13 I visited the scene, accompanied by the Chiefs of Staff and the Russian admiral commanding the Black Sea Fleet, who had had orders from Moscow to be in attendance on me whenever I came ashore. We were a little shy and very tactful with our host. But we need not have worried. As Peake pointed to the line on which the Light Brigade had been drawn up the Russian admiral pointed in almost the same direction and exclaimed, "The German tanks came at us from over there." A little later Peake explained the Russian dispositions, and pointed to the hills where their infantry had stood, whereupon the Russian admiral intervened with obvious pride: "That is where a Russian battery fought and died to the last man." I thought it right at this juncture to explain that we were studying a different war, "a war of dynasties, not of peoples". Our host gave no sign of comprehension, but seemed perfectly satisfied. So all passed off very pleasantly.

Before us lay the valley down which the Light Brigade had charged, and we could see the ridge which had been so gallantly defended by the Highlanders. As the scene lay before us one could grasp the situation which Lord Raglan had faced some ninety years earlier. We had visited his tomb in the morning, and were greatly struck by the care and respect with which it had been treated by the Russians.

*   *   *   *   *

I had much looked forward to the sea voyage through the Dardanelles to Malta, but I felt it my duty to make a lightning trip to Athens and survey the Greek scene after the recent troubles. Early on February 14 accordingly we set off by car for Saki, where our aeroplane awaited us. Eden had already left in advance. As we drove over the winding mountain road we passed a chasm into which the Germans had pitched scores of locomotives. At the airfield a splendid guard of honour of N.K.V.D. troops was drawn up. I inspected them in my usual manner, looking each man straight in the eye. This took some time, as there were at least two hundred of them, but it was commented on in a favourable way by the Soviet Press. I made a farewell speech before entering the plane.

We flew without incident to Athens, making a loop over the island of Skyros to pass over the tomb of Rupert Brooke, and were received at the airfield by the British Ambassador, Mr. Leeper, and General Scobie. Only seven weeks before I had left the Greek capital rent by street-fighting. We now drove into it in an open car, where only a thin line of kilted Greek soldiers held back a vast mob, screaming with enthusiasm, in the very streets where hundreds of men had died in the Christmas days when I had last seen the city. That evening a huge crowd of about fifty thousand people gathered in Constitution Square. The evening light was wonderful as it fell on these classic scenes. I had no time to prepare a speech. Our security services had thought it important that we should arrive with hardly any notice. I addressed them with a short harangue.

Your Beatitude, soldiers and citizens of Athens, and of Greece, these are great days. These are days when dawn is bright, when darkness rolls away. A great future lies before your country.

There has been much misunderstanding and ignorance of our common cause in many parts of the world, and there have been misrepresentations of issues fought out here in Athens. But now these matters are clearing, and there is an understanding of the part Greece has played and will play in the world.

Speaking as an Englishman, I am very proud of the part which the British Army played in protecting this great and immortal city against violence and anarchy. Our two countries have for long marched together along hard, dusty roads in friendship and in loyalty.

Freedom and prosperity and happiness are dear to all nations of the British Commonwealth and Empire. We who have been associated with you in the very long struggle for Greek liberty will march with you till we reach the end of the dark valley, and we will march with you till we reach the broad highlands of justice and peace.

Let no one fail in his duty towards his country. Let no one swerve off the high road of truth and honour. Let no one fail to rise to the occasion of this great moment and of these splendid days. Let the Greek nation stand first in every heart. Let it stand first in every man and woman. Let the future of Greece shine brightly in their eyes.

From the bottom of my heart I wish you prosperity. From the bottom of my heart I hope that Greece will take her proper place in the circle of victorious nations—of nations who have suffered terribly in war. Let right prevail. Let party hatreds die. Let there be unity, let there be resolute comradeship.

Greece for ever! Greece for all!

That evening I dined at our shot-scarred Embassy, and in the early hours of February 15 we took off in my plane for Egypt.

*   *   *   *   *

At Alexandria I went on board H.M.S. *Aurora*. I had taken no part in the discussions between the President and the Middle East sovereigns who had been invited to meet him, King Farouk Haille Selassie, and Ibn Saud. These conversations had taken place on board the *Quincy*, which had been anchored in the Bitter Lake. Later that morning the American cruiser steamed into Alexandria harbour, and shortly before noon I went on board for what was to be my last talk with the President. We gathered afterwards in his cabin for an informal family luncheon. I was accompanied by Sarah and Randolph, and Mr. Roosevelt's daughter, Mrs. Boettiger, joined us, together with Harry Hopkins and Mr. Winant. The President seemed placid and frail. I felt that he had a slender contact with life. I was not to see him again. We bade affectionate farewells. That afternoon the Presidential party sailed for home.

*   *   *   *   *

After the departure of our American friends I had arranged a meeting with Ibn Saud. He had been transported to the conference with the President in the American destroyer *Murphy*, and travelled with all the splendour of an Eastern potentate, with an entourage of some fifty persons, including two sons, his Prime Minister, his Astrologer, and flocks of sheep to be killed according to Moslem rites. On February 17 his reception was organised at the Hôtel du Lac at Fayoum oasis, from which we had temporarily removed all the residents. A number of social problems arose. I had been told that neither smoking nor alcoholic beverages were allowed in the Royal Presence. As I was the host at luncheon I raised the matter at once, and said to the interpreter that if it was the religion of His Majesty to deprive himself of smoking and alcohol I must point out that my rule of life prescribed as an absolutely sacred rite smoking cigars and also the drinking of alcohol before, after, and if need be during all meals and in the intervals between them. The King graciously accepted the position. His own cup-bearer from Mecca offered me a glass of water from its sacred well, the most delicious that I had ever tasted.

It had been indicated to me beforehand that there would be an interchange of presents during the course of our meeting. I had therefore made what I thought were adequate arrangements. "Tommy" Thompson bought for me in Cairo for about a hundred pounds, at the Government's expense, a little case of very choice perfumes, which I presented. We were all given jewelled swords, diamond-hilted, and other splendid gifts. Sarah had an enormous portmanteau which Ibn Saud had provided for "your womenfolk". It appeared that we were rather outclassed in gifts, so I told the King, "What we bring are but tokens. His Majesty's Government have decided to present you with the finest motor-car in the world, with every comfort for peace and every security against hostile action." This was later done.

King Ibn Saud made a striking impression. My admiration for him was deep, because of his unfailing loyalty to us. He was always at his best in the darkest hours. He was now over seventy, but had lost none of his warrior vigour. He still lived the existence of a patriarchal king of the Arabian desert, with his forty living sons and the seventy ladies of his harem, and three of the four official wives, as prescribed by the Prophet, one vacancy being kept.

We returned from Fayoum to Cairo, stopping on the way at the British Ambassador's desert house, where we had tea. I stayed for a few days at the Casey villa, and had interviews both with King Farouk and the President of Syria, at which we talked of the recent difficulties in the Middle East, many of which still continue. Sarah meanwhile unpacked the portmanteau which Ibn Saud had given her. It contained many splendid and beautiful Arab robes and several vessels of rare and delicious perfumes. At the bottom were half a dozen cardboard boxes of different sizes. One of these contained a diamond with the valuation £1,200 attached. There were a number of other gems, and several necklaces of Red Sea pearls. Anthony had a similar set of presents, though in his case the diamond was adjusted to his rank. When eventually I reported on these matters to the Cabinet we told them that of course we should not keep for ourselves any of the valuables. In fact the Treasury, to whom they were given, almost paid with them for the motor-car which I had taken it upon myself to present to Ibn Saud.

On February 19 I flew back to England. Northolt was fog-bound, and our plane was diverted to Lyneham. I drove on to London by car, stopping at Reading to join my wife, who had come to meet me.

*　　*　　*　　*　　*

At noon on February 27 I asked the House of Commons to approve the results of the Crimea Conference. I said:

I am anxious that all parties should be united in this new instrument, so that these supreme affairs shall be, in Mr. Gladstone's words, "high and dry above the ebb and flow of party politics". . . . The Crimea Conference leaves the Allies more closely united than before, both in the military and in the political sphere. Let Germany ever recognise that it is futile to hope for division among the Allies and that nothing can avert her utter defeat. Further resistance will only be the cause of needless suffering. The Allies are resolved that Germany shall be totally disarmed, that Nazism and militarism in Germany shall be destroyed, that war criminals shall be justly and swiftly punished, that all German industry capable of military production shall be eliminated or controlled, and that Germany shall make compensation in kind to the utmost of her ability for damage done to Allied nations. On the other hand, it is not the purpose of the Allies to destroy the people of Germany, or leave them without the necessary means of subsistence. Our policy is not revenge; it is to take such measures as may be necessary to secure the future peace and safety of the world. There will be a place one day for Germans in the comity of nations, but only when all traces of Nazism and militarism have been effectively and finally extirpated.

Poland was the issue which disturbed the House.

The three Powers are agreed that acceptance by the Poles of the provisions on the eastern frontiers, and, so far as can now be ascertained, on the western frontiers, is an essential condition of the establishment and future welfare and security of a strong, independent, homogeneous Polish State. . . . But even more important than the frontiers of Poland, within the limits now disclosed, is the freedom of Poland. The home of the Poles is settled. Are they to be masters in their own house? Are they to be free, as we in Britain and the United

States or France are free? Are their sovereignty and their independence to be untrammelled, or are they to become a mere projection of the Soviet State, forced against their will by an armed minority to adopt a Communist or totalitarian system? I am putting the case in all its bluntness. It is a touchstone far more sensitive and vital than the drawing of frontier lines. Where does Poland stand? Where do we all stand on this?

Most solemn declarations have been made by Marshal Stalin and the Soviet Union that the sovereign independence of Poland is to be maintained, and this decision is now joined in both by Great Britain and the United States. Here also the World Organisation will in due course assume a measure of responsibility. The Poles will have their future in their own hands, with the single limitation that they must honestly follow, in harmony with their Allies, a policy friendly to Russia. That is surely reasonable. . . .

The agreement provides for consultations, with a view to the establishment in Poland of a new Polish Provisional Government of National Unity, with which the three major Powers can all enter into diplomatic relations, instead of some recognising one Polish Government and the rest another. . . . His Majesty's Government intend to do all in their power to ensure that . . . representative Poles of all democratic parties are given full freedom to come and make their views known.

I felt bound to proclaim my confidence in Soviet good faith in the hope of procuring it. In this I was encouraged by Stalin's behaviour about Greece.

The impression I brought back from the Crimea, and from all my other contacts, is that Marshal Stalin and the Soviet leaders wish to live in honourable friendship and equality with the Western democracies. I feel also that their word is their bond. I know of no Government which stands to its obligations, even in its own despite, more solidly than the Russian Soviet Government. I decline absolutely to embark here on a discussion about Russian good faith. It is quite evident that these matters touch the whole future of the world. Sombre indeed would be the fortunes of mankind if some awful schism arose between the Western democracies and the Russian Soviet Union.

I continued:

We are now entering a world of imponderables, and at every stage occasions for self-questioning arise. It is a mistake to look too far ahead. Only one link in the chain of destiny can be handled at a time.

I trust the House will feel that hope has been powerfully strengthened by our meeting in the Crimea. The ties that bind the three Great

Powers together and their mutual comprehension of each other have grown. The United States has entered deeply and constructively into the life and salvation of Europe. We have all three set our hands to far-reaching engagements at once practical and solemn.

*　*　*　*　*

The general reaction of the House was unqualified support for the attitude we had taken at the Crimea Conference. There was however intense moral feeling about our obligations to the Poles, who had suffered so much at German hands and on whose behalf as a last resort we had gone to war. A group of about thirty Members felt so strongly on this matter that some of them spoke in opposition to the motion which I had moved. There was a sense of anguish lest we should have to face the enslavement of a heroic nation. Mr. Eden supported me. In the division on the second day we had an overwhelming majority, but twenty-five Members, most of them Conservatives, voted against the Government, and in addition eleven members of the Government abstained. Mr. H. G. Strauss, who was Parliamentary Secretary to the Ministry of Town and Country Planning, resigned.

It is not permitted to those charged with dealing with events in times of war or crisis to confine themselves purely to the statement of broad general principles on which good people agree. They have to take definite decisions from day to day. They have to adopt postures which must be solidly maintained, otherwise how can any combinations for action be maintained? It is easy, after the Germans are beaten, to condemn those who did their best to hearten the Russian military effort and to keep in harmonious contact with our great Ally, who had suffered so frightfully. What would have happened if we had quarrelled with Russia while the Germans still had two or three hundred divisions on the fighting front? Our hopeful assumptions were soon to be falsified. Still, they were the only ones possible at the time.

CHAPTER XXIV

CROSSING THE RHINE

General Eisenhower's Double Thrust into Germany – British Doubts – Montgomery's Advance to the Rhine – The Enemy are Expelled from the Wesel Bridgehead, March 10 – Cologne Captured, March 7 – A Stroke of Fortune for the Twelfth Army Group – The Last German Stand in the West – Plans and Preparations for Crossing the Rhine – I Visit Montgomery's Headquarters, March 23 – And Watch the Fly-in, March 24 – Heavy Fighting at Wesel and Rees – An Evening in Montgomery's Map Wagon – I Visit Eisenhower, March 25 – And Cross the Rhine – Speedy Progress by the American Armies – The Collapse of Germany's Western Front.

DESPITE their defeat in the Ardennes,* the Germans decided to give battle west of the Rhine, instead of withdrawing across it to gain a breathing-space. General Eisenhower planned three operations. In the first he would destroy the enemy west of the river and close up to it, then he would form bridgeheads, and finally drive deep into Germany. In this last phase there would be two simultaneous thrusts. One would start from the lower Rhine below Duisburg, skirt the northern edge of the Ruhr, which would be enveloped and later subdued, and drive across the North German plain towards Bremen, Hamburg, and the Baltic. The second thrust would be from Karlsruhe to Kassel, whence there were options northwards or eastwards according to circumstances.

We had reviewed this plan at Malta with some concern. We doubted whether we were strong enough for two great simultaneous operations, and felt that the northern advance by Montgomery's Twenty-first Army Group would be much the more important. Only thirty-five divisions probably could take part,

* See Chapter XVII, pp. 238 *et seq.*

353

but we held that the maximum effort should be made here, whatever its size, and that it should not be weakened for the sake of the other thrust. The matter was keenly argued by the Combined Chiefs of Staff. General Bradley* attributes to Montgomery most of the pressure which was brought to bear. This is not a fair assessment. The British view as a whole was that the northern thrust, with its consequences to the Ruhr, was paramount. In a second respect also we questioned the plan. We were anxious that Montgomery should cross the Rhine as soon as possible, and not be held back merely because German forces were still on the near bank at some distant point. General Bedell Smith, Eisenhower's Chief of Staff, came to Malta and gave us assurances. Eisenhower has said in his official report, "The plan of campaign for crossing the Rhine and establishing a strong force on the far bank was, thanks to the success of the operations west of the river, basically the same as that envisaged in our long-term planning in January, and even before D Day. The fundamental features were the launching of a main attack to the north of the Ruhr, supported by a strong secondary thrust from bridgeheads in the Frankfurt area. Subsequently offensives would strike out from the bridgeheads to any remaining organised forces and complete their destruction."†

In terms of divisions we were evenly matched. In early February Eisenhower and the Germans had about eighty-two apiece. But there was a vast difference in quality. Allied morale was high, the Germans were badly shaken. Our ranks were battle-trained and confident. The enemy were scraping up their last reserves, and in January Hitler had sent the ten divisions of his Sixth Panzer Army to try to save the oilfields of Austria and Hungary from the Russians. Our bombing had grievously injured his factories and communications. He was desperately short of petrol, and his Air Force was only a shadow.

✱   ✱   ✱   ✱   ✱

The first task was to clear the enemy from the Colmar pocket. This was completed at the beginning of February by the First French Army, helped by four American divisions. More important, and leading to a long and arduous battle, was Montgomery's advance to the Rhine north of Cologne. A thrust from the

* Omar Bradley, *A Soldier's Story.*
† Eisenhower's report to Combined Chiefs of Staff, p. 118.

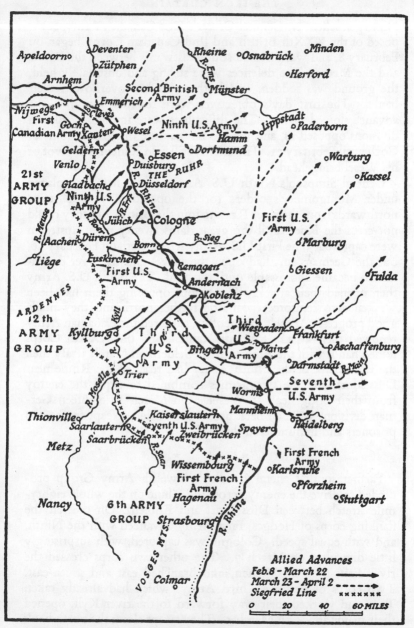

Apeldoorn Deventer   Rheine   Osnabrück   Minden
          Zütphen   Herford
Arnhem
                    Second British   Münster
          Emmerich   Army
Nijmegen   Cleves   Ninth U.S. Army   Lippstadt   Paderborn
First   Goch   Wesel   Hamm
Canadian Army Xanten   Dortmund
          Geldern   Essen
          Venlo   Duisburg THE RUHR   Warburg
21st   Düsseldorf   Kassel
ARMY   Gladbach   Rhine
GROUP   Ninth U.S.
          Army R. Erft
                Jülich   Cologne   First U.S.
          R. Roer   Army
          Düren   R. Sieg
R. Meuse Aachen   Born   Marburg
   Liége   Euskirchen   Remagen   Giessen
            First U.S.   Andernach   Fulda
             Army   Koblenz
ARDENNES   Third
   12th   Wiesbaden   Frankfurt
ARMY   Kyllburg   Third   Bingen   U.S.   Mainz   Aschaffenburg
GROUP   U.S.   Army   Darmstadt   R. Main
           R. Kyll Army
                Trier   Worms   Seventh
         R. Moselle   U.S. Army
Thionville   Mannheim
         Saarlautern Kaiserslautern
                   Seventh U.S. Army   Speyer   Heidelberg
Metz   Saarbrücken Zweibrücken
                 R. Saar   First French
                    Wissembourg   Karlsruhe   Army
                     First French   Pforzheim
Nancy   6th ARMY   Hagenau   Stuttgart
           GROUP   Strasbourg
                              R. Rhine
                   Colmar

Allied Advances
Feb.8 – March 22   ⟶
March 23 – April 2   ⤏
Siegfried Line   ×××××

0   20   40   60 MILES

CROSSING THE RHINE
355

Nijmegen salient by General Crerar's First Canadian Army, composed of the XXXth British and IInd Canadian Corps, began on February 8, and was directed south-eastwards between the Rhine and the Meuse. The defences were strong and obstinately held, the ground was sodden, and both rivers had overflowed their banks. The first day's objectives were reached, but then the advance slowed down. The difficulties were immense. Eleven divisions opposed us, and we did not capture the strong-point of Goch until February 21. The enemy still held Xanten, the pivot of his Wesel bridgehead.

General Simpson's Ninth U.S. Army, which had been placed under Montgomery's orders for the operation, was to strike northwards from the river Roer to join the British, but they could not cross the Roer until the great dams twenty miles upstream were captured. The First U.S. Army seized them on February 10, but the Germans smashed open the valves and the river downstream became uncrossable until the 23rd. The Ninth U.S. Army then opened their attack. The troops opposing them had been weakened to reinforce the battle farther north and they made good progress. As their momentum gathered the Canadian Army renewed their attacks towards Xanten, and the XXXth Corps joined hands with the Americans at Geldern on March 3. By then the right flank of the Ninth Army had reached the Rhine near Düsseldorf, and the two armies combined to expel the enemy from their bridgehead at Wesel. On March 10 eighteen German divisions were all back across the Rhine, except for 53,000 prisoners and unnumbered dead.

* * * * *

Farther south General Bradley's Twelfth Army Group proceeded to drive the enemy across the Rhine on the whole eighty-mile stretch between Düsseldorf and Koblenz. On the left the flanking corps of Hodges' First Army advanced with the Ninth, and with equal speed. Cologne was captured, with surprisingly little difficulty, on March 7. The other two corps crossed the river Erft, took Euskirchen, and branched east and south-east. Two corps of Patton's Third Army, which had already taken Trier and worked their way forward to the river Kyll, opened their major attack on March 5. They swept along the north bank of the Moselle, and three days later joined the First Army on the

Rhine. On the 7th a stroke of fortune was boldly accepted. The 9th Armoured Division of the First U.S. Army found the railway bridge at Remagen partly destroyed but still usable. They promptly threw their advance-guard across, other troops quickly followed, and soon over four divisions were on the far bank and a bridgehead several miles deep established. This was no part of Eisenhower's plan, but it proved an excellent adjunct, and the Germans had to divert considerable forces from farther north to hold the Americans in check. This brief campaign carried the Twelfth Army Group to the Rhine in one swift bound, and 49,000 Germans were taken prisoners. They had fought to the best of their ability, but were largely immobilised for want of petrol.

I sent Eisenhower well-earned congratulations.

Prime Minister to General of the Army Eisenhower   9 Mar 45
Let me offer you my warmest congratulations on the great victory won by the Allied armies under your command, by which the defeat or destruction of all the Germans west of the Rhine will be achieved. No one who studies war can fail to be impressed by the admirable speed and flexibility of the American armies and groups of armies, and the adaptiveness of commanders and their troops to the swiftly changing conditions of modern battles on the greatest scale. I am glad that the British and Canadian armies in the north should have played a part in your far-reaching and triumphant combinations.

*   *   *   *   *

Only one large pocket of Germans now remained west of the Rhine. These stood in a great salient formed by the Moselle from Koblenz to Trier and thence along the Siegfried Line back to the Rhine. Opposite its nose was the XXth Corps of the Third U.S. Army, on its right the Seventh U.S. Army, and, near the Rhine, a French group. The Allies attacked on March 15 against stiff opposition. Good progress was made west of Zweibrücken, but east of it the Germans held firm. It availed them little, for Patton had reached the Rhine north of Koblenz, and he turned five divisions southward across the lower Moselle. The stroke cut in behind the salient. It came as a complete surprise, and met feeble resistance. By March 21 it had reached Worms and joined the XXth Corps, which had burst through the bulge south of Trier.

The defenders of the renowned and dreaded Siegfried Line were

THE NORTHERN CROSSING

Bridgeheads, night March 24-25
Allied line, night March 28-29
Static front

5   10   15 MILES

Dortmund

Dalmen

Coesfeld

Haltern

Recklinghausen   Bochum

Lembeck

Heiden   Dursten   THE RUHR

Borken   Gladbeck

Stadtlohn   Gelsenkirchen   Essen

Ringenberg   Bruckhausen   Oberhausen

Bocholt   Wesel LIPPE CANAL   Holten   Duisburg

Isselburg   6   17   Dinslaken   R. Rhine

Anholt   Diersford

Emmerich   3 Cdn.   Rees   Wesel   30 Div.

Bienen   Div.   15 Div.   Ossenburg   XVI
Div.   Xanten   Geldern Corps   9 Div.
51 Div.   Rheinsberg

Cleves   XII   NINTH
Corps   U.S.
ARMY

R. Rhine   FIRST CANADIAN ARMY

REICHSWALD   XXX
Corps   Goch

Nijmegen   SECOND BRITISH
ARMY   Venlo

Arnhem   R. Meuse

358

thus cut off, and in a few days all organised resistance came to an end. As a by-product of victory the 5th U.S. Division made an unpremeditated crossing of the Rhine fifteen miles south of Mainz, which soon expanded into a deep bridgehead pointing towards Frankfurt.

Thus ended the last great German stand in the West. Six weeks of successive battles along a front of over two hundred and fifty miles had driven the enemy across the Rhine with irreplaceable losses in men and material. The Allied Air Forces played a part of supreme importance. Constant attacks by the tactical air forces aggravated the defeat and disorganisation and freed us from the dwindling Luftwaffe. Frequent patrols over the airfields containing the enemy's new jet-propelled fighters minimised a threat that had caused us anxiety. Continuing raids by our heavy bombers had reduced the German oil output to a critical point, ruined many of their airfields, and so heavily damaged their factories and transportation system as to bring them almost to a standstill.

* * * * *

While the Americans farther south were closing on the Rhine Montgomery had made ready to cross it. Planning and assembling of material had begun several months before. Great quantities of stores, amphibious vehicles, assault craft, and bridging material were now brought up to the fighting zone and troops concentrated on the near bank, behind constant smoke-screens.

Crossing-places were good, the Ruhr was threatened, and Kesselring, who had replaced Rundstedt in the chief command, had no doubt about where the blow would fall. Seven divisions of the First Parachute Army, his best remaining troops, dug themselves into the eastern bank, but, except for the perimeter defences at Wesel and Rees, their field works could not compare with those which the Allies had already overcome. Their artillery however was strong, and anti-aircraft guns had been drawn from the powerful air defences of the Ruhr. The sooner we could strike the better, and the plight of Northern Holland, still in the German grip, heightened the urgency.

Prime Minister to General Ismay, for Chiefs of Staff Committee　　　　　　　　　　　8 Mar 45

This frightful letter from Dr. Gerbrandy and Sir Desmond Morton's comment thereupon require your immediate attention.

Late on Monday night General Bedell Smith volunteered to me at Rheims the statement that he hoped that two divisions might be available to clear Holland immediately after the passage of the Rhine. I understand he contemplated American divisions. I am of opinion that a military plan should now be concerted to prevent the horrors which will befall the Dutch, and incidentally to extirpate the rocket-firing points in Holland at the earliest moment. I consider that if it were inevitable, which I doubt, a certain delay might be accepted in the main advance on Berlin. I am prepared to telegraph to the President on these lines, but I should like first of all to hear your views. Considering how the D.U.K.W.s, Buffaloes, etc., carried forward our northern movement, I do not think too much ought to be made of the water problem. If it be true that the German forces in Holland are now almost entirely static and that all the effective fighting units have left, there is no need to dwell upon the military task and overweight it.

Montgomery pressed forward his preparations and the Allied Air Forces brought all their power to bear. In the last week of February they began a bombardment from Bremen to Koblenz to deny to the enemy the arsenals of the Ruhr and isolate the battle area. As the days passed the attacks intensified. In the fortnight before the assault heavy bombers of the R.A.F. and the 8th and 9th U.S. Air Forces dropped nearly 50,000 tons of bombs. Medium and fighter bombers and an overwhelming force of fighter aircraft joined in to cut off the battlefield and bring chaos and ruin to Western Germany.

*   *   *   *   *

Under Montgomery were the First Canadian, the Second British, and the Ninth U.S. Armies. The two latter would seize bridgeheads north and south of Wesel, while the 1st British Commando Brigade, in the centre, captured Wesel itself. We would cross by night after an hour's bombardment by 2,000 guns, while the Canadians protected the left flank and themselves crossed later to drive northwards. Next morning two airborne divisions, the 6th British and 17th American, would drop behind the enemy's lines north of the town to disrupt his defences from the rear. This arrangement promoted their early junction with the other troops, which had played us false at Arnhem. In support were our heavy bombers and no fewer than 3,000 fighters under Air Marshal Coningham.

I desired to be with our armies at the crossing, and Montgomery made me welcome. Taking only my secretary, Jock Colville, and "Tommy" with me, I flew in the afternoon of March 23 by Dakota from Northolt to the British headquarters near Venlo. The Commander-in-Chief conducted me to the caravan in which he lived and moved. I found myself in the comfortable wagon I had used before. We dined at seven o'clock, and an hour later we repaired with strict punctuality to Montgomery's map wagon. Here were displayed all the maps kept from hour to hour by a select group of officers. The whole plan of our deployment and attack was easily comprehended. We were to force a passage over the river at ten points on a twenty-mile front from Rheinsberg to Rees.

All our resources were to be used. Eighty thousand men, the advance-guard of armies a million strong, were to be hurled forward. Masses of boats and pontoons lay ready. On the far side stood the Germans, entrenched and organised in all the strength of modern fire-power.

Everything I had seen or studied in war, or read, made me doubt that a river could be a good barrier of defence against superior force. In Hamley's *Operations of War*, which I had pondered over ever since Sandhurst days, he argues the truth that a river running parallel to the line of advance is a much more dangerous feature than one which lies squarely athwart it, and he illustrates this theory by Napoleon's marvellous campaign of 1814. I was therefore in good hopes of the battle even before the Field-Marshal explained his plans to me. Moreover, we had now the measureless advantage of mastery in the air. The episode which the Commander-in-Chief particularly wished me to see was the drop next morning of the two airborne divisions, comprising 14,000 men, with artillery and much other offensive equipment, behind the enemy lines. Accordingly we all went to bed before ten o'clock.

I telegraphed to Stalin:

Prime Minister to Marshal Stalin   23 Mar 45
I am with Field-Marshal Montgomery at his H.Q. He has just ordered the launching of the main battle to force the Rhine on a broad front with Wesel at the centre. The operation will be supported by about two thousand guns and by the landing of an airborne corps.

It is hoped to pass the river to-night and to-morrow and to establish bridgeheads. Once the river has been crossed a very large reserve of armour is ready to exploit the assault.

Field-Marshal Montgomery has asked me to present his respects to you. To-morrow I shall send you another message.

*　　*　　*　　*　　*

The honour of leading the attack fell to our 51st and 15th Divisions and the American 30th and 79th. Four battalions of the 51st were the first to set forth, and a few minutes later they had reached the far side. Throughout the night the attacking divisions poured across, meeting little resistance at first, as the bank itself was lightly defended. At dawn bridgeheads, shallow as yet, were firmly held, and the Commandos were already at grips in Wesel.

In the morning Montgomery had arranged for me to witness from a hill-top amid rolling downland the great fly-in. It was full daylight before the subdued but intense roar and rumbling of swarms of aircraft stole upon us. After that in the course of half an hour over 2,000 aircraft streamed overhead in their formations. My view-point had been well chosen. The light was clear enough to enable one to see where the descent on the enemy took place. The aircraft faded from sight, and then almost immediately afterwards returned towards us at a different level. The parachutists were invisible even to the best field-glasses. But now there was a double murmur and roar of reinforcements arriving and of those who had delivered their attacks returning. Soon one saw with a sense of tragedy aircraft in twos and threes coming back askew, asmoke, or even in flames. Also at this time tiny specks came floating to earth. Imagination built on a good deal of experience told a hard and painful tale. It seemed however that nineteen out of every twenty of the aircraft that had started came back in good order, having discharged their mission. This was confirmed by what we heard an hour later when we got back to headquarters.

The assault was now in progress along the whole front, and I was conducted by motor on a long tour from one point to another, and to the various corps headquarters. It was late in the evening when I returned. My private secretary, Jock Colville, had work to do for me, and could not come with me in the car. He had however made a plan of his own, and actually crossed the Rhine in one of the boats during the morning. There was no opposition to the passage, but the lodgments on the other side were under artillery fire. A shell burst near him and an officer with whom he was talking. A soldier of our airborne division standing beside them was severely wounded, and Jock was drenched in his blood. He naturally would have said nothing about this incident but for the fact that he arrived back at head-quarters at precisely the same time as Montgomery and I. His blood-stained tunic caught the Field-Marshal's eye, and he asked what had happened. He then complained that a civil servant should have crossed the river without his personal permission having been obtained. I protected Jock from his wrath, and promised to rebuke him myself, which I did in suitable terms after learning what had passed, and pointing out how much inconvenience he would have caused to the work of my Private Office if he had been killed. Who would have decoded and presented to me the secret telegrams that came in every few hours? He expressed his contrition, and I advised him to keep as far away from the Field-Marshal as possible during mess. This he did by dining elsewhere, and all passed off quietly. He is now forgiven.

* * * * *

Things went well all that day. The four assaulting divisions were safely across and established in bridgeheads 5,000 yards deep. The heaviest fighting was at Wesel and Rees. The airborne divisions were going strong and our air operations were most successful. The strike of the Allied Air Forces had been second only to that of D Day in Normandy. It included not only the strategic air forces in Britain, but also heavy bombers from Italy, who made deep penetrations into Germany.

At 8 p.m. we repaired to the map wagon, and I now had an excellent opportunity of seeing Montgomery's methods of con-ducting a battle on this gigantic scale. For nearly two hours a succession of young officers, of about the rank of major, pre-sented themselves. Each had come back from a different sector of the front. They were the direct personal representatives of the Commander-in-Chief, and could go anywhere and see anything and ask any questions they liked of any commander, whether at the divisional headquarters or with the forward troops. As in turn they made their reports and were searchingly questioned by their chief the whole story of the day's battle was unfolded. This gave Monty a complete account of what had happened by highly competent men whom he knew well and whose eyes he trusted. It afforded an invaluable cross-check to the reports from all the various headquarters and from the commanders, all of which had already been sifted and weighed by General de Guingand, his Chief of Staff, and were known to Montgomery. By this process he was able to form a more vivid, direct, and sometimes more accurate picture. The officers ran great risks, and of the seven or eight to whom I listened on this and succeeding nights two were killed in the next few weeks. I thought the system admirable, and indeed the only way in which a modern Commander-in-Chief could see as well as read what was going on in every part of the front. This process having finished, Montgomery gave a series of directions to de Guingand, which were turned into immediate action by the Staff machine. And so to bed.

* * * * *

The next day, March 25, we went to meet Eisenhower. On our way I told Montgomery how his system resembled that of Marlborough and the conduct of battles in the eighteenth century, where the Commander-in-Chief acted through his lieutenant-generals. Then the Commander in-Chief sat on his horse and directed by word of mouth a battle on a five- or six-mile front, which ended in a day and settled the fortunes of great nations, sometimes for years or generations to come. In order to make his will effective he had four or five lieutenant-generals posted at different points on the front, who knew his whole mind and were concerned with the execution of his plan. These officers commanded no troops and were intended to be off-shoots and expressions of the Supreme Commander. In modern times the general must sit in his office conducting a battle ranging over ten times the front and lasting often for a week or ten days. In these changed conditions Montgomery's method of personal eye-witnesses, who were naturally treated with the utmost consideration by the front-line commanders of every grade, was an interesting though partial revival of old days.

We met Eisenhower before noon. Here a number of American generals were gathered. After various interchanges we had a brief lunch, in the course of which Eisenhower said that there was a house about ten miles away on our side of the Rhine, which the Americans had sandbagged, from which a fine view of the river and of the opposite bank could be obtained. He proposed that we should visit it, and conducted us there himself. The Rhine—here about four hundred yards broad—flowed at our feet. There was a smooth, flat expanse of meadows on the enemy's side. The officers told us that the far bank was unoccupied so far as they knew, and we gazed and gaped at it for a while. With appropriate precautions we were led into the building. Then the Supreme Commander had to depart on other business, and Montgomery and I were about to follow his example when I saw a small launch come close by to moor. So I said to Montgomery, "Why don't we go across and have a look at the other side?" Somewhat to my surprise he answered, "Why not?" After he had made some inquiries we started across the river with three or four American commanders and half a dozen armed men. We landed in brilliant sunshine and perfect peace on the German shore, and walked about for half an hour or so unmolested.

As we came back Montgomery said to the captain of the launch, "Can't we go down the river towards Wesel, where there is something going on?" The captain replied that there was a chain across the river half a mile away to prevent floating mines interfering with our operations, and several of these might be held up by it. Montgomery pressed him hard, but was at length satisfied that the risk was too great. As we landed he said to me, "Let's go down to the railway bridge at Wesel, where we can see what is going on on the spot." So we got into his car, and, accompanied by the Americans, who were delighted at the prospect, we went to the big iron-girder railway bridge, which was broken in the middle but whose twisted ironwork offered good perches. The Germans were replying to our fire, and their shells fell in salvos of four about a mile away. Presently they came nearer. Then one salvo came overhead and plunged in the water on our side of the bridge. The shells seemed to explode on impact with the bottom, and raised great fountains of spray about a hundred yards away. Several other shells fell among the motor-cars which were concealed not far behind us, and it was decided we ought to depart. I clambered down and joined my adventurous host for our two hours' drive back to his headquarters. It seemed to me he had one standard for Jock Colville and another for himself.

★　　★　　★　　★

During the next few days we continued to gain ground east of the Rhine, and by March 28 the Ninth U.S. Army was approaching Duisburg and had entered Gladbeck. The airborne divisions, with the aid of a British armoured brigade, penetrated deep towards Haltern, whence the line was carried to Borken and Bocholt. The fighting on the left flank had been severe, but the 3rd Canadian Division, working down the bank of the Rhine, was closing on Emmerich to join the rest of the Canadian Army on the near bank. Thus by the end of the month we possessed a springboard east of the Rhine from which to launch major operations deep into Northern Germany. Of the engineering feats that were an essential part of this historic battle I need mention but one as an example of many. By the evening of the 26th no fewer than a dozen bridges had been thrown across the great river.

★   ★   ★   ★   ★

All this time the American armies in the south, though not opposed so strongly, had made astonishing progress. The two bridgeheads which were the reward of their boldness were being daily reinforced and enlarged, and more crossings were made south of Koblenz and at Worms. On March 25 the American Third Army was in Darmstadt, and on the 29th in Frankfurt. On the same day the Seventh U.S. Army occupied Mannheim, and the First U.S. Army, breaking out from Remagen, was already in Giessen and heading north. On April 2 the French were also across the Rhine on the right of the Seventh U.S. Army, which was driving east beyond Heidelberg. Kassel fell. The left of the First U.S. Army gained touch with the Ninth U.S. Army east of Hamm. The Ruhr and its 325,000 defenders were encircled. Germany's Western Front had collapsed.

CHAPTER XXV

THE POLISH DISPUTE

Soviet Breaches of the Yalta Agreement – Roosevelt's Failing Health – A Soviet-Nominated Administration Forced on Roumania, March 6 – Dangers and Difficulties of Allied Protest – Molotov Obstructs the Moscow Talks on Poland – My Proposals to the President, March 8 – My Telegram of March 10 – A Direct Demand to Stalin or Negotiations by Our Ambassadors? – Mr. Roosevelt Suggests a Political Truce – His Telegram of March 16, and My Reply – I Send the President a Personal Message, March 18 – Mr. Baruch's Visit – The Deadlock in Moscow Continues – My Telegrams of March 27 – We Agree to Address Stalin Direct – My Telegram of April 1 – Stalin's Reply, April 7 – His Personal Message to Me – Hope of Progress.

AS the weeks passed after Yalta it became clear that the Soviet Government was doing nothing to carry out our agreements about broadening the Polish Government to include all Polish parties and both sides. Molotov steadily refused to give an opinion about the Poles we mentioned, and not one of them was allowed to come even to a preliminary round-table discussion. He had offered to allow us to send observers to Poland, and had been disconcerted by the readiness and speed with which we had accepted. When our Ambassadors raised the point he made difficulties, arguing, among other things, that it might affect the prestige of the Lublin Provisional Government. No progress of any kind was made in the talks at Moscow. Time was on the side of the Russians and their Polish adherents, who were fastening their grip upon the country by all kinds of severe measures, which they did not wish outside observers to see. Every day's delay was a gain to these hard forces.

I therefore appealed to the President in the hope that we could address Stalin jointly on the highest level. The lengthy correspondence which followed sets forth the situation in Poland, as seen by the British and Americans. At this critical time Roosevelt's health and strength had faded. In my long telegrams I thought I was talking to my trusted friend and colleague as I had done all these years. I was no longer being fully heard by him. I did not know how ill he was, or I might have felt it cruel to press him. The President's devoted aides were anxious to keep their knowledge of his condition within the narrowest circle, and various hands drafted in combination the answers which were sent in his name. To these, as his life ebbed, Roosevelt could only give general guidance and approval. This was an heroic effort. The tendency of the State Department was naturally to avoid bringing matters to a head while the President was physically so frail and to leave the burden on the Ambassadors in Moscow. Harry Hopkins, who might have given personal help, was himself seriously ailing, and frequently absent or uninvited. These were costly weeks for all.

*　　*　　*　　*　　*

On the very evening when I was speaking in the House of Commons upon the results of our labours at Yalta the first violation by the Russians both of the spirit and letter of our agreements took place in Roumania. We were all committed by the Declaration on Liberated Europe, so recently signed, to see that both free elections and democratic Governments were established in the countries occupied by Allied armies. On February 27 Vyshinsky, who had appeared in Bucharest without warning on the previous day, demanded an audience of King Michael and insisted that he should dismiss the all-party Government which had been formed after the royal *coup d'état* of August 1944 and had led to the expulsion of the Germans from Roumania. The young monarch, backed by his Foreign Minister, Visoianu, resisted these demands until the following day. Vyshinsky called again, and, brushing aside the King's request at least to be allowed to consult the leaders of the political parties, banged his fist on the table, shouted for an immediate acquiescence, and walked out of the room, slamming the door. At the same time Soviet tanks and troops deployed in the streets of the capital, and on March 6 a Soviet-nominated Administration took office.

I was deeply disturbed by this news, which was to prove a pattern of things to come. The Russians had established the rule of a Communist minority by force and misrepresentation. We were hampered in our protests because Eden and I during our October visit to Moscow had recognised that Russia should have a largely predominant voice in Roumania and Bulgaria while we took the lead in Greece. Stalin had kept very strictly to this understanding during the six weeks' fighting against the Communists and E.L.A.S. in the city of Athens, in spite of the fact that all this was most disagreeable to him and those around him. Peace had now been restored, and, though many difficulties lay before us, I hoped that in a few months we should be able to hold free, unfettered elections, preferably under British, American, and Russian supervision, and that thereafter a constitution and Government would be erected on the indisputable will of the Greek people.

Stalin was now pursuing the opposite course in the two Black Sea Balkan countries, and one which was absolutely contrary to all democratic ideas. He had subscribed on paper to the principles of Yalta, and now they were being trampled down in Roumania. But if I pressed him too much he might say, "I did not interfere with your action in Greece; why do you not give me the same latitude in Roumania?" This would lead to comparisons between his aims and ours. Neither side would convince the other. Having regard to my personal relations with Stalin, I was sure it would be a mistake to embark on such an argument.

Besides this, I was very conscious of the much more important issue of Poland, and I did not want to do anything about Roumania which might harm the prospect of a Polish settlement. Nevertheless I felt we should tell Stalin of our distress at the forceful installation of a Communist minority Government, since this conflicted with the Declaration on Liberated Europe which we had agreed at the Yalta Conference. More especially I was afraid its advent might lead to an indiscriminate purge of anti-Communist Roumanians, who would be accused of Fascism much on the lines of what had been happening in Bulgaria. I therefore suggested to Mr. Roosevelt that Stalin should be asked to ensure that the new Government did not immediately start a purge of all anti-Communists on the ground that they had been encouraged to do so by the Yalta Declaration.

The news from Moscow about Poland was also most disappointing. The Government majorities in Britain bore no relation to the strong undercurrent of opinion among all parties and classes against Soviet domination of Poland. Labour men were as keen as Conservatives, and Socialists as keen as Catholics. I had based myself in Parliament on the belief that the Yalta Declaration would be carried out in the letter and the spirit. Once it was seen that we had been deceived and that the well-known Communist technique was being applied behind closed doors in Poland, either directly by the Russians or through their Lublin puppets, a very grave situation in British public opinion would be reached. Just at the time when everything militarily was going so well in Europe and in the Far East there would come an open rift between us and Russia, not at all confined, in Great Britain at any rate, to Government opinion, but running deep down through the masses of the people.

After a fairly promising start Molotov was now refusing to accept any interpretation of the Crimea proposals except his own extremely rigid and narrow one. He was attempting to bar practically all our candidates for the consultations, was taking the line that he must support the views of Bierut and his gang, and had withdrawn his offer to let us send observers to Poland. He clearly wanted to make a farce of consulting the "non-Lublin" Poles—which meant that the new Government of Poland would be merely the existing one dressed up to look more respectable to the ignorant—and also wanted to stop us seeing the liquidations and deportations and all the manœuvres of setting up a totalitarian régime before elections were held and even before a new Government was installed. If we did not get things right the world would soon see that Mr. Roosevelt and I had underwritten a fraudulent prospectus when we put our signatures to the Crimea settlement.

I was in any case pledged to Parliament to tell them if a new Polish Government could not be set up in the spirit of Yalta. I was sure the only way to stop Molotov's tactics was to send a personal message to Stalin, and make clear what were the essential things we must have if I was to avoid having to tell Parliament we had failed. Far more than Poland was involved. This was the test case between us and the Russians on the meaning of such terms as democracy, sovereignty, independence, representative

Government, and free and unfettered elections. On March 8 I therefore urged these views on the President and proposed that I should send Stalin a message on the lines set out below, and I hoped that he would send a similar one containing the same minimum requirements.

The message which I wished to send Stalin was as follows:

. . . I am bound to tell you that I should have to make a statement of our failure to Parliament if the Commission in Moscow were not in the end able to agree on the following basis:

(a) M. Molotov appears to be contending that the terms of the Crimea communiqué established for the present Warsaw Administration an absolute right of prior consultation on all points. In the English text the passage of the communiqué in question, which was an American draft, cannot bear this interpretation. M. Molotov's construction therefore cannot be accepted.

(b) All Poles nominated by any of the three Governments shall be accepted for the consultations unless ruled out by unanimous decision of the Commission, and every effort made to produce them before the Commission at the earliest possible moment. The Commission should ensure to the Poles invited facilities for communicating with other Poles whom they wish to consult, whether in Poland or outside, and the right to suggest to the Commission the names of other Poles who should be invited to its proceedings. All Poles appearing before the Commission would naturally enjoy complete freedom of movement and of communication among themselves while in Moscow, and would be at liberty to depart whither they chose upon the conclusion of the consultations. M. Molotov has raised objections to inviting M. Mikolajczyk, but his presence would certainly be vital.

(c) The Poles invited for consultations should discuss among themselves with a view to reaching agreement upon the composition of a Government truly representative of the various sections of Polish opinion present before the Commission. The discussions should also cover the question of the exercise of the Presidential functions. The Commission should preside over these discussions in an impartial arbitral capacity.

(d) Pending the conclusion of the Commission's discussions, the Soviet Government should use its utmost influence to prevent the Warsaw Administration from taking any further legal or administrative action of a fundamental character affecting

371

social, constitutional, economic, or political conditions in Poland.

(e) The Soviet Government should make arrangements to enable British and American observers to visit Poland and report upon conditions there in accordance with the offer spontaneously made by M. Molotov at an earlier stage in the Commission's discussions.

We must not let Poland become a source of disagreement and misunderstanding between our two peoples. For this reason I am sure you will understand how important it is for us to reach an early settlement on the basis of the Yalta decision, and it is because I am confident that you will do your utmost to bring this about that I am now telegraphing to you.

Two days later I telegraphed again to Roosevelt.

Prime Minister to President Roosevelt   10 Mar 45

The Lublin Poles may well answer that their Government can alone ensure "the maximum amount of political tranquillity inside", that they already represent the great mass of the "democratic forces in Poland", and that they cannot join hands with *émigré* traitors to Poland or Fascist collaborationists and landlords, and so on, according to the usual technique.

Meanwhile we shall not be allowed inside the country or have any means of informing ourselves upon the position. It suits the Soviets very well to have a long period of delay, so that the process of liquidation of elements unfavourable to them or their puppets may run its full course. This would be furthered by our opening out now into proposals of a very undefined character for a political truce between these Polish parties (whose hatreds would eat into live steel) in the spirit and intent of the Crimea decision, and might well imply the abandonment of all clear-cut requests, such as those suggested in my last telegram to you. Therefore I should find it very difficult to join in this project of a political truce.

I have already mentioned to you that the feeling here is very strong. Four Ministers have abstained from the divisions and two have already resigned. I beg therefore that you will give full consideration to my previous telegram.

The President assured me on March 11 that our objects were identical, namely, to stop the Lublin Poles persecuting their political opponents and *vice versa*. The only difference between us, he claimed, was on tactics. I wanted the demand to be put squarely to the Soviet Government as such, whereas he felt we had a much better chance of success by proposing a general political truce. At Yalta Stalin made quite a point of the terrorist activities of the underground forces of the London Polish Government against the Red Army and the Lublin Poles. Whether this was true was beside the point; it was what the Soviet Government maintained. But if we merely demanded that the Lublin Poles alone be forced to stop persecuting their political opponents Stalin would be certain to refuse. We might also be charged with trying to halt the land reforms, and the Lublin Poles might claim that they and they alone defended the peasants against the landlords.

Mr. Roosevelt agreed about sending in observers; but he preferred to wait until our Ambassadors appealed to Molotov before either of us approached Stalin personally. "I feel," he telegraphed, "that our personal intervention would best be withheld until every other possibility of bringing the Soviet Government into line has been exhausted. I very much hope therefore that you will not send a message to Uncle Joe at this juncture, especially as I feel that certain parts of your proposed text might produce a reaction quite contrary to your intent. We must of course keep in close touch on this question."

As I knew there was a deadlock in Moscow it was only with much reluctance that I deferred to the President's wish, but we could make no progress without American aid, and if we got out of step the doom of Poland was sealed. A month had passed since Yalta and no progress of any kind had occurred. Time was all on the side of Lublin, who were no doubt at work to establish their authority in such a way as to make it impregnable.

On March 13 I accordingly agreed to defer addressing Stalin directly for the time being; but I begged Mr. Roosevelt to allow our Ambassadors to raise the points set forth in my proposed message. I was convinced that unless we could induce the Russians to agree to these fundamental matters of procedure all our work at Yalta would have been in vain.

When the discussions following Yalta began at Moscow we had a perfectly simple object, namely, to bring together for consultation representative Poles from inside Poland and elsewhere and to promote the formation of a new, reorganised Polish Government sufficiently representative of all Poland for us to recognise it. A test case of progress would have been to have invited

Mikolajczyk and two or three of his friends who had resigned from the London Polish Government because they realised that a good understanding must be reached with Russia.

I feared that Mr. Roosevelt's instructions to his Ambassador would lead to little if any progress on all this, as the only definite suggestion they contained was for a truce between the Polish parties. Here we should enter ground of great disadvantage to us both. The Russians would almost at once claim that the truce was being broken by the anti-Lublin Poles, and that Lublin therefore could not be held to it. I had little doubt that some of the supporters of the Polish Government in London, and more particularly the extreme Right Wing Underground Force, the so-called N.S.Z., might give the Russians and Lublin ground for this contention. As we were not allowed to enter the country to see what the truth was we should be at the mercy of assertions. After a fortnight or so of negotiations about the truce we should be farther back than in the days before Yalta, when the President and I agreed that anyhow Mikolajczyk should be invited.

I set this forth in a personal telegram, and concluded:

Prime Minister to President Roosevelt   13 Mar 45
    At Yalta also we agreed to take the Russian view of the frontier line. Poland has lost her frontier. Is she now to lose her freedom? That is the question which will undoubtedly have to be fought out in Parliament and in public here. I do not wish to reveal a divergence between the British and the United States Governments, but it would certainly be necessary for me to make it clear that we are in presence of a great failure and an utter breakdown of what was settled at Yalta, but that we British have not the necessary strength to carry the matter further and that the limits of our capacity to act have been reached. The moment that Molotov sees that he has beaten us away from the whole process of consultations among Poles to form a new Government he will know that we will put up with anything. On the other hand, I believe that combined dogged pressure and persistence along the lines on which we have been working and of my proposed draft message to Stalin would very likely succeed.

    This produced a strongly argued reply, which had no doubt been the work of the State Department since my long telegram of March 8 had been received in Washington.

President Roosevelt to Prime Minister   16 Mar 45
    I cannot but be concerned at the views you expressed in your

message of the 13th. I do not understand what you mean by a divergence between our Governments on the Polish negotiations. From our side there is certainly no evidence of any divergence of policy. We have been merely discussing the most effective tactics, and I cannot agree that we are confronted with a breakdown of the Yalta agreement until we have made the effort to overcome the obstacles incurred in the negotiations at Moscow. I also find puzzling your statement that the only definite suggestion in our instructions to [Ambassador] Harriman is for a political truce in Poland. Those instructions, of which you have a copy, not only set forth our understanding of the Yalta agreement, but they make the definite point that the Commission itself should agree on the list of Poles to be invited for consultation, and that no one of the three groups from which the reorganised Government is to emerge can dictate which individuals from the other two groups ought to be invited to Moscow. . . . Our chief purpose . . . remains, without giving ground, to get the negotiations moving again, and tackle first of all the point on which they had come to a standstill. I cannot urge upon you too strongly the vital importance of agreeing without further delay on instructions to our Ambassadors so that the negotiations may resume. . . . With this in mind I have examined the points which you propose to submit to Stalin in your message of March 8, and have the following comments to make:

We are in agreement on point (a), that the Warsaw Administration is not entitled to an absolute right of prior consultation on all points, and this is covered in our instructions to Harriman.

I cannot believe that Molotov will accept the proposal contained in point (b), that any Pole can be invited unless all three members of the Commission object, and I am opposed to putting forward such a suggestion at this time, as it would in my view almost certainly leave us in a stalemate, which would only redound to the benefit of the Lublin Poles. I also think the demand for freedom of movement and communication would arouse needless discussion at this stage in the negotiations.

On point (c) we are agreed that the Poles invited for consultation should discuss the composition of the Government among themselves, with the Commission presiding in an impartial arbitral capacity so far as possible. Harriman has already been instructed to this effect, but feels, and I agree, that this might be pressed later.

I have covered your point (d) [about stopping any major changes in Poland] in my previous message, and continue to feel that our approach would be better calculated to achieve the desired result.

With reference to point (e) [sending in observers], you will recall that this had been agreed to by Molotov, who took fright when Clark

Kerr revealed that you were thinking of a large special mission. I am willing to include in Averell's instructions the wording you propose in point (e).

Please let me know urgently whether you agree that in the light of the foregoing considerations our Ambassadors may proceed with their instructions. . . .

To this I rejoined:

Prime Minister to President Roosevelt   16 Mar 45

I am most relieved that you do not feel that there is any fundamental divergence between us, and I agree that our differences are only about tactics. You know, I am sure, that our great desire is to keep in step with you, and we realise how hopeless the position would become for Poland if it were ever seen that we were not in full accord. . . .

3. Halifax will explain to you in detail our views upon the various points the inclusion of which I still consider essential. I welcome your agreement on point (a) [that the Warsaw Government is not entitled to prior consultation on all points]. With regard to point (b) [inviting Poles for consultation], what happens if Molotov vetoes every one of our suggestions? And, secondly, what is the use of anyone being invited who has no freedom of movement and communications? We had in fact not understood that Molotov had disputed this latter point when it was raised with him earlier, but Mikolajczyk has made it a condition of going to Moscow, and I gravely doubt whether we could persuade him to leave unless we had some definite assurance to convey to him. Equally it is in order to reassure the anti-Lublin Poles whom we want to see invited that I should like to come to an agreement with Molotov in regard to the character of the discussions and the Commission's arbitral capacity [my point (c)]. If you feel strongly against mentioning the matter of the Presidential functions at this stage I will give way, although it is a matter of great practical importance which the Poles must not be debarred from discussing. As regards point (d) [asking the Soviet Government to prevent the Warsaw Administration from making any more fundamental changes in Poland], I fear I cannot agree that your truce plan would achieve the desired result. How can we guarantee that nothing will be said or done in Poland or by the Polish Government's supporters here which the Russians could not parade as a breach of the truce? I fear that the truce plan will lead us into interminable delays and a dead end in which some at least of the blame may well be earned by the London Polish Government. I fear therefore that it is impossible for us to endorse your truce proposal, for we think it actively dangerous. I beg you once more most earnestly to consider whether you cannot accept [a revised proposal for halting

major changes in Poland]. This would give us something on which to base the work of our observers [point (e)], on which I am very glad to see that we are in agreement.

4. At present all entry into Poland is barred to our representatives. An impenetrable veil has been drawn across the scene. This extends even to the liaison officers, British and American, who were to help in bringing out our rescued prisoners of war. According to our information, the American officers as well as the British who had already reached Lublin have been requested to clear out. There is no doubt in my mind that the Soviets fear very much our seeing what is going on in Poland. It may be that, apart from the Poles, they are being very rough with the Germans. Whatever the reason, we are not to be allowed to see. This is not a position that could be defended by us.

Although I had no exact information about the President's state of health I had the feeling that, except for occasional flashes of courage and insight, the telegrams he was sending us were not his own. I therefore sent him a message in a personal vein to ease the uphill march of official business.

18 Mar 45

I hope that the rather numerous telegrams I have to send you on so many of our difficult and intertwined affairs are not becoming a bore to you. Our friendship is the rock on which I build for the future of the world, so long as I am one of the builders. I always think of those tremendous days when you devised Lend-Lease, when we met at Argentia, when you decided, with my heartfelt agreement, to launch the invasion of Africa, and when you comforted me for the loss of Tobruk by giving me the 300 Shermans of subsequent Alamein fame. I remember the part our personal relations have played in the advance of the World Cause, now nearing its first military goal.

2. I am sending to Washington and San Francisco most of my Ministerial colleagues on one mission or another, and I shall on this occasion stay at home to mind the shop. All the time I shall be looking forward to your long-promised visit. Clemmie is off to Russia next week for a Red Cross tour as far as the Urals to which she has been invited by Uncle Joe (if we may venture to describe him thus), but she will be back in time to welcome you and Eleanor. My thoughts are always with you all.

3. Peace with Germany and Japan on our terms will not bring much rest to you and me (if I am still responsible). As I observed last time, when the war of the giants is over the wars of the pygmies will begin. There will be a torn, ragged, and hungry world to help to its feet; and what will Uncle Joe or his successor say to the way we should both

377

like to do it? It was quite a relief to [me to] talk party politics the other day. It was like working in wood after working in steel. The advantage of this telegram is that it has nothing to do with shop, except that I had a good talk with Rosenman* about our daily bread.

All good wishes.   WINSTON

The President was evidently pleased with this message, for two days later he sent me the following telegram, which he knew I would like:

President Roosevelt to Prime Minister   20 Mar 45
I would very much appreciate it if you would see Bernie Baruch as soon as convenient to you, and also appreciate it if you could wire him, as he counts you one of his oldest friends and would much prefer having your approval before he goes.

Prime Minister to President Roosevelt   21 Mar 45
I am greatly looking forward to seeing Bernie, who is one of my oldest friends. I am telegraphing to him to say how glad I am that he is coming. I should like to know when he will come.

I had often wondered why he did not make more use of Baruch's immense knowledge and experience both of American politics and war production.

Mr. Baruch came over as planned, and we had long and intimate talks, which led to a further agreeable interchange with the President. I had hopes that I might find a new tie of correspondence and communication with my all-important colleague and comrade. Alas, we were very near the end!

Prime Minister to President Roosevelt   30 Mar 45
I am delighted to see from the abundance of messages I have received from you this morning that you are back in Washington and in such vigour. I saw Bernie yesterday, and he is coming to-night for the week-end. He seems in great form. As you know, I think he is a very wise man. Winant is coming to-morrow. Clemmie is in flight for Moscow, and will be flying about there for at least a month, all of which hangs on my mind.

President Roosevelt to Prime Minister   1 Apr 45
I did receive your very pleasing message.
The efforts of Bernie, who is a wise man of wide experience, should be of much assistance to both of us.

* Judge Samuel Rosenman, one of Mr. Roosevelt's close personal advisers, who had helped to draft the President's report to Congress on the Yalta Conference, and who visited me at this time in London.

We hope that Clemmie's long flying tour in Russia will first be safe and next be productive of good, which I am sure it will be. The war business to-day seems to be going very well from our point of view, and we may hope now for the collapse of Hitlerism at an earlier date than had heretofore been anticipated.

* * * * *

Anglo-American tactics and procedure were at length agreed. Meanwhile, as we in London had foreseen, the deadlock in Moscow continued. The Soviet policy became daily more plain, as also did the use they were making of their unbridled and unobserved control of Poland. They asked that Poland should be represented at San Francisco only by the Lublin Government. When the Western Powers would not agree the Soviets refused to let Molotov attend. This threatened to make all progress at San Francisco, and even the Conference itself, impossible. Molotov, in reply to an agreed communication from our Ambassadors on March 19 and in discussion on March 23, returned a series of flat negatives on every point he dealt with and ignored others. He persisted that the Yalta communiqué merely meant adding a few other Poles to the existing Administration of Russian puppets, and that these puppets should be consulted first. He maintained his right to veto Mikolajczyk and other Poles we might suggest, and pretended that he had insufficient information about the names we had put forward long before. Nothing was said about our proposal that the Commission should preside in an arbitral capacity over discussions among the Poles; nothing on our point that measures in Poland affecting the future of the Polish State and action against individuals and groups likely to disturb the peace should be avoided. Molotov ignored his offer about observers and told us to talk to the Warsaw puppets about it. It was as plain as a pikestaff that his tactics were to drag the business out while the Lublin Committee consolidated their power. Negotiations by our Ambassadors held no promise of an honest Polish settlement. They merely meant that our communications would be side-tracked and time would be wasted on finding formulæ which did not decide vital points.

On March 27 I felt bound to renew the discussion.

Prime Minister to President Roosevelt   27 Mar 45
    . . . As you know, if we fail altogether to get a satisfactory solution on Poland and are in fact defrauded by Russia both Eden and I are

pledged to report the fact openly to the House of Commons. There I advised critics of the Yalta settlement to trust Stalin. If I have to make a statement of facts to the House the whole world will draw the deduction that such advice was wrong; all the more so that our failure in Poland will result in a set-up there on the new Roumanian model. In other words, Eastern Europe will be shown to be excluded from the terms of the Declaration on Liberated Europe, and you and we will be excluded from any jot of influence in that area.

Surely we must not be manœuvred into becoming parties to imposing on Poland—and on how much more of Eastern Europe—the Russian version of democracy? . . . There seems to be only one possible alternative to confessing our total failure. That alternative is to stand by our interpretation of the Yalta Declaration. But I am convinced it is no use trying to argue this any further with Molotov. In view of this, is it not the moment now for a message from us both on Poland to Stalin? I will send you our rough idea on this in my immediately following telegram. I hope you can agree.

I see nothing else likely to produce good results. If we are rebuffed it will be a very sinister sign, taken with the other Russian actions at variance with the spirit of Yalta—such as Molotov's rude questioning of our word in the case of "Crossword"*, the unsatisfactory proceedings over our liberated German prisoners, the *coup d'état* in Roumania, the Russian refusal to allow the Declaration on Liberated Europe to operate, and the blocking of all progress in the European Advisory Commission by the Russians.

What also do you make of Molotov's withdrawal from San Francisco? It leaves a bad impression on me. Does it mean that the Russians are going to run out, or are they trying to blackmail us? As we have both understood them, the Dumbarton Oaks proposals, which will form the basis of discussion at San Francisco, are based on the conception of Great Power unity. If no such unity exists on Poland, which is after all a major problem of the post-war settlement—to say nothing of the other matters just mentioned—what, it will legitimately be asked, are the prospects of success of the new World Organisation? And is it not indeed evident that, in the circumstances, we shall be building the whole structure of future world peace on foundations of sand?

I believe therefore that if the success of San Francisco is not to be gravely imperilled we must both of us now make the strongest possible appeal to Stalin about Poland, and if necessary about any other derogations from the harmony of the Crimea. Only so shall we have any real chance of getting the World Organisation established on lines

* See Chapter XXVI, "Soviet Suspicions."

which will commend themselves to our respective public opinions. Indeed, I am not sure that we should not mention to Stalin now the deplorable impression Molotov's absence from San Francisco will cause.

To this I added later in the day a positive proposal.

Prime Minister to President Roosevelt   27 Mar 45

Could we not both tell him [Stalin] that we are distressed that the work of the Polish Commission is held up because misunderstandings have arisen about the interpretation of the Yalta decisions? The agreed purpose of those decisions was that a new Government of National Unity was to be established, after consultations with representatives of Lublin and other democratic Poles, which both our Governments could recognise. We have not got any reply on the various Polish names we have suggested, pleading lack of information. We have given him plenty of information. There ought not to be a veto by one Power on all nominations. We consider that our nominations for the discussions have been made in the spirit of confidence which befits allies; and of course there could be no question of allowing Lublin to bar them. We will accept any nominations he puts forward, being equally confident that the Soviet Government will not suggest pro-Nazi or anti-democratic Poles. The assembled Poles should then discuss the formation of a new Government among themselves. The Commission should preside as arbitrators to see fair play. Molotov wants the Lublinites to be consulted first. The communiqué does not provide for this. But we have no objection to his seeing them first. We cannot authorise our representatives to do so, since we think it contrary to the spirit of the communiqué. Also, to our surprise and regret, Molotov, who suggested at an earlier stage that we might like to send observers, has now withdrawn the offer. Indeed, he appears to suggest it had never been made, and has suggested that we should apply to the present Warsaw Administration. Stalin will understand that the whole point of the Yalta decision was to produce a Polish Government we could recognise, and that we obviously cannot therefore deal with the present Administration. We feel sure he will honour the offer to send observers, and his influence with his Warsaw friends is so great that he will overcome with ease any reluctance they may show in agreeing.

2. Also, Stalin will surely see that while the three Great Allies are arranging for the establishment of the new Government of National Unity those in power in Poland should not prejudice the future. We have asked that the Soviet Government should use their influence with their friends in temporary power there. Stalin will, we feel confident, take steps to this end.

3. Stalin will find all this set out in most reasonable terms in our [Ambassadors'] communication of March 19. Will he cast his eye over it and judge whether our suggestions are not all in line with the spirit of the Yalta decision, and should they not all be met by our Ally in order that the aim of the Yalta settlement of Poland—viz., the setting up of a representative Government which Britain and the U.S.A. can recognise—may be carried out without further delay?

The President replied that he also had been watching "with anxiety and concern the development of the Soviet attitude since the Crimea Conference". He set forth his proposals for further negotiations by our Ambassadors, and then concluded: "I agree with you however that the time has come to take up directly with Stalin the broader aspects of the Soviet attitude (with particular reference to Poland), and my immediately following telegram will contain the text of the message I propose to send. I hope you will let me have your reaction as soon as possible."

I was of course much relieved that we were at length agreed to address Stalin directly. All the time I felt sure that it was only in this way that any practical results could be achieved. "I am glad," I telegraphed on March 30 to the President, "you agree that the time has come for us both to address Stalin directly. Your draft is a grave and weighty document, which, though it does not give full expression to our views, we will wholeheartedly accept. I will also endorse it in my parallel message to Stalin."

On April 1 I addressed Stalin myself.

Prime Minister to Marshal Stalin   1 Apr 45
You will by now, I hope, have received the message from the President of the United States,* which he was good enough to show to me before he sent it. It is now my duty on behalf of His Majesty's Government to assure you that the War Cabinet desire me to express to you our wholehearted endorsement of this message of the President's, and that we associate ourselves with it in its entirety.

2. There are two or three points which I desire specially to emphasise. First, that we do not consider we have retained in the Moscow discussions the spirit of Yalta, nor indeed, at points, the letter. It was never imagined by us that the Commission we all three appointed with so much goodwill would not have been able to carry out their part swiftly and easily in a mood of give and take. We certainly thought that a Polish Government "new" and "reorganised" would by now have been in existence, recognised by all the United Nations.

* See Appendix C, p. 633.

This would have afforded a proof to the world of our capacity and resolve to work together for its future. It is still not too late to achieve this.

3. However, even before forming such a new and reorganised Polish Government it was agreed by the Commission that representative Poles should be summoned from inside Poland and from Poles abroad, not necessarily to take part in the Government, but merely for free and frank consultation. Even this preliminary step cannot be taken because of the claim put forward to veto any invitation, even to the consultation, of which the Soviet or the Lublin Government do not approve. We can never agree to such a veto by any one of us three. This veto reaches its supreme example in the case of M. Mikolajczyk, who is regarded throughout the British and American world as the outstanding Polish figure outside Poland.

4. We also have learned with surprise and regret that M. Molotov's spontaneous offer to allow observers or missions to enter Poland has now been withdrawn. We are therefore deprived of all means of checking for ourselves the information, often of a most painful character, which is sent us almost daily by the Polish Government in London. We do not understand why a veil of secrecy should thus be drawn over the Polish scene. We offer the fullest facilities to the Soviet Government to send missions or individuals to visit any of the territories in our military occupation. In several cases this offer has been accepted by the Soviets and visits have taken place, to mutual satisfaction. We ask that the principle of reciprocity shall be observed in these matters, which would help to make so good a foundation for our enduring partnership.

5. The President has also shown me the messages which have passed between him and you about M. Molotov's inability to be present at the Conference at San Francisco. We had hoped the presence there of the three Foreign Ministers might have led to a clearance of many of the difficulties which have descended upon us in a storm since our happy and hopeful union at Yalta. We do not however question in any way the weight of the public reasons which make it necessary for him to remain in Russia.

6. . . . If our efforts to reach an agreement about Poland are to be doomed to failure I shall be bound to confess the fact to Parliament when they return from the Easter recess. No one has pleaded the cause of Russia with more fervour and conviction than I have tried to do. I was the first to raise my voice on June 22, 1941. It is more than a year since I proclaimed to a startled world the justice of the Curzon Line for Russia's western frontier, and this frontier has now been accepted by both the British Parliament and the President of the United

States. It is as a sincere friend of Russia that I make my personal appeal
to you and to your colleagues to come to a good understanding about
Poland with the Western democracies, and not to smite down the
hands of comradeship in the future guidance of the world which we
now extend.

A week later Stalin replied to us both. He blamed the British
and American Ambassadors in Moscow for getting "the Polish
affair into a blind alley". We had agreed at Yalta to form a new
Polish Government by using the Lublin Government as a nucleus
and reconstructing it. Instead, our Ambassadors were trying to
abolish it and establish a completely new one. At Yalta we had
also agreed to consult five Poles from Poland and about three
from London. Our Ambassadors were now claiming that every
member of the Moscow Commission could invite an unlimited
number from both places. The Soviet Government could not
allow this. The Commission as a whole should decide whom to
ask, and these must only be Poles who accepted the decisions of
Yalta, including the Curzon Line, and who genuinely desired
friendly relations between Poland and the U.S.S.R. "The Soviet
Government," he wrote, "insists on this, since much blood of
Soviet soldiers has been shed for the liberation of Poland, and
since in the course of the last thirty years the territory of Poland
has twice been used by the enemy for an invasion of Russia."
Stalin then summed up the steps we should take to "escape from
the blind alley". The Lublin Government must be reconstructed,
not liquidated, by replacing some of its existing Ministers by new
ones from outside it; only eight Poles should be invited for consultation, five from Poland and three from London, and all must
accept the Yalta decisions and be friendly to the Soviet Government; the Lublin Government must first be consulted because of
its "enormous" influence in Poland and because any other course
might insult the Polish people and make them think we were
trying to impose a Government upon them without consulting
public opinion. "I think," he concluded, "that if the above
observations were taken into account an agreed decision on the
Polish question could be arrived at in a short time."

Stalin also sent me a personal message.

Marshal Stalin to Prime Minister   7 Apr 45
The British and American Ambassadors, who are members of the
Moscow Commission, are unwilling to take account of the Provisional

Polish Government, and insist on inviting Polish personalities for consultation without regard to their attitude to the decisions of the Crimea Conference on Poland and to the Soviet Union. They absolutely insist on summoning to Moscow for consultation, for instance, Mikolajczyk, and this they do in the form of an ultimatum. In this they take no account of the fact that Mikolajczyk has come out openly against the decisions of the Crimea Conference on Poland. However, if you think it necessary, I should be ready to use my influence with the Provisional Polish Government to make them withdraw their objections to inviting Mikolajczyk, if the latter would make a public statement accepting the decisions of the Crimea Conference on the Polish question and declaring that he stands for the establishment of friendly relations between Poland and the Soviet Union.

2. You wonder why the Polish theatre of military operations must be wrapped in mystery. In fact there is no mystery here. You ignore the fact that if British observers or other foreign observers were sent into Poland the Poles would regard this as an insult to their national dignity, bearing in mind the fact moreover that the present attitude of the British Government to the Provisional Polish Government is regarded as unfriendly by the latter. So far as the Soviet Government is concerned, it cannot but take account of the negative attitude of the Provisional Government to the question of sending foreign observers into Poland. Further, you are aware that the Provisional Polish Government puts no obstacles in the way of entrance into Poland by representatives of other States which take up a different attitude towards it, and does not in any way obstruct them. This is the case, for instance, in regard to the representatives of the Czechoslovak Government, the Yugoslav Government, and others.

3. I had an agreeable conversation with Mrs. Churchill, who made a great impression on me. She gave me a present from you. Allow me to express my heartfelt thanks for this present.

These carefully considered documents at least offered some hope of progress. I began at once my painful discussions with Mikolajczyk and other Polish representatives with the object of obtaining their unreserved affirmation of agreement with the Yalta decisions.

"We shall," cabled the President on April 11, "have to consider most carefully the implications of Stalin's attitude and what is to be our next step. I shall of course take no action of any kind nor make any statement without consulting you, and I know you will do the same."

CHAPTER XXVI

SOVIET SUSPICIONS

Unconditional Surrender and Military Capitulation – General Karl Wolff Meets Mr. Allen Dulles in Switzerland, March 8 – A Second Meeting, March 19 – Molotov's Insults – Eisenhower's Anger – My Minutes of March 25 to Eden – Stalin's Telegram to the President of April 3 – Mr. Roosevelt's Reproaches, April 5 – My Telegram to Stalin of April 6 – His Replies, April 7 – The Semblance of an Apology – Mr. Roosevelt's Telegram of April 12.

W HILE all the vexations of the Soviet abandonment of the spirit of Yalta were the subject of the prolonged correspondence set forth in the preceding chapter a far more bitter and important interchange was taking place between the British and American Governments and the Soviets. It has been advisable to recount these issues in separate chapters, but it must not be forgotten that the events recorded reacted continually and forcibly upon each other.

By the middle of February the Nazis realised that defeat was near. The advance of the Soviet armies, Alexander's victories in Italy, the failure of their counter-attack in the Ardennes, and Eisenhower's march to the Rhine convinced all but Hitler and his closest followers that surrender was imminent and unavoidable. The question was, surrender to whom? Germany could no longer make war on two fronts. Peace with the Soviets was evidently impossible. The rulers of Germany were too familiar with totalitarian oppression to invite its importation from the East. There remained the Allies in the West. Might it not be possible, they argued, to make a bargain with Great Britain and the United States? If a truce could be made in the West they could concentrate their troops against the Soviet advance. Hitler alone was obstinate. The Third Reich was finished and he would die with it. But several of his followers tried to make secret approaches to the English-speaking Allies. All these proposals were of course rejected. Our terms were unconditional surrender on all fronts. At the same time our commanders in the field were always fully authorised to accept purely military capitulations of the enemy forces which opposed them, and an attempt to arrange this while we were fighting on the Rhine led to a harsh exchange between the Russians and the President, whom I supported.

In February General Karl Wolff, the commander of the S.S. in Italy, had got into touch through Italian intermediaries with the American Intelligence Service in Switzerland. It was decided to examine the credentials of the persons involved, and the link was given the code-name "Crossword". On March 8 General Wolff himself appeared at Zürich, and met Mr. Allen Dulles, the head of the American organisation. Wolff was bluntly told that there was no question of negotiations, and that if the matter were pursued it could only be on the basis of unconditional surrender. This information was speedily conveyed to Allied Headquarters in Italy and to the American, British, and Soviet Governments. On March 15 General Airey and General Lemnitzer, British and American Chiefs of Staff at Caserta, arrived in Switzerland in disguise, and four days later, on March 19, a second exploratory meeting was held with General Wolff.

I realised at once that the Soviet Government might be suspicious of a separate military surrender in the South, which would enable our armies to advance against reduced opposition as far as Vienna and beyond, or indeed towards the Elbe or Berlin. Moreover, as all our fronts round Germany were part of the whole Allied war the Russians would naturally be affected by anything done on any one of them. If any contacts were made with the enemy, formal or informal, they ought to be told in good time. This rule was scrupulously followed. On March 12 the British Ambassador in Moscow informed the Soviet Government of this link with the German emissaries, and said that no contact would be made until we received the Russian reply. There was at no stage any question of concealing anything from the Russians. The Allied representatives then in Switzerland even explored ways of smuggling a Russian officer in to join them if the Soviet Government wished to send someone. It did not however prove possible to arrange for Soviet representation at the exploratory meeting in Berne, and the Russians were informed on March 13 that if this contact proved to be of serious import we would welcome their representatives at General Alexander's headquarters. Three days later M. Molotov informed the British Ambassador in Moscow that the Soviet Government found the attitude of the British Government "entirely inexplicable and incomprehensible in denying facilities to the Russians to send the representative to Berne". A similar communication was passed to the American Ambassador.

On March 21 our Ambassador in Moscow was instructed to inform the Soviet Government yet again that the only object of the meetings was to make sure that the Germans had authority to negotiate a military surrender and to invite Russian delegates to Allied headquarters at Caserta. This he did. Next day Molotov handed him a written reply, which contained the following expressions:

In Berne for two weeks, behind the backs of the Soviet Union, which is bearing the brunt of the war against Germany, negotiations have been going on between the representatives of the German military command on the one hand and representatives of the English and American commands on the other.

Sir Archibald Clark Kerr of course explained that the Soviets had misunderstood what had occurred and that these "negotiations" were no more than an attempt to test the credentials and authority of General Wolff. Molotov's comment was blunt and insulting. "In this instance," he wrote, "the Soviet Government sees not a misunderstanding, but something worse." He attacked the Americans just as bitterly.

In the face of so astonishing a charge it seemed to me that silence was better than a contest in abuse, and on March 24 I minuted to Mr. Eden:

Prime Minister to Foreign Secretary   24 Mar 45
For the moment these negotiations have dropped. They may be reopened in a far more vital area than Italy. In this military and political questions will be intertwined. The Russians may have a legitimate fear of our doing a deal in the West to hold them well back in the East. On the whole it will be well to send no reply [to Molotov]

till we have checked up with Washington, to whom you should repeat
the Russian message.

*   *   *   *   *

At the same time it was necessary to warn our military commanders in the West. I accordingly showed Molotov's insulting
letter both to Montgomery and to Eisenhower, with whom I at
this time was watching the crossing of the Rhine.

General Eisenhower was much upset, and seemed deeply stirred
with anger at what he considered most unjust and unfounded
charges about our good faith. He said that as a military commander he would accept the unconditional surrender of any body
of enemy troops on his front, from a company to the entire
Army, that he regarded this as a purely military matter, and
that he had full authority to accept such a surrender without
asking anybody's opinion. If however political matters arose
he would immediately consult the Governments. He feared that
if the Russians were brought into a question of the surrender of
Kesselring's forces what could be settled by himself in an hour
might be prolonged for three or four weeks, with heavy losses to
our troops. He made it clear that he would insist upon all the
troops under the officer making the surrender laying down their
arms and standing still until they received further orders, so that
there would be no possibility of their being transferred across
Germany to withstand the Russians. He would also at the same
time advance through these surrendered troops as fast as possible
to the East.

I thought myself that these matters should be left to his discretion, and that the Governments should only intervene if any
political issues arose. I did not see why we should break our
hearts if, owing to mass surrender in the West, we got to the
Elbe, or even farther, before Stalin. Jock Colville reminds me
that I said to him that evening, "I hardly like to consider dismem-
bering Germany until my doubts about Russia's intentions have
been cleared away."

I minuted to Mr. Eden on March 25:

Further reflection convinces me we should send no answer to the
insulting letter from Molotov. I presume you have already sent a
copy of it to the State Department, pointing out, in no spirit of complaint, that it was they who particularly wished that the Russians

should not come to Switzerland and that Alexander should deal with the matter on a purely military basis. I am sure the right thing now is to get absolutely in line with the United States, which should be easy, and meanwhile let Molotov and his master wait.

I agree with you that the whole question of the San Francisco Conference hangs in the balance. The sending of Gromyko instead of Molotov is a grimace. I should suppose the President would be much offended by this.

We have had a jolly day, having crossed the Rhine. To-morrow we go to the 15th Scottish Division, on the other side. I should think it not at all unlikely that the whole German front in the West may collapse and be broken up into blobs. There is still hard fighting going on in the North, and the brunt again seems to come from the left-hand hinge, which, as usual, we form.

And later on the same day:

. . . We should ask the United States where they stand and whether they will now agree to a telegram from the President and me to Stalin, and secondly whether this should, as you say, cover other topics—e.g., access to Poland, treatment of our prisoners, imputations against our good faith about Berne, Roumania, etc.

Molotov's refusal to go to San Francisco is no doubt the expression of the Soviet displeasure. We should put it to Roosevelt that the whole question of going to San Francisco in these conditions is called in question and that quite definite forming up by Britain and the United States against breach of Yalta understandings now is necessary if such a meeting is to have any value.

However, I must say that we cannot press the case against Russia beyond where we can carry the United States. Nothing is more likely to bring them into line with us than any idea of the San Francisco Conference being imperilled. Could you let me have a draft on the above lines on which I can send you back a personal to Roosevelt this time to-morrow? Meanwhile no answer should be sent to any of the Russian messages, even though the loss of time be prejudicial to us. When we come back at them now it must be both together. These matters will not be ripe for debate before Easter.

We have had a glorious day here, and I hope the consequences will be far-reaching. I am to see Eisenhower to-morrow at his request. I showed Montgomery Molotov's rude message, as of course the venue of the negotiations may easily be changed to his theatre. I well understand the Russian anxiety lest we should accept a military surrender in the West or South, which means that our armies will advance against little or no opposition and will reach the Elbe, or even Berlin, before the Bear. Therefore, should military negotiations break out on this front, which is not a secondary front like Italy, it will not be possible to keep the military and the political aspects separate. In my view the Russians should be in from the start, and we should carry on in accordance with our duty, our obvious advantage, and our plain right. They are claiming to have everything yielded to them at every point, and give nothing in return except their military pressure, which has never yet been exerted except in their own interest. They ought to be made to feel that we also have our point of view. In my opinion, the military, in the event of disagreement in negotiations, must refer to their Governments before reaching any conclusion.

My wife was at this time about to visit the Soviet Union on behalf of her "Aid to Russia" fund, but so intense was the Russian suspicion about the conversations at Berne that I even considered postponing her departure.

Prime Minister to Foreign Secretary   25 Mar 45

My immediately preceding minute. I suppose it is all right Clemmie going on her journey in these circumstances. Let me know your unprejudiced opinion whether it would be better to put it off for a few days or weeks, or whether it would be considered as a sign of personal goodwill. I incline to her going as arranged.

In fact she went, and was received with the utmost goodwill. Meanwhile I carefully watched the progress of the negotiations to make sure the Soviets suffered no unfair exclusion.

Prime Minister to Foreign Secretary   30 Mar 45

Have we not told the Russians that the only purpose of the contacts in Switzerland is to arrange a meeting at our military headquarters in Italy, where military questions will be discussed in the presence, if they wish, of a Russian representative, and that if at any moment political affairs are trenched upon the whole matter can be referred to the three Governments? It looks as if the Swiss conversations may go beyond that, if indeed they have not already gone beyond it. We have decided to ignore the insulting telegrams which Molotov has sent. This however does not relieve us from our obligation as Allies on any matter which might involve peace negotiations.

Pray consider this and let me know whether any further information should be conveyed.

* * * * *

On April 5 I received from the President the startling text of his interchanges with Stalin. These were the telegrams:

Marshal Stalin to President Roosevelt   3 Apr 45

I have received your message on the question of negotiations in Berne. You are absolutely right that, in connection with the affair regarding negotiations of the Anglo-American command with the German command, somewhere in Berne or some other place, "has developed an atmosphere of fear and distrust deserving regrets".

You insist that there have been no negotiations yet.

It may be assumed that you have not yet been fully informed. As regards my military colleagues, they, on the basis of data which they have on hand, do not have any doubts that the negotiations have taken place, and that they have ended in an agreement with the Germans, on the basis of which the German commander on the Western Front, Marshal Kesselring, has agreed to open the front and permit the Anglo-American troops to advance to the east, and the Anglo-Americans have promised in return to ease for the Germans the peace terms.

I think that my colleagues are close to the truth. Otherwise one could not have understood the fact that the Anglo-Americans have refused to admit to Berne representatives of the Soviet command for participation in the negotiations with the Germans.

I also cannot understand the silence of the British, who have allowed you to correspond with me on this unpleasant matter, and they themselves remain silent, although it is known that the initiative in this whole affair with the negotiations in Berne belongs to the British.

I understand that there are certain advantages for the Anglo-American troops as a result of these separate negotiations in Berne or some other place, since the Anglo-American troops get the possibility to advance into the heart of Germany almost without resistance on the part of the Germans, but why was it necessary to conceal this from the Russians, and why were your Allies, the Russians, not notified?

As a result of this at the present moment the Germans on the Western Front in fact have ceased the war against England and the United States. At the same time the Germans continue the war with Russia, the Ally of England and the United States. It is understandable that such a situation can in no way serve the cause of preservation of the strengthening of trust between our countries.

I have already written to you in my previous message, and consider it necessary to repeat it here, that I personally and my colleagues would never have made such a risky step, being aware that a momentary advantage, no matter what it would be, is fading before the principal advantage of the preservation and strengthening of the trust among the Allies.

This accusation angered the President deeply. His strength did not allow him to draft his own reply. General Marshall framed the following answer, with Roosevelt's approval. It certainly did not lack vigour.

President Roosevelt to Marshal Stalin   5 Apr 45

I have received with astonishment your message of April 3 containing an allegation that arrangements which were made between Field-Marshal Alexander and Kesselring at Berne "permitted the Anglo-American troops to advance to the east, and the Anglo-Americans promised in return to ease for the Germans the peace terms".

In my previous messages to you in regard to the attempts made in Berne to arrange a conference to discuss a surrender of the German Army in Italy I have told you that (i) no negotiations were held in Berne; (ii) that the meeting had no political implications whatever; (iii) that in any surrender of the enemy Army in Italy there could be no violation of our agreed principle of unconditional surrender; (iv) that Soviet officers would be welcomed at any meeting that might be arranged to discuss surrender.

For the advantage of our common war effort against Germany, which to-day gives excellent promise of an early success in a disintegration of the German armies, I must continue to assume that you have the same high confidence in my truthfulness and reliability that I have always had in yours.

I have also a full appreciation of the effect your gallant Army has had in making possible a crossing of the Rhine by the forces under General Eisenhower, and the effect that your forces will have hereafter on the eventual collapse of the German resistance to our combined attacks.

I have complete confidence in General Eisenhower, and know that he certainly would inform me before entering into any agreement with the Germans. He is instructed to demand, and will demand, unconditional surrender of enemy troops that may be defeated on his front. Our advances on the Western Front are due to military action. Their speed has been attributable mainly to the terrific impact of our air-power, resulting in destruction of German communications, and to the fact that Eisenhower was able to cripple the bulk of the German forces on the Western Front while they were still west of the Rhine.

I am certain that there were no negotiations in Berne at any time, and I feel that your information to that effect must have come from German sources, which have made persistent efforts to create dissension between us in order to escape in some measure for responsibility for their war crimes. If that was Wolff's purpose in Berne your message proves that he has had some success.

With a confidence in your belief in my personal reliability and in

393

my determination to bring about together with you an unconditional surrender of the Nazis, it is astonishing that a belief seems to have reached the Soviet Government that I have entered into an agreement with the enemy without first obtaining your full agreement.

Finally I would say this: it would be one of the great tragedies of history if at the very moment of the victory now within our grasp such distrust, such lack of faith, should prejudice the entire undertaking after the colossal losses of life, material, and treasure involved.

Frankly, I cannot avoid a feeling of bitter resentment toward your informers, whoever they are, for such vile misrepresentations of my actions or those of my trusted subordinates.

I was deeply struck by this last sentence, which I print in italics. I felt that although Mr. Roosevelt did not draft the whole message he might well have added this final stroke himself. It looked like an addition or summing up, and it seemed like Roosevelt himself in anger.

I wrote at once to the President:

Prime Minister to President Roosevelt 5 Apr 45
I am astounded that Stalin should have addressed to you a message so insulting to the honour of the United States and also of Great Britain. His Majesty's Government cordially associate themselves with your reply, and the War Cabinet have instructed me to send to Stalin the message in my immediately following. . . .

Next day I addressed Stalin myself.

Prime Minister to Marshal Stalin 6 Apr 45
The President has sent me his correspondence with you about the contacts made in Switzerland between a British and an American officer on Field-Marshal Alexander's staff and a German general named Wolff relating to possible surrender of Kesselring's army in Northern Italy. I therefore deem it right to send you a precise summary of the action of His Majesty's Government. As soon as we learned of these contacts we immediately informed the Soviet Government on March 12, and we and the United States Government have faithfully reported to you everything that has taken place. The sole and only business mentioned or referred to in any way in Switzerland was to test the credentials of the German emissary and try to arrange a meeting between a nominee of Kesselring's with Field-Marshal Alexander at his headquarters or some convenient point in Northern Italy. There were no negotiations in Switzerland even for a military surrender of Kesselring's army. Still less did any political-military plot, as alleged

in your telegram to the President, enter into our thoughts, which are not, as suggested, of so dishonourable a character.

2. Your representatives were immediately invited to the meeting we attempted to arrange in Italy. Had it taken place and had your representatives come, they could have heard every word that passed.

3. We consider that Field-Marshal Alexander has full right to accept the surrender of the German army of twenty-five divisions on his front in Italy, and to discuss such matters with German envoys who have the power to settle the terms of capitulation. Nevertheless we took especial care to invite your representatives to this purely military discussion at his headquarters should it take place. In fact however nothing resulted from any contacts in Switzerland. Our officers returned from Switzerland without having succeeded in fixing a rendezvous in Italy for Kesselring's emissaries to come to. Of all this the Soviet Government have been fully informed step by step by Field-Marshal Alexander or by Sir Archibald Clark Kerr, as well as through United States channels. I repeat that no negotiations of any kind were entered into or even touched upon, formally or informally, in Switzerland.

4. There is however a possibility that the whole of this request to parley by the German General Wolff was one of those attempts which are made by the enemy with the object of sowing distrust between Allies. Field-Marshal Alexander made this point in a telegram sent on March 11, in which he remarks, "Please note that two of the leading figures are S.S. and Himmler men, which makes me very suspicious." This telegram was repeated to the British Ambassador in Moscow on March 12 for communication to the Soviet Government. If to sow distrust between us was the German intention it has certainly for the moment been successful.

After quoting some of the more insulting phrases from Molotov's letter I continued:

In the interests of Anglo-Russian relations His Majesty's Government decided not to make any reply to this most wounding and unfounded charge, but to ignore it. This is the reason for what you call in your message to the President "the silence of the British". We thought it better to keep silent than to respond to such a message as was sent by M. Molotov, but you may be sure that we were astonished by it and affronted that M. Molotov should impute such conduct to us. This however in no way affected our instructions to Field-Marshal Alexander to keep you fully informed.

6. Neither is it true that the initiative in this matter came, as you state to the President, wholly from the British. In fact the information given to Field-Marshal Alexander that the German General Wolff

wished to make a contact in Switzerland was brought to him by an American agency.

7. There is no connection whatever between any contacts at Berne or elsewhere with the total defeat of the German armies on the Western Front. They have in fact fought with great obstinacy, and inflicted upon us and the American armies since the opening of our February offensive up to March 28 upwards of 87,000 casualties. However, being outnumbered on the ground and literally overwhelmed in the air by the vastly superior Anglo-American Air Forces, which in the month of March alone dropped over 200,000 tons of bombs on Germany, the German armies in the West have been decisively broken. The fact that they were outnumbered on the ground in the West is due to the magnificent attacks and weight of the Soviet armies.

8. With regard to the charges which you have made in your message to the President of April 3, which also asperse His Majesty's Government, I associate myself and my colleagues with the last sentence of the President's reply.

On April 7 Stalin replied to the President's reproach.

Marshal Stalin to President Roosevelt   7 Apr 45
Your message of April 5 received.

In my message of April 3 the point at issue is not that of integrity and trustworthiness. I have never doubted your integrity and trustworthiness or Mr. Churchill's either. My point is that in the course of our correspondence it has become evident that our views differ on the point as to what is admissible and what is inadmissible as between one Ally and another. We Russians think that in the present situation on the fronts, when the enemy is faced with inevitable surrender, if the representatives of any one Ally ever meet the Germans to discuss surrender the representatives of another Ally should be afforded an opportunity of participating in such a meeting. In any case, this is absolutely essential if the Ally in question asks for such participation. The Americans and British however think differently and regard the Russian standpoint as wrong. They have, accordingly, refused the Russians the right to join in meeting the Germans in Switzerland. I have already written you, and I think it should be repeated, that in a similar situation the Russians would never have denied the Americans and British the right to join in such a meeting. I still think the Russian point of view to be the only correct one, as it precludes all possibility of mutual suspicions and makes it impossible for the enemy to sow distrust between us.

2. It is difficult to admit that the lack of resistance by the Germans on the Western Front is due solely to the fact that they have been

defeated. The Germans have 147 divisions on the Eastern Front. They could without prejudicing their own position detach fifteen to twenty divisions from the Eastern Front and transfer them to reinforce their troops on the Western Front. Yet the Germans have not done and are not doing this. They are continuing to wage a crazy struggle with the Russians for an insignificant railway station like Zemlyanitsa in Czechoslovakia, which is as much use to them as hot poultices to a corpse, and yet they yield without the slightest resistance such important towns in the centre of Germany as Osnabrück, Mannheim, and Kassel. You will agree that such behaviour on the part of the Germans is more than curious and unintelligible.

3. As regards my informants, I can assure you that they are extremely honest and modest people who discharge their duties conscientiously and have no intention of offending anyone. We have very often put these people to a practical test. Judge for yourselves. Last February General Marshall sent the Soviet General Staff a number of important reports, in which he warned the Russians, on the basis of data in his possession, that in March there would be two serious counter-attacks by the Germans on the Eastern Front—one would be aimed from Pomerania against Thorn and the other from the region of Moravska Ostrava against Lodz. In actual fact however it turned out that the Germans' main blow was being prepared and was directed not in the directions above-mentioned, but in an entirely different area, namely, in the neighbourhood of Lake Balaton, south-west of Budapest. It is common knowledge now that in this area the Germans had assembled up to thirty-five divisions, including eleven tank divisions. This was one of the heaviest attacks of the whole war, with such a large concentration of tank forces. Marshal Tolbukhin was able to avoid a catastrophe and subsequently inflict a smashing defeat on the Germans, because, among other reasons, my agents discovered, though somewhat tardily, this plan of the Germans for a major attack and immediately warned Marshal Tolbukhin. In this way I was able once again to convince myself how conscientious and well-informed Soviet agents are. . . .

He also sent a copy of his telegram to me, together with the following personal message:

Marshal Stalin to Prime Minister   7 Apr 45
In my message of April 7 to the President, which I am sending to you also, I have already replied to all the fundamental points raised in your message regarding the negotiations in Switzerland. On the other questions raised in your message I consider it necessary to make the following remarks.

1. Neither I nor Molotov had any intention of "blackening" any-one. It is not a matter of wanting to "blacken" anyone, but of our having developed differing points of view as regards the rights and obligations of an Ally. You will see from my message to the President that the Russian point of view on this question is the correct one, as it guarantees each Ally's rights and deprives the enemy of any possibility of sowing discord between us.

2. My messages are personal and strictly confidential. This makes it possible to speak one's mind clearly and frankly. This is the advantage of confidential communications. If however you are going to regard every frank statement of mine as offensive it will make this kind of communication very difficult. I can assure you that I had and have no intention of offending anyone.

I passed this to Roosevelt, with the following comment:

Prime Minister to President Roosevelt   11 Apr 45
I have a feeling that this is about the best we are going to get out of them, and certainly it is as near as they can get to an apology. How-ever, before considering any answer at all from His Majesty's Govern-ment please tell me how you think the matter should be handled so that we may keep in line together.

The President answered next day that he was sending the following message to Stalin:

Thank you for your frank explanation of the Soviet point of view of the Berne incident, which now appears to have faded into the past without having accomplished any useful purpose.
There must not, in any event, be mutual distrust, and minor mis-understandings of this character should not arise in the future. I feel sure that when our armies make contact in Germany and join in a fully co-ordinated offensive the Nazis' armies will disintegrate.

And later:

President Roosevelt to Prime Minister   12 Apr 45
I would minimise the general Soviet problem as much as possible, because these problems, in one form or another, seem to arise every day, and most of them straighten out, as in the case of the Berne meeting.
We must be firm however, and our course thus far is correct.

CHAPTER XXVII

WESTERN STRATEGIC DIVERGENCES

War and Politics – A Deadly Hiatus – Soviet Ambition – Some Practical Points – Eisenhower's Strategy – His Telegram to Stalin – My Minute to the Chiefs of Staff of March 31 – The American Riposte – My Telegram to Eisenhower of March 31 – Telegram to the President – Eisenhower's Answer – Further Correspondence with Eisenhower – The Rescue of Holland.

A S a war waged by a coalition draws to its end political aspects have a mounting importance. In Washington especially longer and wider views should have prevailed. It is true that American thought is at least disinterested in matters which seem to relate to territorial acquisitions, but when wolves are about the shepherd must guard his flock, even if he does not himself care for mutton. At this time the points at issue did not seem to the United States Chiefs of Staff to be of capital importance. They were of course unnoticed by and unknown to the public, and were all soon swamped, and for the time being effaced, by the flowing tide of victory. Nevertheless, as will not now be disputed, they played a dominating part in the destiny of Europe, and may well have denied us all the lasting peace for which we had fought so long and hard. We can now see the deadly hiatus which existed between the fading of President Roosevelt's strength and the growth of President Truman's grip of the vast world problem. In this melancholy void one President could not act and the other could not know. Neither the military chiefs nor the State Department received the guidance they required. The former confined themselves to their professional sphere; the latter did not comprehend the issues involved. The indispensable political direction was lacking at the moment when it was most needed. The United States stood on the scene of victory, master of world fortunes, but without a true and coherent design. Britain, though still very powerful, could not act decisively alone. I could at this stage only warn and plead. Thus this climax of apparently measureless success was to me a most unhappy time. I moved amid cheering crowds, or sat at a table adorned with congratulations and blessings from every part of the Grand Alliance, with an aching heart and a mind oppressed by forebodings.

The destruction of German military power had brought with it a fundamental change in the relations between Communist Russia and the Western democracies. They had lost their common enemy, which was almost their sole bond of union. Henceforward Russian imperialism and the Communist creed saw and set no bounds to their progress and ultimate dominion, and more than two years were to pass before they were confronted again with an equal will-power. I should not tell this tale now when all is plain in glaring light if I had not known it and felt it when all was dim, and when abounding triumph only intensified the inner darkness of human affairs. Of this the reader must be the judge.

The decisive, practical points of strategy and policy with which this narrative seeks to deal were:

First, that Soviet Russia had become a mortal danger to the free world.

Secondly, that a new front must be immediately created against her onward sweep.

Thirdly, that this front in Europe should be as far east as possible.

Fourthly, that Berlin was the prime and true objective of the Anglo–American armies.

Fifthly, that the liberation of Czechoslovakia and the entry into Prague of American troops was of high consequence.

Sixthly, that Vienna, and indeed Austria, must be regulated by the Western Powers, at least upon an equality with the Russian Soviets.

Seventhly, that Marshal Tito's aggressive pretensions against Italy must be curbed.

Finally, and above all, that a settlement must be reached on all major issues between the West and the East in Europe *before the armies of democracy melted*, or the Western Allies yielded any part of the German territories they had conquered, or, as it could soon be written, liberated from totalitarian tyranny.

* * * * *

All kinds of rumours, only slightly sustained by our reports, were rife about Hitler's future plans. I had thought it prudent to have them searchingly examined, because I heard that they were counting for much at Eisenhower's headquarters. Certainly a movement of German administrative departments southwards from Berlin was noticeable.

Prime Minister to General Ismay, for C.O.S. Committee   17 Mar 45

I should like the Intelligence Committee to consider the possibility that Hitler, after losing Berlin and Northern Germany, will retire to the mountainous and wooded parts of Southern Germany and endeavour to prolong the fight there. The strange resistance he made at Budapest and is now making at Lake Balaton, and the retention of Kesselring's army in Italy so long, seem in harmony with such an intention. But of course he is so foolishly obstinate about everything that there may be no meaning behind these moves. Nevertheless the possibilities should be examined.

Although nothing could be positive, the general conclusion of our Chiefs of Staff was that a prolonged German campaign, or even guerrilla, in the mountains was unlikely on any serious scale. The possibility was therefore relegated by us, as it proved rightly, to the shades. On this basis I inquired about the strategy for the advance of the Anglo-American armies as foreseen at Allied Headquarters, and received this reply:

General Eisenhower to Prime Minister   30 Mar 45

As soon as the U.S. Ninth and First Armies join hands and enemy encircled in Ruhr area is incapable of further offensive action I propose driving eastward to join hands with Russians or to attain general line of Elbe. Subject to Russian intentions, the axis Kassel–Leipzig is the best for the drive, as it will ensure the overrunning of that important industrial area, into which German Ministries are believed to be moving; it will cut the German forces approximately in half, and it will not involve us in crossing of Elbe. It is designed to divide and destroy the major part of remaining enemy forces in West.

This will be my main thrust, and until it is quite clear that concentration of all our effort on it alone will not be necessary I am prepared to direct all my forces to ensuring its success. It lies in Bradley's zone, and he will have the Third, First, and Ninth Armies to carry it out, with Fifteenth Army also under his command, following, if possible, mopping up. He will have Montgomery protecting his left flank, with British and Canadian Armies, north of general line Hanover–

Wittenberg, and Devers protecting his right with Seventh and First French Armies.

Once the success of main thrust is assured I propose to take action to clear the northern ports, which in the case of Kiel will entail forcing the Elbe. Montgomery will be responsible for these tasks, and I propose to increase his forces if that should seem necessary for the purpose.

In addition Sixth Army Group will be prepared, when above requirements have been met, to drive to south-east on axis Nuremberg-Regensburg to prevent any possible German consolidation in south and to join hands with Russians in Danube valley.

I trust this added information will make clear my present plans. Naturally they are flexible and subject to changes to meet unexpected situations.

About the same time we learned that Eisenhower had announced his policy in a direct telegram to Marshal Stalin on March 28 without previously mentioning the subject either to his Deputy, Air Chief Marshal Tedder, or the Combined Chiefs of Staff. We all thought that this went beyond the limits of negotiation with the Soviets by the Supreme Commander in Europe as they had previously been understood. General Eisenhower felt justified in this direct correspondence with the head of the Russian State because Stalin was also Commander-in-Chief of the Red Army. Yet it was not with the President of the United States that he corresponded, who was also the head of the military forces, but with General Marshall.

In this telegram Eisenhower said that after isolating the Ruhr he proposed to make his main thrust along the axis Erfurt–Leipzig–Dresden, which, by joining hands with the Russians, would cut in two the remaining German forces. A secondary advance through Regensburg to Linz, where also he expected to meet the Russians, would prevent "the consolidation of German resistance in the redoubt in Southern Germany". Stalin agreed readily. He said that the proposal "entirely coincides with the plan of the Soviet High Command". "Berlin," he added, "has lost its former strategic importance. The Soviet High Command therefore plans to allot secondary forces in the direction of Berlin." This statement was not borne out by events.

The British Chiefs of Staff were concerned both about the merits of the new plan and about the short-circuiting of the highest authorities, both military and constitutional. They drafted a lengthy telegram to their colleagues in Washington, which I did not see till after it had gone. This very often happened in inter-Staff negotiations. I was in full agreement in principle with our Chiefs of Staff, and we thought on the same lines. All the same, I thought that their telegram brought in many minor extraneous matters and did not take the best ground for an argument with the United States Chiefs of Staff. I accordingly sent them the following minute.

Prime Minister to General Ismay, for C.O.S. Committee　　31 Mar 45

I have considered your telegram, and of course it is a good thing for the military points to be placed before the Combined Chiefs of Staff Committee. I hope however we shall realise that we have only a quarter of the forces invading Germany, and that the situation has thus changed remarkably from the days of June 1944. . . .

3. It seems to me that the chief criticism of the new Eisenhower plan is that it shifts the axis of the main advance upon Berlin to the direction through Leipzig to Dresden, and thus raises the question of whether the Twenty-first Army Group will not be so stretched as to lose its offensive power, especially after it has been deprived of the Ninth United States Army. Thus we might be condemned to an almost static rôle in the north and virtually prevented from crossing the Elbe until an altogether later stage in the operations has been reached. All prospect also of the British entering Berlin with the Americans is ruled out.

4. The validity of such criticism depends on the extent of the enemy's resistance. If that resistance is practically collapsing there is no reason why the advances, both of the main Army and of the Twenty-first Army Group, should not take place on a broader front than hitherto. This is a point on which the Supreme Commander must have the final word.

5. It also seems that General Eisenhower may be wrong in supposing Berlin to be largely devoid of military and political importance. Even though German Government departments have to a great extent moved to the south, the dominating fact on German minds of the fall of Berlin should not be overlooked. The idea of neglecting Berlin and leaving it to the Russians to take at a later stage does not appear to me correct. As long as Berlin holds out and withstands a siege in the ruins, as it may easily do, German resistance will be stimulated. The fall of Berlin might cause nearly all Germans to despair.

6. We weaken our case for a stronger concentration between the sea and the Hanover–Berlin flank by suggesting we should like to

turn aside to clean up matters in Denmark, Norway, and along the Baltic shore. . . .

7. In short, I see argumentative possibilities being opened to the United States Chiefs of Staff by our telegram, on which they will riposte heavily. It must be remembered that Eisenhower's credit with them stands very high. He may claim to have correctly estimated so far the resisting strength of the enemy and to have established by deeds (a) the "closing" [i.e., reaching] of the Rhine along its whole length, (b) the power to make the double advance instead of staking all on the northern advance. . . . These events, combined with the continual arrival of American reinforcements, have greatly enhanced General Eisenhower's power and prestige, and the Americans will feel that, as the victorious Supreme Commander, he has a right, and indeed a vital need, to try to elicit from the Russians their views as to the best point for making contact by the armies of the West and of the East.

8. Finally, the capture of Danzig and consequent annihilation of one of the three principal U-boat bases is a new event bringing great relief to the Admiralty. The renewal of the U-boat warfare on the scale which they predicted is plainly now impossible. . . . Therefore I cannot admit a state of urgency in any way justifying left-handed diversions to clear the Baltic ports, etc., if these diversions take anything from the speed or weight of the advance of the Twenty-first Group of Armies.

P.S.—The above was dictated by me *before* I had seen the United States Chiefs of Staff riposte.

The "riposte" had by now arrived. The United States Chiefs replied in substance that Eisenhower's procedure in communicating with the Russians appeared to have been an operational necessity, and that any modification of it should be made by him, and not by them. The course of action outlined in his plan appeared to accord with agreed strategy and with his directive. Eisenhower, they claimed, was deploying across the Rhine in the north the maximum forces which could be used. The secondary effort in the south was achieving an outstanding success, and was being exploited as much as supplies would permit. They were confident that the Supreme Commander's course of action would secure the ports and everything else mentioned by the British more quickly and more decisively than the plan urged by them.

The Battle of Germany, they said, was at a point where it was for the Field Commander to judge the measures which should be taken. To turn away deliberately from the exploitation of the enemy's weakness did not appear sound. The single objective should be quick and complete victory. While recognising that there were factors not of direct concern to the Supreme Commander, the United States Chiefs considered his strategic concept was sound and should receive full support and that he should continue to communicate freely with the Commander-in-Chief of the Soviet Army.

The United States Chiefs of Staff however also suggested that General Eisenhower should be asked to submit to them an amplification of his message to Marshal Stalin, and to delay replying to any new request from Moscow for further information until he had heard from the Combined Chiefs of Staff.

* * * * *

In full agreement with my military colleagues, I repeated the substance of my minute to General Eisenhower.

Prime Minister to General Eisenhower 31 Mar 45

Very many thanks. It seems to me personally that if the enemy's resistance does not collapse the shifting of the main axis of advance so much farther to the southward and the withdrawal of the Ninth U.S. Army from the Twenty-first Army Group may stretch Montgomery's front so widely that the offensive rôle which was assigned to him may peter out. I do not know why it would be an advantage not to cross the Elbe. If the enemy's resistance should weaken, as you evidently expect and which may well be fulfilled, why should we not cross the Elbe and advance as far eastward as possible? This has an important political bearing, as the Russian armies of the South seem certain to enter Vienna and overrun Austria. If we deliberately leave Berlin to them, even if it should be in our grasp, the double event may strengthen their conviction, already apparent, that they have done everything.

2. Further, I do not consider myself that Berlin has yet lost its military and certainly not its political significance. The fall of Berlin would have a profound psychological effect on German resistance in every part of the Reich. While Berlin holds out great masses of Germans will feel it their duty to go down fighting. The idea that the capture of Dresden and junction with the Russians there would be a superior gain does not commend itself to me. The parts of the German Government departments which have moved south can very quickly move southward again. But while Berlin remains under the German flag it cannot, in my opinion, fail to be the most decisive point in Germany.

3. Therefore I should greatly prefer persistence in the plan on which we crossed the Rhine, namely, that the Ninth U.S. Army should

march with the Twenty-first Army Group to the Elbe and beyond Berlin. This would not be in any way inconsistent with the great central thrust which you are now so rightly developing as the result of the brilliant operations of your armies south of the Ruhr. It only shifts the weight of one army to the northern flank.

I also summed up the position in a message to the President.

Prime Minister to President Roosevelt   1 Apr 45
You will have read the telegrams between the British Chiefs of Staff and their United States colleagues. I think there is some misunderstanding on both sides, which I am anxious to disperse without more ado.

2. We are very much obliged to the United States Chiefs of Staff for their paragraph which gives time for a reasonable interchange of views between our two Chiefs of Staff Committees.

3. I am however distressed to read that it should be thought that we wish in the slightest degree to discredit or lower the prestige of General Eisenhower in his increasingly important relations with the Russian commanders in the field. All we sought was a little time to consider the far-reaching changes desired by General Eisenhower in the plans that had been concerted by the Combined Chiefs of Staff at Malta and had received your and my joint approval. The British Chiefs of Staff were naturally concerned by a procedure which apparently left the fortunes of the British Army, which though only a third of yours still amounts to over a million men, to be settled without the slightest reference to any British authority. They also did not fully understand from General Eisenhower's message what actually was intended. In this we may be excused, because General Deane was similarly puzzled and delayed delivery of General Eisenhower's message to Stalin for twenty hours in order to ask for background. I am in full agreement in this instance with the procedure proposed by your Chiefs of Staff, and I am sorry we did not think of it ourselves.

4. At this point I wish to place on record the complete confidence felt by His Majesty's Government in General Eisenhower, our pleasure that our armies are serving under his command, and our admiration of the great and shining qualities of character and personality which he has proved himself to possess in all the difficulties of handling an Allied Command. Moreover, I should like to express to you, Mr. President, as I have already done orally in the field to General Eisenhower, my heartfelt congratulations on the glorious victories and advances by all the armies of the United States Centre in the recent battles on the Rhine and over it. . . .

5. Having dealt with and I trust disposed of these misunderstandings between the truest friends and comrades that ever fought side by side

as allies, I venture to put to you a few considerations upon the merits of the changes in our original plans now desired by General Eisenhower. It seems to me the differences are small, and, as usual, not of principle but of emphasis. Obviously, laying aside every impediment and shunning every diversion, the Allied armies of the North and Centre should now march at the highest speed towards the Elbe. Hitherto the axis has been upon Berlin. General Eisenhower, on his estimate of the enemy's resistance, to which I attach the greatest importance, now wishes to shift the axis somewhat to the southward and strike through Leipzig, even perhaps as far south as Dresden. He withdraws the Ninth United States Army from the northern group of armies, and in consequence stretches its front southwards. I should be sorry if the resistance of the enemy was such as to destroy the weight and momentum of the advance of the British Twenty-first Army Group and to leave them in an almost static condition along the Elbe when and if they reach it. I say quite frankly that Berlin remains of high strategic importance. Nothing will exert a psychological effect of despair upon all German forces of resistance equal to that of the fall of Berlin. It will be the supreme signal of defeat to the German people. On the other hand, if left to itself to maintain a siege by the Russians among its ruins, and as long as the German flag flies there, it will animate the resistance of all Germans under arms.

6. There is moreover another aspect which it is proper for you and me to consider. The Russian armies will no doubt overrun all Austria and enter Vienna. If they also take Berlin will not their impression that they have been the overwhelming contributor to our common victory be unduly imprinted in their minds, and may this not lead them into a mood which will raise grave and formidable difficulties in the future? I therefore consider that from a political standpoint we should march as far east into Germany as possible, and that should Berlin be in our grasp we should certainly take it. This also appears sound on military grounds.

7. To sum up, the difference that might exist between General Eisenhower's new plans and those we advocated, and which were agreed upon beforehand, would seem to be the following, viz., whether the emphasis should be put on an axis directed on Berlin or on one directed on Leipzig and Dresden. This is surely a matter upon which a reasonable latitude of discussion should be allowed to our two Chiefs of Staff Committees before any final commitment involving the Russians is entered into.

8. I need hardly say that I am quite willing that this message, which is my own personal message to you and not a Staff communication, should be shown to General Marshall.

Actually, though I did not realise it, the President's health was now so feeble that it was General Marshall who had to deal with these grave questions.

Eisenhower replied at once to my telegram of March 31.

General Eisenhower to Prime Minister   1 Apr 45

After reading your message dated yesterday I think you still have some misunderstanding of what I intend to do.

In the first place I repeat that I have not changed any plan. I made certain groupings of this force in order to cross the Rhine, with the main deliberate thrust in the north, isolate the Ruhr, and disrupt, surround, or destroy the Germans defending that area. This is as far as strategic objectives of this force have ever been approved by me, because obviously such a victory over the German forces in the West and such a blow to his industrial capacity would necessarily create new situations requiring study and analysis before the next broad pattern of effort could be accurately sketched.

The situation that is now developing is one that I have held before my Staff for more than a year as the one toward which we shall strive, namely, that our forces should be concentrating across the Rhine through avenues of Wesel and Frankfurt and situated roughly in a great triangle with the apex resting in the Kassel area. From there onwards the problem was to determine the direction of the blow that would create the maximum disorganisation to the remaining German forces and the German power of resisting. I had never lost sight of the great importance of the drive to the northernmost coast, although your telegram did introduce a new idea respecting the political importance of the early attainment of particular objectives. I clearly see your point in this matter. The only difference between your suggestions and my plan is one of timing. . . . In order to assure the success of each of my planned efforts, I concentrate first in the Centre to gain the position I need. As it looks to me now, the next move thereafter should be to have Montgomery cross the Elbe, reinforced as necessary by American troops, and reach at least a line including Lübeck on the coast. If German resistance from now on should progressively and definitely crumble you can see that there would be little if any difference in time between gaining central position and crossing the Elbe. On the other hand, if resistance tends to stiffen at all I can see that it is vitally necessary that I concentrate for each effort, and do not allow myself to be dispersed by attempting to do all these projects at once.

Quite naturally, if at any moment collapse should suddenly come about everywhere along the front we would rush forward, and Lübeck and Berlin would be included in our important targets.

And I answered:

Prime Minister to General Eisenhower 2 Apr 45
Thank you again for your most kind telegram. . . . I am however all the more impressed with the importance of entering Berlin, which may well be open to us, by the reply from Moscow to you, which in paragraph 3 says, "Berlin has lost its former strategic importance." This should be read in the light of what I mentioned of the political aspects. I deem it highly important that we should shake hands with the Russians as far to the east as possible. . . .

4. The arrival of your additional information has largely allayed the anxieties of our Staffs, and they have telegraphed in this sense to their opposite numbers in Washington. You will, I am sure, make allowance for the fact that we had heard nothing at all about this either officially or from our Deputy* until we saw your telegram to Stalin, and this telegram made them think that very large changes were proposed.

5. I regard all this business as smoothing itself down quite satisfactorily, though some correspondence is still proceeding between our Chiefs of Staff Committees.

6. Again my congratulations on the great developments. Much may happen in the West before the date of Stalin's main offensive.

I felt it my duty to end this correspondence between friends.

Prime Minister to President Roosevelt 5 Apr 45
I still think it was a pity that Eisenhower's telegram was sent to Stalin without anything being said to our Chiefs of Staff or to our Deputy, Air Chief Marshal Tedder, or to our Commander-in-Chief, Field-Marshal Montgomery. The changes in the main plan have now turned out to be very much less than we at first supposed. My personal relations with General Eisenhower are of the most friendly character. I regard the matter as closed, and to prove my sincerity I will use one of my very few Latin quotations: *Amantium iræ amoris integratio est.*

* * * * *

These discussions had not of course been allowed to hamper our military advance, and indeed it was at about this time that one noteworthy step in the liberation of Europe was accomplished. We had received many terrible accounts of the plight of the Dutch in "Fortress Holland", and the First Canadian Army had been charged with their rescue. Its IInd Corps accordingly drove the enemy from the Wilhelmshaven peninsula and

---

* Air Chief Marshal Tedder, at Eisenhower's headquarters.

North-East Holland, while the Ist Canadian Corps captured Arnhem and swung west towards Amsterdam. Then their advance was halted south of the Zuider Zee. The German commander refused to surrender his forces so long as the German armies were still fighting elsewhere. If we were to turn all this low-lying, partly flooded area into a battlefield the sufferings of the civilians would be unbearably increased. I addressed myself to the President.

Prime Minister to President Roosevelt   10 Apr 45

The plight of the civil population in Occupied Holland is desperate. Between two and three million people are facing starvation. We believe that large numbers are dying daily, and the situation must deteriorate rapidly now that communications between Germany and Holland are virtually cut. I fear we may soon be in the presence of a tragedy.

2. Eisenhower has plans prepared for bringing relief to the civil population when Western Holland is liberated, and we have accumulated the stocks for this purpose in suitable proximity. But if we wait until Holland has been liberated this help may come too late. There is need for action to bring immediate help, on a far larger scale than is afforded by the Swedish relief scheme.

3. I therefore ask you to join me in giving notice to the German Government through the Swiss Government, as the Protecting Power, to the following effect. It is the responsibility of the German Government to sustain the civil population in those parts of Holland which remain in German occupation. As they have failed to discharge that responsibility, we are prepared to send food and medical supplies for distribution to the civil population through the agency of the International Red Cross. We are ready to increase the limited supplies that are already being sent from Sweden, and also to send in further supplies, by sea or direct from areas under military control of the Allies, subject to the necessary safe-conducts being arranged. We invite the German Government to accord the facilities to enable this to be done.

4. In present circumstances I think that the German Government might well accede to this request. If however they should refuse I propose that we should at this stage warn the German commander in Holland and all the troops under his command that by resisting our attempt to bring relief to the civil population in this area they brand themselves as murderers before the world, and we shall hold them responsible with their lives for the fate which overtakes the people of Holland. Full publicity would be given to this warning, so as to bring it home to all German troops stationed in Holland.

5. We must avert this tragedy if we can. But, if we cannot, we must at least make it clear to the world on whose shoulders the responsibility lies.

6. The terms of the communication to be made to the German Government through the Protecting Power are being drafted and will be sent to you to-morrow. In the meantime I hope that you will be able to agree in principle.

This was generally agreed, and parleys began with Seyss-Inquart, the Nazi High Commissioner. It was arranged that we should halt our westward advance. On his part he would stop further flooding, cease all repressive measures against the inhabitants, and help to bring in relief supplies. We had accumulated large quantities of these, and all means were used, by land, sea, and air, to deliver them speedily. This was certainly the best possible arrangement. The Dutch nation has since most gracefully acknowledged by word and deed the help we were so proud to give them after their bravely borne afflictions.

CHAPTER XXVIII

THE CLIMAX: ROOSEVELT'S DEATH

Death of President Roosevelt at the Supreme Climax of the War –
Universal Regret – My Tribute in Parliament – I am Restrained from
Attending the Funeral – Memorial Service in St. Paul's – First Con-
tacts with President Truman – An Informative Message from Lord
Halifax – My Telegram to Stalin, April 14 – Eden's Telegrams from
Washington, April 15 and 16.

PRESIDENT ROOSEVELT died suddenly on Thursday, April 12, at Warm Springs, Georgia. He was sixty-three. In the afternoon, while he was having his portrait painted, he suddenly collapsed, and died a few hours later without regaining consciousness.

Preceding chapters have shown how the problems of impending victory rivalled in their perplexity the worst perils of war. Indeed, it may be said that Roosevelt died at the supreme climax of the war, and at the moment when his authority was most needed to guide the policy of the United States. When I received these tidings early in the morning of Friday, the 13th, I felt as if I had been struck a physical blow. My relations with this shining personality had played so large a part in the long, terrible years we had worked together. Now they had come to an end, and I was overpowered by a sense of deep and irreparable loss. I went down to the House of Commons, which met at eleven o'clock, and in a few sentences proposed that we should pay our respects to the memory of our great friend by immediately adjourning. This unprecedented step on the occasion of the death of the head of a foreign State was in accordance with the unanimous wish of the Members, who filed slowly out of the chamber after a sitting which had lasted only eight minutes.

All the nations paid their tributes in one form or another to

Roosevelt's memory. Black-bordered flags were hung in Moscow, and the Supreme Soviet, when it met, stood in silence. The Japanese Premier expressed "profound sympathy" to the Americans in the loss of their leader, to whom he assigned the responsibility for "the Americans' advantageous position to-day". The German radio said, in contrast, "Roosevelt will go down in history as the man at whose instigation the present war spread into a Second World War, and as the President who finally succeeded in bringing his greatest opponent, the Bolshevik Soviet Union, to power."

In my message to Mrs. Roosevelt I said:

Accept my most profound sympathy in your grievous loss, which is also the loss of the British nation and of the cause of freedom in every land. I feel so deeply for you all. As for myself, I have lost a dear and cherished friendship which was forged in the fire of war. I trust you may find consolation in the magnitude of his work and the glory of his name.

And to Harry Hopkins, who had been my precious link on so many occasions:

I understand how deep your feelings of grief must be. I feel with you that we have lost one of our greatest friends and one of the most valiant champions of the causes for which we fight. I feel a very painful personal loss, quite apart from the ties of public action which bound us so closely together. I had a true affection for Franklin.

When Parliament met on Tuesday, April 17, I moved an address to the King conveying to His Majesty the deep sorrow of the House and their profound sympathy with Mrs. Roosevelt and with the Government and people of the United States. It is customary for the leaders of all parties to speak in support of such a motion, but there developed a spontaneous feeling that it should be left to me alone to speak for the Commons. I cannot find to-day words which I prefer to those I uttered in the emotion of this melancholy event.

"My friendship," I said, "with the great man to whose work and fame we pay our tribute to-day began and ripened during this war. I had met him, but only for a few minutes, after the close of the last war, and as soon as I went to the Admiralty in September 1939 he telegraphed inviting me to correspond with him direct on naval or other matters if at any time I felt inclined.

413

Having obtained the permission of the Prime Minister, I did so. Knowing President Roosevelt's keen interest in sea warfare, I furnished him with a stream of information about our naval affairs, and about the various actions, including especially the action of the Plate River, which lighted the first gloomy winter of the war.

"When I became Prime Minister, and the war broke out in all its hideous fury, when our own life and survival hung in the balance, I was already in a position to telegraph to the President on terms of an association which had become most intimate, and to me most agreeable. This continued through all the ups and downs of the world struggle until Thursday last, when I received my last messages from him. These messages showed no falling-off in his accustomed clear vision and vigour upon perplexing and complicated matters. I may mention that this correspondence, which of course was greatly increased after the United States' entry into the war, comprises, to and fro between us, over 1,700 messages. Many of these were lengthy messages, and the majority dealt with those more difficult points which come to be discussed upon the level of heads of Governments only after official solutions have not been reached at other stages. To this correspondence there must be added our nine meetings—at Argentia, three in Washington, at Casablanca, at Teheran, two at Quebec, and last of all at Yalta—comprising in all about 120 days of close personal contact, during a great part of which I stayed with him at the White House or at his home at Hyde Park or in his retreat in the Blue Mountains, which he called Shangri-La.

"I conceived an admiration for him as a statesman, a man of affairs, and a war leader. I felt the utmost confidence in his upright, inspiring character and outlook, and a personal regard—affection, I must say—for him beyond my power to express to-day. His love of his own country, his respect for its constitution, his power of gauging the tides and currents of its mobile public opinion, were always evident, but added to these were the beatings of that generous heart which was always stirred to anger and to action by spectacles of aggression and oppression by the strong against the weak. It is indeed a loss, a bitter loss to humanity, that those heart-beats are stilled for ever.

"President Roosevelt's physical affliction lay heavily upon him. It was a marvel that he bore up against it through all the many years of tumult and storm. Not one man in ten millions, stricken and crippled as he was, would have attempted to plunge into a life of physical and mental exertion and of hard, ceaseless political controversy. Not one in ten millions would have tried, not one in a generation would have succeeded, not only in entering this sphere, not only in acting vehemently in it, but in becoming indisputable master of the scene. In this extraordinary effort of the spirit over the flesh, of will-power over physical infirmity, he was inspired and sustained by that noble woman his devoted wife, whose high ideals marched with his own, and to whom the deep and respectful sympathy of the House of Commons flows out to-day in all fullness.

"There is no doubt that the President foresaw the great dangers closing in upon the pre-war world with far more prescience than most well-informed people on either side of the Atlantic, and that he urged forward with all his power such precautionary military preparations as peace-time opinion in the United States could be brought to accept. There never was a moment's doubt, as the quarrel opened, upon which side his sympathies lay. The fall of France, and what seemed to most people outside this Island the impending destruction of Great Britain, were to him an agony, although he never lost faith in us. They were an agony to him not only on account of Europe, but because of the serious perils to which the United States herself would have been exposed had we been overwhelmed or the survivors cast down under the German yoke. The bearing of the British nation at that time of stress, when we were all alone, filled him and vast numbers of his countrymen with the warmest sentiments towards our people. He and they felt the Blitz of the stern winter of 1940–41, when Hitler set himself to 'rub out' the cities of our country, as much as any of us did, and perhaps more indeed, for imagination is often more torturing than reality. There is no doubt that the bearing of the British, and above all of the Londoners, kindled fires in American bosoms far harder to quench than the conflagrations from which we were suffering. There was also at that time, in spite of General Wavell's victories—all the more indeed because of the reinforcements which were sent from this country to him—the apprehension widespread in the United States that we should be invaded by Germany after the fullest preparation in the spring of 1941. It was in January that the President sent to

England the late Mr. Wendell Willkie, who, although a political rival and an opposing candidate, felt as he did on many important points. Mr. Willkie brought a letter from Mr. Roosevelt, which the President had written in his own hand, and this letter contained the famous lines of Longfellow:

> Sail on, O ship of State!
> Sail on, O Union, strong and great!
> Humanity with all its fears,
> With all the hopes of future years,
> Is hanging breathless on thy fate.

"At about that same time he devised the extraordinary measure of assistance called Lend-Lease, which will stand forth as the most unselfish and unsordid financial act of any country in all history. The effect of this was greatly to increase British fighting power, and for all the purposes of the war effort to make us, as it were, a much more numerous community. In that autumn I met the President for the first time during the war at Argentia, in Newfoundland, and together we drew up the declaration which has since been called the Atlantic Charter, and which will, I trust, long remain a guide for both our peoples and for other peoples of the world.

"All this time, in deep and dark and deadly secrecy, the Japanese were preparing their act of treachery and greed. When next we met in Washington, Japan, Germany, and Italy had declared war upon the United States, and both our countries were in arms, shoulder to shoulder. Since then we have advanced over the land and over the sea through many difficulties and disappointments, but always with a broadening measure of success. I need not dwell upon the series of great operations which have taken place in the Western Hemisphere, to say nothing of that other immense war proceeding on the other side of the world. Nor need I speak of the plans which we made with our great Ally, Russia, at Teheran, for these have now been carried out for all the world to see.

"But at Yalta I noticed that the President was ailing. His captivating smile, his gay and charming manner, had not deserted him, but his face had a transparency, an air of purification, and often there was a far-away look in his eyes. When I took my leave of him in Alexandria harbour I must confess that I had an indefinable sense of fear that his health and his strength were on the ebb. But nothing altered his inflexible sense of duty. To the end he faced his innumerable tasks unflinching. . . . When death came suddenly upon him he had 'finished his mail'. That portion of his day's work was done. As the saying goes, he died in harness, and we may well say in battle harness, like his soldiers, sailors, and airmen, who side by side with ours are carrying on their task to the end all over the world. What an enviable death was his! He had brought his country through the worst of its perils and the heaviest of its toils. Victory had cast its sure and steady beam upon him.

"In the days of peace he had broadened and stabilised the foundations of American life and union. In war he had raised the strength, might, and glory of the great Republic to a height never attained by any nation in history. With her left hand she was leading the advance of the conquering Allied Armies into the heart of Germany, and with her right, on the other side of the globe, she was irresistibly and swiftly breaking up the power of Japan. And all the time ships, munitions, supplies, and food of every kind were aiding on a gigantic scale her Allies, great and small. . . .

"But all this was no more than worldly power and grandeur, had it not been that the causes of human freedom and social justice, to which so much of his life had been given, added a lustre . . . which will long be discernible among men. He has left behind him a band of resolute and able men handling the numerous interrelated parts of the vast American war machine. He has left a successor who comes forward with firm step and sure conviction to carry on the task to its appointed end. For us it remains only to say that in Franklin Roosevelt there died the greatest American friend we have ever known, and the greatest champion of freedom who has ever brought help and comfort from the New World to the Old."

★   ★   ★   ★   ★

Although Roosevelt's death came as a shock and a surprise, as I have said I had been aware ever since we parted at Alexandria after the Yalta Conference that his strength was fading. I did what I could in personal telegrams to relieve the strain of the divergences on large matters of policy which Soviet antagonism brought into our official correspondence, but I had not fully realised how serious the President's condition had become. I knew that it was not his practice to draft his own telegrams about official business and no change in their style was apparent. Oliver Lyttelton, who saw him on March 29, telegraphed on the 30th that he was "greatly shocked by his appearance".

My first impulse was to fly over to the funeral, and I had already ordered an aeroplane. Lord Halifax telegraphed that both Hopkins and Stettinius were much moved by my thought of possibly coming over, and both warmly agreed with my judgment of the immense effect for good that would be produced; and later that Truman had asked him to say how greatly he would personally value the opportunity of meeting me as early as possible, and that he thought a visit for the funeral, if I had had this in mind, would have been a natural and easy occasion. Mr. Truman's idea was that after the funeral I might have had two or three days' talk with him.

Much pressure was however put on me not to leave the country at this most critical and difficult moment, and I yielded to the wishes of my friends.

I sent the following message to the President:

I very much regret that it is impossible now for me to change my plans, which were approved by the King and the Cabinet this morning, and upon which all arrangements have been made for the conduct of the debates in Parliament next week, including my tribute to the late President on Tuesday, and my attendance upon the King at the memorial service to be held in St. Paul's Cathedral. I am looking forward earnestly to a meeting with you at an early date. Meanwhile the Foreign Secretary knows the whole story of our joint affairs.

In the after-light I regret that I did not adopt the new President's suggestion. I had never met him, and I feel that there were many points on which personal talks would have been of the greatest value, especially if they had been spread over several days and were not hurried or formalised. It seemed to me extraordinary, especially during the last few months, that Roosevelt had not made his deputy and potential successor thoroughly acquainted with the whole story and brought him into the decisions which were being taken. This proved of grave disadvantage to our affairs. There is no comparison between reading about events afterwards and living through them from hour to hour. In Mr. Eden I had a colleague who knew everything and could at any moment take over the entire direction, although I was myself in good health and full activity. But the Vice-President of the United States steps at a bound from a position where he has little information and less power into supreme authority. How could Mr. Truman know and weigh the issues at stake at this climax of the war? Everything that we have learnt about him since shows him to be a resolute and fearless man, capable of taking the greatest decisions. In these early months his position was one of extreme difficulty, and did not enable him to bring his outstanding qualities fully into action.

* * * * *

To my first and formal telegram of condolence and salutation the new President sent a most friendly reply.

I wrote:

13 Apr 45

Pray accept from me the expression of my personal sympathy in the loss which you and the American nation have sustained in the death of our illustrious friend. I hope that I may be privileged to renew with you the intimate comradeship in the great cause we all serve that I enjoyed through these terrible years with him. I offer you my respectful good wishes as you step into the breach in the victorious lines of the United Nations.

Mr. Truman assured me he would do everything in his power to forward the cause for which President Roosevelt gave his life, and to guard and promote the intimate solid relations between our countries which he and I had forged. He hoped to meet me, and promised in the meantime to send me a telegram about Stalin's messages on Poland.*

* * * * *

A very informative telegram from our Ambassador reached me a few days later.

Lord Halifax to Prime Minister   16 Apr 45

Anthony and I saw Harry Hopkins this morning. We both thought that he looked rather frail and fine-drawn. . . . He had not been greatly surprised by the President's death, and was thankful that he had not had a stroke and lost power like Wilson. For some time he had noticed

---

* See Chapter XXV.

how much the President had failed. He had been able to do only very little work.

He judged the President's death to have created a completely new situation in which we should be starting from scratch. One thing we could be certain of would be that the policy would be very much more the concerted action of the Senate. How this would work it was quite impossible to predict. Much would depend on his personal judgments of the people with whom he would be dealing.

Harry thought, on the whole, it had worked out for the best that you had decided not to come over now. To give Truman a few weeks to find his feet was very much to the good. Meanwhile you could be exchanging personal messages, which would make him begin to feel he knew you. Anthony suggested that it would be better, especially if events went the way that looked likely, and if Truman came over to see the United States troops, for him to stop in London *en route*, which we should like also on general grounds, and which F.D.R. had promised to do. Harry liked that idea. . . .

As regards himself, Truman had asked Harry to give him notes on foreign and international policy, which Harry was doing, but he could certainly not carry on in his present job. Truman probably would not want him, and Harry anyhow would not do it. Truman's methods would be quite different from those of F.D.R.: he would conduct his own business, and there would be no room for Harry's particular line of usefulness. They had mentioned the future in a talk they had yesterday, but not much more than to say that they must talk again when Harry was well. . . .

It may be of interest that Truman's hobby is history of military strategy, of which he is reported to have read widely. He certainly betrayed surprising knowledge of Hannibal's campaigns one night here. He venerates Marshall.

To Stalin I said: * * * * *

Prime Minister to Marshal Stalin 14 Apr 45

I have received your message of April 7. I thank you for its reassuring tone, and trust that the "Crossword" misunderstanding may now be considered at an end.*

I have been greatly distressed by the death of President Roosevelt, with whom I had in the last five and a half years established very close personal ties of friendship. This sad event makes it all the more valuable that you and I are linked together by many pleasant courtesies and memories, even in the midst of all the perils and difficulties that we have surmounted.

* Chapter XXVI, pp. 387 *et. seq.*

I must take the occasion to thank you and Molotov for all the kindness with which you have received my wife during her visit to Moscow, and for all the care that is being taken of her on her journey through Russia. We regard it as a great honour that she should receive the Order of the Red Banner of Labour on account of the work she has done to mitigate the terrible sufferings of the wounded soldiers of the heroic Red Army. The amount of money she collected is perhaps not great, but it is a love offering not only of the rich, but mainly of the pennies of the poor, who have been proud to make their small weekly contributions. In the friendship of the masses of our peoples, in the comprehension of their Governments, and in the mutual respect of their armies the future of the world resides.

Marshal Stalin to Prime Minister   15 Apr 45

I have received your message on the occasion of the death of President Roosevelt.

In President Franklin Roosevelt the Soviet people saw a distinguished statesman and a rigid* champion of close co-operation between the three States.

The friendly attitude of President Franklin Roosevelt to the U.S.S.R. will always be most highly valued and remembered by the Soviet people. So far as I personally am concerned I feel exceptionally deeply the burden of the loss of this great man, who was our mutual friend.

Eden, who was now in Washington, wrote:

Foreign Secretary (Washington) to Prime Minister   15 Apr 45

The Ambassador and I had a talk with Stettinius shortly after my arrival this morning. Stettinius said that both Stalin and Molotov had shown signs of being deeply moved by the President's death. Stalin had asked Harriman whether there was any contribution he could make at a moment like this to assist to promote the unity of the great Allies. Stettinius said that fortunately Harriman had not at once replied "Poland", but instead had suggested that it would be a good thing if Molotov would come to San Francisco for the Conference. Stettinius had seized on this, and telegraphed urging not only that Molotov should come to San Francisco, but also that he should come to Washington first for conversations. An hour ago Stettinius rang me up to say that the Russians had agreed to this course and that Molotov was coming by an American aircraft which had been sent to bring him. I suppose therefore that he will be here by Tuesday, when I plan that we should embark upon the Polish issue.

2. This is all good news, but we ought not to build too much on it, for it yet remains to be seen what attitude Molotov adopts when he

* "Unswerving" is, I think, a truer translation.—W.S.C.

gets here. At any rate, it is stimulating to have a chance to get to grips. . . .

3. Stettinius also spoke to me this afternoon about the debate in the House of Commons this week on Poland, and said that he hoped that you would be able to indicate that events have taken a new turn in the light of the meeting of the three Foreign Secretaries. I agreed, but told him that it was my view that it would do the Russians no harm to know how deep was our concern at the failure of the Moscow Commission thus far to make progress on the basis of Yalta decisions. I feel strongly that we must keep a steady pressure on the Russians. There is no justification yet for optimism, and our best chance of success in any of the conversations here is that the Russians should understand to the full the seriousness for us all of the failure.

And the next day:

Foreign Secretary (Washington) to Prime Minister   16 Apr 45

Edward and I paid our first call on the President this morning. He made a good impression. I told him how touched and pleased you had been with his first message to you. I repeated your regret that it had not been possible for you to come to Washington on the occasion of the late President's funeral, but said that you hoped an early meeting would be possible. The President said that he warmly reciprocated these sentiments. We would understand that he had inherited heavy responsibilities. He had to familiarise himself with a wide range of subjects. It was however his intention to continue on exactly the same lines of foreign policy as the late President had followed. . . .

I then reverted to the subject of a meeting between you and the President, and said that the President would probably recall that President Roosevelt had planned to make an early visit to Europe, making London his first port of call. I knew that His Majesty and you and all His Majesty's Government would be much gratified if President Truman felt able to carry through this programme. The President said that he would like this very much, but we should understand for the moment that he had a number of immediate duties to carry through here. He must deal with a number of urgent domestic issues, and he must also familiarise himself with the late President's policies, on a wide range of subjects. It was not therefore possible for him to give a definite answer now. I got the impression however that he would like to come, though it may be that the date will have to be later than President Roosevelt contemplated. . . .

I gave him your message about Mikolajczyk, with the text of what Mikolajczyk said. The President asked me to thank you, and said that it seemed to him that M. Mikolajczyk's comments made "very good sense". The State Department are, I know, much pleased with this development. . . .

My impression from the interview is that the new President is honest and friendly. He is conscious of but not overwhelmed by his new responsibilities. His references to you could not have been warmer. I believe we shall have in him a loyal collaborator, and I am much heartened by this first conversation.

I replied:

Prime Minister to Mr. Eden (Washington)   24 Apr 45

Much though I should like to see the President personally, I am reluctant to go to the United States in the next sixty days. It is very probable that the election will be on foot here before then. This we cannot tell until we can see more clearly the military results which lie before us. The most cordial invitation will, I am sure, be sent to the President by the King and by His Majesty's Government. I should think that ninety days hence would not be an inconvenient date, as the General Election will either have taken place or be relegated to October. This has not yet been settled.

Thus we all resumed our toilsome march under the deep impression of a common loss.

CHAPTER XXIX

GROWING FRICTION WITH RUSSIA

President Truman Proposes a Joint Message to Stalin – My Reply – An Important Declaration by Mikolajczyk, April 16 – His Further Statement About the Curzon Line – Fruitless Discussion with Molotov at Washington – The Treaty Between the U.S.S.R. and the Warsaw Poles – My Telegram to Stalin of April 24, and His Reply – Soviet Security and Western Dictation – A Retrospect – My Appeal to Stalin of April 29 – A Bleak Prospect – The Entrapping of the Sixteen Polish Leaders – A Forbidding Reply from Stalin, May 5 – The Dark Scene in Europe – Urgent Need for a Triple Meeting.

PRESIDENT TRUMAN'S first political act which concerned us was to take up the Polish question from the point where it stood when Roosevelt died, only forty-eight hours earlier. He proposed a joint declaration by us both to Stalin. The document in which this was set forth must of course have been far advanced in preparation by the State Department at the moment when the new President succeeded. Nevertheless it is remarkable that he felt able so promptly to commit himself to it amid the formalities of assuming office and the funeral of his predecessor.

He admitted that Stalin's attitude was not very hopeful, but felt we should "have another go", and he accordingly proposed telling Stalin that our Ambassadors in Moscow had agreed without question to the three leaders of the Warsaw Government being invited to Moscow for consultation and assuring him we had never denied they would play a prominent part in forming the new Provisional Government of National Unity. Our Ambassadors were not demanding the right to invite an unlimited number of Poles from abroad and from within Poland. The real issue was whether the Warsaw Government could veto individual candidates for consultation, and in our opinion the Yalta agreement did not entitle them to do so.

Mr. Truman then suggested asking Stalin to agree to the following plan:

(1) Bierut, Osóbka-Morawski, Rola Zymierski, Bishop Sapieha, one representative Polish leader not connected with the existing Warsaw Government to be proposed by Stalin, and Mikolajczyk, Grabski, and Stanczyk from London should at once be invited to Moscow.

(2) Once the invitations were issued the Warsaw representatives could arrive first if desired.

(3) The Polish leaders should then suggest others from within Poland or abroad who might be brought in for consultation, so that all major Polish groups would be represented in the discussions.

(4) Until the Polish leaders were consulted we could not bind ourselves about the composition of the new Government of National Unity, and did not consider the Yugoslav precedent applied to Poland.

I replied immediately to this important proposal.

Prime Minister to President Truman   15 Apr 45
It gave me great pleasure to receive your message No. 1, and I am thankful indeed for the expressions of friendship and comradeship which it contains. I reciprocate these most cordially.

2. I have just read the draft joint message which you propose we should send to Stalin. In principle I am in complete agreement with its terms, but there is one important point which Mr. Eden will put before you, and as you and he will be able to discuss the text together any points of detail can, I am sure, be adjusted. I will consult the Cabinet on Monday if the final draft reaches me by then, and I hope we may despatch the message with our joint authority on that very day, as I strongly agree with you that our reply is of high urgency. Moreover, it is important to strike the note of our unity of outlook and of action at the earliest moment.

3. Meanwhile Eden will no doubt discuss with you our impressions of what is actually happening in Moscow and Warsaw. As I see it, the Lublin Government are feeling the strong sentiment of the Polish nation, which, though not unfriendly to Russia, is fiercely resolved on independence, and views with increasing disfavour a Polish Provisional Government which is in the main a Soviet puppet. They are

endeavouring, in accord with the Soviet Government, to form a Government more broad-based than the present one by the addition of Polish personalities (including perhaps Witos) whom they have in their power, but whose end they seek and need. This is a step in the right direction, but would not satisfy our requirements or the decisions of the Crimea Conference.

4. Eden saw Mikolajczyk before his departure, and Mikolajczyk promised to make the declaration desired of him in Stalin's private introductory telegram to me dated April 7, which I repeated to President Roosevelt. . . . I hope to have this afternoon the form of his declaration, which he will publish in his own Polish paper here next Thursday. This, if satisfactory, can be telegraphed to Stalin on Monday, either simultaneously with or as part of our joint message, and if it is not satisfactory I will wrestle with him to make it so, and thereafter repeat to you.

Mr. Eden, who was then in Washington, cabled to me next day that he did not consider that we could agree to the first of the President's proposals about invitations to the Polish leaders to come to Moscow for consultations. It was essential that representatives from within Poland should comprise men who really carried weight and could speak on behalf of the Polish parties. We had to have the right to nominate Poles from inside Poland, and could not leave the choice solely to the Russians. Unless the Poles from inside were truly representative he doubted whether Mikolajczyk and his friends would join in the consultations.

The joint message was sent on the 15th in a slightly amended form. Meanwhile I obtained the following declaration from M. Mikolajczyk, whom I saw at Chequers:

16 Apr 45

I consider close and lasting friendship with Russia is the keystone of future Polish policy, within the wider friendship of the United Nations.

2. To remove all doubt as to my attitude, I wish to declare that I accept the Crimea decision in regard to the future of Poland, its sovereign independent position, and the formation of a Provisional Government representative of National Unity.

3. I support the decision arrived at in the Crimea that a conference of leading Polish personalities be called with a view to constituting a Government of National Unity, as widely and fairly representative of the Polish people as possible, and one which will command recognition by the three major Powers.

On receiving this Stalin wrote to me:

17 Apr 15

Mikolajczyk's statement represents of course a great step forward, but it is not clear whether Mikolajczyk also accepts that part of the decisions of the Crimea Conference which deals with the eastern frontiers of Poland. I should be glad, first, to receive a full text of Mikolajczyk's statement, and, secondly, to receive from Mikolajczyk an explanation as to whether he also accepts that part of the decisions of the Crimea Conference on Poland which deals with the eastern frontiers of Poland.

I therefore sent him on the 22nd a public statement by M. Mikolajczyk which had appeared in his newspaper. "There is no doubt," I cabled, "about the answer which he gives in his last sentence to the question you put to me, namely, that he accepts the Curzon Line, including the Lvov cession to the Soviets. I hope this will be satisfactory to you."

Mikolajczyk's statement ran as follows:

On the demand of Russia the three Great Powers have declared themselves in favour of establishing Poland's eastern frontier on the Curzon Line, with the possibility of small rectifications. My own point of view was that at least Lvov and the oil district should be left to Poland. Considering however, firstly, that in this respect there is an absolute demand on the Soviet side, and, secondly, that the existence side by side of our two nations is dependent on the fulfilment of this condition, we Poles are obliged to ask ourselves whether in the name of the so-called integrity of our republic we are to reject it and thereby jeopardise the whole body of our country's interests. The answer to this question must be "No".

As I got no answer to this it may be assumed that the Dictator was for the moment content. Other points were open. Mr. Eden telegraphed from Washington that he and Stettinius agreed that we should renew our demand for the entry of observers into Poland, and that we should once more press the Soviet Government to hold up their negotiations for a treaty with the Lublin Poles. But shortly after deciding this news arrived that the treaty had been concluded.

*　　*　　*　　*　　*

Next day, April 23, Mr. Stettinius and Eden had an hour and a quarter's discussion with Molotov over Poland. They made no progress whatever.

Stettinius opened by asking whether they should discuss Poland or San Francisco. Molotov at once said San Francisco. Mr. Eden said that San Francisco depended on what progress could be made over Poland, and they must start with Poland. This was accepted. Eden then said that on April 15 the President and I had sent a joint message to Stalin about Poland. Could Molotov say what his Government thought about it? Molotov said he was aware of the message, but had not seen the full text, and the Russian Ambassador declared that the Soviet Embassy had not got it. This, if true, augured ill for the attention Stalin paid to it. The text was then read to Molotov. He asked for time to consider it.

He then referred to the treaty between the Soviet Government and the Administration in Warsaw. Eden pointed out that it had been concluded before any progress whatever had been made in setting up the new Provisional Government of National Unity in Poland. Molotov said he would do what was possible, but any new Government must be based on the existing one and be friendly towards the U.S.S.R. He was surprised that the treaty should have caused dissatisfaction, since it was an attempt by the U.S.S.R. to foster pro-Soviet feeling in Poland. The Soviets had made no difficulties about any agreements between Britain or the United States and France or Belgium.

Eden pointed out that all three of us recognised the Governments in France and Belgium, whereas Poland had two Governments, one recognised by the United States and ourselves and most of the world and the other recognised by the Soviet Government. Making a treaty with the Warsaw Government, which we and the Americans did not recognise, was entirely different, and made people think that the Soviet Government was satisfied with the Polish Government as it now was. Stettinius agreed.

Molotov argued that the United States and Britain were not neighbours of Poland and could afford to postpone decisions, but Russia must make her treaties without delay so as to forward the fight against Germany.

"I took a very bad view," wrote Eden to me, "of to-night's meeting with M. Molotov. I could see no sign of any attention having been given to your joint message with the President. Consequently there seems to be no prospect of progress to-morrow. Moreover, the Soviet Government were quite unrepentant about their treaty with the Warsaw Poles.... My impression is that the

Soviet Government is still cavalier in its attitude and will not accept the seriousness of the situation, unless it is brought up sharply against realities. There is only one way in which we can now do this, and that is by postponing the opening of the Conference for some days while we continue to hammer at the Polish issue in Washington. Unless the Russians are prepared to work with us and the Americans on the basis of the Yalta decisions there is no three-Power unity on which San Francisco can be based."

"Seeking as I do," I replied on the 24th, "a lasting friendship with the Russian people, I am sure this can only be founded upon their recognition of Anglo-American strength. My appreciation is that the new President is not to be bullied by the Soviets."

* * * * *

To Stalin I wrote on the same day:

Prime Minister to Marshal Stalin 24 Apr 45

I have seen the message about Poland which the President handed to M. Molotov for transmission to you, and I have consulted the War Cabinet on account of its special importance. It is my duty now to inform you that we are all agreed in associating ourselves with the President in the aforesaid message. I earnestly hope that means will be found to compose these serious difficulties, which if they continue will darken the hour of victory.

Stalin replied in effect that we regarded the Provisional Polish Government, not as the nucleus of a future Polish Government of National Unity, but simply as one of several groups equivalent to any other group of Poles. This was not what we had decided at Yalta. "There," he claimed, "all three of us, including President Roosevelt, proceeded on the assumption that the Provisional Polish Government, functioning now, as it does, in Poland and enjoying the confidence and support of the majority of the Polish people, should be the nucleus—that is to say, the principal part—of a new reorganised Government of National Unity.

"You evidently are not in agreement with such an understanding of the question. In declining to accept the Yugoslav precedent as a model for Poland you confirm that the Provisional Polish Government cannot be considered as a basis and nucleus of a future Government of National Unity."

Stalin also contended that Poland, unlike Great Britain and the United States, had a common frontier with the Soviet Union.

Her security was as important to Russia as that of Belgium and Greece to Great Britain. The Soviet Union had the right to strive for a friendly Government in Poland and could never approve a hostile one. "To this," he wrote, "we are pledged, apart from all else, by the blood of the Soviet people, which has been profusely shed on the fields of Poland in the name of the liberation of Poland. I do not know whether a truly representative Government has been set up in Greece or whether the Government in Belgium is truly democratic." The Soviet Union was not consulted when they were set up and claimed no right to interfere, "as it understands the full significance of Belgium and Greece for the security of Great Britain." For the United States and Great Britain to come to an arrangement together beforehand about Poland, where the U.S.S.R. was concerned above all, was to put the U.S.S.R. in an intolerable position.

He thanked me for sending him Mikolajczyk's statement about the eastern frontiers of Poland, and promised to advise the Provisional Polish Government to withdraw their objections against inviting him for consultations.

"All that is required now," Stalin concluded, "is that the Yugoslav precedent should be recognised as a model for Poland."

This was no answer. We had gone to Yalta with the hope that both the London and Lublin Polish Governments would be swept away and that a new Government would be formed from among Poles of goodwill, among whom the members of Bierut's Government would be prominent. But Stalin had not liked this plan, and we and the Americans had agreed that there was to be no sweeping away of the Bierut Government, but that instead it should become a "new" Government, "reorganised on a broader democratic basis, with the inclusion of democratic leaders from Poland itself and from Poles abroad". For this purpose Molotov and the two Ambassadors were to sit together in Moscow and try to bring such a Government into being by consultations with members of the existing Provisional Government and with other Polish democratic leaders from within Poland and from abroad.

They were then to select the Poles who were to come for consultations. We tried in each case to find representative men, and we were careful to exclude people who we thought were extreme and unfriendly to Russia. We selected from the London Polish Government three good men, namely, Mikolajczyk,

Stanczyk, and Grabski, who accepted the eastern frontiers which Stalin and I had agreed upon.

The names of those from inside and outside Poland were put forward in the same spirit of helpfulness by the Americans and ourselves. But after nine weeks of discussion on the Commission at Moscow no progress had been made. Molotov had steadily refused to give an opinion about the Poles we mentioned, so that not one of them had been allowed to come even to a preliminary round-table discussion.

★ ★ ★ ★ ★

On April 29 I put my whole case to Stalin.

Prime Minister to Marshal Stalin   29 Apr 45

. . . We are all shocked that you should think that we would favour a Polish Government hostile to the Soviet Union. This is the opposite of our policy. But it was on account of Poland that the British went to war with Germany in 1939. We saw in the Nazi treatment of Poland a symbol of Hitler's vile and wicked lust of conquest and subjugation, and his invasion of Poland was the spark that fired the mine. The British people do not, as is sometimes thought, go to war for calculation, but for sentiment. They had a feeling which grew up in years that with all Hitler's encroachments and doctrine he was a danger to our country and to the liberties which we prize in Europe, and when after Munich he broke his word so shamefully about Czechoslovakia even the extremely peace-loving Chamberlain gave our guarantee against Hitler to Poland. When that guarantee was invoked by the German invasion of Poland the whole nation went to war with Hitler, unprepared as we were. There was a flame in the hearts of men like that which swept your people in their noble defence of their country from a treacherous, brutal, and, as at one time it almost seemed, overwhelming German attack. This British flame burns still among all classes and parties in this Island, and in its self-governing Dominions, and they can never feel this war will have ended rightly unless Poland has a fair deal in the full sense of sovereignty, independence, and freedom, on the basis of friendship with Russia. It was on this that I thought we had agreed at Yalta.

Side by side with this keen sentiment for the rights of Poland, which I believe is shared in at least as strong a degree throughout the United States, there has grown up throughout the English-speaking world a very warm and deep desire to be friends on equal and honourable terms with the mighty Russian Soviet Republic and to work with you, making allowances for our different systems of thought and government, in long and bright years for all the world which we three Powers alone can make together. I, who in my years of great responsibility have worked faithfully for this unity, will certainly continue to do so by every means in my power, and in particular I can assure you that we in Great Britain would not work for or tolerate a Polish Government unfriendly to Russia. Neither could we recognise a Polish Government that did not truly correspond to the description in our joint declaration at Yalta, with proper regard for the rights of the individual as we understand these matters in the Western world.

With regard to your reference to Greece and Belgium, I recognise the consideration which you gave me when we had to intervene with heavy armed forces to quell the E.A.M.-E.L.A.S. attack upon the centre of government in Athens. We have given repeated instructions that your interest in Roumania and Bulgaria is to be recognised as predominant. We cannot however be excluded altogether, and we dislike being treated by your subordinates in these countries so differently from the kindly manner in which we at the top are always treated by you. In Greece we seek nothing but friendship, which is of long duration, and desire only her independence and integrity. But we have no intention to try to decide whether she is to be a monarchy or a republic. Our only policy there is to restore matters to the normal as quickly as possible and to hold fair and free elections, I hope within the next four or five months. These elections will decide the régime and later on the constitution. The will of the people, expressed under conditions of freedom and universal franchise, must prevail; that is our root principle. If the Greeks were to decide for a republic it would not affect our relations with them. We will use our influence with the Greek Government to invite Russian representatives to come and see freely what is going on in Greece, and at the elections I hope there will be Russian, American, and British commissioners at large in the country to make sure that there is no intimidation or other frustration of the free choice of the people between the different parties who will be contending. After that our work in Greece may well be done.

As to Belgium, we have no conditions to demand, though . . . we hope they will, under whatever form of government they adopt by popular decision, come into a general system of resistance to prevent Germany striking westward. Belgium, like Poland, is a theatre of war and corridor of communication, and everyone must recognise the force of these considerations, without which great armies cannot operate.

It is quite true that about Poland we have reached a definite line of action with the Americans. This is because we agree naturally upon the subject, and both sincerely feel that we have been rather ill-treated . . . since the Crimea Conference. No doubt these things seem different when looked at from the opposite point of view. But we are absolutely agreed that the pledge we have given for a sovereign, free, independent Poland, with a Government fully and adequately representing all the democratic elements among Poles, is for us a matter of honour and duty. I do not think there is the slightest chance of any change in the attitude of our two Powers, and when we are agreed we are bound to say so. After all, we have joined with you, largely on my original initiative, early in 1944, in proclaiming the Polish-Russian frontier which you desired, namely, the Curzon Line, including Lvov for Russia. We think you ought to meet us with regard to the other half of the policy which you equally with us have proclaimed, namely, the sovereignty, independence, and freedom of Poland, provided it is a Poland friendly to Russia. . . .

Also, difficulties arise at the present moment because all sorts of stories are brought out of Poland which are eagerly listened to by many Members of Parliament, and which at any time may be violently raised in Parliament or the Press in spite of my deprecating such action, and on which M. Molotov will vouchsafe us no information at all in spite of repeated requests. *For instance, there is the talk of the fifteen Poles who were said to have met the Russian authorities for discussion over four weeks ago, and of M. Witos, about whom there has been a similar but more recent report; and there are many other statements of deportations, etc.* How can I contradict such complaints when you give me no information whatever and when neither I nor the Americans are allowed to send anyone into Poland to find out for themselves the true state of affairs? There is no part of our occupied or liberated territory into which you are not free to send delegations, and people do not see why you should have any reasons against similar visits by British delegations to foreign countries liberated by you.

There is not much comfort in looking into a future where you and the countries you dominate, plus the Communist Parties in many other States, are all drawn up on one side, and those who rally to the English-speaking nations and their associates or Dominions are on the other. It is quite obvious that their quarrel would tear the world to pieces and that all of us leading men on either side who had anything to do with that would be shamed before history. Even embarking on a long period of suspicions, of abuse and counter-abuse, and of opposing policies would be a disaster hampering the great developments of world prosperity for the masses which are attainable only by our trinity. I hope there is no word or phrase in this outpouring of my heart to you which unwittingly gives offence. If so, let me know. But do not, I beg you, my friend Stalin, underrate the divergences which

* Author's italics.

are opening about matters which you may think are small to us but which are symbolic of the way the English-speaking democracies look at life.

*   *   *   *   *

The incident of the missing Poles mentioned in my telegram now requires to be recorded, although it carries us somewhat ahead of the general narrative. At the beginning of March 1945 the Polish Underground were invited by the Russian Political Police to send a delegation to Moscow to discuss the formation of a united Polish Government along the lines of the Yalta agreement. This was followed by a written guarantee of personal safety, and it was understood that the party would later be allowed, if the negotiations were successful, to travel to London for talks with the Polish Government in exile. On March 27 General Leopold Okulicki, the successor of General Bor-Komorowski in command of the Underground Army, two other leaders, and an interpreter had a meeting in the suburbs of Warsaw with a Soviet representative. They were joined the following day by eleven leaders representing the major political parties in Poland. One other Polish leader was already in Russian hands. No one returned from the rendezvous. On April 6 the Polish Government in exile issued a statement in London giving the outline of this sinister episode. The most valuable representatives of the Polish Underground had disappeared without a trace in spite of the formal Russian offer of safe-conduct. Questions were asked in Parliament, and stories have since spread of the shooting of local Polish leaders in the areas at this time occupied by the Soviet armies, and particularly of one episode at Siedlce, in Eastern Poland. It was not until May 4 that Molotov admitted at San Francisco that these men were being held in Russia, and an official Russian news agency stated next day that they were awaiting trial on charges of "diversionary tactics in the rear of the Red Army".

On May 18 Stalin publicly denied that the arrested Polish leaders had ever been invited to Moscow, and asserted that they were mere "diversionists" who would be dealt with according to "a law similar to the British Defence of the Realm Act". The Soviet Government refused to move from this position. Nothing more was heard of the victims of the trap until the case against them opened on June 18. It was conducted in the usual Communist manner. The prisoners were accused of subversion,

434

terrorism, and espionage, and all except one admitted wholly or in part the charges against them. Thirteen were found guilty, and sentenced to terms of imprisonment ranging from four months to ten years, and three were acquitted. This was in fact the judicial liquidation of the leadership of the Polish Underground which had fought so heroically against Hitler. The rank and file had already died in the ruins of Warsaw.

* * * * *

To President Truman I said:

5 May 45

I am most concerned about the fate of the fifteen Polish representatives, in view of the statement made by Molotov to Stettinius at San Francisco that they had been arrested by the Red Army, and I think you and I should consult together very carefully upon this matter. If these Poles were enticed into Russian hands and are now no longer alive one cannot quite tell how far such a crime would influence the future. I am in entire agreement with Eden's views and actions. I hope that he will soon pass through Washington on his homeward journey, and that you will talk it all over with him.

* * * * *

I now received a most disheartening reply from Stalin to the lengthy appeal I had made to him on April 29.

Marshal Stalin to Prime Minister   5 May 45

I have received your message of April 29 on the subject of the Polish question.

I am obliged to say that I cannot agree with the arguments which you advance in support of your position.

1. You are inclined to regard the suggestion that the example of Yugoslavia should be taken as a model for Poland as a departure from the procedure agreed between us for the creation of a Polish Government of National Unity. This cannot be admitted. The example of Yugoslavia is important . . . as pointing the way to the most effective and practical solution of the problem of establishing a new united Government there. . . .

2. I am unable to share your views on the subject of Greece in the passage where you suggest that the three Powers should supervise elections. Such supervision in relation to the people of an Allied State could not be regarded otherwise than as an insult to that people and a flagrant interference with its internal life. Such supervision is unnecessary in relation to the former satellite States which have subsequently declared war on Germany and joined the Allies, as has been shown by

the experience of the elections which have taken place, for instance, in Finland; here elections have been held without any outside intervention and have led to constructive results.

Your remarks concerning Belgium and Poland as theatres of war and corridors of communication are entirely unjustified. It is a question of Poland's peculiar position as a neighbour State of the Soviet Union which demands that the future Polish Government should actively strive for friendly relations between Poland and the Soviet Union, which is likewise in the interest of all other peace-loving nations. This is a further argument for following the example of Yugoslavia. The United Nations are concerned that there should be a firm and lasting friendship between the Soviet Union and Poland. Consequently we cannot be satisfied that persons should be associated with the formation of the future Polish Government who, as you express it, "are not fundamentally anti-Soviet", or that only those persons should be excluded from participation in this work who are in your opinion "extremely unfriendly towards Russia". Neither of these criteria can satisfy us. We insist, and shall insist, that there should be brought into consultation on the formation of the future Polish Government only those persons who have actively shown a friendly attitude towards the Soviet Union and who are honestly and sincerely prepared to co-operate with the Soviet State.

3. I must comment especially on [another] point of your message, in which you mention difficulties arising as a result of rumours of the arrest of fifteen Poles, of deportations and so forth.

As to this, I can inform you that the group of Poles to which you refer consists not of fifteen but of sixteen persons, and is headed by the well-known Polish general Okulicki. In view of his especially odious character the British Information Service is careful to be silent on the subject of this Polish general, who "disappeared" together with the fifteen other Poles who are said to have done likewise. But we do not propose to be silent on this subject. This party of sixteen individuals headed by General Okulicki was arrested by the military authorities on the Soviet front and is undergoing investigation in Moscow. General Okulicki's group, and especially the General himself, are accused of planning and carrying out diversionary acts in the rear of the Red Army which resulted in the loss of over 100 fighters and officers of that Army, and are also accused of maintaining illegal wireless transmitting stations in the rear of our troops, which is contrary to law. All or some of them, according to the results of the investigation, will be handed over for trial. This is the manner in which it is necessary for the Red Army to defend its troops and its rear from diversionists and disturbers of order.

The British Information Service is disseminating rumours of the murder or shooting of Poles in Siedlce. These statements of the British Information Service are complete fabrications, and have evidently been suggested to it by agents of Arciszewski.*

4. It appears from your message that you are not prepared to regard the Polish Provisional Government as the foundation of the future Government of National Unity, and that you are not prepared to accord it its rightful position in that Government. I must say frankly that such an attitude excludes the possibility of an agreed solution of the Polish question.

I repeated this forbidding message to President Truman, with the following comment:

6 May 45

It seems to me that matters can hardly be carried further by correspondence, and that as soon as possible there should be a meeting of the three heads of Governments. *Meanwhile we should hold firmly to the existing position obtained or being obtained by our armies in Yugoslavia, in Austria, in Czechoslovakia, on the main central United States front, and on the British front, reaching up to Lübeck, including Denmark.*† There will be plenty to occupy both armies in collecting the prisoners during the next few days, and we hope that the V.E. celebration will also occupy the public mind at home. Thereafter I feel that we must most earnestly consider our attitude towards the Soviets and show them how much we have to offer or withhold.

Stalin had however already sent a copy to the President.

* * * * *

While the Conference at San Francisco was agreeably planning the foundations of a future free, civilised, and united world, and while the rejoicings in our victory over Hitler and the Nazi tyranny transported the peoples of the Grand Alliance, my mind was oppressed with the new and even greater peril which was swiftly unfolding itself to my gaze. There was also at a lower level the worry about a General Election, which, whatever its result, must divide the nation and weaken its expression in this period when all that we had gained in a righteous war might be cast away. It seemed above all vital that Stalin, Truman, and I should meet together at the earliest moment, and that nothing should delay us. On May 4 I drew the European scene as I saw it for Mr. Eden, who was at the San Francisco Conference, in daily

---

* The Prime Minister of the Polish Government in exile.
† Author's italics.

touch with Stettinius and Molotov, and soon to revisit the President at Washington.

I consider that the Polish deadlock can now probably only be resolved at a conference between the three heads of Governments in some unshattered town in Germany, if such can be found. This should take place at latest at the beginning of July. I propose to telegraph a suggestion to President Truman about his visit here and the further indispensable meeting of the three major Powers.

2. The Polish problem may be easier to settle when set in relation to the now numerous outstanding questions of the utmost gravity which require urgent settlement with the Russians. I fear terrible things have happened during the Russian advance through Germany to the Elbe. The proposed withdrawal of the United States Army to the occupational lines which were arranged with the Russians and Americans in Quebec, and which were marked in yellow on the maps we studied there, would mean the tide of Russian domination sweeping forward 120 miles on a front of 300 or 400 miles. This would be an event which, if it occurred, would be one of the most melancholy in history. After it was over and the territory occupied by the Russians Poland would be completely engulfed and buried deep in Russian-occupied lands. What would in fact be the Russian frontier would run from the North Cape in Norway, along the Finnish-Swedish frontier, across the Baltic to a point just east of Lübeck, along the at present agreed line of occupation and along the frontier between Bavaria to Czechoslovakia to the frontiers of Austria, which is nominally to be in quadruple occupation, and half-way across that country to the Isonzo river, behind which Tito and Russia will claim everything to the east. Thus the territories under Russian control would include the Baltic provinces, all of Germany to the occupational line, all Czechoslovakia, a large part of Austria, the whole of Yugoslavia, Hungary, Roumania, Bulgaria, until Greece in her present tottering condition is reached. It would include all the great capitals of Middle Europe, including Berlin, Vienna, Budapest, Belgrade, Bucharest, and Sofia. The position of Turkey and Constantinople will certainly come immediately into discussion.

3. This constitutes an event in the history of Europe to which there has been no parallel, and which has not been faced by the Allies in their long and hazardous struggle. The Russian demands on Germany for reparations alone will be such as to enable her to prolong the occupation almost indefinitely, at any rate for many years, during which time Poland will sink with many other States into the vast zone of Russian-controlled Europe, not necessarily economically Sovietised, but police-governed.

OCCUPATION ZONES IN GERMANY, AS AGREED AT QUEBEC, SEPT. 1944

It is just about time that these formidable issues were examined between the principal Powers as a whole. We have several powerful bargaining counters on our side, the use of which might make for a peaceful agreement. *First, the Allies ought not to retreat from their present positions to the occupational line until we are satisfied about Poland, and also about the temporary character of the Russian occupation of Germany, and the conditions to be established in the Russianised or Russian-controlled countries in the Danube valley, particularly Austria and Czechoslovakia, and the Balkans.* * Secondly, we may be able to please them about the exits from the Black Sea and the Baltic as part of a general statement. All these matters can only be settled before the United States armies in Europe are weakened. If they are not settled before the United States armies withdraw from Europe and the Western World folds up its war machines there are no prospects of a satisfactory solution and very little of preventing a third World War. It is to this early and speedy show-down and settlement with Russia that we must now turn our hopes. Meanwhile I am against weakening our claim against Russia on behalf of Poland in any way. I think it should stand where it was put in the telegrams from the President and me.

"Nothing," I added the next day, "can save us from the great catastrophe but a meeting and a show-down as early as possible at some point in Germany which is under American and British control and affords reasonable accommodation."

* Author's italics.

CHAPTER XXX

THE FINAL ADVANCE

The Situation at Mr. Roosevelt's Death – The Winter Offensive of the Red Army – Fall of Vienna – The Ninth United States Army Crosses the Elbe, April 12 – And the First United States Army Meets the Russians – The Fall of Prague, May 9 – A Retrospect – Early Plans for the Occupation of Germany – Agreement at Quebec, September 1944 – The Change After Yalta – No Withdrawal from the Agreed Zones Without a Meeting with Stalin – My Warning to President Roosevelt of April 5 – My Minute to the Chiefs of Staff, April 7 – I Address Myself to the New President, April 18 – Berlin and Lübeck – A Reply from Mr. Truman – My Telegram to Stalin of April 27 – His Answer, May 2 – Russian Obstruction in Vienna – The Three Fronts Come Together – The Race for Denmark – The End Draws Near.

PRESIDENT ROOSEVELT died at a moment when political and military prizes of the highest consequence hung in the balance. Hitler's Western Front had collapsed; Eisenhower was across the Rhine and driving deep into Germany and Central Europe against an enemy who in places resisted fiercely but was quite unable to stem the onslaught of our triumphant armies; there seemed nothing to stop the Western Allies from taking Berlin. The Russians were only thirty-five miles from the city, on the east, but they were not yet ready to attack. Between them and Berlin lay the Oder. The Germans were entrenched before the river, and hard fighting was to take place before the Red Army could force a crossing and begin their advance. Vienna was another matter. Our chance of forestalling the Russians in this ancient capital by a thrust from Italy had been abandoned eight months before when Alexander's forces had been stripped for the sake of the landing in the south of France. Prague was still within our reach.

In order to understand how this military position had arisen we must look back a few weeks. Their great winter offensive had carried the Russians across the eastern frontier of Germany into Silesia, an industrial area second in importance only to the Ruhr, and into Pomerania. During the next two months they had reached the lower Oder from Stettin to Glogau, and farther south had established themselves firmly across the river. Encircled German garrisons at Oppeln, Posen, and Schneidemühl were reduced, and Danzig fell at the end of March. The modernised fortress of Königsberg proved to be tough, and was not captured until April 9, after a desperate four-day assault. Only at Breslau and in far-off Courland were there considerable German forces holding out behind the Russian lines. On the Danube front the carnage in Budapest ended on February 15, but heavy German counter-attacks at each end of Lake Balaton continued well into March. When these were thrown back the Russians entered Austria. They moved on Vienna from east and south, were in full possession on April 13, and thrust forward up the Danube towards Linz.

Stalin had told Eisenhower that his main blow would be made in "approximately the second half of May", but he was able to advance a whole month earlier. Perhaps the swift approach of the Western armies to the Elbe had something to do with it.

After crossing the Rhine and encircling the Ruhr Eisenhower detached the flanking corps of the First and Ninth U.S. Armies to subdue its garrison. Bradley's Twelfth Army Group, the Ninth, First, and Third Armies, advanced on Magdeburg, Leipzig, and Bayreuth. Opposition was sporadic, and at the two former cities and in the Hartz Mountains it was severe, but by April 19 all had fallen and the head of the Third Army had crossed into Czechoslovakia. The Ninth Army had indeed moved so swiftly that they crossed the Elbe near Magdeburg on April 12 and were about sixty miles from Berlin.

The Russians, in great strength on the Oder, thirty-five miles from the capital, started their attack on April 16 on a 200-mile front, and surrounded Berlin on April 25. On the same day spearheads of the United States First Army from Leipzig met the Russians near Torgau, on the Elbe. Germany was cut in two, and the Ninth and First Armies remained halted facing the Russians on the Elbe and the Mulde. The German Army was disintegrating before our eyes. Over a million prisoners were taken in the first three weeks of April, but Eisenhower believed that fanatical Nazis would attempt to establish themselves in the mountains of Bavaria and Western Austria, and he swung the Third U.S. Army southwards. Its right thrust down the Danube reached Linz on May 5, and later met the Russians coming up from Vienna. Its left penetrated into Czechoslovakia as far as Budě-jovice, Pilsen, and Karlsbad. There was no agreement to debar him from occupying Prague if it were militarily feasible.

I accordingly approached the President.

Prime Minister to President Truman   30 Apr 45

There can be little doubt that the liberation of Prague and as much as possible of the territory of Western Czechoslovakia by your forces might make the whole difference to the post-war situation in Czecho-slovakia, and might well influence that in near-by countries. On the other hand, if the Western Allies play no significant part in Czecho-slovakian liberation that country will go the way of Yugoslavia.

Of course, such a move by Eisenhower must not interfere with his main operations against the Germans, but I think the highly impor-tant political consideration mentioned above should be brought to his attention.

On May 1 President Truman told me that General Eisenhower's immediate plan of operations in Czechoslovakia was expressed as follows:

The Soviet General Staff now contemplates operations into the Vltava valley. My intention, as soon as current operations permit, is to proceed and destroy any remaining organised German forces.

If a move into Czechoslovakia is then desirable, and if conditions here permit, our logical initial move would be on Pilsen and Karlsbad. I shall not attempt any move which I deem militarily unwise.

The President added, "This meets with my approval." This seemed decisive. Nevertheless I returned to the question a week later.

Prime Minister to General Eisenhower   7 May 45

I am hoping that your plan does not inhibit you to advance to Prague if you have the troops and do not meet the Russians earlier. I thought you did not mean to tie yourself down if you had the troops and the country was empty.

Don't bother to reply by wire, but tell me when we next have a talk.

Eisenhower's plan was however to halt his advance generally on the west bank of the Elbe and along the 1937 boundary of Czechoslovakia. If the situation warranted he would cross it to the general line Karlsbad–Pilsen–Budějovice. The Russians agreed to this and the movement was made. But on May 4 the Russians reacted strongly to a fresh proposal to continue the advance of the Third U.S. Army to the river Vltava, which flows through Prague. This would not have suited them at all. So the Americans "halted while the Red Army cleared the east and west banks of the Moldau river and occupied Prague".* The city fell on May 9, two days after the general surrender was signed at Rheims.

* * * * *

At this point a retrospect is necessary. The occupation of Germany by the principal Allies had long been studied. In the summer of 1943 a Cabinet Committee which I had set up under Mr. Attlee, in agreement with the Chiefs of Staff, recommended that the whole country should be occupied if Germany was to be effectively disarmed, and that our forces should be disposed in three main zones of roughly equal size, the British in the north-west, the Americans in the south and south-west, and the Russians in an eastern zone. Berlin should be a separate joint zone, occupied by each of the three major Allies. These recommendations were approved and forwarded to the European Advisory Council, which then consisted of M. Gousev, the Soviet Ambassador, Mr. Winant, the American Ambassador, and Sir William Strang of the Foreign Office.

At this time the subject seemed to be purely theoretical. No one could foresee when or how the end of the war would come. The German armies held immense areas of European Russia. A year was yet to pass before British or American troops set foot in Western Europe, and nearly two years before they entered Germany. The proposals of the European Advisory Council were not thought sufficiently pressing or practical to be brought before the War Cabinet. Like many praiseworthy efforts to make plans for the future, they lay upon the shelves while the war crashed on.

* Eisenhower, "Report to Combined Chiefs of Staff", p. 140.

In those days a common opinion about Russia was that she would not continue the war once she had regained her frontiers, and that when the time came the Western Allies might well have to try to persuade her not to relax her efforts. The question of the Russian zone of occupation in Germany therefore did not bulk in our thoughts or in Anglo-American discussions, nor was it raised by any of the leaders at Teheran.

When we met in Cairo on the way home in November 1943 the United States Chiefs of Staff brought it forward, but not on account of any Russian request. The Russian zone of Germany remained an academic conception, if anything too good to be true. I was however told that President Roosevelt wished the British and American zones to be reversed. He wanted the lines of communication of any American force in Germany to rest directly on the sea and not to run through France. This issue involved a lot of detailed technical argument and had a bearing at many points upon the plans for "Overlord". No decision was reached at Cairo, but later a considerable correspondence began between the President and myself. The British Staff thought the original plan the better, and also saw many inconveniences and complications in making the change. I had the impression that their American colleagues rather shared their view. At the Quebec Conference in September 1944 we reached a firm agreement between us.

The President, evidently convinced by the military view, had a large map unfolded on his knees. One afternoon, most of the Combined Chiefs of Staff being present, he agreed verbally with me that the existing arrangement should stand subject to the United States armies having a near-by direct outlet to the sea across the British zone. Bremen and its subsidiary Bremerhaven seemed to meet the American needs, and their control over this zone was adopted. This decision is illustrated on the accompanying map. We all felt it was too early as yet to provide for a French zone in Germany, and no one so much as mentioned Russia.

At Yalta in February 1945 the Quebec plan was accepted without further consideration as the working basis for the inconclusive discussions about the future eastern frontier of Germany. This was reserved for the Peace Treaty. The Soviet armies were at this very moment swarming over the pre-war frontiers, and we wished them all success. We proposed an agreement about the zones of occupation in Austria. Stalin, after some persuasion, agreed to my strong appeal that the French should be allotted part of the American and British zones and given a seat on the Allied Control Commission. It was well understood by everyone that the agreed occupational zones must not hamper the operational movements of the armies. Berlin, Prague, and Vienna could be taken by whoever got there first. We separated in the Crimea not only as Allies but as friends facing a still mighty foe with whom all our armies were struggling in fierce and ceaseless battle.

The two months that had passed since then had seen tremendous changes cutting to the very roots of thought. Hitler's Germany was doomed and he himself about to perish. The Russians were fighting in Berlin. Vienna and most of Austria was in their hands. The whole relationship of Russia with the Western Allies was in flux. Every question about the future was unsettled between us. The agreements and understandings of Yalta, such as they were, had already been broken or brushed aside by the triumphant Kremlin. New perils, perhaps as terrible as those we had surmounted, loomed and glared upon the torn and harassed world.

My concern at these ominous developments was apparent even before the President's death. He himself, as we have seen, was also anxious and disturbed. His anger at Molotov's accusations over the Berne affair has been recorded. In spite of the victorious advance of Eisenhower's armies, President Truman found himself faced in the last half of April with a formidable crisis. I had for some time past tried my utmost to impress the United States Government with the vast changes which were taking place both in the military and political spheres. Our Western armies would soon be carried well beyond the boundaries of our occupation zones, as both the Western and Eastern Allied fronts approached one another, penning the Germans between them.

The following telegrams show that I never suggested going back on our word over the agreed zones provided other agreements were also respected. I became convinced however that before we halted, or still more withdrew, our troops we ought to seek a meeting with Stalin face to face and make sure that an agreement was reached about the whole front. It would indeed be a disaster if we kept all our agreements in strict good faith while the Soviets laid their hands upon all they could get without the slightest regard for the obligations into which they had entered.

* * * * *

As early as April 5 I had sent a serious warning to Roosevelt.

. . . There is very little doubt in my mind that the Soviet leaders, whoever they may be, are surprised and disconcerted at the rapid advance of the Allied armies in the West and the almost total defeat of the enemy on our front, especially as they say they are themselves in no position to deliver a decisive attack before the middle of May. All this makes it the more important that we should join hands with the Russian armies as far to the east as possible, and, if circumstances allow, enter Berlin.

I may remind you that we proposed and thought we had arranged six weeks ago provisional zones of occupation in Austria, but that since Yalta the Russians have sent no confirmation of these zones. Now that they are on the eve of taking Vienna and very likely will occupy the whole of Austria, it may well be prudent for us to hold as much as possible in the north.

We must always be anxious lest the brutality of the Russian messages does not foreshadow some deep change of policy for which they are preparing. On the whole I incline to think it is no more than their natural expression when vexed or jealous. For that very reason I deem it of the highest importance that a firm and blunt stand should be made at this juncture by our two countries in order that the air may be cleared and they realise that there is a point beyond which we will not tolerate insult. I believe this is the best chance of saving the future. If they are ever convinced that we are afraid of them and can be bullied into submission, then indeed I should despair of our future relations with them and much else.

General Eisenhower had proposed that while the armies in the west and the east should advance irrespective of demarcation lines, in any area where the armies had made contact either side should be free to suggest that the other should withdraw behind the boundaries of their occupation zone. Discretion to request and to order such withdrawals would rest with Army Group commanders. Subject to the dictates of operational necessity, the retirement would then take place. I considered that this proposal was premature and that it exceeded the immediate military needs. I therefore sent the Chiefs of Staff the following minute for their guidance in discussing General Eisenhower's proposal with their American counterparts.

Prime Minister to General Ismay, for C.O.S. Committee 7 Apr 45

When the forces arrive in contact, and after the preliminary salutations have been exchanged, they should rest opposite each other in those positions, except in so far as actual neighbouring military operations require concerted action. Thus, if we crossed the Elbe and advanced to Berlin, or on a line between Berlin and the Baltic, which is all well within the Russian zone, we should not give this up *as a military matter*. It is a matter of State to be considered between the three Governments, and in relation to what the Russians do in the South, where they will soon have occupied not only Vienna but all Austria. There cannot be such a hurry about our withdrawing from a place we have gained that the few days necessary for consulting the Governments in Washington and London cannot be found. I attach great importance to this, and could not agree to proposals of this kind [being decided] *on a Staff level*. They must be referred to the President and me.

I am very glad to see the delaying action proposed in our Chiefs of Staff message. It is entirely in accordance with my thought.

Action was taken accordingly.

*   *   *   *   *

The death of President Roosevelt on April 12 led me to seek the concurrence of the Chiefs of Staff in presenting anew the whole argument about zones to his successor.

Prime Minister to General Ismay, for C.O.S. Committee 14 Apr 45

I should advise the following line:

"We consider that before the Anglo-American armies retire from any ground they have gained from the enemy over and beyond the zones of occupation agreed upon the political issues operative at that time should be discussed between the heads of Governments, and in particular that the situation should be viewed as a whole and in regard to the relations between the Soviet, American, and British Governments. These Governments will have to make sure that there is in fact a friendly and fair execution of the occupation zones as already agreed between Governments. For these reasons we consider the matter is above the sphere of purely military decision by a commander in the field."

I addressed myself to the new President on April 18. Mr. Truman was of course only newly aware at second hand of all the complications that faced us, and had to lean heavily on his advisers. The purely military view therefore received an emphasis beyond its proper proportion.

Prime Minister to President Truman   18 Apr 45

Your armies soon, and presently ours, may come into contact with the Soviet forces. The Supreme Commander should be given instructions by the Combined Chiefs of Staff as soon as possible how to act. In my view there are two zones:

(a) The tactical zone, in which our troops must stand on the line they have reached unless there is agreement for a better tactical deployment against the continuing resistance of the enemy. This should be arranged by the Supreme Commander through our military representatives in Moscow, or if convenient across the lines in the field. The Combined Chiefs of Staff have already taken up the issue of instructions to cover this phase.

(b) The occupational zone, which I agreed with President Roosevelt on the advice of the Combined General Staffs. In my view this zone should be occupied within a certain time from V.E. Day, whenever this is declared, and we should retire with dignity from the much greater gains which the Allied troops have acquired by their audacity and vigour.

2. I am quite prepared to adhere to the occupational zones, but I do not wish our Allied troops or your American troops to be hustled back at any point by some crude assertion of a local Russian general. This must be provided against by an agreement between Governments so as to give Eisenhower a fair chance to settle on the spot in his own admirable way.

3. The occupational zones were decided rather hastily at Quebec in September 1944, when it was not foreseen that General Eisenhower's armies would make such a mighty inroad into Germany. The zones cannot be altered except by agreement with the Russians. But the moment V.E. Day has occurred we should try to set up the Allied Control Commission in Berlin and should insist upon a fair distribution of the food produced in Germany between all parts of Germany. As it stands at present the Russian occupational zone has the smallest proportion of people and grows by far the largest proportion of food, the Americans have a not very satisfactory proportion of food to conquered population, and we poor British are to take over all the ruined Ruhr and large manufacturing districts, which are, like ourselves, in normal times large importers of food. I suggest that this tiresome question should be settled in Berlin by the Allied Control Commission before we move from the tactical positions we have at present achieved. The Russian idea of taking these immense food supplies out of the food-producing areas of Germany to feed themselves is very natural, but I contend that the feeding of the German

population must be treated as a whole and that the available supplies must be divided *pro rata* between the occupational zones.

4. I should be most grateful if you would let me have your views on these points, which, from the information I receive from many sources, are of the highest consequence and urgency.

Mr. Eden was in Washington, and fully agreed with the views I telegraphed to him.

Prime Minister to Mr. Eden (Washington)   19 Apr 45

This is for your eyes alone. It would seem that the Western Allies are not immediately in a position to force their way into Berlin. The Russians have 2½ million troops on the section of the front opposite that city. The Americans have only their spearheads, say twenty-five divisions, which are covering an immense front and are at many points engaged with the Germans. . . .

It is thought most important that Montgomery should take Lübeck as soon as possible, and he has an additional American Army Corps to strengthen his movements if he requires it. Our arrival at Lübeck before our Russian friends from Stettin would save a lot of argument later on. There is no reason why the Russians should occupy Denmark, which is a country to be liberated and to have its sovereignty restored. Our position at Lübeck, if we get it, would be decisive in this matter.

Thereafter, but partly concurrent, it is thought well to push on to Linz to meet the Russians there, and also by an American encircling movement to gain the region south of Stuttgart. In this region are the main German installations connected with their atomic research, and we had better get hold of these in the interests of the special secrecy attaching to this topic.

Mr. Eden replied:

Foreign Secretary (Washington) to Prime Minister   21 Apr 45

I strongly share the view that Montgomery should take Lübeck. A Russian occupation of Denmark would cause us much embarrassment. The fears of Scandinavian countries would be greatly increased, and I seem to remember that one of the causes of dissension between the Russians and Germans in their honeymoon period in 1940 arose out of certain Russian demands for control of the Kattegat.

I am sure that you still have Prague in mind. It might do the Russians much good if the Americans were to occupy the Czech capital, when no doubt they would be willing to invite the Soviet Ambassador to join the United States and ourselves, in contrast to the behaviour the Russians have shown to us. . . .

Mr. Truman's reply however carried us little further. He proposed that the Allied troops should retire to their agreed zones in Germany and Austria as soon as the military situation allowed, and sought my opinion on a draft telegram to Stalin to this effect.

To this I answered:

Prime Minister to President Truman   24 Apr 45

I thank you for your answer to my telegram. I agree with the preamble, but later paragraphs simply allow the Russians to order us back to the occupational zones at any point they might decide, and not necessarily with regard to the position of the fronts as a whole. It is your troops who would suffer most by this, being pushed back about a hundred and twenty miles in the centre and yielding up to the unchecked Russian advance an enormous territory. And this while all questions of our spheres in Vienna or arrangements for triple occupation of Berlin remain unsettled.

On April 27, after discussion with the President, I sent this telegram to Stalin:

Prime Minister to Marshal Stalin   27 Apr 45

The Anglo-American armies will soon make contact in Germany with Soviet forces, and the approaching end of German resistance makes it necessary that the United States, Great Britain, and the Soviet Union decide upon an orderly procedure for the occupation by their forces of the zones which they will occupy in Germany and in Austria.

2. Our immediate task is the final defeat of the German Army. During this period the boundaries between the forces of the three Allies must be decided by commanders in the field, and will be governed by operational considerations and requirements. It is inevitable that our armies will in this phase find themselves in occupation of territory outside the boundaries of the ultimate occupational zones.

3. When the fighting is finished the next task is for the Allied Control Commissions to be set up in Berlin and Vienna, and for the forces of the Allies to be redisposed and to take over their respective occupational zones. The demarcation of the zones in Germany has already been decided upon, and it is necessary that we shall without delay reach an agreement on the zones to be occupied in Austria at the forthcoming meeting proposed by you in Vienna.

4. It appears now that no signed instrument of surrender will be forthcoming. In this event Governments should decide to set up at once the Allied Control Commissions, and to entrust to them the task of making detailed arrangements for the withdrawal of the forces to their agreed occupational zones.

5. In order to meet the requirements of the situation referred to in paragraph 2 above, namely, the emergency and temporary arrangements for the tactical zones, instructions have been sent to General Eisenhower. These are as follows:

(a) To avoid confusion between the two armies and to prevent either of them from expanding into areas already occupied by the other, both sides should halt as and where they meet, subject to such adjustments to the rear or to the flanks as are required, in the opinion of the local commanders on either side, to deal with any remaining opposition.

(b) As to adjustments of forces after cessation of hostilities in an area, your troops should be disposed in accordance with military requirements regardless of zonal boundaries. You will, in so far as permitted by the urgency of the situation, obtain the approval of the Combined Chiefs of Staff prior to any major adjustment, in contrast to local adjustments for operational and administrative reasons.

6. It is requested that you will issue similar instructions to your commanders in the field.

7. I am sending this message to you and to President Truman simultaneously.

The reply was guarded.

Marshal Stalin to Prime Minister   2 May 45
I have received your message of April 27 on the subject of procedure for the occupation of Germany and Austria by the Red Army and the Anglo-American armed forces.

For my part I should inform you that the Soviet High Command has issued instructions that when the Soviet and Allied forces meet the Soviet Command should immediately establish contact with the Command of the American or English forces, and that they should in agreement together

(a) define a provisional tactical demarcation boundary line, and
(b) take measures to suppress any opposition by the German forces within their provisional demarcation line.

*   *   *   *   *

The Russians had not been long in Vienna before we got a foretaste of what would happen in a zone of their occupation. They announced that a Provisional Austrian Government had been formed, and they refused to let our missions fly in. All this made me fear that they were deliberately exploiting their arrival

451

to "organise" the country before we got there. On April 30 I accordingly telegraphed to Mr. Truman as follows:

It seems to me that unless we both take a strong stand now we shall find it very difficult to exercise any influence in Austria during the period of her liberation from the Nazis. Would you be willing to join me in sending Marshal Stalin a message in the following terms:

"We have been much concerned to hear from our Chargé d'Affaires in Moscow that, despite the invitation you extended to Mr. Harriman on April 13, the Soviet Government will not now agree to the Allied missions proceeding to Vienna until agreement has been reached in the European Advisory Commission regarding the respective zones in Vienna and the provisional control machinery. We have also been disagreeably surprised by the announcement of the setting up in Vienna of a Provisional Austrian Government, despite our request for time to consider the matter.

"It has been our understanding that the treatment of Austria, as of Germany, is a matter of common concern to the four Powers who are to occupy and control those countries. We regard it as essential that British, American, and French representatives should be allowed to proceed at once to Vienna in order to report on conditions there before any final settlement is reached in the European Advisory Commission on matters affecting the occupation and control of the country, and especially of Vienna itself. We hope you will issue the necessary instructions to Marshal Tolbukhin, in order that the Allied missions may fly in at once from Italy."

On May 3 President Truman replied that he entirely agreed with my telegram and was himself sending a protest to the Soviet Government. This protest reminded the Russians of Stalin's suggestion that American, British, and French representatives should go to Vienna at once and settle the zones of occupation. Plans had been made for their journey, and now the Soviet Government were saying that their arrival would be "undesirable" until after the zones had been agreed by the European Advisory Commission. The Commission had been unable to agree, partly through lack of information. The only way was to study the problem on the spot, and Soviet unwillingness to let us do so was holding up the Commission's work. Mr. Truman concluded his message by asking the Soviet Government to let the Allied representatives fly to Vienna at once.

These representations were quite ineffectual.

*   *   *   *   *

Meanwhile the advance of the Allied armies continued with increasing momentum. The Seventh U.S. Army of General Devers's Army Group passed through Munich on April 30, and reached Innsbruck on May 3. The First French Army, after skirting the north shore of Lake Constance, turned south and also crossed the Austrian frontier. From Innsbruck a force was sent to the Brenner Pass, and on May 4, a few miles to the south, met the head of the U.S. Fifth Army, which had driven up from Italy as an aftermath of Alexander's victorious campaign. And so all three "fronts", Western, Eastern, and Southern, once thousands of miles apart, at last came together, crushing the life out of the German armies. Their encirclement had been completed by Montgomery in the north. The head of the VIIIth Corps, which led the advance of the Second British Army, reached the Elbe thirty miles above Hamburg on April 19. On their left the XIIth Corps was strongly resisted by hastily organised detachments drawn principally from the officer cadet schools near Rheine, but they took Soltau on April 18 and drove on towards Hamburg. The XXXth Corps also had some hard fighting on their way to Bremen. The whole British advance was delayed by having to replace several hundreds of the bridges across the many waterways, which had been demolished by the enemy. Bremen fell on April 26. The VIIIth Corps, with the XIIth on their left and the XVIIIth U.S. Airborne Corps guarding their right flank, crossed the Elbe on April 29. They headed for the Baltic, so as to place themselves across the land-gate of Denmark. On May 2 the 11th Armoured Division reached Lübeck, and Denmark was liberated by our troops amidst scenes of great rejoicing. Our 6th Airborne Division met the Russians at Wismar. Next day the XIIth Corps entered Hamburg. Everywhere north of the Elbe the country was filled with masses of refugees and disorganised soldiery, fleeing from the Russians to surrender to the Western Allies. The end was near.

CHAPTER XXXI

ALEXANDER'S VICTORY IN ITALY

Our Offensive is Postponed to the Spring – Allied Air Attack – Hitler Forbids a Withdrawal – The Weakness of the German Position – The Fall of Bologna, April 21 – The Allied Pursuit Across the Po – Naval Affairs – A New German Peace Offer – Unconditional Surrender in Italy, April 29 – Mussolini is Murdered – I Send My Victory Congratulations to All Concerned – The End of a Fine Campaign.

GLEAMING successes marked the end of our campaigns in the Mediterranean. In December Alexander had succeeded Wilson as Supreme Commander, while Mark Clark took command of the Fifteenth Army Group. After their strenuous efforts of the autumn our armies in Italy needed a pause to reorganise and restore their offensive power.

The long, obstinate, and unexpected German resistance on all fronts had made us and the Americans very short of artillery ammunition, and our hard experiences of winter campaigning in Italy forced us to postpone a general offensive till the spring. But the Allied Air Forces, under General Eaker, and later under General Cannon, used their thirty-to-one superiority in merciless attacks on the supply lines which nourished the German armies. The most important one, from Verona to the Brenner Pass, where Hitler and Mussolini used to meet in their happier days, was blocked in many places for nearly the whole of March. Other passes were often closed for weeks at a time, and two divisions being transferred to the Russian front were delayed almost a month.

The enemy had enough ammunition and supplies, but lacked fuel. Units were generally up to strength, and their spirit was high in spite of Hitler's reverses on the Rhine and the Oder. In Northern Italy they had twenty-seven divisions, four of them

Italian, against our equivalent of twenty-three drawn from the British Empire, the United States, Poland, Brazil, and Italy.* The German High Command might have had little to fear had it not been for the dominance of our Air Forces, the fact that we had the initiative and could strike where we pleased, and their own ill-chosen defensive position, with the broad Po at their backs. They would have done better to yield Northern Italy and withdraw to the strong defences of the Adige, where they could have held us with much smaller forces, and sent troops to help their over-matched armies elsewhere, or have made a firm southern face for the National Redoubt in the Tyrol mountains, which Hitler may have had in mind as his "last ditch".

But defeat south of the Po spelt disaster. This must have been obvious to Kesselring, and was doubtless one of the reasons for the negotiations recorded in a previous chapter.† Hitler was of course the stumbling-block, and when Vietinghoff, who succeeded Kesselring, proposed a tactical withdrawal he was thus rebuffed: "The Fuehrer expects, now as before, the utmost steadiness in the fulfilment of your present mission to defend every inch of the North Italian areas entrusted to your command."

* * * * *

This eased our problem. If we could break through the Adriatic flank and reach the Po quickly all the German armies would be cut off and forced to surrender, and it was to this that Alexander and Clark bent their efforts when the stage was set for the final battle. The capture of Bologna, which had figured so much in our autumn plans, was no longer a principal object. The plan was for the Eighth Army, under General McCreery, to force a way down the road from Bastia to Argenta, a narrow, strongly defended passage, flooded on both sides, but leading to more open ground beyond. When this was well under way General Truscott's Fifth Army was to strike from the mountainous central front, pass west of Bologna, join hands with the Eighth Army on the Po, and together pursue to the Adige. The Allied naval forces would make the enemy believe that amphibious landings were imminent on both east and west coasts.

In the evening of April 9, after a day of mass air attacks and

* Four Italian "combat groups", each of nearly divisional strength, had been formed, and took an active part in the campaign.
† Chapter XXVI, "Soviet Suspicions."

artillery bombardment, the Eighth Army attacked across the river Senio, led by the Vth and the Polish Corps. On the 11th they reached the next river, the Santerno. The foremost brigade of the 56th Division and Commandos made a surprise landing at Menate, three miles behind the enemy, having been carried across the floods in a new type of amphibious troop-carrying tank called the Buffalo, which had come by sea from an advanced base in the Adriatic. By the 14th there was good news all along the Eighth Army front. The Poles took Imola. The New Zealand Division crossed the Sillaro. The 78th Division, striking north, took the bridge at Bastia and joined the attacks of the 56th on the Argenta road. The Germans knew well that this was their critical hinge and fought desperately.

That same day the Fifth Army began the centre attack west of the Pistoia–Bologna road. After a week of hard fighting, backed by the full weight of the Allied Air Forces, they broke out from the mountains, crossed the main road west of Bologna, and struck north. On the 20th Vietinghoff, despite Hitler's commands, ordered a withdrawal. He tactfully reported that he had "decided to abandon the policy of static defence and adopt a mobile strategy". It was too late. Argenta had already fallen and the 6th British Armoured Division was sweeping towards Ferrara. Bologna was closely threatened from the east by the Poles and from the south by the 34th U.S. Division. It was captured on April 21, and here the Poles destroyed the renowned 1st German Parachute Division. The Fifth Army pressed towards the Po, with the tactical air force making havoc along the roads ahead. Its 10th U.S. Mountain Division crossed the river on the 23rd, and the right flank of the Army, the 6th South African Division, joined the left of the Eighth. Trapped behind them were many thousand Germans, cut off from retreat, pouring into prisoners' cages or being marched to the rear. The offensive was a fine example of concerted land and air effort, wherein the full strength of the strategical and tactical air forces played its part. Fighter-bombers destroyed enemy guns, tanks, and troops; light and medium bombers attacked the lines of supply, and our heavy bombers struck by day and night at the rear installations.

* * * * *

We crossed the Po on a broad front at the heels of the enemy.

THE INVASION OF GERMANY

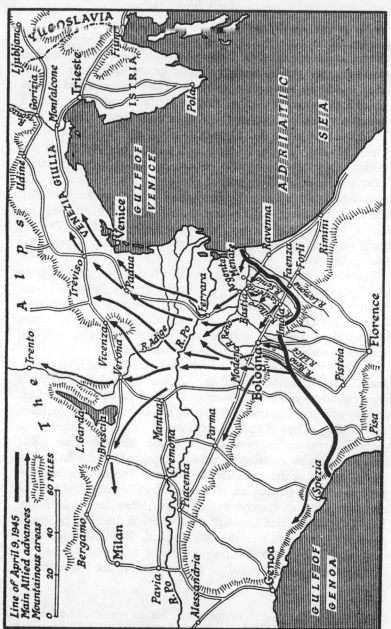

THE BATTLE OF THE RIVER PO

457

All the permanent bridges had been destroyed by our Air Forces, and the ferries and temporary crossings were attacked with such effect that the enemy were thrown into confusion. The remnants who struggled across, leaving all their heavy equipment behind, were unable to reorganise on the far bank. The Allied armies pursued them to the Adige. Italian Partisans had long harassed the enemy in the mountains and their back areas. On April 25 the signal was given for a general rising, and they made widespread attacks. In many cities and towns, notably Milan and Venice, they seized control. Surrenders in North-West Italy became wholesale. The garrison of Genoa, four thousand strong, gave themselves up to a British liaison officer and the Partisans. On the 27th the Eighth Army crossed the Adige, heading for Padua, Treviso, and Venice, while the Fifth, already in Verona, made for Vicenza and Trento, its left extending to Brescia and Alessandria.

The naval campaign, though on a much smaller scale, had gone equally well. In January the ports of Split and Zadar had been occupied by the Partisans, and coastal forces from these bases harassed the Dalmatian shore and helped Tito's steady advance. In April alone at least ten actions were fought at sea, with crippling damage to the enemy and no loss of British ships.

The Navy had operated on both flanks during the final operations. On the west coast British, American, and French forces were continually in action, bombarding and harassing the enemy, driving off persistent attacks by light craft and midget submarines, and clearing mines in the liberated ports. These activities led to the last genuine destroyer action in the Mediterranean. The former Yugoslav destroyer *Premuda*, captured by the Italians at the beginning of the war, left Genoa on the night of March 17, with two Italian destroyers, all manned by Germans, and tried to intercept a British convoy sailing from Marseilles to Leghorn. The British destroyers *Look-out* and *Meteor*, on patrol off the north point of Corsica, got warning and attacked. Both the Italian ships were sunk, the British suffering no loss or damage. By the time our armies reached the Adige the fighting at sea had virtually ended.

* * * * *

Meanwhile the March negotiations for an armistice had probably come to Himmler's ears. Certainly he sent for General

Wolff, the principal envoy and a high S.S. official in Italy, and questioned him closely. There was then a pause before the force of facts overcame German hesitancies, but on April 24 Wolff reappeared in Switzerland with full powers from Vietinghoff. I hastened to tell the Russians.

Prime Minister to Marshal Stalin   26 Apr 45

This is about "Crossword". The German envoys, with whom all contact was broken by us some days ago, have now arrived again on the Lake of Lucerne. They claim to have full powers to surrender the Army in Italy. Field-Marshal Alexander is therefore being told that he is free to permit these envoys to come to A.F.H.Q. in Italy. This they can easily do by going into France and being picked up by our aircraft from there. Will you please send Russian representatives forthwith to Field-Marshal Alexander's headquarters.

Field-Marshal Alexander is free to accept the unconditional surrender of the considerable enemy army on his front, but all political issues are reserved to the three Governments.

2. You will notice that the surrender in Italy was not mentioned in the telegrams I sent you a few hours ago about Himmler's proposed surrender in the West and North.* We have spent a lot of blood in Italy, and the capture of the German armies south of the Alps is a prize dear to the hearts of the British nation, with whom in this matter the United States have shared the costs and perils.

3. All the above is for your personal information. Our Staff have telegraphed to the American Staff in order that the Combined Anglo-American Staff may send instructions in the same sense to Field-Marshal Alexander, who will be told to keep your High Command fully informed through the Anglo-American Military Missions in Moscow.

Two plenipotentiaries were brought to Alexander's headquarters, and on April 29 they signed the instrument of unconditional surrender in the presence of high British, American, and Russian officers.

I duly informed Moscow.

Prime Minister to Marshal Stalin   29 Apr 45

I have just received a telegram from Field-Marshal Alexander that after a meeting at which your officers were present the Germans accepted the terms of unconditional surrender presented to them and are sending the material clauses of the instrument of surrender to

---

* See Chapter XXXII, "The German Surrender."

General von Vietinghoff, with the request to name the date and hour at which conclusion of hostilities can be made effective. It looks therefore as if the entire German forces south of the Alps will almost immediately surrender.

On May 2 nearly a million Germans surrendered as prisoners of war, and the war in Italy ended.

★   ★   ★   ★   ★

For Mussolini also the end had come. Like Hitler he seems to have kept his illusions until almost the last moment. Late in March he had paid a final visit to his German partner, and returned to his headquarters on Lake Garda buoyed up with the thought of the secret weapons which could still lead to victory. But the rapid Allied advance from the Apennines made these hopes vain. There was hectic talk of a last stand in the mountainous areas on the Italo-Swiss frontier. But there was no will to fight left in the Italian Socialist Republic.

On April 25 Mussolini decided to disband the remnants of his armed forces and to ask the Cardinal Archbishop of Milan to arrange a meeting with the underground Military Committee of the Italian National Liberation Movement. That afternoon talks took place in the Archbishop's palace, but with a last furious gesture of independence Mussolini walked out. In the evening, followed by a convoy of thirty vehicles, containing most of the surviving leaders of Italian Fascism, he drove to the prefecture at Como. He had no coherent plan, and as discussion became useless it was each man for himself. Accompanied by a handful of supporters, he attached himself to a small German convoy heading towards the Swiss frontier. The commander of the column was not anxious for trouble with Italian Partisans. The Duce was persuaded to put on a German great-coat and helmet. But the little party was stopped by Partisan patrols; Mussolini was recognised and taken into custody. Other members, including his mistress, Signorina Petacci, were also arrested. On Communist instructions the Duce and his mistress were taken out in a car next day and shot. Their bodies, together with others, were sent to Milan and strung up head downwards on meat-hooks in a petrol station on the Piazzale Loreto, where a group of Italian Partisans had lately been shot in public.

Such was the fate of the Italian dictator.

A photograph of the final scene was sent to me, and I was profoundly shocked.

Prime Minister to Field-Marshal Alexander (Italy)   10 May 45
I have seen the photograph.

The man who murdered Mussolini made a confession, published in the *Daily Express*, gloating over the treacherous and cowardly method of his action. In particular he said he shot Mussolini's mistress. Was she on the list of war criminals? Had he any authority from anybody to shoot this woman? It seems to me the cleansing hand of British military power should make inquiries on these points.

But at least the world was spared an Italian Nuremberg.

*   *   *   *   *

I sent my congratulations to the victorious commanders and their men.

Prime Minister to Field-Marshal Alexander   29 Apr 45
I rejoice in the magnificently planned and executed operations of the Fifteenth Group of Armies, which are resulting in the complete destruction or capture of all the enemy forces south of the Alps. That you and General Mark Clark should have been able to accomplish these tremendous and decisive results against a superior number of enemy divisions, after you have made great sacrifices of whole armies for the Western Front, is indeed another proof of your genius for war and of the intimate brotherhood in arms between the British Commonwealth and Imperial forces and those of the United States. Never, I suppose, have so many nations advanced and manœuvred in one line victoriously. The British, Americans, New Zealanders, South Africans, British-Indians, Poles, Jews, Brazilians, and strong forces of liberated Italians have all marched together in that high comradeship and unity of men fighting for freedom and for the deliverance of mankind. This great final battle in Italy will long stand out in history as one of the most famous episodes in the Second World War. Pray give my heartfelt congratulations to all your commanders and principal officers of all Services, and above all to the valiant and ardent troops whom they have led with so much skill.

Prime Minister to General Clark   3 May 45
Pray let me send you and your gallant men my most heartfelt thanks for all you have done to make this great victory possible.

To President Truman I telegraphed:

Prime Minister to President Truman   3 May 45
I have received the generous tribute to Field-Marshal Alexander and

the Allied forces under his command which you, Mr. President, have sent me. I forwarded it to him immediately, as you desired, and have requested him to reply direct to you. I know that he will value profoundly, as also do indeed the nations of the British Commonwealth represented in these campaigns, the warmth of the feelings which you express. May I also, in my turn, express British gratitude for the services of the highest quality, both in counsel and on the battle-line, of General Mark Clark of the United States Army, who commanded the fighting front with its magnificent United States divisions, and whose comradeship with Field-Marshal Alexander, shared by this Army of many States and races, will long be cherished in both our countries and commended by history.

I also sent a message to Signor Bonomi.

Prime Minister to Signor Bonomi   3 May 45
On the occasion of the surrender of the German armed forces in Italy I send Your Excellency, on behalf of His Majesty's Government in the United Kingdom, a message of warm congratulation on the final liberation of Italian territory from our common enemy, and in particular on the part played by the Italian regular forces and Patriots behind the lines.

2. The knowledge that they have contributed to this unprecedented victory and have materially accelerated the cleansing of their country's soil will, I trust, be a source of strength to the Italian people in the no less strenuous days which lie ahead. . . .

3. I extend to Your Excellency the good wishes of His Majesty's Government for the great work of reconstruction which now faces the Italian Government and people.

4. I look forward to the time, which cannot be long delayed, when Italy, whose forces have co-operated in war with those of the United Nations, will work with the United Nations in the more fruitful labours of peace.

Thus ended our twenty months' campaign in Italy. Our losses had been grievous, but those of the enemy, even before the final surrender, far heavier. The principal task of our armies had been to draw off and contain the greatest possible number of Germans. This had been admirably fulfilled. Except for a short period in the summer of 1944, the enemy had always outnumbered us. At the time of their crisis in August of that year no fewer than fifty-five German divisions were deployed along the Mediterranean fronts. Nor was this all. Our forces rounded off their task by devouring the larger army they had been ordered to contain. There have been few campaigns with a finer culmination.

CHAPTER XXXII

THE GERMAN SURRENDER

Hitler Resolves to Make His Last Stand in Berlin, April 22 – His Suicide in the Bunker, April 29 – Himmler's Peace Offer – My Telephone Conversation with President Truman, April 25 – A Cordial Message from Stalin – Himmler's Death – The German Surrender to Field-Marshal Montgomery, May 4 – The Instrument of General Capitulation is Signed at Rheims, May 7 – The End of the Luftwaffe – Goering is Taken Prisoner in the Tyrol – The U-Boats Capitulate – The Fate of Germany's Surface Fleet – Allied Convoys to Russia – President Truman's Telegram, and My Reply – Mrs. Churchill in Moscow – Stalin's Message – My Victory Warning.

B Y the middle of April it had been clear that Hitler's Germany would soon be utterly destroyed. The invading armies drove onwards in their might and the space between them narrowed daily. Hitler had pondered where to make his last stand. As late as April 20 he still thought of leaving Berlin for the "Southern Redoubt" in the Bavarian Alps. That day he held a meeting of the principal Nazi leaders. As the German double front, East and West, was in imminent danger of being cut in twain by the spear-point thrust of the Allies, he agreed to set up two separate commands. Admiral Doenitz was to take charge in the North both of the military and civil authorities, with the particular task of bringing back to German soil nearly two million refugees from the East. In the South General Kesselring was to command the remaining German armies. These arrangements were to take effect if Berlin fell.

Two days later, on April 22, Hitler made his final and supreme decision to stay in Berlin to the end. The capital was soon completely encircled by the Russians and the Fuehrer had lost all power to control events. It remained for him to organise his own death amid the ruins of the city. He announced to the Nazi leaders who remained with him that he would die in Berlin. Goering and Himmler had both left after the conference of the 20th, with thoughts of peace negotiations in their minds. Goering, who had gone south, assumed that Hitler had in fact abdicated by his resolve to stay in Berlin, and asked for confirmation that he should act formally as the successor to the Fuehrer. The reply was his instant dismissal from all his offices.

* * * * *

The last scenes at Hitler's headquarters have been described elsewhere in much detail. Of the personalities of his régime only Goebbels and Bormann remained with him to the end. The Russian troops were now fighting in the streets of Berlin. In the early hours of April 29 Hitler made his will. The day opened with the normal routine of work in the air-raid shelter under the Chancellery. News arrived of Mussolini's end. The timing was grimly appropriate. On the 30th Hitler lunched quietly with his suite, and at the end of the meal shook hands with those present and retired to his private room. At half-past three a shot was heard, and members of his personal staff entered the room to find him lying on the sofa with a revolver by his side. He had shot himself through the mouth. Eva Braun, whom he had married secretly during these last days, lay dead beside him. She had taken poison. The bodies were burnt in the courtyard, and Hitler's funeral pyre, with the din of the Russian guns growing ever louder, made a lurid end to the Third Reich.

The leaders who were left held a final conference. Last-minute attempts were made to negotiate with the Russians, but Zhukov demanded unconditional surrender. Bormann tried to break through the Russian lines, and disappeared without trace. Goebbels poisoned his six children and then ordered an S.S. guard to shoot his wife and himself. The remaining staff of Hitler's headquarters fell into Russian hands.

That evening a telegram reached Admiral Doenitz at his headquarters in Holstein:

In place of the former Reich-Marshal Goering the Fuehrer appoints you, Herr Grand Admiral, as his successor. Written authority is on its way. You will immediately take all such measures as the situation requires. BORMANN.

Chaos descended. Doenitz had been in touch with Himmler, who, he assumed, would be nominated as Hitler's successor if Berlin fell, and now supreme responsibility was suddenly thrust upon him without warning and he faced the task of organising the surrender.

* * * * *

Himmler had for some months been urged to make personal contact with the Western Allies on his own initiative in the hope of negotiating a separate surrender. A General Schellenberg of the S.S. had proposed to him as an intermediary Count Bernadotte, the head of the Swedish Red Cross, who had occasion from time to time to visit Berlin. There had been secret meetings between Bernadotte and Himmler in February, and again in April, when Bernadotte visited the German capital. But the Nazi leader felt too deeply committed in his loyalty to Hitler to make any move. The Fuehrer's announcement on April 22 of the last stand in Berlin led him to act.

In the early hours of April 25 a telegram arrived in London from Sir Victor Mallet, British Minister to Sweden. He reported that at 11 p.m. on April 24 he and his American colleague, Mr. Herschel Johnson, had been asked to call upon the Swedish Minister for Foreign Affairs, Mr. Boheman. The purpose of the interview was to meet Count Bernadotte, who had an urgent mission. Bernadotte told them that Himmler was on the Eastern Front, and had asked that he might meet him urgently in North Germany. Bernadotte suggested Lübeck, and they had met the previous evening. Himmler, though tired and admitting Germany was finished, was still calm and coherent. He said that Hitler was so desperately ill that he might be dead already, and in any case would be so within the next few days. Himmler stated that while the Fuehrer was still active he would not have been able to do what he now proposed, but as Hitler was finished he could act with full authority. He then asked if the Swedish Government would arrange for him to meet General Eisenhower and capitulate on the whole Western Front. Bernadotte said there was no need for this as he could simply order his troops to surrender, and in any case he would not forward the request unless Norway and Denmark were included in the capitulation. If this were done there might be some point in a meeting, because special arrangements might be necessary as to how and to whom the

Germans there were to lay down their arms. Himmler thereupon said he was prepared to order the German forces in Denmark and Norway to surrender to either British, American, or Swedish troops. When asked what he proposed to do if the Western Allies refused his offer, Himmler replied that he would take command of the Eastern Front and die in battle. Himmler said he hoped that the Western Allies rather than the Russians would be the first to enter Mecklenburg, in order to save the civilian population.

Count Bernadotte ended by saying that General Schellenberg was now in Flensburg, near the Danish border, eagerly waiting for news, and could make sure that any message would reach Himmler immediately. Both Ministers remarked that Himmler's refusal to surrender on the Eastern Front looked like a last attempt to make trouble between the Western Allies and Russia. Obviously the Nazis would have to surrender to all the Allies simultaneously. The Swedish Minister admitted this might be so, but pointed out that if the troops on the whole of the Western Front and in Norway and Denmark laid down their arms it would be a great help for all the Allies, including Russia, and would lead to an early and total capitulation. In any case, he thought that Bernadotte's information should be passed to the British and United States Governments. As far as his own Government was concerned, we were completely at liberty to tell the Soviets, as the Swedes would in no way be, or be thought to be, promoting discord between the Allies. The only reason why the Swedish Government could not inform the Soviets direct was that Himmler had stipulated that his information was exclusively for the Western Powers.*

* * * * *

I received this news on the morning of April 25, and immediately summoned the War Cabinet. Our reactions are shown in the message which I dispatched to President Truman:

You will no doubt have received some hours ago the report from Stockholm by your Ambassador on the Bernadotte-Himmler talks. I called the War Cabinet together at once, and they approved the immediately following telegram, which we are sending to Marshal Stalin and repeating through the usual channels to you. We hope you

* A slightly different version of this episode is given by Count Bernadotte in his book *The Fall of the Curtain*, pp. 54 ff.

will find it possible to telegraph to Marshal Stalin and to us in the same sense. As Himmler is evidently speaking for the German State, as much as anybody can, the reply that should be sent him through the Swedish Government is in principle a matter for the triple Powers, since no one of us can enter into separate negotiations. This fact however in no way abrogates General Eisenhower's or Field-Marshal Alexander's authority to accept local surrenders as they occur.

In view of the importance of this German peace offer and of our experience of Russian suspicions over "Crossword",* I think it well to record our attitude in detail.

That evening I telephoned the President, and then dictated the following note for the next meeting of the Cabinet:

I spoke to President Truman at 8.10 p.m. He knew nothing of what had happened at Stockholm, except that when I asked to speak to him he inquired what it was about, and I told him about the important message from Stockholm. He had not received any report from the American Ambassador there. I therefore read him the full text of Mallet's telegram. I also told him that we were convinced the surrender should be unconditional and simultaneous to the three major Powers. He expressed strong agreement with this. I then read him the telegram I had sent, in accordance with the Cabinet's decision, to Marshal Stalin, and he expressed strong agreement with this also. He asked me to read it out to him a second time, which I did, so that he could send a similar message at once to the Russians. I gave him the substance of the covering note which I had prepared on our Stalin telegram, which I attach. An hour and a half before this talk I had sent both the Stalin message and my covering note, so that he should have the written text of them by now.

2. He also stated to me that he hoped to see me soon, to which I replied that we were telegraphing him proposals for a meeting, preferably here. I also told him that we strongly approved of the lead he was taking on the Polish issue. These, together with compliments, formed the whole of our conversation.

W. S. C., 25.4.45

* * * * *

Here is the text of my covering message to Stalin:

Prime Minister to Marshal Stalin   25 Apr 45
The President of the United States has the news also. There can be no question, as far as His Majesty's Government is concerned, of anything less than unconditional surrender simultaneously to the three

---

* See Chapter XXVI, "Soviet Suspicions."

major Powers. We consider Himmler should be told that German forces, either as individuals or in units, should everywhere surrender themselves to the Allied troops or representatives on the spot. Until this happens the attack of the Allies upon them on all sides and in all theatres where resistance continues will be prosecuted with the utmost vigour. . . .

His reply was the most cordial message I ever had from him.

Marshal Stalin to Prime Minister   25 Apr 45
I thank you for your communication of April 25 regarding the intention of Himmler to surrender on the Western Front.

I consider your proposal to present to Himmler a demand for unconditional surrender on all fronts, including the Soviet front, the only correct one. *Knowing you, I had no doubt that you would act in this way.** I beg you to act in the sense of your proposal, and the Red Army will maintain its pressure on Berlin in the interests of our common cause.

I have to state, for your information, that I have given a similar reply to President Truman, who also addressed to me the same inquiry.

I answered:

Prime Minister to Marshal Stalin   27 Apr 45
I am extremely pleased to know that you had no doubt how I would act, and always will act, towards your glorious country and yourself. British and I am sure American action on this matter will go forward on the lines you approve, and we all three will continually keep each other fully informed.

Count Bernadotte conveyed our demand to Himmler. No more was heard of the Nazi leader till May 21, when he was arrested by a British control post at Bremervörde. He was disguised and was not recognised, but his papers made the sentries suspicious and he was taken to a camp near Second Army Headquarters. He then told the commandant who he was. He was put under armed guard, stripped, and searched for poison by a doctor. During the final stage of the examination he bit open a phial of cyanide, which he had apparently hidden in his mouth for some hours. He died almost instantly, just after eleven o'clock at night on Wednesday, May 23.

*　　*　　*　　*　　*

In the north-west the drama closed less sensationally. On May 2 news arrived of the surrender in Italy. On the same day our troops

* Author's italics.

reached Lübeck, on the Baltic, making contact with the Russians and cutting off all the Germans in Denmark and Norway. On the 3rd we entered Hamburg without opposition and the garrison surrendered unconditionally. A German delegation came to Montgomery's headquarters on Luneburg Heath. It was headed by Admiral Friedeburg, Doenitz's emissary, who sought a surrender agreement to include German troops in the North who were facing the Russians. This was rejected as being beyond the authority of an Army Group commander, who could deal only with his own front. Next day, having received fresh instructions from his superiors, Friedeburg signed the surrender of all German forces in North-West Germany, Holland, the Islands, Schleswig-Holstein, and Denmark.

In the course of a telegram to Mr. Eden at San Francisco, dated May 5, I told him:

In the North Eisenhower threw in an American corps with great dexterity to help Montgomery in his advance on Lübeck. He got there with twelve hours to spare. There were reports from the British Naval Attaché at Stockholm, which we are testing, that, according to Swedish information, the Russians have dropped parachutists a few miles south of Copenhagen and that Communist activities have appeared there. It now appears there were only two parachutists. We are sending in a moderate holding force to Copenhagen by air, and the rest of Denmark is being rapidly occupied from henceforward by our fast-moving armoured columns. I think therefore, having regard to the joyous feeling of the Danes and the abject submission and would-be partisanship of the surrendered Huns, we shall head our Soviet friends off at this point too.

You will by now have heard the news of the tremendous surrender that has been made to Montgomery of all North-West Germany, Holland, and Denmark, both as regards men and ships. The men alone must be more than a million. Thus in three successive days 2,500,000 Germans have surrendered to our British commanders. This is quite a satisfactory incident in our military history. Ike has been splendid throughout. We must vie with him in sportsmanship.

Friedeburg went on to Eisenhower's headquarters at Rheims, where he was joined by General Jodl on May 6. They played for time to allow as many soldiers and refugees as possible to disentangle themselves from the Russians and come over to the Western Allies, and they tried to surrender the Western Front separately. Eisenhower imposed a time-limit and insisted on a general capitulation. Jodl reported to Doenitz: "General Eisenhower insists that we sign to-day. If not, the Allied fronts will be closed to persons seeking to surrender individually. I see no alternative—chaos or signature. I ask you to confirm to me immediately by wireless that I have full powers to sign capitulation."

The instrument of total, unconditional surrender was signed by Lieut.-General Bedell Smith and General Jodl, with French and Russian officers as witnesses, at 2.41 a.m. on May 7. Thereby all hostilities ceased at midnight on May 8. The formal ratification by the German High Command took place in Berlin, under Russian arrangements, in the early hours of May 9. Air Chief Marshal Tedder signed on behalf of Eisenhower, Marshal Zhukov for the Russians, and Field-Marshal Keitel for Germany.

* * * * *

The final destruction of the German Army has been related; it remains to describe the end of Hitler's other fighting forces. During the previous autumn the German Air Force, by a remarkable feat of organisation, but at the cost of its long-range bomber output, had greatly increased the numbers of its fighter aircraft. Our strategic bombing had thrown it on to the defensive and 70 per cent. of its fighters had to be used for home defence. Although greater in numbers their effectiveness was less, largely owing to fuel shortage caused by our attacks on oil installations, which it became their principal duty to prevent. German high-performance jet fighters perturbed us for a time, but special raids on their centres of production and their airfields averted the threat. Throughout January and February our bombers continued to attack, and we made a heavy raid in the latter month on Dresden, then a centre of communications of Germany's Eastern Front. The enemy air was fading. As our troops advanced the airfields of the Luftwaffe were more and more squeezed into a diminishing area, and provided excellent targets.

I felt the time had come to reconsider our policy of bombing industrial areas. Victory was close and we had to think ahead. "If we come into control of an entirely ruined land," I wrote on April 1, "there will be a great shortage of accommodation for ourselves and our Allies. We must see to it that our attacks do not do more harm to ourselves than they do to the enemy's immediate war effort." Hitler felt differently, and wanted factories and utility installations of every kind to be destroyed, but the able Speer and the German generals ignored his order. Our Chiefs of Staff instructed Bomber Command on April 6: "No great or additional advantage can be expected from attacks on remaining industrial centres, since the full effects would not be likely to mature before hostilities ceased." It soon became difficult to bomb ahead of our troops without risk to the Russians, but British and United States aircraft did much other useful work. Advancing troops were sustained from the air; Holland was saved from famine; our released prisoners of war and wounded were carried home.

In judging the contribution to victory of strategic air-power it should be remembered that this was the first war in which it was fully used. We had to learn from hard-won experience. Success depends on sound deductions from a mass of intelligence, often specialised and highly technical, on every aspect of the enemy's national life, and much of this information has to be gathered in peace-time. We certainly under-estimated the strong latent reserve in Germany's industry and the great resources she had gained from Occupied Europe. Thanks to well-organised relief measures, strict police action, and innate discipline and courage, the German people endured more than we had thought possible. But although the results of the early years fell short of our aims we forced on the enemy an elaborate, ever-growing but finally insufficient air defence system which absorbed a large proportion of their total war effort. Before the end we and the United States had developed striking forces so powerful that they played a major part in the economic collapse of Germany. Great efforts were made by the sister nations of the Commonwealth, especially Canada, in the Empire Training Scheme, which turned out 200,000 men for air crews, and in 1945 nearly half the operational pilots of British Bomber Command were men from overseas.

The final Russian attack, which began on April 16, provoked the Luftwaffe to a last dying effort, but in a few days the great Berlin airfields, with many intact aircraft, were in Soviet hands, and, like the German Army, the Air Force was split in two. Disruption and disintegration spread fast. It had no more power to recover and fell to pieces. Part of its headquarters escaped south from Berlin, and for a few days tried to operate from a lunatic asylum near Munich. Thence it scattered into Austria. In a remote mountain village of the Tyrol nearly a hundred of the more senior officers, including Goering himself, were taken prisoners by the Americans. Retribution had come at last.

*　　*　　*　　*　　*

The immense scale of events on land and in the air has tended to obscure the no less impressive victory at sea. The whole Anglo-American campaign in Europe depended upon the movement of convoys across the Atlantic, and we may here carry the story of the U-boats to its conclusion. In spite of appalling losses to themselves they continued to attack, but with diminishing success, and the flow of shipping was unchecked. Even after the autumn of 1944, when they were forced to abandon their bases in the Bay of Biscay, they did not despair. The Schnorkel-fitted boats now in service, breathing through a tube while charging their batteries submerged, were but an introduction to the new pattern of U-boat warfare which Doenitz had planned. He was counting on the advent of the new type of boat, of which very many were now being built. The first of these were already under trial. Real success for Germany depended on their early arrival on service in large numbers. Their high submerged speed threatened us with new problems, and would indeed, as Doenitz predicted, have revolutionised U-boat warfare. His plans failed mainly because the special materials needed to construct these vessels became very scarce and their design had constantly to be changed. But ordinary U-boats were still being made piecemeal all over Germany and assembled in bomb-proof shelters at the ports, and in spite of the intense and continuing efforts of Allied bombers the Germans built more submarines in November 1944 than in any other month of the war. By stupendous efforts and in spite of all losses about sixty or seventy U-boats remained in action until almost the end. Their achievements were not large, but they carried the undying hope of stalemate at sea. The new revolutionary submarines never played their part in the Second World War. It had been planned to complete 350 of them during 1945, but only a few came into service before the capitulation. This weapon in Soviet hands lies among the hazards of the future.

The final phase of our onslaught lay in German coastal waters and the exits from the Baltic, and Allied air attacks on Kiel,

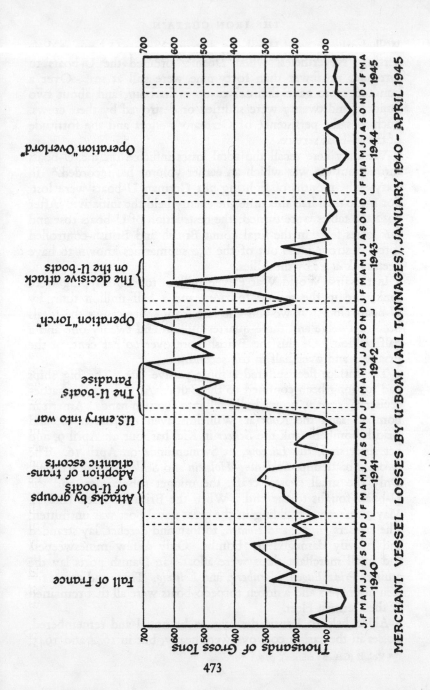

MERCHANT VESSEL LOSSES BY U-BOAT (ALL TONNAGES), JANUARY 1940 – APRIL 1945

473

Wilhelmshaven, and Hamburg destroyed many U-boats at their berths. Nevertheless when Doenitz ordered the U-boats to surrender no fewer than forty-nine were still at sea. Over a hundred more gave themselves up in harbour, and about two hundred and twenty were scuttled or destroyed by their crews. Such was the persistence of Germany's effort and the fortitude of the U-boat service.

We may here recall the total losses inflicted on the U-boats throughout the war which an earlier volume has recorded.* In sixty-eight months of fighting 781 German U-boats were lost. For more than half this time the enemy held the initiative. After 1942 the tables were turned; the destruction of U-boats rose and our losses fell. In the final count British and British-controlled forces destroyed 500 out of the 632 submarines known to have been sunk *at sea* by the Allies.

In the first World War eleven million tons of shipping were sunk, and in the second fourteen and a half million tons, by U-boats alone. If we add the loss from other causes the totals become twelve and three-quarter million and twenty-one and a half million. Of this the British bore over 60 per cent. in the first war and over half in the second.

The surface fleet suffered a more passive fate. The big ships had for long been confined to the Baltic. At Gdynia the battle-cruiser *Gneisenau*, now a hulk, fell into Russian hands. American bombers sank the *Köln* at Wilhelmshaven on March 30, and British bombers sank the *Scheer* in Kiel harbour on April 9, and her sister ship, the *Lützow*, at Swinemünde on April 16. The two old battleships *Schleswig-Holstein* and *Schlesien* were scuttled. Only the small coastal craft, the midget submarines, and the U-boats fought to the end. When the British entered Kiel on May 3 scarcely a building in the great naval port was unsmitten. The cruisers *Hipper* and *Emden*, forlorn and derelict, lay stranded and heavily damaged by bombs. Only a few minesweepers and small merchant ships were afloat. In Danish ports lay the cruisers *Prinz Eugen*, *Nürnberg*, and *Leipzig*. These and about fifteen destroyers and a dozen torpedo-boats were all that remained of the German Fleet.

Allied help to Russia deserves to be noted and remembered. Losses in the earlier convoys were heavy, but in 1944 and 1945,

* Vol. II (Cassell's edition), p. 7.

when convoys sailed only during the dark winter months they were small. In the whole of the war ninety-one merchant ships were lost on the Arctic route, amounting to 7.8 per cent. of the loaded vessels outward bound and 3.8 per cent. of those returning. Only fifty-five of these were in escorted convoys. In this arduous work the Merchant Navy lost 829 lives, while the Royal Navy paid a still heavier price. Two cruisers and seventeen other warships were sunk and 1,840 officers and men died.

The forty convoys to Russia carried the huge total of £428,000,000 worth of material, including 5,000 tanks and over 7,000 aircraft from Britain alone. The approximate figures are:

Year	Approximate Amount of Cargo dispatched from U.K. or U.S.A.	Approximate Amount of Cargo lost en route
1941	300,000 tons	10,000 tons
1942	1,350,000 ,,	270,000 ,,
1943	450,000 ,,	Nil ,,
1944	1,250,000 ,,	10,000 ,,
1945	650,000 ,,	10,000 ,,
	4,000,000 tons	300,000 tons

Thus we redeemed our promise, despite the many hard words of the Soviet leaders and their harsh attitude towards our rescuing sailors.

* * * * *

In the hour of overwhelming victory I was only too well aware of the difficulties and perils that lay ahead, but here at least there could be a brief moment for rejoicing. The President sent me a telegram of congratulation and warmly recorded his Government's appreciation of our contribution to victory.

I replied:

Prime Minister to President Truman   9 May 45

Your message is cherished by the British nation, and will be regarded as if it were a battle honour by all His Majesty's Armed Forces, of all the races in all the lands. Particularly will this be true throughout the great armies which have fought together in France and Germany under General of the Army Eisenhower, and in Italy under Field-Marshal Alexander. In all theatres the men of our two countries were brothers-in-arms, and this was also true in the air, on the oceans, and in the narrow seas. In all our victorious armies in Europe we have fought as one. Looking at the staffs of General Eisenhower and Field-Marshal Alexander, anyone would suppose that they were the organisation of one country, and certainly a band of men with one high purpose. Field-Marshal Montgomery's Twenty-first Army Group, with its gallant Canadian Army, has played its part both in our glorious landing last June and in all the battles which it has fought, either as the hinge on which supreme operations turned or in guarding the northern flank or advancing northward at the climax. All were together, heart and soul.

You sent a few days ago your message to Field-Marshal Alexander, under whom, in command of the Army front in Italy, is serving your doughty General, Mark Clark.

Let me tell you what General Eisenhower has meant to us. In him we have had a man who set the unity of the Allied Armies above all nationalistic thoughts. In his headquarters unity and strategy were the only reigning spirits. The unity reached such a point that British and American troops could be mixed in the line of battle and large masses could be transferred from one command to the other without the slightest difficulty. At no time has the principle of alliance between noble races been carried and maintained at so high a pitch. In the name of the British Empire and Commonwealth I express to you our admiration of the firm, far-sighted, and illuminating character and qualities of General of the Army Eisenhower.

I must also give expression to our British sentiments about all the valiant and magnanimous deeds of the United States of America under the leadership of President Roosevelt, so steadfastly carried forward by you, Mr. President, since his death in action. They will for ever stir the hearts of Britons in all quarters of the world in which they dwell, and will, I am certain, lead to even closer affections and ties than those that have been fanned into flame by the two World Wars through which we have passed with harmony and elevation of mind.

My wife was in Moscow at this time, and I therefore asked her to deliver my message there.

Prime Minister to Mrs. Churchill (Moscow)   8 May 45

It would be a good thing if you broadcast to the Russian people to-morrow, Wednesday, provided that were agreeable to the Kremlin. If so you might give them the following message from me, on which of course our Embassy would obtain approval:

"Prime Minister to Marshal Stalin, to the Red Army, and to the Russian people. From the British nation I send you heartfelt greetings on the splendid victories you have won in driving the invader from your soil and laying the Nazi tyrant low. It is my firm belief that on the friendship and understanding between the British and Russian peoples depends the future of mankind. Here in our island home we are thinking to-day very often about you all, and we send you from the bottom of our hearts our wishes for your happiness and well-being, and that, after all the sacrifices and sufferings of the Dark Valley through which we have marched together, we may also in loyal comradeship and sympathy walk in the sunshine of victorious peace. I have asked my wife to speak these few words of friendship and admiration to you all."

Let me know what you will do. Much love. W.

In this atmosphere of general goodwill Stalin sent his answer.

Marshal Stalin to Prime Minister   10 May 45

A MESSAGE TO THE ARMED FORCES AND THE PEOPLES OF GREAT BRITAIN
FROM THE PEOPLES OF THE SOVIET UNION

I send my personal greetings to you, the stout-hearted British Armed Forces and the whole British people, and I congratulate you with all my heart on the great victory over our common enemy, German imperialism. This historic victory has been achieved by the joint struggle of the Soviet, British, and American Armies for the liberation of Europe.

I express my confidence in the further successful and happy development in the post-war period of the friendly relations which have grown up between our countries in the period of the war.

I have instructed our Ambassador in London to convey my congratulations to you all on the victory we have won and to give you my very best wishes.

*   *   *   *   *

The unconditional surrender of our enemies was the signal for the greatest outburst of joy in the history of mankind. The Second World War had indeed been fought to the bitter end in Europe. The vanquished as well as the victors felt inexpressible relief. But for us in Britain and the British Empire, who had alone been in the struggle from the first day to the last and staked our existence on the result, there was a meaning beyond what even our most powerful and most valiant Allies could feel. Weary and worn, impoverished but undaunted and now triumphant, we had a moment that was sublime. We gave thanks to God for the noblest of all His blessings, the sense that we had done our duty.

When in these tumultuous days of rejoicing I was asked to

speak to the nation I had borne the chief responsibility in our Island for almost exactly five years. Yet it may well be there were few whose hearts were more heavily burdened with anxiety than mine. After reviewing the varied tale of our fortunes I struck a sombre note which may be recorded here.

"I wish," I said, "I could tell you to-night that all our toils and troubles were over. Then indeed I could end my five years' service happily, and if you thought that you had had enough of me and that I ought to be put out to grass I would take it with the best of grace. But, on the contrary, I must warn you, as I did when I began this five years' task—and no one knew then that it would last so long—that there is still a lot to do, and that you must be prepared for further efforts of mind and body and further sacrifices to great causes if you are not to fall back into the rut of inertia, the confusion of aim, and the craven fear of being great. You must not weaken in any way in your alert and vigilant frame of mind. Though holiday rejoicing is necessary to the human spirit, yet it must add to the strength and resilience with which every man and woman turns again to the work they have to do, and also to the outlook and watch they have to keep on public affairs.

"On the continent of Europe we have yet to make sure that the simple and honourable purposes for which we entered the war are not brushed aside or overlooked in the months following our success, and that the words 'freedom', 'democracy', and 'liberation' are not distorted from their true meaning as we have understood them. There would be little use in punishing the Hitlerites for their crimes if law and justice did not rule, and if totalitarian or police Governments were to take the place of the German invaders. We seek nothing for ourselves. But we must make sure that those causes which we fought for find recognition at the peace table in facts as well as words, and above all we must labour to ensure that the World Organisation which the United Nations are creating at San Francisco does not become an idle name, does not become a shield for the strong and a mockery for the weak. It is the victors who must search their hearts in their glowing hours, and be worthy by their nobility of the immense forces that they wield.

"We must never forget that beyond all lurks Japan, harassed and failing, but still a people of a hundred millions, for whose warriors death has few terrors. I cannot tell you to-night how much time or what exertions will be required to compel the Japanese to make amends for their odious treachery and cruelty. We, like China, so long undaunted, have received horrible injuries from them ourselves, and we are bound by the ties of honour and fraternal loyalty to the United States to fight this great war at the other end of the world at their side without flagging or failing. We must remember that Australia and New Zealand and Canada were and are all directly menaced by this evil Power. These Dominions came to our aid in our dark times, and we must not leave unfinished any task which concerns their safety and their future. I told you hard things at the beginning of these last five years; you did not shrink, and I should be unworthy of your confidence and generosity if I did not still cry: Forward, unflinching, unswerving, indomitable, till the whole task is done and the whole world is safe and clean."

CHAPTER XXXIII

AN UNEASY INTERLUDE

Tito's Troops Enter Trieste – Correspondence with President Truman – The German Garrison Surrenders to General Freyberg – My Instructions to Field-Marshal Alexander – A Strong and Welcome Telegram from the President, May 12 – I Urge a Standfast Order for the American Forces in Europe – Hesitation in Washington – More Difficulties in Trieste – The President and I Send Joint Instructions to Eisenhower and Alexander – Stalin's Telegram of June 21, and My Reply – Crisis in the Levant – My Speech to the Commons of February 27 – Fighting in Beirut, Aleppo, and Damascus – The British Commander-in-Chief Restores Order, May 31 – Difficulties with France in the Alpes-Maritimes – Mr. Truman's Indignation and General de Gaulle's Reply.

AS the German armies in Italy retreated Tito's forces had pushed rapidly into Italian territory in the north-east. They hoped to seize the lands which they claimed in this area, and in particular to capture Trieste before the Anglo-American troops arrived. Both the Americans and ourselves were not only determined to prevent any frontiers being settled in this manner before the Peace Treaty, but also intended to secure Trieste, with its splendid port, as the essential supply point for future occupation zones in Austria. We were clear on these issues, and General Alexander, who had visited Tito at Belgrade in February, was accordingly authorised to take the steps necessary to secure the position.

Even before the surrender of the German armies I had raised the question of Trieste with President Truman. "It seems to me vital," I said on April 27, "to get Trieste if we can do so in the easy manner proposed, and to run the risks inherent in these political-military operations. The late President always attached great importance to Trieste, which he thought should be an international port forming an outlet into the Adriatic from all the regions of the Danube basin. There are many points to consider about this, but that there should be an outlet to the south seems of interest to the trade of many States involved. The great thing is to be there before Tito's guerrillas are in occupation. Therefore it does not seem to me there is a minute to wait. The actual status of Trieste can be determined at leisure. . . . I should be most grateful if you would give your personal attention to this."

On the 30th Mr. Truman said he agreed there was no need to ask the Russians beforehand while operations were going on. Before entering Venezia Giulia Alexander would explain his intentions to Tito and make it clear that any Yugoslav forces in the area must come under our command. Alexander's instructions were to communicate with the Combined Chiefs of Staff before taking further action if the Yugoslavs did not co-operate. The President thought this was important, for he wished to avoid having Americans fighting Yugoslav forces or being used in Balkan combats.

Alexander told me on May 1 that he expected troops of the Eighth Army to reach Trieste within the next twenty-four hours. Their orders were to secure Trieste, the anchorage at Pola, and the lines of communication between Italy and Austria. In linking up with regular Yugoslavs great care was to be used to avoid armed clashes.

At the same time Alexander telegraphed to Tito informing him of his plans. "They are similar," he said, "to those we discussed at Belgrade. . . . I presume that any of your forces which may be in the area affected by my operations will come under my command, as you suggested during our recent discussions in Belgrade, and that you will now issue orders to that effect."

To me he reported:

1 May 45

Tito's regular forces are now fighting in Trieste, and have already occupied most of Istria. I am quite certain that he will not withdraw his troops if ordered to do so unless the Russians tell him to.

If I am ordered by the Combined Chiefs of Staff to occupy the whole of Venezia Giulia by force if necessary, we shall certainly be committed to a fight with the Yugoslav Army, who will have at least the moral backing of the Russians. Before we are committed I think it as well to consider the feelings of our own troops in this matter. They have a profound admiration for Tito's Partisan Army, and a great sympathy

481

for them in their struggle for freedom. We must be very careful therefore before we ask them to turn away from the common enemy to fight an Ally. Of course I should not presume to gauge the reaction of our people at home, whom you know so well.

*   *   *   *   *

Tito's troops had in fact entered Trieste on April 30 in the hope not only of securing the city and the surrounding area, but also of obtaining the surrender of the German garrison of seven thousand men with all its equipment. It was not until the afternoon of the following day that the Yugoslav forces made contact with the advance-guard of the 2nd New Zealand Division just west of Monfalcone. On May 2 General Freyberg and his New Zealand troops entered Trieste, took the surrender of the German garrison, and occupied the dock areas.

On May 5 Alexander telegraphed:

Tito ... now finds himself in a much stronger military position than he foresaw when I was in Belgrade, and wants to cash in on it. Then he hoped to step into Trieste when finally I stepped out. Now he wants to be installed there and only allow me user's right.

We must bear in mind that since our meeting he has been to Moscow. I believe he will hold to our original agreement if he can be assured that when I no longer require Trieste as a base for my forces in Austria he will be allowed to incorporate it in his New Yugoslavia.

The last sentence of Alexander's message made it necessary for me to make our political view clear.

"I like all your correspondence with Tito," I replied on May 6. "I am very glad you got into Trieste, Gorizia, and Monfalcone in time to put your foot in the door. Tito, backed by Russia, will push hard, but I do not think that they will dare attack you in your present position. Unless you can make a satisfactory working arrangement with Tito the argument must be taken up by the Governments. *There is no question of your making any agreement with him about incorporating Istria, or any part of the pre-war Italy, in his 'New Yugoslavia'. The destiny of this part of the world is reserved for the peace table, and you should certainly make him aware of this.*"*
I added:

In order to avoid leading Tito or the Yugoslav commanders into any temptation, it would be wise to have a solid mass of troops in this

* Author's subsequent italics.

area, with a great superiority of modern weapons and frequent demonstrations of the Air Force, as far as possible without hurting your
advance in the direction of Vienna, which I am sure you are pressing
with all possible speed.

I suppose you have cleared the approaches to Trieste so that you can
soon have some strong naval forces there. Strength is safety and peace.

You have no doubt seen our telegrams of complaint on the way we
are being treated [by the Russians] about Vienna. You are clearly
entitled to advance as far and as fast as you can into former enemy
territory until you form contact with Russian or Yugoslav forces,
when the same method of friendly recognition should be adopted as
has proved a success on the Western Front.

*　　*　　*　　*　　*

A week later, on May 12, after the great events had happened
in the Western theatre, there arrived from President Truman a
most welcome and strong message. He said he was becoming
increasingly concerned at Tito's actions in Venezia Giulia. Tito
seemed to have no intention of abandoning the territory or letting
this ancient problem await a general post-war settlement. We
must now decide, said the President, whether to uphold the
fundamental principles of territorial settlement by orderly process against force, intimidation, or blackmail. If Tito succeeded
he would probably claim parts of South Austria, Hungary, and
Greece. Although the stability of Italy and her relations with
Russia might be at stake, the present issue was not a question of
taking sides in a dispute between Italy and Yugoslavia or becoming involved in Balkan politics, but of deciding whether Britain
and America were going to allow their Allies to engage in un-
controlled land-grabbing or tactics which were all too reminis-
cent of Hitler and Japan. Yugoslav occupation of Trieste would
have more far-reaching consequences than the immediate territory
involved. We should insist, he urged, on Field-Marshal Alexander
obtaining complete and exclusive control of Trieste and Pola, the
line of communication through Gorizia and Monfalcone, and of a
big enough area to the east to ensure proper administration. Mr.
Truman said we should be prepared to consider any necessary
steps to effect Tito's withdrawal.

He also added a draft message for our Ambassadors to deliver to
Belgrade.

He suggested informing Stalin of our plans in accordance with the Yalta agreement, and concluded as follows: "If we stand firm on this issue, as we are doing on Poland, we can hope to avoid a host of other similar encroachments."

I need not say how relieved I was to receive this invaluable support from my new companion.

Prime Minister to President Truman   12 May 45

I agree with every word you say, and will work with all my strength on the line you propose. . . . If the situation is handled firmly before our strength is dispersed Europe may be saved from another bloodbath. Otherwise the whole fruits of our victory may be cast away and none of the purposes of World Organisation to prevent territorial aggression and future wars will be attained.

I trust that a standstill order can be given on the movements of the American armies and Air Forces from Europe [to the Far East], at any rate for a few weeks. We will also conform in our demobilisation. Even if this standstill order should become known it would do nothing but good. . . .

In accordance with your suggestion, I am instructing our Ambassador at Belgrade to address Tito on the lines which you have set forth, and to keep in step with your Ambassador at every stage, whether in oral representations or the delivery of identical or parallel notes or of a joint note.

I hastened to tell the good news to Alexander.

Prime Minister to Field-Marshal Alexander   12 May 45

You will have seen the most robust and encouraging telegram I have just received from the President about Tito. I have assured him that we shall support his policy, and instructions are being given to Stevenson at Belgrade to take identical action with the American Ambassador there. Of the eighteen divisions concerned, you could, I should think, count on all. The six British and British-Indian divisions are under Imperial orders. I should think it likely that the Brazilian division would act with the seven Americans. I should imagine there is nothing the two Poles [Polish divisions] would like better. The fact that Great Britain and the United States are acting together should make the matter clearly comprehensible to the troops.

You must indeed rejoice at the prospects of so much help being given by our great Allies and by the new President. This action if pursued with firmness may well prevent a renewal of the World War. I recognise of course that it affects every theatre, and so, I am sure, does President Truman.

*   *   *   *   *

484

It seems probable that a somewhat violent internal reaction at Washington followed the new President's bold telegram to me. The argument "Don't let us get tied up in Europe" had always been formidable. It had undoubtedly led to the Second World War through the ruin of the League of Nations by the withdrawal of the United States. It was now to play almost as deadly a part at a moment when the future hung in the balance. There was also at this time the desire to finish off Japan by concentrating all available and suitable forces in the Far East. This was supported by the powerful school which had from the beginning set the Far East before Europe. My suggestion of a "standstill" or a "stand-fast" order seems to have raised this issue abruptly in the President's circle. At any rate, his replies seemed couched in a somewhat different mood to that expressed in his telegram about Trieste.

On May 14 he said he preferred to see what happened before considering a continued though temporary occupation of the Soviet Zone in Germany by the Western Allies. As for the Yugoslavs, Mr. Truman declared that we should await reports about our messages to Belgrade before deciding what forces to use if our troops were attacked. Unless Tito did attack it was impossible to involve the United States in another war. Two days later he declared he was unable and unwilling to involve his country in a war with the Yugoslavs unless they attacked us, in which case we should be justified in using Allied troops to throw them back far enough to stop any further aggression.

Meanwhile the local situation around Trieste sharpened. At first Alexander would have been content if Tito had put his fighting and administrative troops under Allied command, at any rate in the areas where we were operating, though of course we would rather they withdrew completely. But Yugoslav posts and sentries were restricting our movements. Their behaviour, both in Austria and Venezia Giulia, made a bad impression on the Allied troops, both United States and British. Our men were obliged to look on without power to intervene at actions which offended their sense of justice, and felt that they were condoning wrongdoing. "As a result," cabled Alexander, "feeling against Yugoslavia is now strong, and is getting stronger. It is now certain that any solution by which we shared an area with Yugoslav troops or Partisans or permitted Yugoslav administration to function would not work."

485

On May 19 I replied to the President.

Prime Minister to President Truman   19 May 45

I hope you will not mind my putting to you, with great respect, the need for some further consideration of the words "a war with the Yugoslavs", and secondly "attack us". I do not envisage a war with the Yugoslavs, and, short of war, I do not consider Ambassadors should be withdrawn. It is at critical junctures that Ambassadors should be on the spot. Meanwhile Tito's answer has arrived, and is completely negative. We clearly cannot leave matters in this state, and immediate action will now be necessary. Otherwise we shall merely appear to have been bluffing, and will in fact be bluffed out.

I think we should prevent the rough handling of our front-line troops, or infiltrations ostensibly peaceful but contrary to the directions of the Allied commanders and on a scale to endanger the position of our forces where they now stand. For instance, supposing they take up positions all round a British or American unit until they have it at their mercy, are we to wait till they open fire before asking them to move back beyond the lines you have indicated as desirable? I am sure this is not what you mean, but it is just the sort of incident which I think may arise. . . .

In these conditions I should not consider action by Alexander to ensure the proper functioning of his Military Government as constituting "a war with the Yugoslavs". But I certainly think that pressure should be put upon them to quit Trieste and Pola and return to the lines marked out, and that this pressure should be regarded as in the nature of frontier incidents rather than as principal diplomatic decisions. I cannot allow our own troops to be knocked about and mishandled inside the zone which we both consider they are entitled to occupy, on the basis that they are in no circumstances to open fire. A great many of the Yugoslavs have been filtering back to-day over the Isonzo, and their truculent attitude is already somewhat abated. I rest myself on your telegram to me of May 12.

On May 21 Mr. Truman said he agreed we could not leave matters as they were. We should reject Tito's answer and reinforce our troops at once so that the Yugoslavs should have no doubts about our intentions. He proposed that Eisenhower and Alexander should make a show of force by land and air, and time it to coincide with our rejection of Tito's demands. The President thought that a heavy demonstration might bring Tito to his senses, but if hostilities began he doubted if they could be treated as frontier incidents.

He accordingly sent me the text of appropriate instructions for Alexander and Eisenhower, but he ended his telegram with a revealing sentence: "I must not have any avoidable interference with the re-deployment of American forces to the Pacific."

Prime Minister to President Truman   21 May 45

I am in entire agreement with the message you are issuing to Alexander and Eisenhower, and our Chiefs of Staff will notify yours accordingly so that Combined Chiefs of Staff can give the necessary directions. To save time I am notifying Field-Marshal Alexander privately.

2. I think there is a very good chance that, if our deployment is formidable, a solution may be reached without fighting. Our firm attitude in this matter will, I believe, be of value in our discussions with Stalin. It seems to me that the need for our triple meeting at the earliest moment is very great. There will probably be a General Election campaign here during June, but as all parties are agreed on foreign policy it need not make any postponement necessary. Could you give me any idea of the date and place which would be suitable, so that we can make our several requests to Stalin. I have a fear he may play for time in order to remain all-powerful in Europe when our forces have melted. . . .

Lieut.-General Morgan, Alexander's Chief of Staff, eventually agreed with the Yugoslavs upon a line of demarcation around Trieste.

*   *   *   *   *

It was not until a month had passed of increasing friction with the Soviets and with Tito that Stalin himself addressed me on the Yugoslav problem.

Marshal Stalin to Prime Minister   21 June 45

Notwithstanding the fact that the Yugoslav Government accepted the proposal of the American and British Governments with regard to the Istria-Trieste region, the conversations in Trieste seem to have reached a deadlock. This is principally to be explained by the fact that the representatives of the Allied Command in the Mediterranean are unwilling to take account even of the minimum wishes of the Yugoslavs. Yet the Yugoslavs earned the merit of liberating this territory from the German invaders, and in that territory moreover the Yugoslav population is in a majority. This position cannot be regarded as satisfactory from the point of view of the Allies.

In my desire not to make matters worse, I have hitherto not drawn attention in our correspondence to the behaviour of Field-Marshal

Alexander, but now it is time to emphasise that I cannot accept the supercilious tone with regard to the Yugoslavs which Field-Marshal Alexander has occasionally adopted in these conversations. It is absolutely unacceptable that Field-Marshal Alexander in an official and public message allowed himself to compare Marshal Tito with Hitler and Mussolini. Such a comparison is unjustified and offensive to Yugoslavia.

The Soviet Government also found unexpected the tone of ultimatum in the declaration which the Anglo-American representatives presented to the Yugoslav Government on June 2. How is it possible with such methods to secure solid and positive results?

All this compels me to draw your attention to the situation which has developed.

As before, I hope that with regard to Trieste-Istria legitimate Yugoslav interests will be satisfied, especially since on the main question the Yugoslavs have met the Allies half-way.

I replied:

Prime Minister to Marshal Stalin   23 June 45

I thank you for your message of June 21. Our joint idea at the Kremlin in October was that the Yugoslav business should work out around 50-50 Russian and British influence. In fact it is at present more like 90-10, and even in that poor 10 we have been subjected to violent pressure by Marshal Tito. So violent was this pressure that the United States and His Majesty's Government had to put in motion many hundreds of thousands of troops in order to prevent themselves from being attacked by Marshal Tito.

2. Great cruelties have been inflicted by the Yugoslavs on the Italians in this part of the world, particularly in Trieste and Fiume, and generally they have shown a disposition to grasp all the territory into which their light forces have penetrated. The movement of these light forces could not have been made unless you for your part had made immense and welcome advances from the East and in the North, and unless Field-Marshal Alexander had held twenty-seven enemy divisions on his front in Italy and finally reduced them to surrender. I do not consider that it can be said that Marshal Tito has conquered all this territory. It has been conquered by the movements of far greater forces both in the West and in the East which compelled the strategic retreat of the Germans from the Balkans.

3. At any rate, we have reached an agreement which it is proposed to enforce. We think that any permanent territorial changes should be settled at the peace table, and Marshal Tito is in no way prejudiced by accepting the present line which we demand until that meeting takes

place. In the interval we can talk all these matters over together at Berlin.

4. The actual wording of Field-Marshal Alexander's telegram has been largely taken from the President's draft. We do not see why we should be pushed about everywhere, especially by people we have helped, and helped before you were able to make any contact with them. Therefore I do not see any reason to make excuses for Field-Marshal Alexander, although I was not aware that he was going to draft his telegram exactly in this way.

5. It seems to me that a Russianised frontier running from Lübeck through Eisenach to Trieste and down to Albania is a matter which requires a very great deal of argument conducted between good friends.

6. These are just the things we have to talk over together at our meeting, which is not long now.

Here for the time being in these pages we may leave the problem of Tito and Trieste.

*   *   *   *   *

In the uneasy interlude between the German surrender and the Tripartite Conference in Berlin General de Gaulle was also determined to assert the position of France both in Syria, where he ran counter to the policy we had consistently pursued of Syrian independence, and in Italy, where he affronted the United States.

As early as February 27 I had stated our policy in plain terms to the House of Commons.

I must make clear, once and for all, the position of His Majesty's Government in respect of Syria and the Lebanon, and in relation to our French Allies. That position is governed by the statements made in 1941, in which the independence of these Levant States was definitely declared by Great Britain and France. At that time, and ever since, His Majesty's Government have made it clear that they would never seek to supplant French influence by British influence in the Levant States. We are determined also to respect the independence of these States and to use our best endeavours to preserve a special position for France, in view of the many cultural and historic connections which France has so long established with Syria. We hope that it may be possible for the French to preserve that special position. We trust that these States will be firmly established by the authority of the World Organisation, and that French privilege will also be recognised.

However, I must make it clear that it is not for us alone to defend by force either Syrian or Lebanese independence or French privilege.

We seek both, and we do not believe that they are incompatible. Too much must not be placed therefore upon the shoulders of Great Britain alone. We have to take note of the fact that Russia and the United States have recognised and favour Syrian and Lebanese independence, but do not favour any special position for any other foreign country.

The liberation of France led to a serious crisis in the Levant. It had been evident for some time that a new treaty would be needed to define French rights in this area, and on my way home from Yalta I had met the President of Syria in Cairo and urged him to make a peaceful settlement with France. The Levant States had been unwilling to start negotiations, but we had persuaded them to do so and conversations had begun. The French delegate, General Beynet, went to Paris for instructions, and his proposals were awaited with anxiety and excitement throughout Syria. Delay occurred; no proposals arrived; and then news spread that French reinforcements were on their way. On May 4 I had sent a friendly message to de Gaulle explaining that we had no ambitions of any kind in the Levant States and would withdraw all our troops from Syria and the Lebanon as soon as the new treaty was concluded and in operation, but I also mentioned that we had to keep our war communications throughout the Middle East free from disturbance and interruption. We represented to him that the arrival of reinforcements, however small, was bound to be looked upon as a means of pressure, and might have serious consequences. This advice was not accepted, and on May 17 French troops landed at Beirut.

An explosion followed. The Syrian and Lebanese Governments broke off negotiations and said that now the war was over the Allies would be asked to evacuate all foreign troops. Anti-French strikes and demonstrations began. Eight people were killed and twenty-five injured in Aleppo. The Syrian Chamber of Deputies ordered conscription. A Foreign Office announcement of May 26 regretting the arrival of French reinforcements drew a reply from Paris next day that the disturbances were artificially provoked and that many more British troops had also been moved in without protest by the Syrians or the Lebanese and without agreement by the French. We had in fact appealed to the Syrian Government on May 25 to keep control of the situation, but on the 28th they told us that events were too much for them and they could no longer be responsible for internal order. French shelling had begun in Homs and Hama; French armoured cars were patrolling the streets of Damascus and Aleppo; French aircraft were flying low over the mosques during the hour of prayer, and machine-guns were mounted on the roofs of buildings.

At about seven o'clock in the evening of May 29 fierce fighting began in Damascus between French troops and Syrians, and continued for several hours into the night. French artillery opened fire, with serious loss of life and damage to property, and French troops occupied the Syrian Parliament buildings. Shelling continued on and off till the morning of May 31, and about two thousand people were killed and injured.

The Governor of Homs had already appealed to the British Ninth Army to arrange a truce. It was now impossible for us to stand aside, and on May 31 General Sir Bernard Paget, Commander-in-Chief Middle East, was told to restore order. He communicated our request to the French commander, and the latter, on instructions from Paris, proclaimed the "Cease fire". I sent the following message to General de Gaulle:

Prime Minister to General de Gaulle (Paris)   31 May 45

In view of the grave situation which has arisen between your troops and the Levant States, and the severe fighting which has broken out, we have with profound regret ordered the Commander-in-Chief Middle East to intervene to prevent the further effusion of blood in the interests of the security of the whole Middle East, which involves communications for the war against Japan. In order to avoid collision between British and French forces, we request you immediately to order the French troops to cease fire and to withdraw to their barracks.

Once firing has ceased and order has been restored we shall be prepared to begin tripartite discussions in London.

By an error in transmission, and with no intentional discourtesy, this message was read to the House of Commons by Mr. Eden about three-quarters of an hour before it reached the General. He felt obliged to issue a public reply in Paris on June 1, saying in effect that the French troops had been attacked by the Syrians, but had everywhere gained control, and that the French Government had themselves ordered a "Cease fire" on May 31.

A vehement protest reached me from the President of the Syrian Republic. But the action we had already taken proved effective. I was most anxious not to vex the French more than was inevitable, and I understood de Gaulle's view and mood about a cause for which he felt passionately. But he also struck a states-manlike note. "We feel," he said, "not the slightest rancour or anger towards the British. France and myself have the highest regard and affection for them. But there are opposing interests, and these must be reconciled. I hope all this will not have too far-reaching consequences. There are too many common interests at stake. There must be peace."

I was in accord with this view, and when I gave an account of these regrettable incidents to the House of Commons on June 5 I said it was a case of "the less said the better".

Prime Minister to General Paget   3 June 45

As soon as you are master of the situation you should show full consideration to the French. We are very intimately linked with France in Europe, and your greatest triumph will be to produce a peace without rancour. Pray ask for advice on any point you may need, apart from military operations.

In view of reports that French soldiers have been killed, pray take the utmost pains to protect them.

And to the Syrian President, whom I deemed a sensible and competent man:

3 June 45

Now that we have come to your aid I hope you will not make our task harder by fury and exaggeration. The French have got to have fair treatment as well as you, and we British, who do not covet any-thing that you possess, expect from you that moderation and helpful-ness which are due to our disinterested exertions.

Our intervention was immediately effective. On June 3 the French garrison at Damascus was withdrawn to a camp outside the city, and a British detachment which had been landed at Beirut from H.M.S. *Arethusa* arrived in the Syrian capital on the same day.

On June 4 Mr. Shone, our Minister at Damascus, delivered my message to the Syrian President, who took it well and sent the following reply:

I sent my message of May 31 to Your Excellency under stress of bombardment and of deep emotion at the sufferings which the Syrian people were undergoing, and which I assure you were no exaggera-tion. Your Excellency will since have received my message of June 1 expressing the gratitude of the Syrian people for the intervention of the British Government, and I and my Government have assured His Majesty's Minister and the Commander-in-Chief that our one desire is to co-operate with the British authorities in their task of restoring order and security in Syria. Your Excellency can be sure that this co-operation with the British authorities will soon have its good results.

"The President," said Mr. Shone, "who was ill in bed when he sent his message of May 31, is now up again and seems fully composed. He is in full accord with you and deeply grateful. As regards fair treatment for the French, he said they could have their schools (if any Syrians still wanted to go to them) and their commercial interests, but neither the Syrian Government nor the Chamber nor the people could ever give them any privilege in this country after what had happened."

General Paget handled the situation with much discretion. All passed off smoothly, and this difficult and untoward Syrian episode came to an end.

* * * * *

A smaller though not less vexatious dispute arose between de Gaulle and President Truman.

In the closing days of the fighting troops of the First French Army in the Alpine region crossed the frontier and moved forward into North-Western Italy, in the province of Cuneo. Orders were presently issued by General Eisenhower for their withdrawal. These orders were ignored by the French units concerned under the authority of their Government.

On May 30 General Doyen, commanding the French Army in the Alps, sent a letter to Major-General Crittenberger, commander of the United States IVth Corps in North-Western Italy, referring to an attempt to establish Allied military government in the province of Cuneo. The letter ended with the following paragraph: "France cannot consent that a modification against her will should be made in the existing state of affairs in the Alpes-Maritimes. This would be contrary to her honour and her security. I have been ordered by the Provisional Government of the French Republic to occupy and administer this territory. This mission being incompatible with the installation of an Allied military agency in the same region, I find myself obliged to oppose it. *Any insistence in this direction would assume a clearly unfriendly character, even a hostile character, and could have grave consequences.*"*

* Author's subsequent italics.

On June 2 General Crittenberger received another letter from General Doyen:

General de Gaulle has instructed me to make as clear as possible to the Allied Command that I have received the order to prevent the setting up of Allied military government in territories occupied by our troops and administered by us *by all necessary means without exception*.

This was astonishing language to use in all the circumstances. "Is it not rather disagreeable," I wrote to the President, when Alexander reported the facts, "for us to be addressed in these terms by General de Gaulle, whom we have reinstated in liberated France at some expense of American and British blood and treasure? Our policy with France is one of friendship."

Mr. Truman was indignant. He wrote to de Gaulle, pointing out that the messages contained the almost unbelievable threat that French troops bearing American arms would fight American and Allied soldiers, whose efforts and sacrifices had so recently and successfully helped to liberate France. The President said that as long as this threat remained no more equipment or ammunition would be issued to the French forces.

This produced immediate results. De Gaulle wrote through his Foreign Minister:

Obviously there has never been an intention either in the orders of the French Government or in those of General Doyen, who commands the Army detachment of the Alps, to oppose by force the presence of American troops in the small areas which French troops occupy to the east of the 1939 frontier between France and Italy. Besides, American troops are now in these areas side by side with French troops, and here as elsewhere good comradeship prevails. . . . To-morrow morning General Juin will proceed to Field-Marshal Alexander's headquarters to deal with this matter in the broadest spirit of conciliation in order that a solution may be found.

Thus the matter ended, if not pleasantly, at least without another quarrel. The British public, whose attention had been distracted from world events by the General Election, were not troubled by these affairs.

CHAPTER XXXIV

THE CHASM OPENS

The Soviet Menace – Pressures and Policies in Washington – The Need for a Conference with Stalin – My "Iron Curtain" Telegram of May 12 – I Take Steps to Preserve the Military Strength of the Western Democracies – A Message from Smuts – A Visit from Mr. Joseph E. Davies, May 26 – "Ganging Up" – My Minute of May 27, and the President's Friendly Reply – Stalin Suggests a Meeting in Berlin – Harry Hopkins Goes to Moscow – His Efforts to Break the Deadlock over Poland – Britain and the United States Recognise a New Polish Provisional Government, July 5.

APPREHENSION for the future and many perplexities had filled my mind as I moved about among the cheering crowds of Londoners in their hour of well-won rejoicing after all they had gone through. The Hitler peril, with its ordeals and privations, seemed to most of them to have vanished in a blaze of glory. The tremendous foe they had fought for more than five years had surrendered unconditionally. All that remained for the three victorious Powers was to make a just and durable peace, guarded by a World Instrument, to bring the soldiers home to their longing loved ones, and to enter upon a Golden Age of prosperity and progress. No more, and surely, thought their peoples, no less.

However, there was another side to the picture. Japan was still unconquered. The atomic bomb was still unborn. The world was in confusion. The main bond of common danger which had united the Great Allies had vanished overnight. The Soviet menace, to my eyes, had already replaced the Nazi foe. But no comradeship against it existed. At home the foundations of national unity, upon which the war-time Government had stood so firmly, were also gone. Our strength, which had overcome so many storms, would no longer continue in the sunshine. How then could we reach that final settlement which alone could reward the toils and sufferings of the struggle? I could not rid my mind of the fear that the victorious armies of democracy would soon disperse and that the real and hardest test still lay before us. I had seen it all before. I remembered that other joy-day nearly thirty years before, when I had driven with my wife from the Ministry of Munitions through similar multitudes convulsed with enthusiasm to Downing Street to congratulate the Prime Minister. Then, as at this time, I understood the world situation as a whole. But then at least there was no mighty army that we need fear.

* * * * *

My prime thought was a meeting of the three Great Powers, and I hoped that President Truman would come through London on his way. As will be seen, very different ideas were being pressed upon the new President from influential quarters in Washington. The sort of mood and outlook which had been noticed at Yalta had been strengthened. The United States, it was argued, must be careful not to let herself be drawn into any antagonism with Soviet Russia. This, it was thought, would stimulate British ambition and would make a new gulf in Europe. The right policy, on the other hand, should be for the United States to stand between Britain and Russia as a friendly mediator, or even arbiter, trying to reduce their differences about Poland or Austria and make things settle down into a quiet and happy peace, enabling American forces to be concentrated against Japan. These pressures must have been very strong upon Truman. His natural instinct, as his historic actions have shown, may well have been different. I could not of course measure the forces at work in the brain-centre of our closest Ally, though I was soon conscious of them. I could only feel the vast manifestation of Soviet and Russian imperialism rolling forward over helpless lands.

Obviously the first aim must be a conference with Stalin. Within three days of the German surrender I cabled the President:

Prime Minister to President Truman   11 May 45

I think we should offer an invitation jointly or severally at the same moment to Stalin to meet us at some agreed unshattered town in Germany for a tripartite meeting in July. We should not rendezvous at any place within the present Russian military Zone. Twice running we have come to meet him. They are concerned about us on account of our civilisation and various instrumentalities. But this will be greatly diminished when our armies are dispersed.

2. I do not know at the moment when our General Election will be, but I do not see any reason why it should influence your movements or mine where public duty calls. If you will entertain the idea of coming over here in the early days of July His Majesty will send you the most cordial invitation, and you will have a great reception from the British nation. I would have suggested the middle of June but for your reference to your fiscal year [June 30], because I feel that every minute counts. Thereafter we might move to the rendezvous fixed in Germany and have the grave discussions on which the immediate future of the world depends. I should of course bring with me representatives of both parties in our State, and both would use exactly the same language about foreign affairs as we are closely agreed. Therefore I urge your coming here in the earliest days of July, and that we leave together to meet U.J. at wherever is the best point outside Russian-occupied territory to which he can be induced to come. *Meanwhile I earnestly hope that the American front will not recede from the now agreed tactical lines.**

3. I doubt very much whether any enticements will get a proposal for a tripartite meeting out of Stalin. But I think he would respond to an invitation. If not what are we to do?

4. I rejoice that your present intention is to adhere to our rightful interpretation of the Yalta agreements and to stand firmly on our present announced attitude towards all the questions at issue. Mr. President, in these next two months the gravest matters in the world will be decided. May I add that I have derived a great feeling of confidence from the correspondence we have interchanged.

5. We are drawing up, as you desire, a list of subjects for discussion among us three, which will take a few days but will be forwarded to you immediately.

He replied at once that he would rather have Stalin propose the meeting, and he hoped our Ambassadors would induce him to suggest it. Mr. Truman then declared that he and I ought to go to the meeting separately, so as to avoid any suspicion of "ganging up". When the Conference ended he hoped to visit England if his duties in America permitted.

I did not fail to notice the difference of view which this telegram conveyed, but I accepted the procedure the President proposed.

* Author's subsequent italics.

497

Prime Minister to President Truman   13 May 45

F. D. R. promised me he would visit England before he went to France—or, as it has now become, Germany. We should feel disappointed if you did not come to us. But, having regard to the gravity of the next few months, no question of ceremonial should intervene with the organised sequence of events. Therefore I am for the conference of the Three as soon as possible and wherever possible.

2. In this case I consider that we should try to bring the meeting off some time in June, and I hope your fiscal year will not delay it. We greatly hope you will come to England later.

3. I agree that our Ambassadors should do their utmost to induce Stalin to propose the meeting, and instructions will be given accordingly by us. I doubt very much whether he will accede. Time is on his side if he digs in while we melt away.

4. I look forward to your meeting with Eden.

In these same days I also sent what may be called the "Iron Curtain" telegram to President Truman. Of all the public documents I have written on this issue I would rather be judged by this.

Prime Minister to President Truman   12 May 45

I am profoundly concerned about the European situation. I learn that half the American Air Force in Europe has already begun to move to the Pacific theatre. The newspapers are full of the great movements of the American armies out of Europe. Our armies also are, under previous arrangements, likely to undergo a marked reduction. The Canadian Army will certainly leave. The French are weak and difficult to deal with. Anyone can see that in a very short space of time our armed power on the Continent will have vanished, except for moderate forces to hold down Germany.

2. Meanwhile what is to happen about Russia? I have always worked for friendship with Russia, but, like you, I feel deep anxiety because of their misinterpretation of the Yalta decisions, their attitude towards Poland, their overwhelming influence in the Balkans, excepting Greece, the difficulties they make about Vienna, the combination of Russian power and the territories under their control or occupied, coupled with the Communist technique in so many other countries, and above all their power to maintain very large armies in the field for a long time. What will be the position in a year or two, when the British and American Armies have melted and the French has not yet been formed on any major scale, when we may have a handful of divisions, mostly French, and when Russia may choose to keep two or three hundred on active service?

3. An iron curtain is drawn down upon their front. We do not

know what is going on behind. There seems little doubt that the whole of the regions east of the line Lübeck–Trieste–Corfu will soon be completely in their hands. To this must be added the further enormous area conquered by the American armies between Eisenach and the Elbe, which will, I suppose, in a few weeks be occupied, when the Americans retreat, by the Russian power. All kinds of arrangements will have to be made by General Eisenhower to prevent another immense flight of the German population westward as this enormous Muscovite advance into the centre of Europe takes place. And then the curtain will descend again to a very large extent, if not entirely. Thus a broad band of many hundreds of miles of Russian-occupied territory will isolate us from Poland.

4. Meanwhile the attention of our peoples will be occupied in inflicting severities upon Germany, which is ruined and prostrate, and it would be open to the Russians in a very short time to advance if they chose to the waters of the North Sea and the Atlantic.

5. Surely it is vital now to come to an understanding with Russia, or see where we are with her, before we weaken our armies mortally or retire to the zones of occupation. This can only be done by a personal meeting. I should be most grateful for your opinion and advice. Of course we may take the view that Russia will behave impeccably, and no doubt that offers the most convenient solution. To sum up, this issue of a settlement with Russia before our strength has gone seems to me to dwarf all others.

*　　*　　*　　*　　*

From the very first moment I took whatever practical steps were in my power to hold the position and prevent the melting of the Western armies.

Prime Minister to General Eisenhower (France)　　　　　　　9 May 45

I have heard with some concern that the Germans are to destroy all their aircraft *in situ*. I hope that this policy will not be adopted in regard to weapons and other forms of equipment. We may have great need of these some day, and even now they might be of use, both in France and especially in Italy. I think we ought to keep everything worth keeping. The heavy cannon I preserved from the last war fired constantly from the heights of Dover in this war.

There is great joy here.

General Eisenhower to Prime Minister　　　　　　　　　　10 May 45

Our policy as laid down in the Act of Surrender is that the Germans will not destroy aircraft, and this policy applies to German action in respect of all their equipment. If Germans are destroying equipment it is in violation of the Act of Surrender, and I should be glad to have any particulars which would enable me to punish the offenders.

There is great joy [here] too.

Prime Minister to Mr. Eden (San Francisco)   11 May 45

To-day there are announcements in the newspapers of the large withdrawals of American troops now to begin month by month. What are we to do? Great pressure will soon be put on us [at home] to demobilise partially. In a very short time our armies will have melted, but the Russians may remain with hundreds of divisions in possession of Europe from Lübeck to Trieste, and to the Greek frontier on the Adriatic. All these things are far more vital than the amendments to a World Constitution which may well never come into being till it is superseded after a period of appeasement by a third World War.

In Washington, at my desire, Mr. Eden on the 14th asked General Marshall and Mr. Stimson about the withdrawal of American troops from Europe. The General was on the whole reassuring. Actual figures of withdrawals for the next few months would not reach fifty thousand a month out of a total of three million. Eden then turned to the question of the withdrawal of Anglo-American forces within their previously agreed zones. Marshall, who had seen my telegram to the President, seemed sympathetic to what I had proposed. The Russians, he said, had however withdrawn in Austria from a small area of the American Zone which they had occupied. He thought that this had been done deliberately in order to strengthen their case when they asked that we should withdraw from areas of the Russian Zone which we were occupying.

*   *   *   *   *

I could at least keep the British Air Force in effective strength.

Prime Minister to General Ismay, for C.O.S. Committee   17 May 45

All reduction of Bomber Command is to be stopped. All reduction of the Metropolitan Air Force, except Coastal Command, is to be stopped. In both cases leave may be given where necessary, but the structure and number of squadrons is not to be cut down till further orders are received from the War Cabinet.

Prime Minister to Secretary of State for Air and Chief of the   17 May 45
Air Staff

No weakening of the Air Force in Italy or demobilisation must take place at present.

Prime Minister to Chief of the Air Staff and General Ismay,   17 May 45
for all concerned

No German aircraft in British control which has a serviceable war value, including spares, is to be destroyed by the Germans or by us without Cabinet sanction being obtained beforehand.

Prime Minister to General Ismay, for Chiefs of Staff Committee   20 May 45

It is about a week since I sent instructions against the further de-mobilisation of the Air Force, and for "Steady on" in that of the Army. Of course I am relying on the C.O.S. Committee to show the best and least obtrusive measures in which these necessary procedures can be carried out.

★   ★   ★   ★   ★

Smuts, who was at San Francisco, and whom I had apprised of all, was in full accord with my mood and actions. He cabled on May 14:

The correspondence confirms forebodings which have been forming in my mind for some time as Russia has been showing her hand in Poland, Roumania, Bulgaria, Czechoslovakia, and Austria. Now the same thing is happening in more drastic form in Yugoslavia, where it is the voice of Tito but the hand of Russia. To this must be added Stalin's surly and truculent reply to your friendly letter. It looks as if the elimination of Germany is looked upon as Russia's opportunity, or as if she intends to exact a heavy price for co-operation against Japan.

I did not think that the point about Japan was valid, but I thought it might weigh unduly with the State Department. I therefore telegraphed to our Ambassador:

Prime Minister to Lord Halifax (Washington)   14 May 45

We desire the entry of the Soviets into the war against Japan at the earliest moment. Having regard to their own great interests in the Far East, they will not need to be begged, nor should their entry be purchased at the cost of concessions prejudicing a reign of freedom and justice in Central Europe or the Balkans.

★   ★   ★   ★   ★

A week passed before I heard again from Mr. Truman on the major issues. Then on May 22 he cabled that he had asked Mr. Joseph E. Davies to come to see me before the Triple Conference, about a number of matters he preferred not to handle by cable.

Mr. Davies had been the American Ambassador in Russia before the war, and was known to be most sympathetic to the régime. He had in fact written a book on his mission to Moscow which was produced also as a film which seemed in many ways to palliate the Soviet system. I of course made immediate arrangements to receive him, and he spent the night of the 26th at Chequers. I had a very long talk with him. The crux of what he had to propose was that the President should meet Stalin first somewhere in Europe before he saw me. I was indeed astonished at this suggestion. I had not liked the President's use in an earlier message of the term "ganging up" as applied to any meeting between him and me. Britain and the United States were united by bonds of principle and by agreement upon policy in many directions, and we were both at profound difference with the Soviets on many of the greatest issues. For the President and the British Prime Minister to talk together upon this common ground, as we had so often done in Roosevelt's day, could not now deserve the disparaging expression "ganging up". On the other hand, for the President to by-pass Great Britain and meet the head of the Soviet State alone would have been, not indeed a case of "ganging up"—for that was impossible—but an attempt to reach a single-handed understanding with Russia on the main issues upon which we and the Americans were united. I would not agree in any circumstances to what seemed to be an affront, however unintentional, to our country after its faithful service in the cause of freedom from the first day of the war. I objected to the implicit idea that the new disputes now opening with the Soviets lay between Britain and Russia. The United States was as fully concerned and committed as ourselves. I made this quite clear to Mr. Davies in our conversation, which also ranged over the whole field of Eastern and Southern European affairs.

In order that there should be no misconception I drafted a formal minute which I gave to Mr. Davies, after cordial agreement with the Foreign Secretary, who had now returned to London.

NOTE BY THE PRIME MINISTER ON MR. DAVIES'S MESSAGE

27 May 45

It is imperative to hold a Conference of the three major Powers at the earliest possible date. The Prime Minister is prepared to attend on any date at any place agreeable to the other two Powers. He hopes however that the United States and Great Britain will not find it necessary to go into Russian territory or the Russian zone of occupation. Many visits have been paid to Moscow, and the last meeting, at Yalta, was held upon Russian soil. The Prime Minister declares that London, the greatest city in the world, and very heavily battered during the war, is the natural and appropriate place for the Victory meeting of the three Great Powers. However, if this is refused His Majesty's Government will none the less discuss with the United States and with Soviet Russia what is the best place to be appointed.

2. The Prime Minister received with some surprise the suggestion conveyed by Mr. Davies that a meeting between President Truman and Premier Stalin should take place at some agreed point, and that the representatives of His Majesty's Government should be invited to join a few days later. *It must be understood that the representatives of His Majesty's Government would not be able to attend any meeting except as equal partners from its opening. This would be undoubtedly regrettable. The Prime Minister does not see that there is any need to raise an issue so wounding to Britain, to the British Empire and Commonwealth of Nations.** Meetings like these always require two or three days of preliminary discussions, when the agenda is framed and where complimentary contacts are made between the three heads of States. In such circumstances all three Great Powers are obviously free to make what contacts they wish and when they please.

3. The Prime Minister realises that President Truman would no doubt like to make the acquaintance of Premier Stalin, the pleasure of which he has not previously enjoyed. None of the Allies at these meetings has sought to put the slightest restraint upon the most free intercourse between the heads of Governments or between their Foreign Secretaries. The Prime Minister has himself been looking forward to making the personal acquaintance for the first time of President Truman, and he has indulged the hope that he might have some private talks with the President before the general sittings commence. However, at such meetings everything is perfectly free and the principals meet together how they like, when they like, and for as long as they like, and discuss any questions that they may consider desirable. This does not of course prevent certain lunches and dinners at which the strong bonds of unity which have hitherto united the three major Powers are vivified by agreeable intercourse and often form the subject of congenial toasts. The Prime Minister's experience has been that these matters work out quite easily on the spot.

4. It would of course be more convenient to the Prime Minister if the meeting of the three major Powers took place after July 5, when the British pollings will be over. But he does not rate such a consideration as comparable at all to the vital importance of having a meeting at the earliest possible moment, before the United States forces in

* Author's subsequent italics.

Europe are to a large extent dissolved. He would therefore be quite ready, if Premier Stalin's consent can be obtained, to meet as early as June 15.

5. It must be remembered that Britain and the United States are united at this time upon the same ideologies, namely, freedom, and the principles set out in the American Constitution and humbly reproduced with modern variations in the Atlantic Charter. The Soviet Government have a different philosophy, namely, Communism, and use to the full the methods of police government, which they are applying in every State which has fallen a victim to their liberating arms. The Prime Minister cannot readily bring himself to accept the idea that the position of the United States is that Britain and Soviet Russia are just two foreign Powers, six of one and half a dozen of the other, with whom the troubles of the late war have to be adjusted. Except in so far as force is concerned, there is no equality between right and wrong. The great causes and principles for which Britain and the United States have suffered and triumphed are not mere matters of the balance of power. They in fact involve the salvation of the world.

6. The Prime Minister has for many years now gone by striven night and day to obtain a real friendship between the peoples of Russia and those of Great Britain, and, as far as he was entitled to do so, of the United States. It is his resolve to persevere against the greatest difficulties in this endeavour. He does not by any means despair of a happy solution conferring great advantages upon Soviet Russia, and at the same time securing the sovereign independence and domestic liberties of the many States and nations which have now been overrun by the Red Army. The freedom, independence, and sovereignty of Poland was a matter for which the British people went to war, ill-prepared as they were. It has now become a matter of honour with the nation and Empire, which is now better armed. The rights of Czechoslovakia are very dear to the hearts of the British people. The position of the Magyars in Hungary has been maintained over many centuries and many misfortunes, and must ever be regarded as a precious European entity. Its submergence in the Russian flood could not fail to be either the source of future conflicts or the scene of a national obliteration horrifying to every generous heart. Austria, with its culture and its historic capital of Vienna, ought to be a free centre for the life and progress of Europe.

7. The Balkan countries, which are the survivors of so many centuries of war, have built up hard civilisations of their own. Yugoslavia is at present dominated by the Communist-trained leader Tito, whose power has been mainly gained by the advances of the British and American armies in Italy. Roumania and Bulgaria are largely

swamped by the fact of their proximity to Soviet Russia and their having taken the wrong side in several wars. Nevertheless these countries have a right to live. As for Greece, by hard fighting by Greeks and by the British Army the right has been obtained for the Greek' people to express at an early approaching election, without fear of obstruction, on the basis of universal suffrage and secret ballot, their free, unfettered choice alike of régime and Government.

8. The Prime Minister cannot feel it would be wise to dismiss all these topics in the desire to placate the imperialistic demands of Soviet Communist Russia. Much as he hopes that a good, friendly, and lasting arrangement may be made and that the World Organisation will come into being and act with some reality, the Prime Minister is sure that the great causes involved in the above epitome of some of the European relationships cannot be ignored. He therefore urges (a) a meeting at the earliest moment, and (b) that the three major Powers shall be invited thereto as equals. He emphasises the fact that Great Britain would not be able to attend any meeting of a different character, and that of course the resulting controversy would compel him to defend in public the policy to which His Majesty's Government is vowed.

The President received this note in a kindly and understanding spirit, and replied on May 29 that he was considering possible dates for the Triple Conference.

I was very glad to learn that all was well and that the justice of our view was not unrecognised by our cherished friends.

On May 27 Stalin suggested that "the Three" should meet in Berlin in "the very near future". I replied that I should be very glad to meet him and the President in what was left of the city, and that I hoped this meeting would take place about the middle of June. I now received the following message:

Marshal Stalin to Prime Minister   30 May 45
A few hours after I had received your telegram Mr. Hopkins was with me, and told me that President Truman thinks the most convenient date for the meeting of the Three would be July 15. I have no objections to that date if you also agree to it.

I send you my best wishes.

*   *   *   *   *

About the same time as President Truman sent Mr. Davies to see me he had asked Harry Hopkins to go as his special envoy to Moscow to make another attempt to reach a working agreement on the Polish question. Although far from well, Hopkins, taking his bride with him, set out gallantly for Moscow. His friendship for Russia was well known, and he received a most friendly welcome. Certainly for the first time some progress was made. Stalin agreed to invite Mikolajczyk and two of his colleagues to Moscow from London for consultation, in conformity with our interpretation of the Yalta agreement. He also agreed to invite some important non-Lublin Poles from inside Poland.

In a telegram to me the President said he felt this was a very encouraging, positive stage in the negotiations. Most of the arrested Polish leaders were apparently only charged with operating illegal radio transmitters, and Hopkins was pressing Stalin to grant them an amnesty so that consultations could be conducted in the most favourable atmosphere possible. He asked me to urge Mikolajczyk to accept Stalin's invitation.

We of course concurred in these proposals for what they were worth.

Prime Minister to President Truman   4 June 45

... I agree with you that Hopkins' devoted efforts have produced a breaking of the deadlock. I am willing that the invitation should be issued to the non-Lublin Poles on that basis if nothing more can be gained at this moment. I also agree that the question of the fifteen or sixteen arrested Poles should not hamper the opening of these discussions. We cannot however cease our efforts on their behalf. I will therefore join with you, either jointly or separately, in a message to Stalin accepting the best that Hopkins can get, provided of course that our Ambassadors are not debarred from pressing for further improvements in the invitations once conversations have begun again.

While it is prudent and right to act in this way at this moment, I am sure you will agree with me that these proposals are no advance on Yalta. They are an advance upon the deadlock, but we ought by now, according to Yalta and its spirit, to have had a representative Polish Government formed. All we have got is a certain number of concessions for outside Poles to take part in preliminary discussions, out of which some improvements in the Lublin Government may be made. I cannot feel therefore that we can regard this as more than a milestone in a long hill we ought never to have been asked to climb. I think we ought to guard against any newspaper assumptions that the Polish problem has been solved or that the difficulties between the Western democracies and the Soviet Government on this matter have been more than relieved. Renewed hope and not rejoicing is all we can indulge in at the moment. ...

*   *   *   *   *

I sent Hopkins my congratulations. He replied next day:

Thanks so much for your nice personal message.

I hope that you can agree to list as proposed and will not make released prisoners [release of the sixteen Polish prisoners] a condition to getting started with consultations here. I am doing everything under Heaven to get these people out of jug, but the more important thing, it seems to me, is to get these Poles together in Moscow right away.

And later, in characteristic vein:

I am leaving Moscow in the morning. Am going to do a little sight-seeing in Berlin, then home. The business here came off reasonably well, and both Averell and I are very hopeful that at least some of these prisoners are going to be sprung.

I want you to know I have not the vaguest notion what the word "amnesty" means, and I hope British Cabinet did not spend too much time debating this one. The only thing I ever said to Stalin was to let these poor Poles out of jug. If you should find out what technical definition of "amnesty" is won't you please let me know.

I persuaded Mikolajczyk to go to Moscow, and in the upshot a new Polish Provisional Government was set up. At Truman's request this was recognised by both Britain and the United States on July 5.

It is difficult to see what more we could have done. For five months the Soviets had fought every inch of the road. They had gained their object by delay. During all this time the Lublin Administration, under Bierut, sustained by the might of the Russian armies, had given them a complete control of Poland, enforced by the usual deportations and liquidations. They had denied us all the access for our observers which they had promised. All the Polish parties, except their own Communist puppets, were in a hopeless minority in the new recognised Polish Provisional Government. We were as far as ever from any real and fair attempt to obtain the will of the Polish nation by free elections. There was still a hope—and it was the only hope—that the meeting of "the Three", now impending, would enable a genuine and honourable settlement to be achieved. So far only dust and ashes had been gathered, and these are all that remain to us to-day of Polish national freedom.

CHAPTER XXXV

THE END OF THE COALITION

The Achievements of the National Government – Strength and Weaknesses of the Rival Party Organisations – My Speech to the Commons, October 31, 1944 – The Need for an Appeal to the Electorate Once Germany was Defeated – The Choice Between June and October – My Desire to Postpone a General Election until Japan Surrendered – Correspondence with Mr. Eden – I Suggest that the Coalition should Continue until Victory over Japan had been Gained – Mr. Attlee Rejects This – I Tender My Resignation to the King, May 23 – The "Caretaker Government" – Declaration of the Poll Fixed for July 26.

FEW questions, national or personal, have so perplexed my mind as did fixing the date of the General Election. The war-time Parliament had lasted nearly ten years, or double the normal span. The supreme task for which the parties had come together in May 1940 was already accomplished. Nothing could have enabled Britain to evoke her gigantic latent strength and endurance except an all-party National Government, strong enough to withstand long years of peril, misfortune, and the disappointment resulting from the errors and chances of war. Now the task in Europe for which we had come together was done. The fruits were yet to be gathered. This process involved a range of less violent but no less vital problems affecting all that we had fought for. If not done well and with war-time strength there could be no fruitful, and even perhaps no lasting, peace.

No Prime Minister could ever have wished for more loyal and steadfast colleagues than I had found in the Labour Party. Nevertheless, as the total defeat of Germany drew ever nearer their party machine began to work, as was certainly its right, with far-reaching and ever-increasing activity. As the war deepened and darkened practically all the agents of the Conservative Party had found war work. Many were young enough to join the forces. The core of the Labour, or, as we call them in our controversial moods, the Socialist Party was at that time the trade unions. Many of the trade union leaders wanted of course to go to the front, but the whole process of organising our production and getting the highest results from day to day forbade their release. They all did work on the home front which no one else could have done, and at the same time they maintained—and who could blame them?—their party affiliations; and once our mortal danger had passed these increasingly took on a partisan character. Thus on the one side there had been complete effacement of party activities, while on the other they ran forward unresisted. This is not a reproach, but a fact. Party conflict and party government should not be disparaged. It is in time of peace, and when national safety is not threatened, one of those conditions of a free Parliamentary democracy for which no permanent substitute is known.

On the Conservative side we became acutely conscious, as the war danger waned and as victory brightened on our horizon, that we stood at an unusual disadvantage so far as political organisation was concerned, and now very sharply up against us came the constitutional need for an appeal to the people by an election on universal suffrage. As this drew nearer the members of the Government felt themselves dividing off in opposite directions, and a whole new set of values became apparent. Instead of being comrades-in-arms we became rivals for power. In Britain, where party differences are now in practice mainly those of emphasis, all points of vantage are contested, and whole hives of men and women are busy night and day in canvassing support for their views and organisations.

When I had moved the prolongation of Parliament on October 31, 1944, I said to the House of Commons:

Let us assume that the German war ends in March, April, or May, and that some or all the other parties in the Coalition recall their Ministers out of the Government, or wish to bring it to an end from such dates. That would be a matter of regret, both on public and on personal grounds, to a great many people, but it would not be a matter of reproach or bitterness between us in this Government or in this House once Germany has been defeated. . . .

When the whole of the Japanese problem is examined, on military grounds alone it would certainly not be prudent to assume that a shorter period than eighteen months after the destruction of Hitler would be required for the final destruction of the Japanese will or capacity to fight, and this period must be continually revised every few months by the Combined Chiefs of Staff.

The prolongation of the life of the existing Parliament by another two or three years would be a very serious constitutional lapse. Even now no one under thirty has ever cast a vote at a General Election, or even at a by-election, since the registers fell out of action at the beginning of the war. *Therefore it seems to me that unless all political parties resolve to maintain the present Coalition until the Japanese are defeated we must look to the termination of the war against Nazism as a pointer which will fix the date of the General Election.** I should regret the break-up of the present highly efficient Government, which has waged war with unsurpassed success and has shaped or carried out within the last two years a programme of reform and social progress which might well have occupied a whole Parliament under the ordinary conditions of peace for five or six years. In fact, I may say—and I will indeed be quite candid on this point—that, having served for forty-two years in this House, I have never seen any Government to which I have been able to give a more loyal, confident, and consistent support. [Laughter.] But while I should regret and deplore the break-up of these forces, so knit together by personal goodwill, by the comradeship of fighting in a great cause, and by the sense of growing success arising from that comradeship, yet I could not blame anyone who claimed that there should be an appeal to the people once the German peril is removed. *Indeed I have myself a clear view that it would be wrong to continue this Parliament beyond the period of the German war.** . . .

I can assure the House that, in the absence of most earnest representations by the Labour and Liberal Parties, I could not refrain from making a submission to the Crown in respect of a dissolution after the German war is effectively and officially finished. I am sure this is a straightforward, fair, and constitutional method of dealing with what is in many ways an unprecedented situation, though not one which need in any way baffle our flexible British system. Meanwhile I must confess that the position will not become increasingly easy. The odour of dissolution is in the air, and parties are inclined to look at each other across the House with an increasing sense of impending division. . . .

The announcement of the dissolution would necessarily mark the close of the present Administration. The Conservative Party have a majority of more than a hundred above all parties and Independents in the present House, and it would therefore fall to us to make arrange-

* Author's subsequent italics.

ments for the inevitable General Election. I cannot conceive that anyone would wish that election to be held in a violent hurry or while
we were all rejoicing together and rendering thanks to God for our
deliverance. There must be an interval. Moreover, we have above all
things to be careful that practically everybody entitled to vote has a
fair chance to do so. This applies above all to the soldiers, many of
whom are serving at great distances from this country. . . .

It may therefore be taken as certain that from the moment the King
gives his consent to a dissolution a period of between two and three
months would be required. This also would be fair to the political
parties and candidates, who have to set about one another in the usual
lusty manner. . . . Finally, it is contrary to precedent for Governments
to hold on to office until the last moment of their legal tenure, or
legally extended tenure, and it would be very unwholesome for any
practice of that kind to be introduced.

* * * * *

It will be seen how decisively I had committed myself in the
previous autumn to an election at the end of the German war.
Looking back, it would have been prudent to claim more latitude
than I did. This could easily have been obtained. I had not done
so, and from the moment of the German surrender the public
mind turned swiftly from national rejoicing to party strife. The
choice lay between June and October. I now hoped and urged
that we should stay together till Japan was defeated, the peace
settlement made, and the armies brought home. Mr. Herbert
Morrison, who stood in the forefront of party affairs, eventually
made the offer that the Labour Ministers should remain in the
Government until the end of October. The sense of deliverance
from the German peril would have abated, the burden of the
re-deployment of our forces against Japan would have become
heavy, and the new register, which came into force on October 15, was thought to give the Labour Party greater advantage.
As we had assimilated the local and national franchise, and thereby
doubled the municipal electorate, they might expect a victory in
the local elections as an encouraging prelude to the Parliamentary
conflict. These same reasons in reverse inclined the Conservative
managers to June. At a meeting of the principal Conservative
Ministers I took the unusual course of asking everyone to write
his opinion on a slip of paper. All but two were for June. This
of course did not govern. The right of recommending a dissolution to the Crown rests solely with the Prime Minister. Apart

from this, my colleagues, when they saw from my demeanour how unwelcome a party struggle was to me, at once assured me they would support whatever choice I made.

I did not like either June or October. Six months before, when the defeat of Germany dwarfed all other issues, it had been easy to speak in a detached manner about what should happen after that. But now, with all the new and grave issues which foregoing chapters have described pressing upon me, I earnestly desired that the national comradeship and unity should be preserved till the Japanese war was ended. This might well have required a year, or even eighteen months, more of Coalition. Was this too much to ask of a nation we had not served ill? It certainly seemed to be in accord with the national interest. But only a friendly agreement between the two parties would render it possible. Having regard to what I had said in the autumn of 1944, I had the feeling that we ought to ask the electors to approve by a referendum, or in some other way, this limited but reasonable prolongation of our tenure. The electioneering atmosphere, which had oppressed us since the defeat of Germany drew near, would have been dissipated at any rate till the end of the year, and we might all have worked together at the great tasks which still lay ahead and required our combined strength. The worst of all solutions of our problem seemed to me an October election. This was too short to give any effective relief to the political tension, which must increase with every one of the four or five months that were to pass, and must vitiate our thoughts and work at home and abroad. If there must be an election in 1945, the sooner it came the better.

No one could tell what the result would be. For the reasons which have been given, the Conservative organisation was far less prepared than that of Labour. On the other hand, many had confidence that the nation would not be likely to take the helm out of my hands. On both sides opinion about the result was divided, and contradictory guesswork reigned. I was myself deeply distressed at the prospect of sinking from a national to a party leader. Naturally I hoped that power would be accorded to me to try to make the settlement in Europe, to end the Japanese war, and to bring the soldiers home. This was not because it seemed less pleasant to live a private life than to conduct great affairs. At this time I was very tired and physically so feeble that I had to be carried upstairs in a chair by the Marines from the

Cabinet meetings under the Annexe. Still, I had the world posi tion as a whole in my mind, and I deemed myself to possess knowledge, influence, and even authority, which might be of service. I therefore saw it as my duty to try, and at the same time as my right. I could not believe this would be denied me.

I put the June or October issue bluntly to Mr. Eden at San Francisco.

11 May 45

Home politics. I have not finally settled between June and October. May 17 is the latest for June 28 poll, and a decision must be made within the next three or four days. There is a consensus of opinion on our side that June is better for our party; that October would prolong the present uneasy electioneering atmosphere, in which many questions requiring settlement are looked at from party angles, and Government may be paralysed. When you left you were in favour of June. Let me know if you have changed. If you and Cranborne arrive no sooner than 16th this matter will have to be settled without you.

2. On the other hand, the Russian peril, which I regard as enormous, could be better faced if we remain united. I expect the Labour Party will offer to stay on till October, no doubt to their party advantage.

His reply, after stating the case impartially, was decided.

Foreign Secretary (San Francisco) to Prime Minister　　　12 May 45
I agree that a June election would probably be better for our party than an October one, though Labour Party will no doubt blame us for ending the Coalition, which the nation, I believe, would like to retain for a while yet. But any advantage they might derive from this would be lost as the campaign developed.

2. It is also inevitable that to continue the Coalition until October, with certainty of parting then, will be an uncomfortable business. Against this it would presumably be advantageous from national standpoint that present Minister of Labour [Mr. Ernest Bevin] should continue to handle demobilisation plans during that period.

3. Big question mark is foreign affairs. Dangers of present situation need no emphasis. I sometimes feel that we are entering period like that of Second Balkan War transferred on to world stage. We are clearly in a stronger position to handle foreign affairs as a National Government, and we shall have need of all our strength in the next few months. Against this, will international situation be any easier in October or need for National Government any less then? As far as I can judge, all signs point to greater difficulties in October than to-day. An election in that month is likely to be even more harmful in relation to the international situation than an election in June. Yet in October

no further postponement will be possible. What therefore we have to balance is the limited advantage of prolonging Coalition on an uneasy basis until October against risk of election at an even more dangerous period in international affairs than now and increased chances of a Socialist victory in October. After carefully weighing all these conflicting considerations, I hold to the opinion I had previously expressed that from the national point of view the balance of argument is in favour of an election in June.

Prime Minister to Mr. Eden (San Francisco)   13 May 45

I have received your message, which is in general harmony with my own opinion and most of us here. May I say how admirably you have balanced the situation.

2. However, since then I have received the President's telegram of May 12 about Trieste*, which has been transmitted to you. I must regard this as one of the most far-sighted, surefooted, and resolute telegrams which it has ever been my fortune to read. I have telegraphed to him supporting him in every way. I send you in my immediately following the text of my reply, and there is also another telegram, of which I have already sent you a copy, about not dispersing our armies while all the fruits of our conquests are still ungathered. I am sure you will do justice to all these topics in conversations with the President, especially about a standstill in the dispersal of our armies.

3. All this brings forth a new factor in the election question. We can hardly ask for the support in so serious a venture of our Labour colleagues and then immediately break up the Government. If there is going to be trouble of this kind the support of men like Attlee, Bevin, Morrison, and George Hall is indispensable to the national presentation of the case. In that event I should on no account agree to an election in October, but simply say that we must prolong our joint tenure. It is common objectives, not fixed dates, which should determine the end of such an alliance as ours. However, do not bother about this for the next two or three days, and I will watch every step. An election on July 5 would be quite possible, thus giving us an extra week to consider all matters together.

The Labour Party were in conference at Blackpool and Mr. Morrison was the reigning spirit in party manœuvre. Mr. Bevin did not want the Coalition to break up, for reasons which may not have been very different from those which guided me. Mr. Attlee had now returned from America, and before going to Blackpool he came to see me at Downing Street. I had a long talk with him, in which I urged most strongly that in one way or

* See Chapter XXXIII, page 483.

another we should postpone an election, not merely till October,
but in some way or other till the end of the Japanese war. He too
was not looking at the issue from a narrow party point of view,
and listened with much apparent sympathy to the appeal which I
made. I certainly had the impression when he left that he would
do his best to keep us together, and so reported to my colleagues.
However, the tide of party feeling proved too strong.

*   *   *   *   *

No satisfactory news arriving, I sent the following letter to
Mr. Attlee:

18 May 45

My dear Attlee,
From the talks I have had with you and your principal Labour
colleagues I have gathered the impression that the Labour Party, instead
of leaving the Government on the defeat of Germany, would be willing to continue the Coalition until the autumn.

I have given the most careful and anxious thought to this suggestion,
and I regret to say that in its present form I cannot feel it would be in
the public interest. A union of parties like that which now exists
should come together and work together, not for a particular date
without regard to world events, but for the achievement of some great
national purpose transcending all party differences. For the last five
or six months our Ministerial and Parliamentary affairs have been increasingly affected by the assumed approach of a General Election at
the end of the German war. This has not conduced to the national
interest so far as domestic affairs are concerned.

I therefore make you the following proposal, which I earnestly hope
you will not readily reject—namely, that we should fix upon another
object for our joint endeavours and adjourn the question of our separa-
tion until it is gained. The First Lord of the Admiralty [Mr. Alexander]
has already expressed in his speech in the City of London his regret that
a General Election should be held before the Japanese war was finished.
It would give me great relief if you and your friends were found
resolved to carry on with us until a decisive victory has been gained
over Japan. In the meanwhile we would together do our utmost to
implement the proposals for social security and full employment contained in the White Papers which we have laid before Parliament. On
this basis we could work together with all the energy and comradeship which has marked our long and honourable association.

I am conscious however in the highest degree of our duty to
strengthen ourselves by a direct expression of the nation's will. If
you should decide to stand on with us, all united together, until the

Japanese surrender is compelled, let us discuss means of taking the nation's opinion—for example, a referendum—on the issue whether in these conditions the life of this Parliament should be further prolonged.

I am sending letters in similar terms to Sir Archibald Sinclair and to Mr. Ernest Brown.

Yours sincerely,
WINSTON S. CHURCHILL

To this letter Mr. Attlee replied rejecting my proposal for the continuation of the Coalition, and I then sent him a second letter as follows:

My dear Attlee,

I am sorry to receive your letter of May 21, in which you reject my proposal that we should work together until the defeat of Japan is achieved and the job is finished.

In this letter you tell me that our only course is to prolong the present Coalition till a General Election in October. This would mean that from now until October, outside the Government, and even within it, we should be continually preparing for an election. We have already suffered several months of this electioneering atmosphere, which, I am sure, is already affecting administrative efficiency, and might soon weaken the country before the world at a time when, above all others, it should be strongest.

I agree with what you say in your letter that it is "on the problems of the reconstruction of the economic life of the country that party differences are most acute". "What is required," you say, "is decisive action. This can only be forthcoming from a Government united on principle and policy." I agree also with your statement, "My colleagues and I do not believe that it would be possible to lay aside political controversy now that the expectation of an election has engaged the attention of the country." For my part, I am sure that a continuance of uncertainty and agitation would be harmful to the whole process of the recovery of our trade and the change-over in industry. It is not good for any country, and it is impossible for any Coalition, to live for so long a time under the spell of an approaching General Election. Least of all is this possible in a world where events are so tumultuous and dangerous as now. . . .

I regret that you should speak of "rushing" an election. Foreseeing what might arise at the close of the German war, we discussed, as you will remember, the whole question of procedure in detail in the War Cabinet. The normal period between a dissolution and the poll is seventeen days, and it was you and your colleagues who proposed that there should be at least a three weeks' additional interval, in view of the special circumstances prevailing. We gladly accepted this reasonable request, and the unanimous decision of the Cabinet was made known by you on January 17, when you announced in the House of Commons that the King had been graciously willing for this occasion to announce his intention to dissolve Parliament at least three weeks beforehand.

Yours very sincerely,

WINSTON S. CHURCHILL

*   *   *   *   *

On May 23, being confronted by a definite breach between the parties, I tendered my resignation to the King. This is almost the only constitutional privilege which a British Prime Minister possesses. But since it involves the end of the Government it is a fairly solid foundation of power. His Majesty, whom I had of course kept fully informed of all that had passed, was graciously pleased to accept my resignation, and asked me if I could form another Government. As the Conservatives still possessed a majority in the House of Commons of one hundred over all parties combined, I undertook this task, and proceeded to form what I regarded as a National Administration, but which was in fact called "the Caretaker Government". Its main structure and core was of course my Conservative and National Liberal colleagues, but in addition all those non-political, or non-party, figures who had played so important a part in the war-time Government without exception remained at their posts. These included Sir John Anderson, Chancellor of the Exchequer, Lord Leathers, Sir Andrew Duncan, Sir James Grigg, Mr. Gwilym Lloyd George, and others.

The formation of a modern British Administration is a complex affair, involving nearly eighty persons and offices. When I thought of the elaborate processes of personal correspondence or interviews with which in Gladstonian days Governments had been formed I felt that only extreme emergency could excuse the use I made of the telephone. In forty-eight hours the new Administration was complete. No one impugned its character or quality. It commanded a majority in the House of Commons, and was able to pass the necessary financial and other measures which were required. I gave an "At Home" to the principal Ministers of the former Government at Downing Street on the

28th. The temper was friendly, but electric. Many were genuinely sorry to give up their work, and none more sorry than I to lose their help. We had been through so much together, and all regarded the past five years as a famous period in their lives. History will endorse this estimate.

The following were the "Caretakers":

THE CABINET

Prime Minister and Minister of Defence	Mr. WINSTON CHURCHILL
Foreign Secretary	Mr. ANTHONY EDEN
Chancellor of the Exchequer	SIR JOHN ANDERSON
Lord President of the Council	LORD WOOLTON
Lord Privy Seal	LORD BEAVERBROOK
President of the Board of Trade and Minister of Production	Mr. OLIVER LYTTELTON
Minister of Labour and National Service	Mr. R. A. BUTLER
Home Secretary	SIR DONALD SOMERVELL
Dominions Secretary	LORD CRANBORNE
Secretary for India and Burma	Mr. L. S. AMERY
Colonial Secretary	Mr. OLIVER STANLEY
First Lord of the Admiralty	Mr. BRENDAN BRACKEN
Secretary for War	SIR JAMES GRIGG
Secretary for Air	Mr. HAROLD MACMILLAN
Secretary for Scotland	LORD ROSEBERY
Minister of Agriculture and Fisheries	Mr. R. S. HUDSON

OTHER MINISTERS OF CABINET RANK

Lord Chancellor	VISCOUNT SIMON
Minister of Education	Mr. RICHARD LAW
Minister of Health	Mr. H. U. WILLINK
Minister of Supply	SIR ANDREW DUNCAN
Minister of Aircraft Production	Mr. ERNEST BROWN
Minister of Works	Mr. DUNCAN SANDYS
Minister of Food	COLONEL J. J. LLEWELLIN
Minister of War Transport	LORD LEATHERS
Minister of Fuel and Power	MAJOR GWILYM LLOYD GEORGE
Minister of Town and Country Planning	Mr. W. S. MORRISON, K.C.
Minister of National Insurance	Mr. LESLIE HORE-BELISHA
Minister of Civil Aviation	LORD SWINTON

Minister of Information	MR. GEOFFREY LLOYD
Postmaster-General	CAPTAIN H. F. C. CROOKSHANK
Minister of State	MR. WILLIAM MABANE
Minister Resident in the Middle East	SIR EDWARD GRIGG
Minister Resident in West Africa	CAPTAIN H. H. BALFOUR
Chancellor of the Duchy of Lancaster	SIR ARTHUR SALTER
Paymaster-General	LORD CHERWELL
Minister of Pensions	SIR WALTER WOMERSLEY
Attorney-General	SIR DAVID MAXWELL FYFE, K.C.
Advocate-General	MR. J. S. C. REID, K.C.
Solicitor-General	SIR WALTER MONCKTON, K.C.
Solicitor-General for Scotland	SIR DAVID KING MURRAY, K.C.

Everything had been arranged to the satisfaction of the other parties about the dates and timings of the various election stages. The King had allowed it to be known that he would consent to a dissolution after three weeks' interval from my receiving his new commission. Accordingly on June 15 Parliament was dissolved. Ten days were to elapse before the nominations of candidates, and ten more before polling day, July 5. Every arrangement had been made on strictly equal terms about bringing candidates home from the front, about uniforms and petrol rations, and not the slightest reproach was ever levelled at those who held the executive power. Owing to the fact that the soldiers' votes must come home to be counted, a further twenty-one days had to elapse between the polling in the United Kingdom and the counting of votes and declaration of results. This final act was fixed for July 26. In several Continental countries, when it was known that the ballot-boxes would be in charge of the British Government for three weeks, astonishment was expressed that there could be any doubt about the result. However, in our country these matters are treated exactly as if they were a cricket match, or other sporting event. Long may it so continue.

CHAPTER XXXVI

A FATEFUL DECISION

Stalin Agrees to a Triple Conference in Berlin in Mid-July – I Try to Arrange an Earlier Meeting – I Invite Mr. Attlee to Go to Potsdam – He Accepts – The Impending Withdrawal of the American Army to Its Zone of Occupation – The Need for a Prior Settlement in Europe – The Position To-day – My Telegrams of June 4 and 9 to President Truman – His Proposed Message to Stalin – My Answer to the President, June 14 – My Telegram to Stalin of June 15 – His Reply – The Armies of the Western Allies Withdraw – The Burden of the General Election – The Soviet Armies Occupy Their Allotted Zones – End of the San Francisco Conference – My Views on the Composition of a World Instrument – Correspondence with Lord Halifax – A Holiday at Hendaye.

ON June 1 President Truman told me that Marshal Stalin was agreeable to a meeting of what he called "the Three" in Berlin about July 15. I replied at once that I would gladly go to Berlin with a British delegation, but I thought that July 15, which Truman had suggested, was much too late for the urgent questions demanding attention between us, and that we should do an injury to world hopes and unity if we allowed personal or national requirements to stand in the way of an earlier meeting. "Although," I cabled, "I am in the midst of a hotly contested election I would not consider my tasks here as comparable to a meeting between the three of us. If June 15 is not possible why not July 1, 2, or 3?" Mr. Truman replied that after full consideration July 15 was the earliest for him, and that arrangements were being made accordingly. Stalin did not wish to hasten the date.

I could not press the matter further.

Prime Minister to President Truman   9 June 45

While I have agreed in principle to our triple meeting in Berlin on July 15, I hope you will agree with me that the British, American, and Russian delegations shall have entirely separate quarters assigned to them and have their own guards, and that there shall be a fourth place prepared in which we meet to confer. I could not accept, as at Yalta, the principle that we go to Berlin, over which it is agreed we are to have triple, or, with the French, quadruple parity, merely as guests of the Soviet Government and armies. We should provide everything for ourselves and be able to meet on equal terms. I should like to know how you stand about this.

Stalin agreed that the delegations should be housed as I proposed. Each would have its own closed territory, under a régime regulated at the discretion of its head. The palace of the German Crown Prince in Potsdam would be used for the joint sessions. There was a good airfield hard by.

*   *   *   *   *

I have already mentioned how strongly I felt that every head of Government in periods of crisis should have a deputy who knows everything and can thus preserve continuity should accidents occur. During the war-time Parliament, with its large Conservative majority, I had always looked upon Mr. Eden as my successor, and had, when invited, so advised the King. But now a new Parliament had been elected and the results were as yet unknown. I therefore felt it right to invite the Leader of the Opposition, Mr. Attlee, to attend the Conference at Potsdam, so that there should be no break in his knowledge of affairs. On June 15 I wrote to him:

15 June 45

I now send you a formal invitation to come with us to the forthcoming Tripartite Conference in the near future.

Since I announced this intention to Parliament I observe that a statement was made last night by Professor Harold Laski, the chairman of the Labour Party, in which he said, "It is of course essential that if Mr. Attlee attends this gathering he shall do so in the rôle of an observer only."

His Majesty's Government must of course bear the responsibility for all decisions, but my idea was that you should come as a friend and counsellor, and help us on all the subjects on which we have been so long agreed, and have been known to be agreed by public declaration. In practice I thought the British delegation would work just as they did at San Francisco, except that, as I have already stated, you would not have official responsibility to the Crown otherwise than as a Privy Counsellor.

Merely to come as a mute observer would, I think, be derogatory to your position as the Leader of your party, and I should not have a right to throw this burden upon you in such circumstances.

I hope however I may have your assurance that you accept my invitation.

Mr. Attlee accepted the invitation in a letter in which he said he had consulted his principal colleagues in the House of Commons, and that they agreed that my offer should be accepted on the basis set out in his letter. Mr. Attlee added that there was never any suggestion that he should go as a mere observer.

The Conference was called "Terminal".

* * * * *

The main reason why I had been anxious to hasten the date of the meeting was of course the impending retirement of the American Army from the line which it had gained in the fighting to the zone prescribed in the occupation agreement. The story of the agreement about the zones and the arguments for and against changing them are recorded in an earlier chapter.* I feared that any day a decision might be taken in Washington to yield up this enormous area—400 miles long and 120 at its greatest depth. It contained many millions of Germans and Czechs. Its abandonment would place a broader gulf of territory between us and Poland, and practically end our power to influence her fate. The changed demeanour of Russia towards us, the constant breaches of the understandings reached at Yalta, the dart for Denmark, happily frustrated by Montgomery's timely action, the encroachments in Austria, Marshal Tito's menacing pressure at Trieste, all seemed to me and my advisers to create an entirely different situation from that in which the zones of occupation had been prescribed two years earlier. Surely all these issues should be considered as a whole, and *now* was the time. Now, while the British and American Armies and Air Forces were still a mighty armed power, and before they melted away under demobilisation and the heavy claims of the Japanese war—now, at the very latest, was the time for a general settlement.

* See Chapter XXX.

A month earlier would have been better. But it was not yet too late. On the other hand, to give up the whole centre and heart of Germany—nay, the centre and key-stone of Europe—as an isolated act seemed to me to be a grave and improvident decision. If it were done at all it could only be as part of a general and lasting settlement. We should go to Potsdam with nothing to bargain with, and all the prospects of the future peace of Europe might well go by default. The matter however did not rest with me. Our own retirement to the occupation frontier was inconsiderable. The American Army was three millions to our one. All I could do was to plead, first, for advancing the date of the meeting of "the Three", and, secondly, when that failed, to postpone the withdrawal until we could confront all our problems as a whole, together, face to face, and on equal terms.

How stands the scene after eight years have passed? The Russian occupation line in Europe runs from Lübeck to Linz. Czechoslovakia has been engulfed. The Baltic States, Poland, Roumania, and Bulgaria have been reduced to satellite States under totalitarian Communist rule. Austria is denied all settlement. Yugoslavia has broken loose. Greece alone is saved. Our armies are gone, and it will be a long time before even sixty divisions can be once again assembled opposite Russian forces, which in armour and man-power are in overwhelming strength. This also takes no account of all that has happened in the Far East. Only the atomic bomb stretches its sinister shield before us. The danger of a third World War, under conditions at the outset of grave disadvantage except in this new, terrible weapon, casts its lurid shadow over the free nations of the world. Thus in the moment of victory was our best, and what might prove to have been our last, chance of durable world peace allowed composedly to fade away. On June 4 I cabled to the President these words, which few would now dispute:

Prime Minister to President Truman   4 June 45

I am sure you understand the reason why I am anxious for an earlier date, say the 3rd or 4th [of July]. I view with profound misgivings the retreat of the American Army to our line of occupation in the central sector, thus bringing Soviet power into the heart of Western Europe and the descent of an iron curtain between us and everything to the eastward. I hoped that this retreat, if it has to be made, would be accompanied by the settlement of many great things which would be the true foundation of world peace. Nothing really important has been settled yet, and you and I will have to bear great responsibility for the future. I still hope therefore that the date will be advanced.

I reinforced this argument by referring to the high-handed behaviour of the Russians in Vienna.

Prime Minister to President Truman   9 June 45
Our missions to Vienna have been ordered by Marshal Tolbukhin to leave by 10th or 11th June. They have not been allowed to see anything outside the strict city limits, and only one airfield can be permitted for the Allies. Here is the capital of Austria, which by agreement is to be divided, like the country itself, into four zones; but no one has any power there except the Russians, and not even ordinary diplomatic rights are allowed. If we give way in this matter we must regard Austria as in the Sovietised half of Europe.

2. On the other hand, the Russians demand the withdrawal of the American and British forces in Germany to the occupation line, fixed so long ago in circumstances so different, and Berlin of course is so far completely Sovietised.

3. Would it not be better to refuse to withdraw on the main European front until a settlement has been reached about Austria? Surely at the very least the whole agreement about zones should be carried out at the same time?

4. A telegram has been despatched to the State Department showing the actual situation of our missions in Vienna, which, as ordered, will, I presume, depart on June 10 or 11, after making their protests.

Two days later I minuted to the Foreign Office, which in Mr. Eden's absence in Washington I at this time directed:

11 June 45
I am still hoping that the retreat of the American centre to the occupation line can be staved off till "the Three" meet, and I take the view that large movements to enable France to assume her agreed part of her zone will stimulate the Russian demand to occupy the heart of Germany. Of course at any moment the Americans may give way to the Russian demand, and we shall have to conform. That will be the moment for making this partial arrangement with the French [about their zone], but not before, even if delay causes some inconvenience in the re-deployments. We ought not to let ourselves be hurried into a decision which touches issues so vast and fateful. There is no objection to the matter being considered by the British Chiefs of Staff meanwhile.

*   *   *   *   *

On June 12 the President replied to my message of June 4., He said that the tripartite agreement about the occupation of Germany, approved by President Roosevelt after "long consideration and detailed discussion" with me, made it impossible to delay the withdrawal of American troops from the Soviet Zone in order to press the settlement of other problems. The Allied Control Council could not begin to function until they left, and the military government exercised by the Allied Supreme Commander should be terminated without delay and divided between Eisenhower and Montgomery. He had been advised, he said, that it would harm our relations with the Soviet to postpone action until our meeting in July, and he accordingly proposed sending a message to Stalin.

This document suggested that we should at once instruct our armies to occupy their respective zones. As for Germany, he was ready to order all American troops to begin withdrawing on June 21. The military commanders should arrange for the simultaneous occupation of Berlin and for free access thereto by road, rail, and air from Frankfurt and Bremen for the United States forces. In Austria arrangements could be completed more quickly and satisfactorily by making the local commanders responsible for defining the zones both there and in Vienna, only referring to their Governments such matters as they were unable to resolve themselves.

This struck a knell in my breast. But I had no choice but to submit.

Prime Minister to President Truman 14 June 45

Obviously we are obliged to conform to your decision, and the necessary instructions will be issued.

2. It is not correct to state that the tripartite agreement about zones of occupation in Germany was the subject of "long consideration and detailed discussion" between me and President Roosevelt. References made to them at Quebec were brief, and concerned only the Anglo-American arrangements which the President did not wish to raise by correspondence beforehand. These were remitted to the Combined Chiefs of Staff, and were certainly acceptable to them.

3. As to Austria, I do not think we can make the commanders on the spot responsible for settling the outstanding questions. Marshal Stalin made it quite plain in his message of May 18 that agreement on the occupation and control of Austria must be settled by the European

Advisory Commission. I do not believe that he would agree to change, and in any case our missions may already have left Vienna. I suggest for your consideration the following redraft of the penultimate paragraph of your message to Marshal Stalin:

"I consider the settlement of the Austrian problem is of equal urgency to the German matter. The redistribution of forces into occupation zones which have been agreed in principle by the European Advisory Commission, the movement of the national garrisons into Vienna, and the establishment of the Allied Commission for Austria should take place simultaneously with these developments in Germany. I therefore attach the utmost importance to settling the outstanding Austrian problems, in order that the whole arrangement of German and Austrian affairs may be put into operation simultaneously. I hope that the recent visit of American, British, and French missions to Vienna will result in the European Advisory Commission being able to take the necessary remaining decisions to this end without delay."

4. I for my part attach particular importance to the Russians evacuating the part of the British Zone in Austria that they are now occupying at the same time as the British and American forces evacuate the Russian Zone in Germany.

5. I sincerely hope that your action will in the long run make for a lasting peace in Europe.

The President accepted textually my suggested paragraph about Austria and sent his telegram to Stalin on June 14. There was nothing more that I could do. I replied, "I am grateful to you for meeting our views about Austria. As I have already told you, we are conforming to your wishes, and I have so informed Stalin."

To Stalin I wrote next day:

I have seen a copy of President Truman's message to you of June 14 regarding the withdrawal of all American troops into their own occupation zone, beginning on June 21, in accordance with arrangements to be made between the respective commanders.

I also am ready to issue instructions to Field-Marshal Montgomery to make the necessary arrangements in conjunction with his colleagues for the similar withdrawal of British troops into their zone in Germany, for the simultaneous movement of Allied garrisons into Greater Berlin, and for the provision of free movement for British forces by air, rail, and road to and from the British Zone to Berlin.

I entirely endorse what President Truman says about Austria. In particular I trust that you will issue instructions that Russian troops should begin to withdraw from that part of Austria which the Euro-

THE WITHDRAWAL OF THE WESTERN ALLIES, JULY 1945

pean Advisory Commission has agreed in principle should form part of the British Zone on the same date as movements begin in Germany.

It must not be overlooked that President Truman had not been concerned or consulted in the original fixing of the zones. The case as presented to him so soon after his accession to power was whether or not to depart from and in a sense repudiate the policy of the American and British Governments agreed under his illustrious predecessor. He was, I have no doubt, supported in his action by his advisers, military and civil. His responsibility at this point was limited to deciding whether circumstances had changed so fundamentally that an entirely different procedure should be adopted, with the likelihood of having to face accusations of breach of faith. Those who are only wise after the event should hold their peace.

Stalin's reply to me completed matters.

Marshal Stalin to Prime Minister   17 June 45

I have received your message concerning the withdrawal of the Allied troops to their respective zones in Germany and Austria.

I regret to have to tell you that there are difficulties in the way of beginning the withdrawal of the British and American troops to their zones and of introducing British and American troops into Berlin on June 21, in that from June 19 onwards Marshal Zhukov and all our other commanders in the field have been invited to Moscow for the session of the Supreme Soviet, and also to organise and participate in a parade on June 24. They will be able to return to Berlin by June 28-30. It must also be borne in mind that work on clearing Berlin of land mines is still not complete, and that it may only be completed towards the end of June.

With regard to Austria, I have to repeat what I have already told you about the summons of Soviet commanders to Moscow and the date of their return to Vienna. It is also necessary that in the very near future the European Advisory Commission should complete its work on establishing the zones of occupation in Austria and Vienna, which is still outstanding.

In view of the facts presented above, I would propose to postpone the withdrawal and replacement by the appropriate forces in the respective zones until July 1, both in Germany and in Austria.

Further, as regards Germany and Austria, it would be desirable to establish now the zones of occupation for the French troops.

We will take all necessary measures in Germany and Austria in accordance with the above plan.

I have written to President Truman too about the whole matter.

On July 1 the United States and British Armies began their withdrawal to their allotted zones, followed by masses of refugees. Soviet Russia was established in the heart of Europe. This was a fateful milestone for mankind.

★　★　★　★　★

While all this was passing I was plunged into the turmoil of the General Election, which began in earnest in the first week of June. This month was therefore hard to live through. Strenuous motor tours to the greatest cities of England and Scotland, with three or four speeches a day to enormous and, it seemed, enthusiastic crowds, and, above all, four laboriously prepared broadcasts, consumed my time and strength. All the while I felt that much we had fought for in our long struggle in Europe was slipping away and that the hopes of an early and lasting peace were receding. The days were passed amid the clamour of multitudes, and when at night, tired out, I got back to my headquarters train, where a considerable staff and all the incoming telegrams awaited me, I had to toil for many hours. The incongruity of party excitement and clatter with the sombre background which filled my mind was in itself an affront to reality and proportion. I was glad indeed when polling day at last arrived and the ballot papers were safely sealed for three weeks in their boxes.

★　★　★　★　★

While the Soviet armies flooded into their allotted zones without incident the San Francisco Conference, which had sought to frame the World Instrument for Peace on which our hearts were set, had reached the conclusion of its labours. Mr. Eden and Mr. Attlee had had to leave on account of the election, but on June 26 I sent my congratulations to Lord Halifax, Lord Cranborne, and all the members of our delegation on the success of their labours and on the quality of the results which, under extraordinary difficulties, had been achieved. "By wisdom in counsel and sincerity of conviction the United Kingdom delegates did much to secure the unity of views without which a World Organisation can have no reality. You have made an invaluable contribution to the re-establishment of a hopeful basis for the future." Unhappily these hopes have so far been very imperfectly fulfilled.

I have always held the view that the foundation of a World Instrument should be sought on a regional basis. Most of the principal regions suggest themselves—the United States, United Europe, the British Commonwealth and Empire, the Soviet Union, South America. Others are more difficult at present to define—like the Asian group or groups, or the African group—but could be developed with study. But the object would be to have many issues of fierce local controversy thrashed out in the Regional Council, which would then send three or four representatives to the Supreme Body, choosing men of the greatest eminence. This would make a Supreme Group of thirty or forty world statesmen, each responsible not only for representing their own region, but for dealing with world causes, and primarily the prevention of war. What we have now is not effective for that outstanding purpose. The summoning of all nations, great and small, powerful or powerless, on even terms to the central body may be compared with the organisation of an army without any division between the High Command and the divisional and brigade commanders. All are invited to the headquarters. Babel, tempered by skilful lobbying, is all that has resulted up to the present. But we must persevere.

* * * * *

A few days later I sent Halifax a cable on details about which the President and his colleagues might be glad to be informed.

Prime Minister to Lord Halifax (Washington) 6 July 45
Naturally I am looking forward very much to meeting the President. The political members of the British delegation will quit the [Potsdam] Conference on July 25 in order to await the poll in England. This will avoid embarrassment when the results are made known. I am led to believe that the present Government will obtain a majority, but, as the President knows, electioneering is full of surprises. It is most unlikely in any event that I should resign on an adverse declaration of the poll, unless it amounted to a very extreme expression of national displeasure. I should await the result of a confidence vote in the House of Commons on the King's Speech, and take my dismissal from the House. This would enable the various parties and individuals to define their position by a vote.

2. The British delegation could therefore return to Berlin on the 27th, and I should personally be able to stay there if necessary till about the 5th or 6th August. Parliament meets on the 1st to elect a Speaker and to swear in Members. But it is not till Wednesday, 8th, that the King opens Parliament, and the decisive division would not take place

before Friday, 10th. I thought all these details, some of which are extremely private, would be of interest to the President.

3. I am delighted to hear that the President contemplates two or even three weeks, as I think it of the utmost importance that whatever happens in England the Conference should not be hurried. It was somewhat abruptly curtailed in the Crimea. We have here to try to reach settlements on a great number of questions of the greatest consequence, and to prepare the way for a Peace Conference, which presumably will be held later in the year or in the early spring.

He replied next day with the following telegram, which shows how well he understood the Washington view.

Lord Halifax (Washington) to Prime Minister   7 July 45

The President had already left for Potsdam when your telegram reached me. Your message will be relayed to him on board ship.

I am sure you will find Truman most anxious to work with us, and fully alive to the long-range implications as well as short-term difficulties of the decisions we have to make. I judge that American tactics with the Russians will be to display at the outset confidence in Russian willingness to co-operate. I should also expect the Americans in dealing with us to be more responsive to arguments based upon the danger of economic chaos in European countries than to *the balder pleas about the risks of extreme Left Governments or of the spread of Communism.** They showed some signs of nervousness in my portrayal of Europe (whatever the facts) as the scene of a clash of ideas in which the Soviet and Western influences are likely to be hostile and conflicting. At the back of their minds there are still lingering suspicions that we want to back Right Wing Governments or monarchies for their own sake. This does not in the least mean that they will be unwilling to stand up with us against the Russians when necessary. But they are likely to pick their occasions with care, and are half expecting to play, or at any rate to represent themselves as playing, a moderating rôle between ourselves and the Russians.

A few years later it was Britain and Western Europe who were urged in many quarters to play the "moderating rôle" between the U.S.A. and the U.S.S.R. Such are the antics of fortune.

*   *   *   *   *

I was resolved to have a week of sunshine to myself between the General Election and the Conference. On July 7, two days after polling day, I flew to Bordeaux with Mrs. Churchill and Mary,

* Author's italics.

and found myself agreeably installed at General Brutinel's villa
near the Spanish frontier at Hendaye, with lovely bathing and
beautiful surroundings. I spent most of the mornings in bed
reading a very good account, by an excellent French writer, of the
Bordeaux armistice and its tragic sequel at Oran. It was strange
to revive my own memories of five years before and to learn of
many things which I had not known at the time. In the after-
noons I even sallied forth with my elaborate painting outfit, and
found attractive subjects on the river Nive and the Bay of St.
Jean de Luz. I found a gifted companion of the brush in Mrs.
Nairn, the wife of the British Consul at Bordeaux, with whom I
had made friends at Marrakesh a year before. I dealt only with a
few telegrams about the impending Conference, and strove to
put party politics out of my head. And yet I must confess the
mystery of the ballot-boxes and their contents had an ugly trick
of knocking on the door and peering in at the windows. When
the palette was spread and I had a paint-brush in my hand it was
easy to drive these intruders away.

The Basque people were everywhere warm in their welcome.
They had endured a long spell of German occupation and were
joyful to breathe freely again. I did not need to prepare myself
for the Conference, for I carried so much of it in my head, and
was happy to cast it off, if only for these few fleeting days. The
President was at sea in the United States cruiser *Augusta*, the same
ship which had carried Roosevelt to our Atlantic meeting in 1941.
On the 15th I motored through the forests to the Bordeaux airfield, and my Skymaster took me to Berlin.

CHAPTER XXXVII

THE DEFEAT OF JAPAN

Plans and Perplexities in South-East Asia – The Fourteenth Army Crosses the Irrawaddy – General Slim Wins the Battle for Meiktila – Chiang Kai-shek Recalls the Chinese Divisions – My Message to General Marshall About the Withdrawal of the American Transport Squadrons – His Disquieting Reply – The Fall of Mandalay, March 20 – The Race for Rangoon – The Amphibious Assault, May 2 – The End of a Long Struggle – My Telegram to Admiral Mountbatten, May 9 – Climax in the Pacific – A British Fleet Arrives in Australia – The American Attack on Iwo Jima – General Lumsden Killed – The Fall of Okinawa, June 22 – My Congratulations to the President – The Liberation of the East Indies – British Earnestness to Aid the Final Assault on Japan – A Merciful Reprieve.

WINTER operations in Burma have already been described,* and hard strategic decisions confronted Admiral Mountbatten when the decisive battle across the Irrawaddy began in February 1945. His instructions were to liberate Burma, for which purpose he was not to expect greater resources than he already had, and then to occupy Malaya and open the Malacca Straits. Weather was dominant. The first task was to occupy the central plain of Burma and capture Rangoon before the monsoon, and the monsoon was due in early May. He could either concentrate the whole Fourteenth Army on a decisive battle in the Mandalay plain and make a swift advance to the south, or use some of his troops for an amphibious operation against Rangoon and a northward stroke against the Japanese rear. An amphibious attack on Rangoon would mean deferring the capture of Puket Island, off the coast of the Kra Isthmus, which was a valuable stepping-stone towards Malaya. Important but uncertain

* See Chapter XI, "Advance in Burma."

factors complicated his choice. Success depended greatly on air supply, in which United States aircraft played a big part, and we also hoped that General Sultan's Chinese-American forces, who had been fighting two Japanese divisions north of Lashio, would remain with us in the struggle. But aid to China was still the overriding American policy, and this help might be withdrawn and the Admiral's plans ruined.

In face of these dangers, which were soon to become acute, Mountbatten decided on the single, fully supported operation by the Fourteenth Army, commanded by General Slim, against the main enemy body west of Mandalay, and a subsequent advance on Rangoon, which, he was advised, could be reached by April 15. At the same time he ordered the XVth Corps in the Arakan to enlarge the air bases at Akyab and on Ramree Island, and extend its hold along the coast, and over the only two passes leading towards the lower Irrawaddy. Despite a large reduction in its supplying aircraft, the corps completed its task, and stopped a Japanese division from joining the decisive battle farther east.

$$\star \quad \star \quad \star \quad \star \quad \star$$

Events here moved swiftly. The 19th Division had already seized bridgeheads across the Irrawaddy about forty miles north of Mandalay, and throughout February they beat off a series of fierce counter-attacks. On February 12 the 20th Division crossed the river lower down and to the west of Mandalay. For a fortnight they had a hard fight to hold their gains, but by then they were joined by the 2nd British Division. This convinced the Japanese High Command that a decisive battle was imminent, and they sent heavy reinforcements. They did not believe that a serious flank attack was also possible, and even dispatched to Siam a division they could ill spare. This however was precisely the stroke which General Slim had prepared. On February 13 the 7th Division crossed the Irrawaddy south of Pakokku and formed a bridgehead. The enemy thought this was a mere diversion, but he was soon to be better informed. On the 21st two motorised brigades of the 17th Division and a brigade of tanks, having crossed the river, broke out from the bridgehead, and by the 28th they reached Meiktila. Here was the principal administrative centre of the Japanese main front, a nodal point of their communications and the focus of several airfields. It was strongly defended, and the enemy sent two divisions post-haste to aid the garrison, but they were held at a distance until reinforcements came to our 17th Division, its airborne brigade, and the 5th Division. After a week of bitter fighting the town was in our hands, and all attempts to recapture it were repulsed. The Japanese admit losing five thousand men and as many wounded in a battle which their Commander-in-Chief has since described as "the master-stroke of Allied strategy".

* * * * *

Far off to the north-east General Sultan was also on the move. His American "Mars" Brigade, three Chinese divisions, and the 36th British Division had opened the road to China at the end of January and advanced southwards. By mid-March they reached the road from Lashio to Mandalay. But Chiang Kai-shek now imposed a halt. He would not allow his Chinese divisions to continue. He had already demanded the American brigade, all the Chinese divisions, and the U.S. air squadrons which sustained them, so that he could build up an army in China and free the rice-producing areas from the Japanese. He suggested that General Slim should halt his advance when Mandalay was taken. This was precisely what Mountbatten had feared when he made his plans a month before. But Chiang Kai-shek insisted on removing his ground forces, and the Japanese were able to take two of their three divisions from this front and march them against our Fourteenth Army.

There remained the American aircraft. On March 30 I urged General Marshall to let them stay.

Prime Minister to Field-Marshal Wilson (*Washington*)   30 Mar 45
Please convey orally and unofficially to General Marshall the following views which I hold:

As General Marshall will remember from our talks at Quebec, we greatly disliked the prospect of a large-scale campaign in the jungles of Burma, and I have always had other ideas myself. But the United States Chiefs of Staff attached the greatest importance to this campaign against the Japanese, and especially to the opening of the Burma Road. We therefore threw ourselves into the campaign with the utmost vigour. Although the prolongation of the German war has withheld from Mountbatten the three British-Indian divisions on which all his hopes were built, he has succeeded far beyond our hopes. . . . The very considerable battle upon such difficult communications which is now being fought with the main Japanese army in Burma is important not only for Burma and as a preliminary to the capture of Rangoon, but plays its part in the general wearing down of the military and particularly of the air-power of Japan. Moreover, once Rangoon is taken the powerful forces which we have on foot there will be set free for further operations in 1946, and even earlier, in combination with the general American onslaught. . . . I feel therefore entitled to appeal to General Marshall's sense of what is fair and right between us, in which I have the highest confidence, that he do all in his power to let Mountbatten have the comparatively small additional support which his air force now requires to enable the decisive battle now raging in Burma to be won. . . .

General Marshall assured us that no planes would be taken away until Rangoon was captured or June 1 was reached, whichever happened first. This sharpened the urgency. If we could not seize Rangoon by June 1 and the transport planes were then reduced the Fourteenth Army would have to make a long retreat until they could be supplied mostly by land. The whole campaign would have failed.

*　　*　　*　　*　　*

The conjoint battles of Mandalay and Meiktila raged through March. The 19th Division broke out of its bridgeheads, fought down the east bank of the Irrawaddy, and entered Mandalay on March 9. The Japanese resisted strongly. Mandalay Hill, 780 feet above the surrounding country, was taken in two days, but the massive walls of Fort Dufferin were impenetrable to ordinary missiles. Finally a breach was made with 2,000-pound bombs, and on March 20 the enemy fled.

The rest of the XXXIIIrd Corps meanwhile fought on to Meiktila. They met great opposition, as the Japanese Commander-in-Chief, in spite of the intervention of the 17th Division behind his front, showed no signs as yet of withdrawing, and the armies were well matched. General Slim had six divisions and two armoured brigades, against over eight, under-strength, Japanese divisions and a division of the "Indian National Army".* But at the end of the month the enemy gave up the struggle and began to fall back down the main road to Toungoo and Rangoon, and through the mountains to the east. After many weeks of

* Indian prisoners of war who had been equipped by and were fighting for the Japanese.

battle our supplies were badly strained, but the enemy's must have been much worse. Long-sustained attacks by the Strategical Air Force on communications and rear installations and constant harrying of their retiring troops by the tactical air forces produced a crisis in their affairs.

★ ★ ★ ★ ★

There was also a crisis in our own affairs. The battles had lasted much longer than we expected. General Sultan was now halted on the Lashio road, and the advent of two Japanese divisions which had opposed him aggravated the delay. There was now no prospect of the Fourteenth Army reaching Rangoon by mid-April, and it was very doubtful if they could get there before the monsoon. Mountbatten accordingly decided to make an amphibious assault on the town after all. This would have to be much smaller than had been anticipated, and the attack on Puket Island would have to be cancelled. Even so the attack could not be launched before the first week in May, and by then it might be too late.

General Slim meanwhile resolved that the IVth Corps should pursue the enemy down the road and railway, while the XXXIIIrd Corps worked their way down the Irrawaddy. He was determined not only to reach Rangoon, but to draw a double net down Southern Burma and trap the enemy within it. The 7th and 20th Divisions accordingly drove down the river with overlapping thrusts and reached Prome on May 2. After a stiff fight with the remnants of three Japanese divisions at Pyawbe the advance went even more swiftly along the road and the railway. The troops were on half-rations to enable them to carry more petrol. An armoured column, and the mechanised brigades of the 5th and 17th Divisions, leap-frogging over each other, reached Toungoo on April 22. The airfields here were badly needed to further the air supply on which all depended, and also to bring the fighters of 221 Group R.A.F. within range of Rangoon. The next bound was to Pegu, whose capture would close the enemy's southernmost escape route from Lower Burma. Our advance troops reached it on April 29. That afternoon torrential rain fell, heralding an early monsoon. Forward air-strips were out of action; tanks and vehicles could not move off the roads. The Japanese mustered every possible man to hold the town and the bridges over the river. On May 2 the 17th Division finally broke

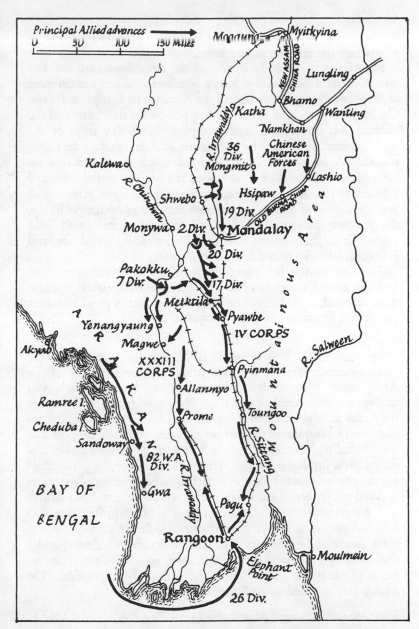

Principal Allied advances →

0   50   100   150 Miles

Moaung   Myitkyina

NEW ASSAM-CHINA ROAD

Lungling

Katha   Bhamo   Wanting

Namkham

36 Div.   Chinese American Forces

R. Irrawaddy

Kalewa   Mongmit   Lashio

Hsipaw

R. Chindwin   Shwebo   OLD BURMA-CHINA ROAD

19 Div.

Monywa 2 Div.   Mandalay

20 Div.

Pakokku   17 Div.

7 Div.

Meiktila

Pyawbe

Yenangyaung   IV CORPS

Magwe

Akyab   XXXIII CORPS   Pyinmana

Ramree I.   Allanmyo   R. Salween

Cheduba I.   Prome   Toungoo

Sandoway   R. Sittang

82 W.A. Div.

BAY OF   Gwa   Pegu

BENGAL   Rangoon   Moulmein

Elephant Point

26 Div.

CENTRAL   BURMA

through, and, hoping to be first in Rangoon, prepared themselves to advance the few remaining miles.

But May 2 was also the D Day of the amphibious assault. For two days beforehand Allied heavy bombers attacked the defences at Elephant Point, which barred the entrance to Rangoon River. On May 1 a parachute battalion dropped on the defenders and the channel was opened for minesweeping. Next day ships of the 26th Division, supported by 224 Group R.A.F., reached the river-mouth. A Mosquito aircraft flew over Rangoon and saw no signs of the enemy. The crew landed at a near-by airfield, walked into the city, and were greeted by a number of our prisoners of war. In the belief that an amphibious attack was no longer likely, the Japanese garrison had departed some days before to hold Pegu. That afternoon the monsoon broke in all its violence, and Rangoon fell with only a few hours to spare.

This amphibious force soon joined the 17th Division at Pegu and the 20th Division at Prome. Many thousands of Japanese were trapped, and during the next three months great numbers perished in attempts to escape eastwards.

*   *   *   *   *

Thus ended the long struggle in Burma, but a tribute is due to those other Services without whose aid the struggles of the Army would have availed little. The Royal Navy had achieved undisputed command of the sea. They could, and did, convey the Army in safety wherever it was needed. The Allied Air Forces had utterly vanquished the Japanese planes, and their support had been unfailing. Airborne supply had been developed and maintained on a prodigious scale. Under General Snelling, the chief administrative officer of the Fourteenth Army, the supply services worked admirably. Last, but not least, the Engineers, both British and American, wrought many wonders of improvisation and achievement, such as laying nearly 3,000 miles of pipe-line across river, forest, and mountain. The famous Fourteenth Army, under the masterly command of General Slim, fought valiantly, overcame all obstacles, and achieved the seemingly impossible. On May 9 I telegraphed to the Supreme Commander:

Prime Minister to Admiral Mountbatten (South-East Asia)   9 May 45
I send you my most heartfelt congratulations upon the culminating victory at Rangoon of your Burma campaigns. The hard fighting at

Imphal and Kohima in 1944 prepared the way for the brilliant operations, conducted over a vast range of territory, which have crowned the exertions of the South-East Asia Command in 1945. When these matters were considered at Quebec last September it was thought both by your High Command and by the Combined Chiefs of Staff, reporting to the President and me, that about six British and British-Indian divisions, together with much shipping and landing vessels, all of which, and more, were asked for by you, would be required for enterprises less far-reaching than those you and your gallant forces and Allies have in fact accomplished. The prolongation of the German war made it impossible to send the British and British-Indian divisions which you needed, and a good many other units on which you were counting had to be retained in the decisive European theatre. In spite of this diminution and disappointment you and your men have done all and more than your directive required. Pray convey to everyone under your command or associated with you the sense of admiration and gratitude felt by all at home at the splendid close of the Burma campaign.

In honour of these great deeds of South-East Asia Command His Majesty the King has commanded that a special decoration, the "Burma Star", should be struck, and the ribbons will be flown out to you at the earliest moment.

* * * * *

The struggle in the Pacific moved no less swiftly to its climax. We had promised at Quebec to send British forces of all arms to the Far East as soon as Germany was defeated, and on my return to London I stated in the House of Commons that the United States had accepted our offer of a fleet. Our effort on land and in the air would be limited only by the available shipping, and plans went ahead accordingly.

In December 1944 Admiral Fraser arrived in Sydney with his flag in the battleship *Howe*. For the first time our main Fleet was to deploy in the Pacific, and under the operational control of an American officer. Our chief difficulties were supply and maintenance. In three years of fighting the Americans had built up an immense supply organisation and a network of island bases. We could not hope to match this achievement, but it was essential that our Fleet should not wholly depend on our Allies for logistic support.

We had been studying the problem throughout 1944. A mission had been sent in June to consult the Australian Government about establishing a base, but Australia's man-power was already fully engaged in General MacArthur's campaigns and in supplying both themselves and the Americans, and it was evident that much material and skilled labour would have to be found from the United Kingdom. The fine port of Sydney was four thousand miles from the fighting. To serve the Fleet we needed a train of fuel and store ships, depot and repair ships, hospital ships, and many other types, and huge supplies would have to be transported from the British Isles. This naturally raised misgivings in the mind of Lord Leathers, the Minister of War Transport, but plans were made, the first essentials were provided, and expansion was still going on when the war ended.

*   *   *   *   *

Soon after his arrival Admiral Fraser went by air to visit both General MacArthur and Admiral Nimitz. He, and later the Fleet, were most cordially received, and from the outset there grew up a spirit of comradeship which overcame all difficulties and led to the most intimate co-operation at all levels. A message from Admiral Nimitz ran:

The British force will greatly increase our striking power and demonstrate our unity of purpose against Japan. The United States Pacific Fleet welcomes you.

Admiral Fraser's seniority however made it difficult for him to command afloat, where he would have outranked Nimitz's immediate subordinates. Vice-Admiral Rawlings, who had a distinguished fighting record in the Mediterranean, had accordingly been selected as second-in-command, and to command at sea. Early in February 1945 he reached Australia with the main Fleet, many of whose ships had for some time been operating in the Indian Ocean. By the beginning of March the Fleet and the elements of the fleet train were assembled at the American base at Manus Island in the Admiralty Group, and on the 18th it sailed for its first campaign in the Pacific under Admiral Spruance.

Here much was happening. The time at last had come to strike at the enemy's homeland. On February 19 Spruance had attacked Iwo Jima, in the Bonin Islands, whence American fighters would be able to escort bombers from the Marianas in attacking Honshu. The struggle was severe and lasted over a month, but victory was won. Meanwhile the British Fleet, now known as Task Force 57, comprising the battleships *King George V* and *Howe*, four fleet carriers mustering nearly 250 aircraft, five cruisers, and eleven destroyers, reached its battle area east of Formosa on March 26. That day its bombers made their first strike at airfields and installations among the islands south of Okinawa. Spruance himself was engaged in full-scale air operations as a prelude to the amphibious attack on Okinawa itself, due to be launched on April 1. On March 18 his fast carrier groups attacked enemy bases near the coast of Japan, and from March 23 they switched to Okinawa. The task of the British Fleet was to stop the enemy using the airfields in the islands to the south and in Northern Formosa.

From March 26 until April 20 the Fleet, refuelling at sea, continued its mission. Wastage in aircraft and exhaustion of supplies then forced a brief withdrawal to Leyte. The opposition had not been heavy. The *Indefatigable* had been struck by a suicide bomber on April 1, causing casualties, and one destroyer was damaged and had to be withdrawn.

*   *   *   *   *

In January, as I have already recounted, we sustained a heavy loss in the death of Lieutenant-General Lumsden.* He was my personal liaison officer with General MacArthur, whose confidence he had completely gained. Lumsden's war record was magnificent. In the very first contacts in Belgium, when commanding the 12th Lancers, he had brought the armoured car into its own again and had played a distinguished part in the fighting which ended at Dunkirk. Later he had commanded the 1st Armoured Division in many months of desert warfare. It was for this record that I selected him to serve with General MacArthur, and the reports which he made to me enabled me to comprehend the distant fierce war with all its novel features by which Japan was defeated in the Far East. On January 6 he was standing on the bridge of the *New Mexico*, talking to Admiral Fraser. The Admiral happened by chance to move to the other side of the bridge. A Japanese suicide bomber suddenly swirled down. General Lumsden and Fraser's aide-de-camp were both killed instantly. The pure chance of a casual walk across the bridge saved our Commander-in-Chief.

*   *   *   *   *

Meanwhile there had been fierce fighting in Okinawa island, and its capture was the largest and most prolonged amphibious

* See Vol. V (Cassell's edition), p. 84.

operation of the Pacific war. Four American divisions made the first landings. The rugged island provided ample facilities for defence and the Japanese garrison of over a hundred thousand fought desperately. The whole of Japan's remaining sea- and air-power was committed. Her last surviving modern battleship, the *Yamato*, supported by cruisers and destroyers, tried to intervene on April 7, but the expedition was intercepted by Spruance's air carrier fleet and almost annihilated. Only a few destroyers survived.

Attacks by suicide bombers reached astonishing proportions. No fewer than 1,900 were made before the island was conquered, and according to Admiral King thirty-four destroyers and small craft were sunk and about two hundred other ships were hit. These attacks and several thousand ordinary sorties constituted the most furious onslaught ever launched by the Japanese. But all was in vain. On June 22, after nearly three months' fighting, the island was subdued. The battle had occupied the full strength of Admiral Nimitz's Central Pacific forces, including an army of 450,000 men.

In the midst of my election and other preoccupations I had followed these moving struggles with day-to-day interest, and I realised at once the magnitude of the American achievement.

Prime Minister to President Truman   22 June 45

I wish to offer my sincere congratulations upon the splendid victory gained by the United States Army, Fleet, and Air Force in Okinawa. The strength of will-power, devotion, and technical resources applied by the United States to this task, joined with the death-struggle of the enemy, of whom 90,000 are reported to be killed, places this battle among the most intense and famous in military history. It is in profound admiration of American valour and resolve to conquer at whatever cost might be necessary that I send you this tribute from your faithful Ally and all your British comrades-in-arms, who watch these memorable victories from this Island and all its camps abroad. We make our salute to all your troops and their commanders engaged.

⋆   ⋆   ⋆   ⋆   ⋆

The British Fleet had sailed again from Leyte on May 1. Between May 4 and 25 our air groups struck the same area as before, and on May 4 our ships bombarded the island of Miyako. The enemy mostly fought back with suicide attacks. The carriers *Formidable* and *Victorious* were severely damaged, the former with

THE LAST PHASE IN THE PACIFIC WAR

heavy casualties, but their armoured decks saved them from disaster and both were able to carry on. By May 25 supplies were running low, and the ships withdrew to Manus Island, much heartened by the following message from Admiral Spruance:

I would express to you, to your officers and to your men, after two months' operations as a Fifth Fleet Task Force, my appreciation of your fine work and co-operative spirit. Task Force 57 has mirrored the great traditions of the Royal Navy to the American Task Forces.

* * * * *

Farther south the liberation of the East Indies was proceeding. On May 1 the 9th Australian Division, supported by United States and Australian naval and air forces, landed at Tarakan, in Dutch Borneo. In June the Australians recaptured Brunei and Sarawak. This was followed on July 1 by a landing at Balikpapan by the 7th Australian Division, supported by Dutch, American, and Australian naval forces. But these heartening events were soon to be overshadowed by the climax of the Far Eastern war, which was now approaching.

We were still determined to send troops and aircraft to invade Japan itself, but we had also to liberate Malaya, Singapore, and the territories beyond. The most we could contribute was three divisions for the main assault, and perhaps another two later on. General MacArthur promised the most generous help, and even offered to arm our forces with American weapons and equipment and supply them from the United States. This was far more than we had ever hoped for, and would ease the strain on our shipping, but it would have been very difficult to carry out. In the air we planned to build up twenty squadrons, comprising four hundred heavy bombers, half from Britain and the rest from the Dominions bordering the Pacific, but here also there were difficulties. It transpired after the Yalta Conference that this force would have to be self-contained, providing its own airfields and installations, ports, roads, and pipe-lines.

As soon as it became clear that Okinawa would be captured General Marshall offered us a base there for the development of our air-power. I welcomed this proof that we were to play our part in the main attack upon Japan.

Prime Minister to General Marshall (Washington)   12 June 45
I am very pleased indeed with your offer to us of a base in Okinawa,

from which our first instalment of ten squadrons can take part in the air bombardment of Japan. This is a very handsome gesture on your part, and in full accordance with all the kindnesses we have received from the United States Chiefs of Staff. Our contribution will help, though nothing like what we should like to give you in your tremendous effort to crush Japan speedily.

We could however only hope to have two squadrons in Okinawa by October 1945 and ten early in 1946. But all these projects were overtaken by events. Japan surrendered before our planes and soldiers could arrive, and only our Fleet and the combined forces of Australia and New Zealand saw action in the final stages of the Pacific war.

The Americans intended to seize Kyushu, the most westerly island of Japan, early in November 1945, and from there to invade the main island of Honshu. Here stood an army of more than a million men, well trained, well equipped, and fanatically determined to fight to the last. What remained of the Japanese Navy and Air Force was just as resolute. These two great operations would have entailed bitter fighting and great loss of life, but they were never required. We may well be thankful.

CHAPTER XXXVIII

POTSDAM: THE ATOMIC BOMB

My First Meeting with President Truman — I Make a Tour of Berlin — Luncheon with the President — Tariffs and Bases — Dinner with Stalin — His Forecast of the British Election Result — Russia's Access to the Oceans — Balkan Troubles and Soviet Policy — The Future of Europe — The Message from the Mexican Desert — Prospect of a Speedy End to the Japanese War Without Soviet Help — The Decision to Use the New Weapon — Discussions with the President — The Onslaught on Japan Continues — "Unconditional Surrender" — We Send an Ultimatum, July 26 — The Bombing of Hiroshima and Nagasaki, August 6 and 9, 1945 — Japan Capitulates, August 14 — Sea-Power Decisive Against Japan.

PRESIDENT TRUMAN arrived in Berlin the same day as I did. I was eager to meet a potentate with whom my cordial relations, in spite of differences, had been established by the correspondence included in this volume. I called on him the morning after our arrival, and was impressed with his gay, precise, sparkling manner and obvious power of decision.

On July 16 both the President and I made separate tours of Berlin. The city was nothing but a chaos of ruins. No notice had of course been given of our visit and the streets had only the ordinary passers-by. In the square in front of the Chancellery there was however a considerable crowd. When I got out of the car and walked about among them, except for one old man who shook his head disapprovingly, they all began to cheer. My hate had died with their surrender and I was much moved by their demonstrations, and also by their haggard looks and threadbare clothes. Then we entered the Chancellery, and for quite a long time walked through its shattered galleries and halls. Our Russian guides then took us to Hitler's air-raid shelter. I went

down to the bottom and saw the room in which he and h
mistress had committed suicide, and when we came up agai
they showed us the place where his body had been burned. W
were given the best first-hand accounts available at that time c
what had happened in these final scenes.

The course Hitler had taken was much more convenient for u
than the one I had feared. At any time in the last few months o
the war he could have flown to England and surrendered himself
saying, "Do what you will with me, but spare my misguidec
people." I have no doubt that he would have shared the fate
of the Nuremberg criminals. The moral principles of moderr
civilisation seem to prescribe that the leaders of a nation defeatec
in war shall be put to death by the victors. This will certainly
stir them to fight to the bitter end in any future war, and nc
matter how many lives are needlessly sacrificed it costs them nc
more. It is the masses of the people who have so little to say abo ut
the starting or ending of wars who pay the additional cost. The
Romans followed the opposite principle, and their conquests were
due almost as much to their clemency as to their prowess.

* * * * *

On another occasion I inspected a four-mile line of American
armour drawn up in impressive array, and also many British
troops and tanks. I opened a soldiers' club for the 7th Armoured
Division, whose extraordinary voyages and marches from Cairo
to the goal of victory have to some extent been mentioned in
previous volumes. Three or four hundred of them were gathered
in the club. They all sang "For he's a Jolly Good Fellow", and
were entirely friendly. I thought I detected a certain air of
sheepishness, which might be due to most of them having voted
adversely.

On July 18 I lunched alone with the President, and we touched
on many topics. I spoke of the melancholy position of Great
Britain, who had spent more than half her foreign investments for
the common cause when we were all alone, and now emerged
from the war with a great external debt of three thousand million
pounds. This had grown up through buying supplies from India,
Egypt, and elsewhere, with no Lend-Lease arrangement, and
would impose upon us an annual exportation without any com-
pensatory import to nourish the wages fund. He followed this

attentively and with sympathy, and declared that the United States owed Great Britain an immense debt for having held the fort at the beginning. "If you had gone down like France," he said, "we might be fighting the Germans on the American coast at the present time. This justifies us in regarding these matters as above the purely financial plane." I said I had told the election crowds that we were living to a large extent upon American imported food, for which we could not pay, but we had no intention of being kept by any country, however near to us in friendship. We should have to ask for help to become a going concern again, and until we got our wheels turning properly we could be of little use to world security or any of the high purposes of San Francisco. The President said he would do his very utmost, but of course I knew all the difficulties he might have in his own country.

I then spoke about Imperial Preference, and explained that it might cause a split in the Conservative Party if it were not wisely handled. I had heard that America was making great reductions in her tariff. The President said it had been reduced by 50 per cent., and he now had authority to reduce it by another 50 per cent., leaving it at one-quarter of its pre-war height. I replied that this was a great factor, and would have a powerful influence on our Dominions, especially Canada and Australia.

The President raised the subject of air and communications. He had great difficulties to face about airfields in British territory, especially in Africa, which the Americans had built at enormous cost. We ought to meet them on this and arrange a fair plan for common use. I assured him that if I continued to be responsible I would reopen the question with him personally. It would be a great pity if the Americans got worked up about bases and air traffic and set themselves to make a win of it at all costs. We must come to the best arrangement in our common interests. President Roosevelt knew well that I wished to go much further on this matter of airfields and other bases, and would have liked to have a reciprocal arrangement between our two countries all over the world. Britain was a smaller Power than the United States, but she had much to give. Why should not an American battleship calling at Gibraltar be able to get the torpedoes to fit her tubes and the shells to fit her guns? Why should we not share facilities for defence all over the world? We could add 50 per cent. to the mobility of the American Fleet.

Mr. Truman replied that all these sentiments were very near his own heart. Any plan would have to be fitted in, in some way, with the policy of the United Nations. I said that was all right so long as the facilities were shared between Britain and the United States. There was nothing in it if they were made common to everybody. A man might propose marriage to a young lady, but it was not much use if he were told that she would always be a sister to him. I wanted, under whatever form or cloak, a continuation of the existing war-time system of reciprocal facilities between Britain and the United States about bases and fuelling points.

The President seemed in full accord with this, if it could be presented in a suitable fashion and did not appear to take crudely the form of a military alliance *à deux*. These last were not his words, but give the impression I got of his mind. Encouraged by this, I went on with my long-cherished idea of keeping the organisation of the Combined Chiefs of Staff in being, at any rate until the world calmed down after the great storm and until there was a world structure of such proved strength and capacity that we could safely confide ourselves to it.

The President was replying to this in an encouraging way when we were interrupted by his officers reminding him that he must now start off to see Marshal Stalin. He was good enough to say that this had been the most enjoyable luncheon he had had for many years, and how earnestly he hoped the relations I had had with President Roosevelt would be continued between him and me. He invited personal friendship and comradeship, and used many expressions at intervals in our discussion which I could not easily hear unmoved. I felt that here was a man of exceptional character and ability, with an outlook exactly along the line of Anglo-American relations as they had developed, simple and direct methods of speech, and a great deal of self-confidence and resolution.

*   *   *   *   *

That night, July 18, I dined with Stalin. We were alone except for Birse and Pavlov. We conversed agreeably from half-past eight in the evening to half-past one next morning without reaching any crucial topic. Major Birse produced a fairly long note which I summarise here. My host seemed indeed to be physically rather oppressed, but his easy friendliness was most agreeable.

About the British election, he said that all his information from Communist and other sources confirmed his belief that I should be returned by a majority of about eighty. He thought the Labour Party would obtain between 220 and 230 seats. I did not attempt to prophesy, but I said I was not sure how the soldiers had voted. He said that the Army preferred a strong Government and would therefore vote for Conservatives. It seemed plain that he hoped that his contacts with me and Eden would not be broken.

He asked why the King was not coming to Berlin, and I said it was because his visit would complicate our security problems. He then affirmed that no country needed a monarchy so much as Great Britain, because the Crown was the unifying force through-out the Empire, and no one who was a friend of Britain would do anything to weaken the respect shown to the Monarchy.

Our conversation continued. I said that it was my policy to welcome Russia as a Great Power on the sea. I wished to see Russian ships sailing across the oceans of the world. Russia had been like a giant with his nostrils pinched by the narrow exit from the Baltic and the Black Sea. I then brought up the question of Turkey and the Dardanelles. The Turks were naturally anxious. Stalin explained what had happened. The Turks had approached the Russians about a treaty of alliance. In reply the Russians had said that there could only be a treaty if neither side had any claims. Russia however wanted Kars and Ardahan, which had been taken away from her at the end of the last war. The Turks said that they could not consider this. Russia then raised the question of the Montreux Convention. Turkey said she could not discuss that either, so Russia replied that she could not discuss a treaty of alliance.

I said that I personally would support an amendment to the Montreux Convention, throwing out Japan and giving Russia access to the Mediterranean. I repeated that I welcomed Russia's appearance on the oceans, and this referred not only to the Dardanelles, but also to the Kiel Canal, which should have a régime like the Suez Canal, and to the warm waters of the Pacific. This was not out of gratitude for anything Russia had done, but was my settled policy.

He then asked me about the German Fleet. He said that a share of it would be most useful to Russia, who had suffered severe losses at sea. He was grateful for the ships we had delivered to him after the surrender of the Italian Navy, but he would like his share of the German ships as well. I did not dissent.

He then spoke of Greek aggression on the Bulgarian and Albanian frontiers. He said there were elements in Greece which were stirring up trouble. I replied that the situation on the frontiers was confused, and the Greeks were grievously alarmed about Yugoslavia and Bulgaria, but I had not heard of any fighting worthy of the name. The Conference should make its will plain to these smaller Powers, and none should be allowed to trespass or fight. They should be told this plainly, and made to understand that any alteration to the frontier lines could only be settled at the Peace Conference. Greece was to have a plebiscite and free elections, and I suggested that the Great Powers should send observers to Athens. Stalin thought this would show a want of confidence in the honesty of the Greek people. He thought that the Ambassadors of the Great Powers should report on the elections.

He then asked what I thought about Hungary. I said I was not sufficiently informed to give a view on the immediate situation, but I would inquire of the Foreign Secretary.

Stalin said that in all the countries liberated by the Red Army the Russian policy was to see a strong, independent, sovereign State. He was against Sovietisation of any of those countries. They would have free elections, and all except Fascist parties would participate.

I then spoke of the difficulties in Yugoslavia, where we had no material ambitions, but there had been the fifty-fifty arrangement. It was now ninety-nine to one against Britain. Stalin protested that the proportions were 90 per cent. British, 10 per cent. Yugoslav, and 0 per cent. Russian interests. The Soviet Government often did not know what Tito was about to do.

Stalin also said that he had been hurt by the American demand for a change of Government in Roumania and Bulgaria. He was not meddling in Greek affairs, and it was unjust of them. I said I had not yet seen the American proposals. He explained that in countries where there had been an *émigré* Government he had found it necessary to assist in the creation of a home Government. This of course did not apply to Roumania and Bulgaria, where everything was peaceful. When I asked why the Soviet Government had given an award to King Michael he said he thought the King had acted bravely and wisely at the time of the *coup d'état*.

I then said how anxious people were about Russia's intentions. I drew a line from the North Cape to Albania, and named the capitals east of that line which were in Russian hands. It looked as if Russia were rolling on westwards. Stalin said he had no such intention. On the contrary, he was withdrawing troops from the West. Two million men would be demobilised and sent home within the next four months. Further demobilisation was only a question of sufficient railway transport. Russian losses during the war had amounted to five million killed and missing. The Germans had mobilised eighteen million men, apart from industry, and the Russians twelve million.

I said I hoped that before the Conference ended we should be able to agree about the frontiers of all the European countries, as well as Russia's access to the seas and the division of the German Fleet. The three Powers gathered round the table were the strongest the world had ever seen, and it was their task to maintain the peace of the world. Although satisfactory to us, the German defeat had been a great tragedy. But the Germans were like sheep. Stalin spoke again of his experience in Germany in 1907, when two hundred Germans missed a Communist meeting because there was no one to take their railway tickets at the station barrier.* He then apologised for not having officially thanked Great Britain for her help in sending supplies during the war. Russia would make her acknowledgment.

In reply to my questioning, he explained the working of Collective and State farms. We agreed that both in Russia and Britain there was no fear of unemployment. He said that Russia was ready to talk about trade with Britain. I said that the best publicity for Soviet Russia abroad would be the happiness and well-being of her people. Stalin spoke of the continuity of Soviet policy. If anything were to happen to him there would be good men ready to step into his shoes. He was thinking thirty years ahead.

* * * * *

On July 17 world-shaking news had arrived. In the afternoon Stimson called at my abode and laid before me a sheet of paper on which was written, "Babies satisfactorily born." By his manner I saw something extraordinary had happened. "It

* See Chapter XXIII, p. 344.

means," he said, "that the experiment in the Mexican desert has come off. The atomic bomb is a reality." Although we had followed this dire quest with every scrap of information imparted to us, we had not been told beforehand, or at any rate I did not know, the date of the decisive trial. No responsible scientist would predict what would happen when the first full-scale atomic explosion was tried. Were these bombs useless or were they annihilating? Now we knew. The "babies" had been "satisfactorily born". No one could yet measure the immediate military consequences of the discovery, and no one has yet measured anything else about it.

Next morning a plane arrived with a full description of this tremendous event in the human story. Stimson brought me the report. I tell the tale as I recall it. The bomb, or its equivalent, had been detonated at the top of a pylon 100 feet high. Everyone had been cleared away for ten miles round, and the scientists and their staffs crouched behind massive concrete shields and shelters at about that distance. The blast had been terrific. An enormous column of flame and smoke shot up to the fringe of the atmosphere of our poor earth. Devastation inside a one-mile circle was absolute. Here then was a speedy end to the Second World War, and perhaps to much else besides.

The President invited me to confer with him forthwith. He had with him General Marshall and Admiral Leahy. Up to this moment we had shaped our ideas towards an assault upon the homeland of Japan by terrific air bombing and by the invasion of very large armies. We had contemplated the desperate resistance of the Japanese fighting to the death with Samurai devotion, not only in pitched battles, but in every cave and dugout. I had in my mind the spectacle of Okinawa island, where many thousands of Japanese, rather than surrender, had drawn up in line and destroyed themselves by hand-grenades after their leaders had solemnly performed the rite of *hara-kiri*. To quell the Japanese resistance man by man and conquer the country yard by yard might well require the loss of a million American lives and half that number of British—or more if we could get them there: for we were resolved to share the agony. Now all this nightmare picture had vanished. In its place was the vision—fair and bright indeed it seemed—of the end of the whole war in one or two violent shocks. I thought immediately myself of how the

Japanese people, whose courage I had always admired, might find in the apparition of this almost supernatural weapon an excuse which would save their honour and release them from their obligation of being killed to the last fighting man.

Moreover, we should not need the Russians. The end of the Japanese war no longer depended upon the pouring in of their armies for the final and perhaps protracted slaughter. We had no need to ask favours of them. A few days later I minuted to Mr. Eden: "It is quite clear that the United States do not at the present time desire Russian participation in the war against Japan." The array of European problems could therefore be faced on their merits and according to the broad principles of the United Nations. We seemed suddenly to have become possessed of a merciful abridgment of the slaughter in the East and of a far happier prospect in Europe. I have no doubt that these thoughts were present in the minds of my American friends. At any rate, there never was a moment's discussion as to whether the atomic bomb should be used or not. To avert a vast, indefinite butchery, to bring the war to an end, to give peace to the world, to lay healing hands upon its tortured peoples by a manifestation of overwhelming power at the cost of a few explosions, seemed, after all our toils and perils, a miracle of deliverance.

British consent in principle to the use of the weapon had been given on July 4, before the test had taken place. The final decision now lay in the main with President Truman, who had the weapon; but I never doubted what it would be, nor have I ever doubted since that he was right. The historic fact remains, and must be judged in the after-time, that the decision whether or not to use the atomic bomb to compel the surrender of Japan was never even an issue. There was unanimous, automatic, unquestioned agreement around our table; nor did I ever hear the slightest suggestion that we should do otherwise.

It appeared that the American Air Force had prepared an immense assault by ordinary air-bombing on Japanese cities and harbours. These could certainly have been destroyed in a few weeks or a few months, and no one could say with what very heavy loss of life to the civilian population. But now, by using this new agency, we might not merely destroy cities, but save the lives alike of friend and foe.

* * * * *

A more intricate question was what to tell Stalin. The President and I no longer felt that we needed his aid to conquer Japan. His word had been given at Teheran and Yalta that Soviet Russia would attack Japan as soon as the German Army was defeated, and in fulfilment of this a continuous movement of Russian troops to the Far East had been in progress over the Siberian Railway since the beginning of May. In our opinion they were not likely to be needed, and Stalin's bargaining power, which he had used with such effect upon the Americans at Yalta, was therefore gone. Still, he had been a magnificent ally in the war against Hitler, and we both felt that he must be informed of the great New Fact which now dominated the scene, but not of any particulars. How should this news be imparted to him? Should it be in writing or by word of mouth? Should it be at a formal and special meeting, or in the course of our daily conferences, or after one of them? The conclusion which the President came to was the last of these alternatives. "I think," he said, "I had best just tell him after one of our meetings that we have an entirely novel form of bomb, something quite out of the ordinary, which we think will have decisive effects upon the Japanese will to continue the war." I agreed to this procedure.

The following is a note which I made for the Cabinet at the time:

18 July 45

The President showed me telegrams about the recent experiment, and asked what I thought should be done about telling the Russians. He seemed determined to do this, but asked about the timing, and said he thought that the end of the Conference would be best. I replied that if he were resolved to tell it might well be better to hang it on the experiment, which was a new fact on which he and we had only just had knowledge. Therefore he would have a good answer to any question, "Why did you not tell us this before?" He seemed impressed with this idea, and will consider it.

On behalf of His Majesty's Government I did not resist his proposed disclosure of the simple fact that we have this weapon. He reiterated his resolve at all costs to refuse to divulge any particulars. . . .

*　　*　　*　　*　　*

Meanwhile the devastating attack on Japan had continued from the air and the sea. Among the principal targets were the remnants of the Japanese Fleet, now dispersed for shelter in the island sea.

One by one the big ships were picked out, and by the end of July the Japanese Navy had virtually ceased to exist.

The homeland was in chaos and on the verge of collapse. The professional diplomats were convinced that only immediate surrender under the authority of the Emperor could save Japan from complete disintegration, but power still lay almost entirely in the hands of a military clique determined to commit the nation to mass suicide rather than accept defeat. The appalling destruction confronting them made no impression on this fanatical hierarchy, who continued to profess belief in some miracle which would turn the scale in their favour.

In several lengthy talks with the President alone, or with his advisers present, I discussed what to do. Earlier in the week Stalin had told me privately that as his party was leaving Moscow an unaddressed message had been delivered to him through the Japanese Ambassador. It was presumably meant for either himself or President Kalinin or other members of the Soviet Government, and was from the Japanese Emperor. It stated that Japan could not accept "unconditional surrender", but might be prepared to compromise on other terms. Stalin had replied that as the message contained no definite proposals the Soviet Government could take no action. I explained to the President that Stalin had not wished to tell him direct lest he might think the Russians were trying to influence him towards peace. In the same way I thought we should abstain from saying anything which would make us seem at all reluctant to go on with the war against Japan for as long as the United States thought fit. However, I dwelt upon the tremendous cost in American and to a smaller extent in British life if we enforced "unconditional surrender" upon the Japanese. It was for him to consider whether this might not be expressed in some other way, so that we got all the essentials for future peace and security and yet left them some show of saving their military honour and some assurance of their national existence, after they had complied with all safeguards necessary for the conqueror. The President replied bluntly that he did not think the Japanese had any military honour after Pearl Harbour. I contented myself with saying that at any rate they had something for which they were ready to face certain death in very large numbers, and this might not be so important to us as it was to them. He then became quite sympathetic, and spoke, as had Mr. Stimson, of the terrible responsibilities that rested upon him for the unlimited effusion of American blood.

I felt there would be no rigid insistence upon "unconditional surrender", apart from what was necessary for world peace and future security and for the punishment of a guilty and treacherous deed. Mr. Stimson, General Marshall, and the President were evidently searching their hearts, and we had no need to press them. We knew of course that the Japanese were ready to give up all conquests made in the war.

Eventually it was decided to send an ultimatum calling for an immediate unconditional surrender of the armed forces of Japan. This document was published on July 26.

26 July 45

We, the President of the United States, the President of the National Government of the Republic of China, and the Prime Minister of Great Britain, representing the hundreds of millions of our countrymen, have conferred and agree that Japan shall be given an opportunity to end the war.

2. The prodigious land, sea, and air forces of the United States, the British Empire, and China, many times reinforced by their armies and air fleets from the West, are poised to strike the final blows upon Japan. This military power is sustained and inspired by the determination of all the Allied nations to prosecute the war against Japan until she ceases to resist.

3. The result of the futile and senseless German resistance to the might of the aroused free peoples of the world stands forth in awful clarity as an example to the people of Japan.

The might that now converges on Japan is immeasurably greater than that which, when applied to the resisting Nazis, necessarily laid waste the lands, the industry, and the methods of life of the whole German people. The full application of our military power, backed by our resolve, will mean the inevitable and complete destruction of the Japanese forces, and just as inevitably the utter devastation of the Japanese homeland.

4. The time has come for Japan to decide whether she will continue to be controlled by those self-willed militaristic advisers whose unintelligent calculations have brought the Empire of Japan to the threshold of annihilation, or whether she will follow the path of reason.

5. The following are our terms. We shall not deviate from them. There are no alternatives. We shall brook no delay.

6. There must be eliminated for all time the authority and influence of those who have deceived and misled the people of Japan into

embarking on world conquest, for we insist that a new order of peace, security, and justice will be impossible until irresponsible militarism is driven from the world.

7. Until such a new order is established and until there is convincing proof that Japan's war-making power is destroyed points in Japanese territory designated by the Allies will be occupied to secure the achievement of the basic objectives we are here setting forth.

8. The terms of the Cairo Declaration shall be carried out, and Japanese sovereignty shall be limited to the islands of Honshu, Hokkaido, Kyushu, Shikoku, and such minor islands as we determine.

9. The Japanese military forces after being completely disarmed shall be permitted to return to their homes, with the opportunity of leading peaceful and productive lives.

10. We do not intend that the Japanese shall be enslaved as a race nor destroyed as a nation, but stern justice will be meted out to all war criminals, including those who have visited cruelties upon our prisoners. The Japanese Government shall remove all obstacles to the revival and strengthening of democratic tendencies among the Japanese people. Freedom of speech, of religion, and of thought, as well as respect for fundamental human rights, shall be established.

11. Japan shall be permitted to maintain such industries as will sustain her economy and allow of the exaction of just reparations in kind, but not those industries which would enable her to rearm for war.

To this end access to, as distinguished from control of, raw materials shall be permitted. Eventual Japanese participation in world trade relations shall be permitted.

12. The occupying forces of the Allies shall be withdrawn from Japan as soon as these objectives have been accomplished, and there has been established, in accordance with the freely expressed will of the Japanese people, a peacefully inclined and responsible Government.

13. We call upon the Government of Japan to proclaim now the unconditional surrender of all the Japanese armed forces, and to provide proper and adequate assurances of their good faith in such action. The alternative for Japan is complete and utter destruction.

These terms were rejected by the military rulers of Japan, and the United States Air Force made its plans accordingly to cast one atomic bomb on Hiroshima and one on Nagasaki.

We agreed to give every chance to the inhabitants. The procedure was developed in detail. In order to minimise the loss of life eleven Japanese cities were warned by leaflets on July 27 that they would be subjected to intensive air bombardment. Next day six of them were attacked. Twelve more were warned on July 31, and four were bombed on August 1. The last warning was given on August 5. By then the Superfortresses claimed to have dropped a million and a half leaflets every day and three million copies of the ultimatum. The first atomic bomb was not cast till August 6.

*　　*　　*　　*　　*

The closing scenes of the war against Japan took place after I left office, and I record them only briefly. On August 9 the Hiroshima bomb was followed by a second, this time on the city of Nagasaki. Next day, despite an insurrection by some military extremists, the Japanese Government agreed to accept the ultimatum, provided this did not prejudice the prerogative of the Emperor as a sovereign ruler. The Allied Governments, including France, replied that the Emperor would be subject to the Supreme Command of the Allied Powers, that he should authorise and ensure the signature of the surrender, and that the armed forces of the Allies would remain in Japan until the purposes set forth at Potsdam had been achieved. These terms were accepted on August 14, and Mr. Attlee broadcast the news at midnight.

The Allied fleets entered Tokyo Bay, and on the morning of September 2 the formal instrument of surrender was signed on board the United States battleship *Missouri*. Russia had declared war on August 8, only a week before the enemy's collapse. None the less she claimed her full rights as a belligerent.

We could brook no delay in enforcing the capitulation. Malaya, Hong Kong, and the greater part of the Dutch East Indies still remained in enemy hands, and elsewhere there were isolated forces who might ignore the Emperor's command and fight on. The occupation of these vast territories was thus a matter of urgency. After his Burma campaign Mountbatten had been preparing to liberate Malaya, and everything was in train for a landing near Port Swettenham. This took place on September 9. Other ports were occupied early in September, without fighting, and on September 12 Mountbatten held a surrender ceremony at Singapore.

A British officer, Admiral Harcourt, reached Hong Kong on August 30, and accepted the formal surrender of the island on September 16.

*　　*　　*　　*　　*

There were some in America who believed that Japan's downfall could have been achieved more economically by a greater use of air-power from bases in China, and possibly Siberia. They maintained that her sea communications could have been severed and her power of resistance in the homeland destroyed just as effectively by air action alone, without a long and costly approach by sea as a prelude to invasion. The more advanced exponents of air-power maintained that political objectives elsewhere, in Burma, Malaya, and the East Indies, might have been renounced for the time being and could have been attained without fighting once the air battle had been won. The American Chiefs of Staff had rejected these ideas.

It would be a mistake to suppose that the fate of Japan was settled by the atomic bomb. Her defeat was certain before the first bomb fell, and was brought about by overwhelming maritime power. This alone had made it possible to seize ocean bases from which to launch the final attack and force her metropolitan Army to capitulate without striking a blow. Her shipping had been destroyed. She had entered the war with over five and a half million tons, later much augmented by captures and new construction, but her convoy system and escorts were inadequate and ill-organised. Over eight and a half million tons of Japanese shipping were sunk, of which five million fell to submarines. We, an island Power, equally dependent on the sea, can read the lesson and understand our own fate had we failed to master the U-boats.

CHAPTER XXXIX

POTSDAM: THE POLISH FRONTIERS

Poland, Germany, and the U.S.S.R. – Polish Compensation for the Curzon Line – Transfer of Populations – The First Plenary Session of the Potsdam Conference, July 17 – Second Session, July 18 – Trouble with the Press – A Plan for Drafting the Peace Treaties – "What is Meant by Germany?" – My Appeal for the Poles in Exile – Discussion About Poland's Frontier in the West – Germany's Food and the Eastern Provinces – I Stress the Urgency of a Settlement – I Meet the Polish Provisional Government, July 24 – My Plea for Free Elections – A Talk with Bierut, July 25.

VICTORY over Japan was neither the most difficult nor perhaps the most far-reaching of the problems which confronted us at the Potsdam Conference. Germany had collapsed; Europe must be rebuilt. The soldier must go home and the refugee return, if he could, to his country. Above all, the nations must make a peace in which they could live together, if not in comfort, at any rate in freedom and safety. I do not intend to recount our detailed exchanges in formal conference and in private conversation on all the urgent and multitudinous questions which pressed upon us. Many of them are still unsolved. Poland, for whom Britain went to war, is neither free nor quiet; Germany is still divided; there is no peace with Russia. Russia's share of Poland, Poland's share of Germany, and the place of Germany and the Soviet Union in the world, such were the topics which dominated our discussions, and to which, for reasons of space, this account must be limited.

We had agreed at Yalta that Russia should advance her western frontier into Poland as far as the Curzon Line. We had always recognised that Poland in her turn should receive substantial accessions of German territory. The question was, how much?

How far into Germany should she go? There had been much disagreement. Stalin had wanted to extend the western frontier of Poland along the river Oder to where it joined the Western Neisse; Roosevelt, Eden, and I had insisted it should stop at the Eastern Neisse. All three heads of Governments had publicly bound themselves at Yalta to consult the Polish Government, and to leave it to the Peace Conference for final settlement. This was the best we had been able to do. But in July 1945 we faced a new situation. Russia had advanced her frontier to the Curzon Line. This meant, as Roosevelt and I had realised, that the three or four million Poles who lived on the wrong side of the line would have to be moved to the west. Now we were confronted with something much worse. The Soviet-dominated Government of Poland had also pressed forward, not to the Eastern Neisse, but to the Western. Much of this territory was inhabited by Germans, and although several millions had fled many had stayed behind. What was to be done with them? Moving three or four million Poles was bad enough. Were we to move more than eight million Germans as well? Even if such a transfer could be contemplated, there was not enough food for them in what was left of Germany. Much of Germany's grain came from the very land which the Poles had seized, and if this was denied us the Western Allies would be left with wrecked industrial zones and a starved and swollen population. For the future peace of Europe here was a wrong beside which Alsace-Lorraine and the Danzig Corridor were trifles. One day the Germans would want their territory back, and the Poles would not be able to stop them.

* * * * *

The first plenary session of the Conference was held at five o'clock on the afternoon of Tuesday, July 17. Stalin proposed that the President should take the chair. I supported this, and Mr. Truman accepted our invitation. A number of lesser problems then appeared. Mr. Truman proposed that Italy should join the United Nations, and that the Foreign Ministers of Great Britain, Russia, China, France, and the United States should draft the peace treaties and boundary settlements of Europe. I was doubtful about both these suggestions. Although we had suffered heavy naval losses in the Mediterranean we had much goodwill to Italy, and had provided fourteen out of the fifteen ships which Russia claimed from the Italian Fleet. But I said bluntly that the British people would not easily forget that Italy declared war on the Commonwealth in the hour of her greatest peril, when French resistance was on the point of collapse; nor could they overlook the long struggle against her in North Africa before America came into the war.

Stalin was just as doubtful about asking China to join the Council of Foreign Ministers. Why should she deal with questions which were primarily European ones? And why have this new body at all? We had the European Advisory Commission, and we had agreed at Yalta to regular meetings of the three Foreign Secretaries. Another organisation would only complicate matters, and anyway when would the Peace Conference be held? The President maintained that as China was a member of the World Security Council she ought to have a say in the European settlement, and he admitted that the new United Nations organisation would leave little scope for meetings of the Foreign Secretaries of the "Big Three". All this seemed to me somewhat premature. I feared a dissolution of the Grand Alliance. A World Organisation, open to all and all-forgiving, might be both diffuse and powerless. Free elections in Poland were more to the point, and I reminded my colleagues that this practical problem still lay before us. On this we parted.

*   *   *   *   *

When the Conference met for its second session at five o'clock on the afternoon of July 18 I at once raised another matter which, though outside the agenda, was of immediate importance. At Teheran it had been very difficult for the Press to get near the meeting-place, and at Yalta it had been impossible. But now, immediately outside the delegation area, there were a hundred and eighty journalists prowling around in a state of furious indignation. They carried very powerful weapons and were making a great outcry in the world Press about the lack of facilities accorded to them. Stalin asked who had let them in. I explained that they were not within the delegation area, but mostly in Berlin. The Conference could only do its work in quiet and secrecy, which must be protected at all costs, and I offered to see the Pressmen myself and explain why they had to be excluded and why nothing could be divulged until the Conference ended. I hoped that Mr. Truman would see them too. The plumage of the Press needed to be smoothed down, and I thought that if the importance of secrecy and quiet for those engaged in the Conference were explained to them they would take their exclusion with a good grace.

Stalin irritably asked what all the journalists wanted, and Mr. Truman said that each of us had his own representative to stand between him and the Press. We had agreed to exclude them and matters should be left as they were. I submitted to the majority, but I thought and still think that a public explanation would have been better.

The Foreign Secretaries then produced their plan for drafting the European peace treaties. The Council would still consist of the Foreign Ministers of the five Powers enumerated by the President, but only those who had signed the articles of surrender imposed on the enemy State concerned would draw up the terms of settlement. This we agreed to, but I was concerned at an American proposal to submit the terms to the United Nations. I pointed out that if this meant consulting every member of the United Nations it would be a lengthy and laborious process, and I should be sorry to agree to it. Mr. Byrnes said we were so bound by the United Nations Declaration, but both he and Stalin admitted that reference to the United Nations could only be made after the five Powers had agreed among themselves. I left it at that.

Then there was Germany. The exact powers of the Control Council, economic questions, the disposal of the Nazi Fleet, none were ready for discussion. "What," I asked, "is meant by Germany?" "What she has become after the war," said Stalin. "The Germany of 1937," said Mr. Truman. Stalin said it was impossible to get away from the war. The country no longer existed. There were no definite frontiers, no frontier guards, no troops, merely four occupied zones. At length we agreed to take the Germany of 1937 as a starting-point. This shelved the problem, and we turned to Poland.

* * * * *

Stalin thereupon proposed the immediate transfer to the Lublin Poles "of all stocks, assets, and all other property belonging to Poland which is still at the disposal of the Polish Government in London, in whatever form this property may be and no matter where or at whose disposal this property may prove to be at the present moment". He also wanted the Polish armed forces, including the Navy and Merchant Marine, to be subordinated to the Lublin Poles. This led me to speak at some length.

The burden lay on British shoulders. When their homeland had been overrun and they had been driven from France many Poles had sheltered upon our shores. There was no worth-while property belonging to the Polish Government in London. I said I believed there was about £20,000,000 in gold in London and Canada. This had been frozen by us, since it was an asset of the Central Bank of Poland. Unfreezing and moving it to a Central Polish Bank must follow the normal channels for such transfers. It was not the property of the Polish Government in London and they had no power to draw upon it. There was of course the Polish Embassy in London, which was open and available for a Polish Ambassador as soon as the new Polish Government cared to send one—and the sooner the better.

In view of this one might well ask how the Polish Government had been financed during its five and a half years in the United Kingdom. The answer was that it had been supported by the British Government; we had paid the Poles about £120,000,000 to finance their Army and diplomatic service, and to enable them to look after Poles who had sought refuge on our shores from the German scourge. When we had disavowed the Polish Government in London and recognised the new Provisional Polish Government it was arranged that three months' salary should be paid to all employees and that they should then be dismissed. It would have been improper to have dismissed them without this payment, and the expense had fallen upon Great Britain.

I then asked the indulgence of the President to unfold an important matter, because our position with regard to it was unique—namely, the demobilisation or transfer to their homeland of the Polish forces that had fought with us in the war. When France fell we had evacuated all Poles who wished to come—about 45,000 men—and built up from these men, and from others who had come through Switzerland and elsewhere, a Polish Army, which had finally reached the strength of some five divisions. There were about 30,000 Polish troops in Germany, and a Polish Corps of three divisions in Italy in a highly excited state of mind and grave moral distress. This army, totalling, from front to rear, more than 180,000 men, had fought with great bravery and good discipline, both in Germany and, on a larger scale, in Italy. There they had suffered severe losses, and had held their positions as steadfastly as any troops on the Italian front. The honour of His Majesty's Government was thus involved. These troops had fought gallantly side by side with ours, at a time when trained troops had been scarce. Many had died, and even if I had not given pledges in Parliament we should wish to treat them honourably.

Stalin said he agreed with this, and I continued that our policy was to persuade as many as possible, not only of the soldiers but also of the civilian employees of the late Polish Government, to go back to their country. But we must have a little time to get over our difficulties.

There had been great improvements in Poland in the last two months, and I cordially hoped for the success of the new Government, which, although not all we could wish, marked a great advance and was the result of patient work by the three Great Powers. I had told the House of Commons that if there were Polish soldiers who had fought at our side and did not want to go back we would take them into the British Empire. Of course, the better the conditions in Poland the more Poles would go back, and it would help if the new Polish Government would assure them their livelihood and freedom and would not victimise them for their former allegiance. I hoped that, with continued improvement in Poland, most of these people would return and become good citizens of the land of their fathers, which had been liberated by the bravery of Russian armies.

Stalin said he appreciated our problems. We had sheltered the former rulers of Poland, and in spite of our hospitality they had caused us many difficulties. But the London Polish Government still existed. They had means of continuing their activities in the Press and elsewhere, and they had their agents. This made a bad impression on all the Allies.

I said we must face facts. The London Government was liquidated in the official and diplomatic sense, but it was impossible to stop its individual members living and talking to people, including journalists and former sympathisers. Moreover, we had to be careful about the Polish Army, for if the situation was mishandled there might be a mutiny. I asked Stalin to put his trust and confidence in His Majesty's Government and give us reasonable time. In return everything possible should be done to make Poland an encouraging place for the Poles to go back to.

Mr. Truman declared that he saw no fundamental differences between us. I had asked for a reasonable amount of time, and Stalin had undertaken to drop any of his proposals which would complicate the issue. The best thing was for the Foreign Secretaries to discuss these points; but he hoped the Yalta agreement would be carried out as soon as possible.

Stalin then suggested referring the whole matter to the Foreign Secretaries.

"Including elections," I said.

"The Provisional Government have never refused to hold free elections," Stalin replied.

This ended the second meeting.

* * * * *

The third and fourth meetings of the Potsdam Conference were occupied with a variety of questions, none of which were pushed to any definite conclusions. Stalin wanted the United Nations to break off all relations with Franco "and help the democratic forces in Spain" to establish a régime "agreeable to the Spanish people". I resisted this suggestion, and eventually the subject was dismissed. The disposal of the German Navy and Merchant Marine, peace terms for Italy, and the Allied occupation of Vienna and Austria also raised discussion without reaching any result. Most of the problems were remitted to our Foreign Secretaries for examination and report. My own policy was to let these points accumulate and then bring matters to a head after the result of our election was known.

* * * * *

We did not recur to Poland till our fifth meeting, on July 21. The Soviet delegation wanted Poland's western frontier to run to the west of Swinemünde, as far as the Oder river, leaving Stettin on the Polish side, then up the river Oder to the estuary of the Western Neisse, and from there along its course to Czechoslovakia.

Mr. Truman recalled that we had agreed to divide Germany

OCCUPATION ZONES IN GERMANY AND AUSTRIA, AS FINALLY ADOPTED, JULY 1945

into four zones of occupation, based on our 1937 frontiers. The British and the Americans had moved their troops back into their new zones, but apparently the Soviet Government had given the Poles a zone of their own without consulting us. Unless this zone counted as part of Germany how could we settle reparations and all the other German questions?

Stalin denied giving the Poles a zone of their own. He declared that the Soviet Government had not been able to stop them. The German population had retreated westwards with the German armies. Only the Poles remained. The Soviet armies needed someone to administer their rear areas. They were not accustomed to fight and clear territory and set up their own administration at the same time. Why not let the Poles do it?

"We ought to keep to the zones we agreed at Yalta," said the President. "If we don't reparations and all sorts of other matters will be difficult to settle."

"We are not worried about reparations," said Stalin.

"The United States will get none anyhow," answered Mr. Truman, "but they will also try to avoid paying anything."

"Nothing definite was fixed at Yalta about the western frontier," said Stalin. "None of us are bound."

This was true. The President said he did not think we could settle the matter now. It would have to wait for the Peace Conference.

"It will be still more difficult," said Stalin, "to restore a German Administration."

"You can use a Polish one in your own zone of occupation in Germany," said the President.

"That is all very well," was Stalin's answer, "but the Germans have fled and the natural and indeed the only solution is to set up a friendly administration of Poles. This does not commit us to any particular boundary, and if the Conference cannot agree about one it can remain in suspense."

"Can it?" I interrupted. "These are very important areas for feeding Germany."

"Who will produce the grain?" countered Stalin. "There is nobody left to plough the land except Poles."

"What has become of the Germans?" we both asked.

"They have fled."

I had taken little part in these interchanges, but now I spoke.

567

How, I asked, were we to feed the Germans who had fled? A quarter of Germany's arable land would be lost. If the area suggested by Britain and America was given to Poland about three or four million Poles would have to be moved from east of the Curzon Line; but the Soviet plan would mean shifting more than eight million Germans. Would there be room for them in what was left of Germany? I was not even sure that Stalin was right that all the Germans had fled. Some people thought that more than two millions were still there.

Stalin thereupon challenged my figures, saying that the Germans had called up many men from these regions. The rest had fled. Not a single German remained in the area which he proposed to give to the Poles. The Germans had quitted their lands between the Oder and the Vistula. The Poles were cultivating them, and they were not likely to let the Germans come back.

The President still wanted us to leave the western frontier to the Peace Conference, but I persisted.

Poland, I said, deserved compensation for the land east of the Curzon Line which she was going to lose to Russia, but she was now claiming more than she had given up. If there were three or four million Poles east of the Curzon Line then room should be made in the west. So considerable a movement of population would shock the people of Great Britain, but a move of eight and a quarter millions would be more than I could defend. Compensation should bear some relation to loss. It would do Poland no good to acquire so much extra territory. If the Germans had run away from it they should be allowed to go back. The Poles had no right to risk a catastrophe in feeding Germany. We did not want to be left with a vast German population who were cut off from their sources of food. The Ruhr was in our zone, and if enough food could not be found for the inhabitants we should have conditions like the German concentration camps.

"Germany has always had to import food," said Stalin. "Let her buy it from Poland." "His Majesty's Government," I answered, "can never admit that East German territory overrun in the war has become Polish." "But Poles inhabit it," said Stalin, "and cultivate the land. We can't compel them to produce bread and give it away to the Germans." I protested that these were not normal times. The Poles were apparently selling Silesian coal to Sweden while Great Britain was having the worst fuel shortage of any time during the war. Food and fuel from the Germany of the 1937 frontiers should be available for all Germans within them, irrespective of the zone in which they lived. Stalin asked who was to produce the coal. The Germans were not producing it, but the Poles were. The German proprietors of the Silesian coalfield had fled. If they came back the Poles would probably hang them. I reminded him of his remark at a previous meeting about not allowing memories of injuries or feelings of retribution to govern our policy, and I asked him to realise what we were faced with, namely, a large number of Germans dumped in our zone who could only be fed from the area which the Poles had occupied.

Stalin said that his remarks before did not apply to war criminals. "But not all the eight and a quarter millions who have fled are war criminals," I answered. He then said he meant the German owners of the Silesian coal-mines. Russia herself was short of coal and was buying it from Poland. Here Mr. Truman supported me. It seemed, he said, to be an accomplished fact that East Germany had been given to Poland, but it could not be treated separately when it came to reparations and supplies. He was quite ready to discuss Poland's western boundary, even though it could only be settled at the Peace Conference, but he was not prepared to see sections of Germany given away piecemeal. Stalin persisted that only the Poles could cultivate these areas. The Russians were short of labour and there were no Germans. We could either stop all production or let the Poles do it. The Poles had lost a valuable coal basin to Russia, and had taken the Silesian one in its place. I pointed out that Poles had always worked in the Silesian mines, and I did not object to their doing so, as agents of the Russian Government, but I did object to Silesia being treated as though it were already part of Poland. Stalin persisted that it was impossible to upset the present state of affairs. The Germans themselves had been short of labour. As the Russians advanced into Germany they had found industries employing forcibly deported Italians, Bulgarians, and other nationalities, including Russians and Ukrainians. When the Red Army arrived these foreign labourers had gone home. Enormous numbers of men had been mobilised in Germany, and most of them were either killed or captured. The vast German industries had had few German workers, but depended on foreign labour, which had now melted away. They must either be closed down or the Poles must be given a chance to run them. What had happened was not the result of deliberate policy, but a spontaneous course of events. And only the Germans were to blame for it. He agreed that the Polish Government's proposals would make difficulties for Germany. "And for the British as well," I interjected. But Stalin said he did not mind making difficulties for the Germans. It was his policy, and it would stop them starting another war. It was better to make difficulties for Germans than for Poles, and the less industry in Germany the more markets for Britain.

* * * * *

When we met next day, on Sunday, July 22, we were no nearer agreement. I repeated and emphasised the more important reasons why His Majesty's Government could not accept the Polish demands, and I set them forth as follows:

(i) The final decision on all boundary questions could only be reached at the Peace Conference. (Stalin said he agreed with this.)

(ii) It would not be advantageous for the Polish nation to take over so large an area as they were now asking for.

(iii) It would rupture the economic unity of Germany, and throw too heavy a burden on the Powers occupying the western zones, particularly as to food and fuel.

(iv) The British had grave moral scruples about vast movements of population. We could accept a transfer of Germans from Eastern Germany equal in number to the Poles from Eastern Poland transferred from east of the Curzon Line—say two to three millions; but a transfer of eight or nine million Germans, which was what the Polish request involved, was too many and would be entirely wrong.

(v) The information about the number of Germans in the disputed areas was not agreed. The Soviet Government said that they had all gone. The British Government believed that great numbers, running into millions, were still there. We of course had not been able to check these figures on the spot, but we must accept them until they were shown to be wrong.

Stalin still insisted that Germany could get enough fuel from the Ruhr and the Rhineland and that there were no Germans left in the territory which the Poles had occupied.

Considerable discussion followed about referring the whole matter to the Council of Foreign Ministers. The President said he could not understand why it was so urgent. It could not be finally settled till the Peace Conference. We had had a most useful and helpful discussion, and the best course was to remit the question to the Foreign Ministers. I protested that it was very urgent. Grievances would remain unremedied. The Poles who had assigned to themselves or had been assigned to this area would be digging themselves in and making themselves masters. The Conference ought to make some sort of a decision, or at least we should know where we stood. It was no use asking the Poles to a discussion with the Council of Foreign Ministers in London if the three Powers could not agree now. In the meantime the whole burden of fuel and food problems would remain, and would fall particularly on the British, whose zone had poor food supplies and the largest population.

Suppose the Council of Foreign Ministers, after hearing the Poles, could not agree—and it appeared unlikely that they would —the winter would be coming on with all its difficulties and it would be impossible to settle the matter without another meeting of the heads of Governments. I was most anxious to tackle the practical difficulties which Stalin had explained the day before, difficulties which sprang from the movement of armies and the march of events. Why not have a line which the Polish authorities could provisionally occupy as Poles, and agree that west of that line any Poles would be working as the agents of the Soviet Government?

We agreed that the new Poland should advance its western frontier to what might be called the line of the Oder. The difference between Stalin and myself was how far this extension should reach. The words "line of the Oder" had been used at Teheran. This was not an exact expression, but the British delegation had a line which could be considered in some detail by the Foreign Secretaries. I pointed out that I had only used the words "line of the Oder" as a general expression, and that it could not be properly explained without a map. But I begged my colleagues to persevere. What would happen if the Foreign Secretaries met in September and discussed Poland and again reached a deadlock just when the winter was upon us? Berlin, for instance, used to get some of its fuel from Silesia.

"No, from Saxony," said Stalin.

"About forty per cent. of its hard coal came from Silesia," I answered.

At this point Mr. Truman read us the crucial passage of the Yalta Declaration, namely:

The three heads of Governments consider that the eastern frontier of Poland should follow the Curzon Line, with digressions from it in some regions of five to eight kilometres in favour of Poland. They recognise that Poland must receive substantial accessions of territory in the north and west. They feel that the opinion of the new Polish Provisional Government of National Unity should be sought in due course on the extent of these accessions, and that the final delimitation of the western frontier of Poland should thereafter await the Peace Conference.

This, he said, was what President Roosevelt, Stalin, and I had decided, and he himself was in complete accord with it. Five countries were now occupying Germany instead of four. It would have been easy enough to have agreed upon a zone for Poland, but he did not like the way the Poles had occupied this area without consulting the "Big Three". He understood Stalin's difficulties, and he understood mine. It was the way in which it had been done that mattered.

"Very well," said Stalin. "We bound ourselves at Yalta to consult the Polish Government. This has been done. We can either approve their proposals or summon them to the Conference to hear what they have to say. We ought to settle the matter here, but as we cannot agree then it had better go to the Council of Foreign Ministers."

At Teheran, he said, Roosevelt and I had wanted the frontier to run along the river Oder to where the Eastern Neisse joined it, while he had insisted on the line of the Western Neisse. Moreover, Mr. Roosevelt and I had planned to leave Stettin and Breslau on the German side of the frontier. Were we to settle the question or put it off?

"If the President," he added, "thinks anyone is to blame, it is not so much the Poles as the Russians and the circumstances."

"I understand your point, and that is exactly what I meant," answered Mr. Truman.

Meanwhile I had pondered over these questions, and I now said that we should invite the Poles to come to the Conference at once. Stalin and the President agreed, and we resolved to send them an invitation.

*　　*　　*　　*　　*

Accordingly, at a quarter-past three on the afternoon of July 24 the representatives of the Polish Provisional Government, headed by their President, Bierut, came to my house in the Ringstrasse. Mr. Eden, Sir Archibald Clark Kerr, our Ambassador in Moscow, and Field-Marshal Alexander were with me.

I began by reminding them that Great Britain had entered the war because Poland had been invaded, and we had always taken the greatest interest in her, but the frontiers which she had now been offered and apparently wished to take meant that Germany would lose one quarter of the arable land she possessed in 1937. Eight or nine million persons would have to be moved, and such great shiftings of population not only shocked the Western democracies, but also imperilled the British Zone in Germany itself, where we had to support the people who had sought refuge there. The result would be that the Poles and the Russians had the food and the fuel, while we had the mouths and the hearths. We would oppose such a division, and were convinced that it was just as dangerous for the Poles to press too far to the west as they had once pressed too far to the east.

I told them that there were other matters which troubled us. If British opinion were to be reassured about Poland the elections should be genuinely free and unfettered, and all the main democratic parties should have full opportunity to participate and proclaim their programmes. What was the definition of democratic parties? I did not believe that only Communists were democrats. It was easy to call everyone who was not a Communist a Fascist beast; but between these two extremes there lay great and powerful forces which were neither one nor the other, and had no intention of being one or the other. Poland should admit as many as possible of these moderate elements into her political life instead of branding everyone who did not fit the preconceived definitions of the extremists.

In the present distracted state of Europe anyone with power could strike at his opponents and condemn them, but the only result was to exclude the moderate elements from political life.

It took all sorts to make a nation. Could Poland afford to divide herself? She should seek as broad a unity as possible and join hands with the West as well as with her Russian friends. For example, the Christian Democratic Party and all those of the National Democratic Party who had not actively collaborated with the enemy should take part in the elections. We should also expect full freedom for the Press, and for our Embassy, to see and report what was happening before and during the poll. Only by tolerance, and even on occasion mutual forgiveness, could Poland preserve the regard and support of the Western democracies, and especially of Great Britain, who had something to give and also something to withhold.

Bierut protested that it would be a terrible mistake if Great Britain, having entered the war for Poland's sake, now showed no understanding of her claims. They were modest and took account of the need of peace in Europe. Poland asked for no more than she had lost. Only a million and a half Germans would have to be shifted (including those in East Prussia). These were all that remained. New land was needed to settle four million Poles from east of the Curzon Line, and about three million who would return from abroad, but even then Poland would have less territory than before the war. She had lost rich agricultural land round Vilna, valuable forests (she had always been poor in timber), and the oil-fields of Galicia. Before the war about eight hundred thousand Polish farm-hands used to go to Eastern Germany as seasonal workers. Most of the inhabitants of the areas the Poles claimed, especially Silesia, were really Poles, though attempts had been made to Germanise them. These territories were historically Polish, and East Prussia still had a large Polish population in the Masurians.

I reminded Bierut that there was no dispute about giving Poland the portions of East Prussia which were south and west of Königsberg, but he persisted that Germany, who had lost the war, would lose only 18 per cent. of her territory, while Poland would still lose 20 per cent. Before the war Poland's population was so dense (about eighty-three per square kilometre) that many Poles had had to emigrate. The Poles only asked for their claims to be closely examined. The boundary they proposed was the shortest possible line between Poland and Germany. It would give Poland just compensation for her losses and for her contribution to Allied victory, and she believed that the British would wish her wrongs to be righted.

I reminded him that till now it had been impossible for us to find out for ourselves what was going on in Poland, since it was a closed area. Could we not send people into Poland with full freedom to move about and tell us what was happening? I favoured ample compensation for his country, but I warned him they were wrong to ask for so much.

*　*　*　*　*

Mr. Eden saw the Poles again at his house late that night. Many subjects were touched upon. At ten o'clock next morning I had a stern talk with Bierut alone.

War, he said, provided an opportunity for "new social developments". I asked whether this meant that Poland was to plunge into Communism, to which I was opposed, though of course it was purely a matter for the Poles. Bierut assured me that according to his ideas Poland would be far from Communist. She wanted to be friendly with the Soviet Union and learn from her, but she had her own traditions and did not wish to copy the Soviet system, and if anyone tried to impose it by force the Poles would probably resist. I said that internal questions were their own affair, but would affect relations between our two countries. Of course there was room for reform, especially on the great landed estates.

"Poland will develop on the principles of Western democracy," he answered. She was not small; she was in the centre of Europe; she would have twenty-six million Polish inhabitants. The Great Powers could not be indifferent about her development, and if this was to be on democratic lines, particularly on the English model, some changes would be inevitable.

I once more impressed on him the importance of free elections. It was no good if only one side could put up candidates. There must be free speech, so that everyone could argue matters out and everyone could vote, as was the case in Great Britain. I hoped that Poland would follow the British example and be proud of it. I would do all in my power to persuade Poles abroad to return to Poland at the right time. But his Provisional Government must encourage them. They must be able to start life again on honourable terms with their fellow-countrymen. I was certainly not

575

satisfied with the behaviour of some Polish officers who suggested that all Poles who returned would be sent to Siberia, though it was true that many Poles had been deported in the past.

Bierut assured me that none were being deported now.

I continued that Poland must have courts of law independent of the Executive. The latest development in the Balkans had been not towards Sovietisation but to police government. The political police arrested people on the orders of the Government. The Western democracies deplored this. Would Poland improve? Was the N.K.V.D. leaving the country?

Bierut replied that, generally speaking, the whole Russian Army was leaving. The N.K.V.D. played no rôle in Poland. The Polish Security Police were independent of them and under the Polish Government. The Soviet Union could no longer be accused of attempting to impose such "forms of assistance" upon Poland. Conditions were returning to normal now that the war was over. He professed to agree with me about elections and democracy, and assured me that Poland would be one of the most democratic countries in Europe. The Poles did not favour police régimes, though exceptional measures had had to be taken to heal the serious rifts of war. About 99 per cent. of the population were Catholics. There was no intention to oppress them, and the clergy, generally speaking, were satisfied.

I replied that Great Britain wanted nothing for herself in Poland, but only to see Poland strong, happy, prosperous, and free. There had been no progress after Yalta, but matters had improved greatly in the last few weeks. There was now a recognised Polish Government. I hoped that it would make itself as broad as possible, or at least make sure that the elections were as broad as possible. Not everyone had been equal to the terrible events of the German occupation. The strong resisted, but many average folk bowed their heads. Not all men could be martyrs or heroes. It would be wise to bring all back into the main stream of political life.

Bierut said his Government did not want to stop people expressing political views, but they were anxious to avoid a lot of small parties. As many small parties as wished could take part in the elections, but normally there would be only a few large groups, probably not more than four or five. Such was the present trend. Elections in Poland would be even more democratic than

English ones, and home politics would develop more and more harmoniously.

In reply I said there was no question of our standing in the way of Poland's future, but the frontier question was entangled with the problems of reparations and supply. We had had a great mass of Germans thrown upon our hands, while the Poles had the rich territories from which they had been fed. They were asking too much. We and the Americans might pursue one policy and the Russians another. That would have serious consequences.

My appeal came to nothing. The world has yet to measure the "serious consequences" which I forecast.

CHAPTER XL

THE END OF MY ACCOUNT

Frustration – Social Contacts – I Give a Final Banquet, July 23 – Stalin is Told About the Atom Bomb, July 24 – His Reaction – I Attend the Conference for the Last Time, July 25 – More Discussion on Poland – My Policy at Potsdam – I Fly to London – The Result of the General Election – My Farewell Message to the British People, July 26, 1945.

FRUSTRATION was the fate of this final Conference of "the Three". I have not attempted to describe all the questions which were raised though not settled at our various meetings. I content myself with telling the tale, so far as I was then aware of it, of the atomic bomb and outlining the terrible issue of the German-Polish frontiers. These events dwell with us to-day.

It remains for me only to mention some of the social and personal contacts which relieved our sombre debates. Each of the three great delegations entertained the other two. First was the United States. When it came to my turn I proposed the toast of "The Leader of the Opposition", adding "whoever he may be". Mr. Attlee, and indeed the company, were much amused by this. The Soviets' dinner was equally agreeable, and a very fine concert, at which leading Russian artistes performed, carried the proceedings so late that I slipped away.

It fell to me to give the final banquet on the night of the 23rd. I planned this on a larger scale, inviting the chief commanders as well as the delegates. I placed the President on my right and Stalin on my left. There were many speeches, and Stalin, without even ensuring that all the waiters and orderlies had left the room, proposed that our next meeting should be in Tokyo. There was no doubt that the Russian declaration of war upon Japan would come at any moment, and already their large armies were massed upon the frontier ready to overrun the much weaker Japanese front line in Manchuria. To lighten the proceedings we changed places from time to time, and the President sat opposite me. I had another very friendly talk with Stalin, who was in the best of tempers and seemed to have no inkling of the momentous information about the new bomb the President had given me. He spoke with enthusiasm about the Russian intervention against Japan, and seemed to expect a good many months of war, which Russia would wage on an ever-increasing scale, governed only by the Trans-Siberian Railway.

Then a very odd thing happened. My formidable guest got up from his seat with the bill-of-fare card in his hand and went round the table collecting the signatures of many of those who were present. I never thought to see him as an autograph-hunter! When he came back to me I wrote my name as he desired, and we both looked at each other and laughed. Stalin's eyes twinkled with mirth and good-humour. I have mentioned before how the toasts at these banquets were always drunk by the Soviet representatives out of tiny glasses, and Stalin had never varied from this practice. But now I thought I would take him on a step. So I filled a small-sized claret glass with brandy for him and another for myself. I looked at him significantly. We both drained our glasses at a stroke and gazed approvingly at one another. After a pause Stalin said, "If you find it impossible to give us a fortified position in the Marmora, could we not have a base at Dedeagatch?" I contented myself with saying, "I will always support Russia in her claim to the freedom of the seas all the year round."

* * * * *

Next day, July 24, after our plenary meeting had ended and we all got up from the round table and stood about in twos and threes before dispersing, I saw the President go up to Stalin, and the two conversed alone with only their interpreters. I was perhaps five yards away, and I watched with the closest attention the momentous talk. I knew what the President was going to do. What was vital to measure was its effect on Stalin. I can see it all as if it were yesterday. He seemed to be delighted. A new bomb! Of extraordinary power! Probably decisive on the whole Japanese war! What a bit of luck! This was my impression at the moment, and I was sure that he had no idea of the significance of what he was being told. Evidently in his intense toils and stresses the atomic bomb had played no part. If he had had the slightest idea of the revolution in world affairs which was in progress his reactions would have been obvious. Nothing would have been easier than for him to say, "Thank you so much for telling me about your new bomb. I of course have no technical knowledge. May I send my expert in these nuclear sciences to see your expert to-morrow morning?" But his face remained gay and genial and the talk between these two potentates soon came to an end. As we were waiting for our cars I found myself near Truman. "How did it go?" I asked. "He never asked a question," he replied. I was certain therefore that at that date Stalin had no special knowledge of the vast process of research upon which the United States and Britain had been engaged for so long, and of the production for which the United States had spent over four hundred million pounds in an heroic gamble.

This was the end of the story so far as the Potsdam Conference was concerned. No futher reference to the matter was made by or to the Soviet delegation.

*   *   *   *   *

On the morning of the 25th the Conference met again. This was the last meeting I attended. I urged once more that Poland's western frontier could not be settled without taking into account the million and a quarter Germans who were still in the area, and the President emphasised that any Peace Treaty could only be ratified with the advice and consent of the Senate. We must, he said, find a solution which he could honestly recommend to the American people. I said that if the Poles were allowed to assume the position of a fifth occupying Power without arrangements being made for spreading the food produced in Germany equally over the whole German population, and without our agreeing about reparations or war booty, the Conference would have failed. This network of problems lay at the very heart of our work, and so far we had come to no agreement. The wrangle went on. Stalin said that getting coal and metal from the Ruhr was more important than food. I said they would have to be bartered against supplies from the East. How else could the miners win coal? "They have imported food from abroad before, and can do so again," was the answer. And how could they pay reparations? "There is still a good deal of fat left in Germany," was the grim reply. I refused to accept starvation in the Ruhr because the Poles held all the grain-lands in the east. Britain herself was short of coal. "Then use German prisoners in the mines; that is what I am doing," said Stalin. "There are forty thousand German troops still in Norway, and you can get them from there." "We are exporting our own coal," I said, "to France, Holland, and Belgium. Why should the Poles sell coal to Sweden while Britain is denying herself for the liberated countries?" "But that is Russian coal," Stalin answered. "Our position is even more difficult than yours. We lost over five million men in the war, and we are desperately short of labour." I put my point once again. "We will send coal from the Ruhr to Poland or anywhere else providing we get in exchange food for the miners who produce it."

This seemed to make Stalin pause. He said the whole problem needed consideration. I agreed, and said I only wanted to point out the difficulties in front of us. Here, so far as I am concerned, was the end of the matter.

★　★　★　★　★

I take no responsibility beyond what is here set forth for any of the conclusions reached at Potsdam. During the course of the Conference I allowed differences that could not be adjusted either round the table or by the Foreign Ministers at their daily meetings to stand over. A formidable body of questions on which there was disagreement was in consequence piled upon the shelves. I intended, if I were returned by the electorate, as was generally expected, to come to grips with the Soviet Government on this catalogue of decisions. For instance, neither I nor Mr. Eden would ever have agreed to the Western Neisse being the frontier line. The line of the Oder and the Eastern Neisse had already been recognised as the Polish compensation for retiring to the Curzon Line, but the overrunning by the Russian armies of the territory up to and even beyond the Western Neisse was never and would never have been agreed to by any Government of which I was the head. Here was no point of principle only, but rather an enormous matter of fact affecting about three additional millions of displaced people.

There were many other matters on which it was right to confront the Soviet Government, and also the Poles, who, gulping down immense chunks of German territory, had obviously become their ardent puppets. All this negotiation was cut in twain and brought to an untimely conclusion by the result of the General Election. To say this is not to blame the Ministers of the new Government, who were forced to go over without any serious preparation, and who naturally were unacquainted with the ideas and plans I had in view, namely, to have a show-down at the end of the Conference, and, if necessary, to have a public break rather than allow anything beyond the Oder and the Eastern Neisse to be ceded to Poland.

However, the real time to deal with these issues was, as has been explained in earlier chapters, when the fronts of the mighty Allies faced each other in the field, and before the Americans, and to a lesser extent the British, made their vast retirement on a four-hundred-mile front to a depth in some places of 120 miles, thus giving the heart and a great mass of Germany over to the Russians. At that time I desired to have the matter settled before we had made this tremendous retirement and while the Allied armies were still in being. The American view was that we were committed to a definite line of occupation, and I held strongly that this line of occupation could only be taken up when we were satisfied that the whole front, from north to south, was being settled in accordance with the desires and spirit in which our engagements had been made. However, it was impossible to gather American support for this, and the Russians, pushing the Poles in front of them, wended on, driving the Germans before them and de-populating large areas of Germany, whose food supplies they had seized, while chasing a multitude of mouths into the over-crowded British and American zones. Even at Potsdam the matter might perhaps have been recovered, but the destruction of the British National Government and my removal from the scene at the time when I still had much influence and power rendered it impossible for satisfactory solutions to be reached.

* * * * *

I flew home with Mary on the afternoon of July 25. My wife met me at Northolt, and we all dined quietly together.

Excellent arrangements had been made by Captain Pim and the

staff of the Map Room to present a continuous tale of election results as they came in the next day. The latest view of the Conservative Central Office was that we should retain a substantial majority. I had not burdened myself unduly with the subject while occupied with the grave business of the Conference. On the whole I accepted the view of the party managers, and went to bed in the belief that the British people would wish me to continue my work. My hope was that it would be possible to reconstitute the National Coalition Government in the proportions of the new House of Commons. Thus slumber. However, just before dawn I woke suddenly with a sharp stab of almost physical pain. A hitherto subconscious conviction that we were beaten broke forth and dominated my mind. All the pressure of great events, on and against which I had mentally so long maintained my "flying speed", would cease and I should fall. The power to shape the future would be denied me. The knowledge and experience I had gathered, the authority and goodwill I had gained in so many countries, would vanish. I was discontented at the prospect, and turned over at once to sleep again. I did not wake till nine o'clock, and when I went into the Map Room the first results had begun to come in. They were, as I now expected, unfavourable. By noon it was clear that the Socialists would have a majority. At luncheon my wife said to me, "It may well be a blessing in disguise." I replied, "At the moment it seems quite effectively disguised."

In ordinary circumstances I should have felt free to take a few days to wind up the affairs of the Government in the usual manner. Constitutionally I could have awaited the meeting of Parliament in a few days' time, and taken my dismissal from the House of Commons. This would have enabled me to present before resignation the unconditional surrender of Japan to the nation. The need for Britain being immediately represented with proper authority at the Conference, where all the great issues we had discussed were now to come to a head, made all delay contrary to the public interest. Moreover, the verdict of the electors had been so overwhelmingly expressed that I did not wish to remain even for an hour responsible for their affairs. At seven o'clock therefore, having asked for an audience, I drove to the Palace, tendered my resignation to the King, and advised His Majesty to send for Mr. Attlee.

I issued to the nation the following message, with which this account may close:

26 July 45

The decision of the British people has been recorded in the votes counted to-day. I have therefore laid down the charge which was placed upon me in darker times. I regret that I have not been permitted to finish the work against Japan. For this however all plans and preparations have been made, and the results may come much quicker than we have hitherto been entitled to expect. Immense responsibilities abroad and at home fall upon the new Government, and we must all hope that they will be successful in bearing them.

It only remains for me to express to the British people, for whom I have acted in these perilous years, my profound gratitude for the unflinching, unswerving support which they have given me during my task, and for the many expressions of kindness which they have shown towards their servant.

FINIS

APPENDICES

APPENDIX A

LIST OF ABBREVIATIONS

A.D.G.B.	Air Defence of Great Britain
A.F.H.Q.	Allied Force Headquarters (Mediterranean Command)
A.K.	The Polish Underground Army
A.R.P.	Air Raid Precautions
A.T.S.	(Women's) Auxiliary Territorial Service
C.A.S.	Chief of the Air Staff
C.I.G.S.	Chief of the Imperial General Staff
C.-in-C.	Commander-in-Chief
C.O.S.	Chiefs of Staff
D.D. tanks	"Swimming" tanks
D.U.K.W.	Amphibious load-carrying vehicle
E.A.C.	European Advisory Committee
E.A.M.	The Greek "National Liberation Front"
E.D.E.S.	The Greek "National Democratic Army"
E.L.A.S.	The Greek "People's National Army of Liberation"
G.H.Q.	General Headquarters
G.O.C.-in-C.	General Officer Commanding-in-Chief
M.V.D.	The Polish "Ministry of Internal Affairs"
N.K.V.D.	The Russian Secret Police
N.S.Z.	The Polish Right Wing Underground Force
O.K.W.	Supreme Command of the German Armed Forces
P.M.	Prime Minister
S.C.A.E.F.	Supreme Commander Allied Expeditionary Force
S.E.A.C.	South-East Asia Command
S.H.A.E.F.	Supreme Headquarters Allied Expeditionary Force
S.S.	The *corps d'élite* of the Nazi Party
U.N.R.R.A.	The United Nations Relief and Rehabilitation Administration
V1.	The flying bomb (the "doodle-bug")
V2.	The jet-propelled rocket bomb
V.C.A.S.	Vice-Chief of the Air Staff
V.C.I.G.S.	Vice-Chief of the Imperial General Staff
V.C.N.S.	Vice-Chief of the Naval Staff
V.E.	Victory in Europe
V.J.	Victory in the Far East

APPENDIX B

LIST OF CODE-NAMES

ADMIRAL Q: President Roosevelt.

ANVIL: Allied landings in the south of France, 1944. Later called "Dragoon".

ARGONAUT: The Conference of "the Three" at Yalta, February 1945.

BUFFALO: An amphibious troop-carrying tank.

CAPITAL: The advance into Central Burma from the north.

COLONEL WARDEN: The Prime Minister.

CROSSBOW: The committee set up to deal with measures against pilotless weapons.

CROSSWORD: The German approach to the Allies through Italian intermediaries.

CULVERIN: Operations against Northern Sumatra.

DRACULA: The capture of Rangoon and the cutting off of the Japanese from their bases and lines of communication in Siam.

DRAGOON: The Allied landings in Southern France. Originally called "Anvil".

MANNA: The British expedition to Greece, 1944.

MULBERRY: Artificial harbour.

OCTAGON: The second Conference at Quebec, 1944.

OMAHA BEACH: A beach north-west of Bayeux, site of an American landing on D Day.

OVERLORD: The landing in Normandy in 1944.

PHŒNIX: Reinforced concrete caissons for use as breakwaters.

PLUTO: The submarine pipe-line taking petrol across the Channel.

QUADRANT: The Quebec Conference, 1943.

TERMINAL: The Potsdam Conference, July 1945.

TUBE ALLOYS: Atom bomb research.

WHALE: Floating roadway used in pier construction.

WINDOW: Tinfoil strips used to confuse German Radar.

APPENDIX C

PRIME MINISTER'S DIRECTIVES, PERSONAL MINUTES, AND TELEGRAMS

JUNE 1944 – JULY 1945

JUNE

Prime Minister to Secretary of State for War   1 June 44

Pray let me have a statement setting out the full establishment of the service of psychiatrists in the Army, including all dependent and ancillary personnel.

Prime Minister to Chief of the Air Staff   5 June 44

Thank you for your minute of May 17, enclosing photographs taken from the air with the latest camera modification. They show a remarkable improvement in technique, and I hope we have ample special cameras available for use in operations.

Lord Cherwell also showed me a night photograph taken by similar technique which was very much better defined than any night photos I have previously seen. No doubt the enemy will do much of his movement by night, so that we should make sure cameras for this sort of work are ready.

Prime Minister to Secretary of State for War and   6 June 44
C.I.G.S.

At the Cabinet the other day we were told that the impending shortage of man-power in the Army, equalling 90,000 men deficit, would lead to the destruction of five divisions. Why do you cut down divisions in this way without considering all the reactions which the destruction of divisions implies? For instance, what happens to the corps and army troops, who for five divisions amount to well over 100,000 men? Surely they could be re-trained for infantry duties in less time than fresh entries, or even young men from the R.A.F.? When new divisions are formed about 40,000 men are required. When existing divisions are cannibalised they only work out at an economy of 18,000 men apiece. It is this kind of thing which makes it so difficult for one to help you in keeping up the Army.

2. We have over 1,600,000 men in England. Even after our men have crossed the sea there ought to be a great many in this country from whom it ought to be possible to find sufficient drafts to make good a shortage of 90,000.

3. The above in no way means that I shall relax my pressure on the R.A.F. Regiment, on the Royal Marines, and all other sources. But to go and feed up to the Cabinet that a deficit of 90,000 men means the loss of five divisions cannot be accepted.

Prime Minister to Minister of Production, First Lord of   7 June 44
the Admiralty, Secretary of State for War, and
Minister of Supply

I am glad to hear that the production of "Mulberry" was successfully completed in time. This was a fine feat. The final operational requirement for "Phœnix" had been met by May 23, and the minimum operational requirement for "Whale" by the following day. As a result the whole of the equipment is now in the hands of the Admiralty, and was ready and waiting in the assembly areas on D Day.

The production of this novel and complicated equipment and its towing to the final erection sites and thence to the assembly areas has been a very considerable undertaking. I feel that all the departments concerned are to be congratulated on the completion of the work.

Prime Minister to Minister of Production, First Lord,   7 June 44
First Sea Lord, Minister of War Transport
(Foreign Secretary to see)

Since the losses in the bombardment for "Overlord" appear to be so much less than was anticipated, I am expecting you to organise a convoy to Russia in July, and to run them regularly thereafter as long as the Americans will send anything or there is anything due from us on the existing protocol. It may be necessary to make a new protocol as far as we are concerned.

Prime Minister to Minister of Production and   10 June 44
Sir Edward Bridges

The most obnoxious and burdensome of all "Overlord" security restrictions is the thirty-mile coastal ban, which should be removed at once. This should be done as far as possible unostentatiously, and by local arrangement rather than by proclamation. It would be best if it could just fade away.

2. It is agreed that the diplomatic ban should also be removed privately and unostentatiously from D + 7. There should be no mitigation towards Ireland, except in compassionate cases. Otherwise I am in agreement with the proposal that the removal of other bans should be discussed with S.H.A.E.F. after D + 12.

Prime Minister to Chancellor of the Exchequer   15 June 44

I attach a memorandum prepared by officials about the man-power position in the second half of 1944. Pray reassemble your Ministerial committee which dealt with these matters last November, and prepare a scheme for consideration by the War Cabinet on the following basis:

For the present we must base our plans on the continuance of the war in Europe throughout the first half of 1945; and we cannot assume that it will not continue thereafter. By the end of August 1944 it may be possible to take a clearer view of the position, and for the present the position had best be dealt with by short-term adjustments.

I consider that a temporary increase should be made in the Army intake during July and August (say, 15,000 instead of 6,000 a month). This increase should be at the expense of the other Services, and not of munitions and other essential industries.

The Ministry of Supply claim for an increased allotment cannot, in my view, be accepted, and the Ministry must conform with the allocation approved for them.

There will still remain a shortage of labour amounting to about 100,000. This should in the main be shared between the industrial staffs of the Admiralty and the Ministry of Aircraft Production and from Civil Defence. Something more might also be found from industries and elsewhere.

"Overlord" casualties to date are much lower than anticipated. Your recommendations should take this into account.

Prime Minister to Sir Edward Bridges, and to　　　　18 June 44
General Ismay for C.O.S. Committee

Ministers who wish to visit Normandy for any purpose should inform me of what they propose. I will then consult General Eisenhower's wishes. No other visitors should be permitted to go, except with General Eisenhower's concurrence.

2. This should be brought to the notice of all concerned.

Prime Minister to Major Morton　　　　22 June 44

Remind me later on to make a note about a hostel in France for correspondents, as in the last war, with facilities for distinguished visitors to be entertained and taken to proper places in the front line without in the slightest degree affecting the command posts or the Commanders-in-Chief. Let me have a report on the château which was then formed. Major Neville Lytton played an important part in this. The Minister of Information should be consulted.

Prime Minister to First Lord and First Sea Lord　　　　22 June 44

Please let me have without delay the measures you are taking to sustain the bombarding fleet. *Warspite* should certainly be used as long as she can swim and her guns can fire. *Malaya*, I presume, is ready to take over gadgets from *Warspite* in good time, or is she properly equipped already? It is much better to rely on the 15-inch than the

16-inch, because of the larger stocks of ammunition and replacement mountings and tubes. I understand you are using *Revenge* and *Resolution* as stoker training-ships. These vessels should be put to a higher use. We have great need to sustain our bombarding fleet, which may have to deal with Cherbourg, and will certainly be required for the flanks of the liberating armies.

2. Let me see your estimate of losses of the British fleet employed in "Overlord" and the losses which have actually occurred.

3. Let me have the stocks of ammunition for the *Nelson-Rodney* class and the *Malaya-Ramillies* class. Let me also have the story about retubing of the guns in both the *Malaya* and *Ramillies* types. How long does it take to shift 15-inch guns from one ship to another? It would surely be wrong to use important fighting vessels as stoker training-schools in the height of a great battle, when either they or their armament or equipment are urgently needed.

4. I had hoped to hear something from you before now about replenishment of the bombarding fleet. I am quite ready to complain about the Americans, but let us make sure that our own house is in order.

Prime Minister to First Sea Lord   22 June 44

Pray let me have an immediate report on the synthetic harbours. How have they stood up to the recent rough weather, and how much has their construction been delayed? How many units have we lost on the way over or in any other way? And have we replacements?

I had been expecting a report from you.

Prime Minister to General Ismay   23 June 44

I am concerned to hear that the War Office have circulated a report which raises doubt whether we shall be able to destroy the U-boat and E-boat pens in Occupied Europe before handing over the ports to the Allied Governments.

Of course the battle must not be hampered by precipitate attempts at demolition. But we must reserve our right to deal with such threats to our security when leisure serves. It would be intolerable if any Allied Government objected after their failure to defend their country had exposed us to so much danger and after we had shed so much blood in the liberation of their people.

Pray let me have the views of the Chiefs of Staff.

Prime Minister to General Hollis   23 June 44

I consider that the Polish Parachute Brigade should not be lightly cast away. It may have a value in Poland itself far out of proportion to its actual military power. I trust that these views may be conveyed to

Generals Eisenhower and Montgomery before the brigade is definitely established in France.

Prime Minister to General Hollis   29 June 44

Thank you for your analysis of the United States troops arriving in the United Kingdom up to August 1944.

I am still by no means satisfied that the Americans could not bring over more fighting divisions and less ancillary troops. I am astonished, for example, that the total number of corps and army troops (131,243) exceeds the number of men in the divisions (87,689) by 43,554. And what is the function of the S.O.S. troops, who amount to about one-fifth of the total? What is the meaning of the sinister phrase "excluding casuals"? How many casuals?

Prime Minister to Foreign Secretary   29 June 44

I do not advise any decision at the present time on the Palestine policy. It is well known I am determined not to break the pledges of the British Government to the Zionists expressed in the Balfour Declaration, as modified by my subsequent statement at the Colonial Office in 1921. No change can be made in policy without full discussion in Cabinet. We have so little to do now, it should be easy to find an opportunity to do this.

JULY

Prime Minister to Home Secretary   1 July 44

The use of the air-raid sirens requires immediate consideration. Undoubtedly they cause a great deal of needless unrest. Vast numbers of people hear the sirens who never hear or see the bomb. The question presents itself differently in the country and in the town, and particularly of course on the bomb highway, where I spent last Friday. Here the "Alert" and the "All clear" were almost continuous, with intervals of perhaps a quarter of an hour.

2. One ought to be able to answer the question, "What do you want each given class of person to do on hearing the alarm?" Is the ploughman to stop ploughing, or the shopkeeper to retire into the cellar? Are people who are congregated for any purpose to disperse? What action, if any, do you wish them to perform? If there is nothing for them to do different from what they would be doing, as is the case with almost everyone, what is the use of sounding the sirens? In the case of the old-fashioned air raids, falling now here, now there, all over the country, the sirens were very useful, and should be retained for this purpose.

3. I have watched from my windows a great many people in the park to see what they do when the sirens sound. They do nothing at all. The smallest shower of rain will make them melt away, but they take not the slightest notice of the banshee. Nor with these bombs is it possible in most cases to give any *local* warning, and even so I do not know what the people can do in the day-time. At night they ought always to go to whatever is the safest place, and forget their cares. Once there they require no sirens to disturb their slumber.

4. I have been this afternoon to see several scenes of incidents. In one case the greater part of the explosion was within four or five feet of the shelter. I asked whether it was occupied, and they assured me it was full and that was why there had been so little loss of life. Not one person in the shelter had been hurt in any way. But this is not the case in many places. I should like a return of the London shelters and the use made of them. For how many is there accommodation—by which I mean bunks—and how many of them are being used? Considering that you began strengthening the street shelters in good time, it seems a great pity not to reap where you have sown.

Prime Minister to Paymaster-General   3 July 44

The point I object to is this. The casualties [in Italy] are presented with various uneven dates at totals from September 3, 1943, or January 22, 1944. Each week the casualties of the week are added, making cumulative totals. This is not what one wants to know. One wants to know how many have been killed, wounded, and missing in the week, and then look back to the general totals for reflection on the campaign. In the present circumstances it is necessary to add up each week the totals of the British, American, and other forces killed, wounded, and missing, and subtract from them the totals given in the return the week before. This can of course be done; but why should it be done, and why should the War Office impose it upon us? All that is needed surely is to present the total casualties, killed, wounded, and missing, armies and nations, and on the lower part of the page the addition of·the week or fortnight, I care not which. I certainly do not intend to accept the present lay-out. Pray arm me for a further attack.

Prime Minister to General Ismay, for C.O.S. Committee   5 July 44

I cannot agree with this melting down of the Polish 2nd Division. The few divisions they have embody the life of Poland. They are not to be treated as ordinary masses of men in the pool for replacements. Had not this Polish Division better go and join the other two in Italy, where it will be welcomed, to make a strong corps? I am not prepared to agree to S.H.A.E.F.'s proposals.

Prime Minister to General Ismay   5 July 44

Pray thank General Bedell Smith from me for his paper on build-up, and tell him with what great interest I have read it.

2. It certainly seems a very serious thing that there should be no increase in the United States forces between D 90 and D 120, and an increase of only four [divisions] between D 90 and D 150, which makes it all the more necessary to capture the ports of St. Nazaire and Havre, to develop the small ports, and not to dissipate the landing-craft which so much increase the effectiveness of the small ports.

3. We are told that there are more than forty trained divisions in the United States, and certainly a far larger army than we now possess or have in prospect will be needed to advance through France into Germany, unless there is a total collapse on the Russian or German front.

4. It is for this reason, among others, that I deprecate the sending away of L.S.T.s from the decisive theatre to operations so remote as those now projected in Southern France. It seems to me the main object should be to open the French ports and maintain at its full or even enlarge the synthetic harbour which remains to us, so that the immense armed strength of America can be applied. He would be a bold man who would say that the existing forces assigned have it in their power to deliver decisive blows. The greatest importance should be attached to the direct landing of United States forces across the Atlantic. This will only be possible when ports like the two additional ones I have mentioned have been got into working order, as well as Cherbourg.

I have ventured to put down my first thoughts on reading General Bedell Smith's paper, but I am keeping it by me for further study.

The War Office should let me have a similar statement on their build-up.

Prime Minister to General Ismay, for C.O.S. Committee   6 July 44

The one thing to fight for now is a clean cut, so that Alexander knows what he has and we know what we have a right to give him. Let him take their seven divisions—three American and four French. Let them monopolise all the landing-craft they can reach. But let us at least have a chance to launch a decisive strategic stroke with what is entirely British and under British command. I am not going to give way about this for anybody. Alexander is to have his campaign. If the Americans try to withdraw the two divisions still left with him I shall ask you to send the 52nd Division from the United Kingdom to bridge the gap. I hope you realise that an intense impression must be made upon the Americans that we have been ill-treated and are furious. Do not let any smoothings or smirchings cover up this fact. After a little we shall get together again; but if we take everything lying down there will be no end to what will be put upon us.

Prime Minister to Secretary of State for War　　　　　7 July 44
How is it that the 36th Indian Division consists of two British brigades? There is much to be said for humility in the world, but to call a British division an Indian division is really going below the level of grovelling to which we have been subject. If they are British troops let them be called British troops.

Prime Minister to Minister of Aircraft Production　　　　8 July 44
Thank you for your minute of June 5 [about centralising jet-propulsion research under one man, with the best possible advice available from within and without the Government service].

In research and development there is great risk in too much centralisation, and the present move seems likely to run this risk. But if you are anxious to take it I leave the matter in your hands.

As you know, I have always taken a great personal interest in this question of jet-propulsion, and I should be obliged if you would let me have a note on the progress made at, say, two-monthly intervals.

Prime Minister to Secretary of State for War　　　　　10 July 44
Army psychiatric services.

I am very glad to receive your report of the deficiencies you mention. Could you let me know how much is the cost of running this service, which, I see, amounts to over 2,000 officers, nurses, and attendants?

Prime Minister to First Sea Lord　　　　　　　　　10 July 44
At one time in the last war we had a great development of anti-U-boat nets of quite a light description which enwrapped the U-boat and towed a buoy on the surface. Can anything like this be adapted to the human torpedo? Surely these light webs could be spread about in the harbour, buoyed so as not to impede the navigation, and yet always giving a tell-tale buoy or flare for counter-attack.

Prime Minister to First Sea Lord　　　　　　　　　10 July 44
Please let me have a short report on the capacity of Caen harbour.
I see figures stated much larger than any which have previously been mentioned.

Prime Minister to Sir Edward Bridges　　　　　　　10 July 44
In your report [about the War Cabinet meeting on forming a Jewish fighting force] you say that it was decided that a brigade group would be carefully examined. I certainly understood and hold very strongly the view that a brigade group should be made. When the War Office say they will carefully examine a thing they mean they will do it in.

The matter must therefore be set down for an early meeting of the War Cabinet only this week, and the Secretary of State for War should be warned of my objection. A copy of the further letter from Dr. Weizmann may also be forwarded to the War Office.

Prime Minister to General Ismay, for C.O.S. Committee　　　11 July 44

You must surely give the Turks that small outfit of Radar equipment and one or two night fighter squadrons to defend Constantinople. Pray let me know what can be done. The matter is urgent.

Prime Minister to Foreign Secretary　　　11 July 44

There is no doubt that this [persecution of Jews in Hungary and their expulsion from enemy territory] is probably the greatest and most horrible crime ever committed in the whole history of the world, and it has been done by scientific machinery by nominally civilised men in the name of a great State and one of the leading races of Europe. It is quite clear that all concerned in this crime who may fall into our hands, including the people who only obeyed orders by carrying out the butcheries, should be put to death after their association with the murders has been proved.

I cannot therefore feel that this is the kind of ordinary case which is put through the Protecting Power, as, for instance, the lack of feeding or sanitary conditions in some particular prisoners' camp. There should therefore, in my opinion, be no negotiations of any kind on this subject. Declarations should be made in public, so that everyone connected with it will be hunted down and put to death.

Prime Minister to Secretary of State for War　　　12 July 44

I am anxious to reply promptly to Dr. Weizmann's request for the formation of a Jewish fighting force put forward in his letter of July 4, of which you have been given a copy. I understand that you wish to have the views of Generals Wilson and Paget before submitting to the Cabinet a scheme for the formation of a Jewish brigade force. As this matter has now been under consideration for some time I should be glad if you would arrange for a report setting out your proposals to be submitted to the Cabinet early next week.

Prime Minister to Secretary of State for War　　　13 July 44

I have had most disturbing news from my old regiment, the Oxfordshire Hussars, of which I am now Honorary Colonel. Apparently its rôle is to find drafts for the Twenty-first Army Group and to be a holding unit for wounded, trainees, etc. This means that it can never serve as a fighting unit, and will in fact disappear in all but name. It seems very wrong that a regiment with such a fine history and record should be treated in this shabby fashion. Surely they deserve their chance in the field?

Pray go into this and let me know what can be done.

Prime Minister to Foreign Secretary   13 July 44

We must not forget that both we and the French have promised independence to the people of Syria and the Lebanon. I have frequently interpreted this as meaning that the French have the same primacy in Syria and the Lebanon, and the same sort of relations, as we have in Iraq—so much and no more. We cannot go back on this.

Prime Minister to Secretary of State for Air   13 July 44

These figures [comparing the weight of high explosive delivered by German aircraft and flying bombs] require your attention, and should be compared with the figures of bombs dropped from British and American aircraft over Germany.

It is a puzzle to me and to many others how, with perhaps thirty or forty tons of bombs, including containers, dropped over London by the robot, such very noticeable damage occurs; whereas instead of fifty tons of bombs there may well be two or three thousand tons dropped over Berlin or Munich, and yet the German people seem to get away with it all right. You should yourself, with some of your experts, go and look at some of the damage done here. Why is it that such a small weight in German robot bombs creates results which seem eight to ten times greater than their equal quantity on German cities?

Air Chief Marshal Harris should also be asked to express an opinion. Big claims are made about the destruction wrought in Germany, and we wonder why the effect is so severe here, or whether, *per contra*, very much less hitting and useful effect is done in Germany.

Prime Minister to Foreign Secretary   14 July 44

ESCAPE OF JEWS FROM GREECE

This requires careful handling. It is quite possible that rich Jews will pay large sums of money to escape being murdered by the Huns. It is tiresome that this money should get into the hands of E.L.A.S., but why on earth we should go and argue with the United States about it I cannot conceive. We should take a great responsibility if we prevented the escape of Jews, even if they should be rich Jews. I know it is the modern view that all rich people should be put to death wherever found, but it is a pity that we should take up that attitude at the present time. After all, they have no doubt paid for their liberation so high that in future they will only be poor Jews, and therefore have the ordinary rights of human beings.

Prime Minister to Secretary of State for War   14 July 44

I hope that now that the initial needs of "Overlord" have been met and security considerations no longer apply the armies in Italy will receive a reasonable proportion of the latest types of equipment, such as 17-pounder Shermans, Heavy Churchills, flame-throwers, special assault vehicles, and sabot ammunition.

Pray let me know what arrangements have been made and what are your present intentions.

Prime Minister to General Ismay and Mr. Peck   16 July 44

I wish the British casualties, dating from the outset of the operations [in Normandy], to be issued, subject to military considerations, at regular monthly intervals, covering the same period as the fortnightly Allied casualties. I am particularly anxious that the Canadian casualties, although stated separately, should be included in the British publication of casualties; otherwise they will be very readily assumed to be part of the American casualties. The point is of Imperial consequence. The matter should be put to the Dominions Office.

It is a great mistake to whittle away the British share in these battles by a form of presentation and to have the Canadians incorporated directly or by implication with the Americans. It is of course understood that military considerations may at any time override the publication of any casualties.

I am sure demands will be made in America for the publication of the American casualties, which are at present greatly in excess of our own. If General Eisenhower's broadminded plan of publishing only "Allied casualties" could survive I should be content with it at any rate for a good many months; but I do not think there is any chance of this.

However, I should like to see the case more carefully studied. There will be time enough for this, I think, and perhaps I will have a talk with General Eisenhower myself.

Prime Minister to First Lord and First Sea Lord   17 July 44

There seems to be a sort of idea that the Navy will retain much of its present strength and man-power after the defeat of Hitler. I thought I had better let you have a few general ideas on the subject which seem worthy of your consideration. The Japanese are already outnumbered two to one by the Americans. We shall of course send a powerful fleet, consisting of our most modern ships, with the necessary train and attendant vessels. I should imagine however that Vote A will not exceed 400,000 men as long as the Japanese war lasts, after which you will have to return to at least pre-war strength.

Prime Minister to Secretary to the Cabinet and General Ismay   25 July 44

Pray issue to all three Service departments and to the Postmaster-General a request from me to furnish a weekly report of the average time taken for letters to reach the Navy, Army, and Air Force in Normandy, and for their similar correspondence to reach destinations here.

Prime Minister to First Lord   26 July 44

I have heard from various quarters that the delays in letters reaching the Navy on the other side, or *vice versa*, are serious. The Army is now getting a far better service. With all the facilities of boats going to and fro, it should be the easiest thing in the world to solve. Pray make me a report on the past and present position, and what you are going to do about it.

I am sending a copy of this minute to the Postmaster-General.

Prime Minister to Home Secretary   26 July 44

I agree with this report [of a meeting on the institution of a system of imminent danger warnings in connection with flying bombs], subject to the following comments:

1. There should be some indication that this system cannot be made universal, but will be expanded steadily as circumstances and resources allow.

2. "Particular classes," like bus-drivers and others, must be given perfectly clear directions. The responsibility in the case of bus-drivers lies with the Minister of War Transport. You cannot leave every bus-driver to solve this difficult problem for himself, especially as his passengers may disagree with his solution. It ought to be quite easy to put on a single sheet of paper what he is to do in various circumstances. I should have thought that, in principle, wherever there was any possibility of cover the bus should stop and people be allowed to get out. A bus, with all its glass, full of passengers, is a very painful object when we are under bombardment of this kind.

3. When you give a signal of any kind you should attach to it a perfectly clear recommendation—or, if need be, order—to the public to take what you think is the best action. I have seen some very good papers drawn up already on this subject. These should be amplified for town and simplified for country as you now think fit in the light of greater experience.

Prime Minister to Secretary of State for War   26 July 44

I am in general agreement with your proposals [for a Jewish fighting force], but I think the brigade should be formed and sent to Italy as soon as convenient, and worked up to a brigade group there as time goes on by the attachment of the other units.

2. I like the idea of the Jews trying to get at the murderers of their fellow-countrymen in Central Europe, and I think it would give a great deal of satisfaction in the United States.

3. The points of detail which occur to me are:

I do not think this brigade should be any more liable to be split by serious military emergencies than any other unit in the Middle East. On the contrary, only a serious emergency should affect it, considering what it represents.

I believe it is the wish of the Jews themselves to fight the Germans anywhere. It is with the Germans they have their quarrel. There is no need to put the conditions in such a form as to imply that the War Office in its infinite wisdom might wish to send the Jews to fight the Japanese, and that otherwise there would be no use in having the brigade group.

Surely political as well as military considerations govern the demobilisation or disposal of any of the forces under British command. In the case of a contingent of this kind there certainly might be political reasons either for dispersing it or for maintaining it after the war. . . .

I will consult the King about this [proposal that the force should have its own flag]. I cannot conceive why this martyred race, scattered about the world and suffering as no other race has done at this juncture, should be denied the satisfaction of having a flag. However, not only the King but the Cabinet might have views on this.

4. Should I be able to visit Italy I will discuss the details of this with General Wilson, and also very likely I shall see General Paget. Meanwhile please go ahead within the lines proposed and negotiate with the Jewish Agency. Remember the object of this is to give pleasure and an expression to rightful sentiments. and that it certainly will be welcomed widely in the United States. Let me see the form of any announcement that is made.

Prime Minister to First Lord and First Sea Lord　　　　　29 July 44
Further to my minute of July 17 [about man-power for the Navy], you should prepare a plan showing the size and the composition of the fleet which you could maintain on the assumption that at the end of twelve months after the defeat of Germany the total man-power of the Navy had been reduced to 400,000 men. On this hypothesis you should set out the details of the fleet which could be maintained here and in the Far East, together with the comparative figures for ships and Vote A on the completion of mobilisation at the outbreak of war in 1939.

Prime Minister to Secretary of State for War and C.I.G.S.   29 July 44

General Montgomery spoke to me last week about the Queen's Own Oxfordshire Hussars and other ancient Yeomanry regiments which are being used merely as holding units. I have pointed out to you the great importance of not destroying these permanent elements in our defensive system.

2. General Montgomery proposed to me that when a "Hostilities Only" or war-time-raised unit at the front was so depleted that it had to be broken up, that unit should be sent home to form part of the pool of reinforcements, and one of these now perfectly trained, permanent Yeomanry units should be sent out in its place. They are of course actually at the present time trained as artillery or anti-tank. This proposal seems quite satisfactory to me.

3. I have your minute of July 18 informing me that you have given instructions that men are not to be taken from the Oxfordshire Hussars for the time being. What about the Gloucestershire and other similar units—*i.e.*, those playing a permanent part in the military defences for generations in good times and bad? Can they not be treated in this way? Perhaps you would be so kind as to let me have a list.

Prime Minister to Home Secretary   29 July 44

Your minute of July 28 [enclosing a revised draft statement on the subject of an imminent danger warning for flying bombs].

I cannot see that the new paragraph is defensible. The bus-drivers are to use their discretion, and the Government rely on the good sense of the people. What happens when the bus-driver decides one way and the good sense of the people in the bus decides the other way? Surely something more intelligible than that can be provided. How do the people in the bus bring their good sense to bear on the conductor or the driver? You had better bring this question up at the Cabinet on Monday.

Otherwise I like your paper.*

AUGUST

Prime Minister to First Sea Lord   4 Aug 44

What are the Navy doing on the western flank of the armies? I should have thought that they would be very lively all along the Atlantic shores of the Brest peninsula, driving off all enemy vessels, isolating the Channel Islands from all food or escape of the German garrison, being ready at Quiberon Bay or elsewhere to join hands with the advancing American columns. We shall soon be possessed of

* See minute of July 26 to Home Secretary.

harbours or inlets at which bases for U-boats and destroyers could be established, dominating the waters round the Brest peninsula and greatly helping the movements of the land forces. As it is, they seem to be doing very little except to fight on the north-eastern flank. There are plums to be picked in the Brest peninsula. Admiral Ramsay must not weary of well-doing.

2. When I have heard from you I will address General Eisenhower on the subject. It is not the least use telling me that General Eisenhower has not asked for anything. He is very busy with the land battle and knows very little about the sea. I am convinced that opportunities are passing.

3. I shall be very glad to see you about this at any time.

Prime Minister to Minister of Production   4 Aug 44

Thank you for your report of the effect of flying-bomb attacks on production in London. Such a loss of man-hours, particularly in the radio industry, would certainly be serious if it continued. I am glad to see that, together with the Minister of Labour, you are doing everything possible to improve the local warning systems and the general efficiency of factory A.R.P. organisation. I hope that your efforts, together with the easing off in the attacks, will soon yield some improvement.

Please continue for the time being to send me a fortnightly report.

A steady drift of radio production *out* of London should be promoted.

Prime Minister to Foreign Secretary   6 Aug 44

This seems to be a rather doubtful business [the case of the Hungarian Jews]. These unhappy families, mainly women and children, have purchased their lives with probably nine-tenths of their wealth. I should not like England to seem to be wanting to hunt them down. By all means tell the Russians anything that is necessary, but please do not let us prevent them from escaping.

I cannot see how any suspicion of peace negotiations could be fixed on this miserable affair.

Prime Minister to General Ismay, for C.O.S. Committee   10 Aug 44

We could not agree that "Dragoon" in its relation to "Overlord" can be considered a "major operation" for the sake of which, if necessary, General Alexander's army and its prospects are to be ruined. This army, after the withdrawals made from it of the French and half the United States forces, still consists of about twenty divisions, of which sixteen are British, British Empire, or British-controlled. We are not prepared in any circumstances to have it regarded as a lesser operation than "Dragoon" or that "Dragoon" should have priority over its essential needs. Moreover, when we come to the strategic aspect, we are not prepared at the present time, in the absence of further discussion, to agree that it is more probable that an advance to the westward would be more desirable for General Alexander's army than one towards Trieste, which might be accompanied by action in the Istrian peninsula in conjunction with the Partisans of Yugoslavia under Marshal Tito. It is obvious that different views may be taken on these matters. These can only be resolved at a conference not only between the Chiefs of Staff but between the heads of Governments.

Prime Minister to C.I.G.S.   10 Aug 44
How is it that the 6th Guards Tank Brigade has got no Churchills with the heavy armour? Where are the heavy armoured tanks? Let me have a return.

Prime Minister to Secretary of State for War and   18 Aug 44
V.C.I.G.S.
General Clark mentioned to me that about six months ago he proposed to the War Office that 2,000 A.T.S., etc., should be sent out here in order to replace men of the Allied Force Headquarters and other rearward services for active war work. Only 250 were sent however.
2. In view of the urgent need to sustain General Alexander's army every device must be adopted.
The Americans are making far more use of women out here than we are.

Prime Minister to Secretary of State for War   18 Aug 44
I understand that the managing director of the *Continental Daily Mail* has been asking for facilities to re-start the paper in France and that S.H.A.E.F. favour this proposal. I should like to see the *Daily Mail*, and any other London newspaper which so wished, start Continental editions for the forces in France if this could be managed. It would however be necessary to obtain the consent of the Newspaper Proprietors Association to the allocation of newsprint, and if you see no objection to this project I should be glad if you would take the matter up with them. Speed is essential.

Prime Minister to Foreign Secretary, First Lord, and   18 Aug 44
Secretary of State for War
At my request the Chiefs of Staff have considered a report by the War Office which suggested that it would not be possible to destroy the U-boat and E-boat pens in captured Continental ports before the British and American Governments ceased to control them.
It is certainly our policy to hand over the control of occupied terri-

tory to the national authorities as soon as possible, and it will therefore be difficult to complete the destruction of these installations before that happens.

I am sure however that the Governments concerned (namely, the French Committee of National Liberation, the Norwegian, Belgian, and Dutch Governments) should be told that when we hand back any occupied territory to them we intend to reserve control over all military installations, including U-boat and E-boat pens, which have been erected by the enemy, until they have been completely dismantled or destroyed. I should be glad if the War Office and the Admiralty would investigate and report on the possibility of having this done by civilian contractors.

Prime Minister to General Ismay   23 Aug 44

The best way to bring about the fall of the present High Command in Germany is to draw up a list of war criminals who will be executed if they fall into the hands of the Allies. This list need not be more than fifty to a hundred long (apart from the punishment of local offenders). This would open a gulf between the persons named in it and the rest of the population. At the present moment none of the German leaders has any interest but fighting to the last man, hoping he will be that last man. It is very important to show the German people that they are not on the same footing as Hitler, Goering, Himmler, and other monsters, who will infallibly be destroyed.

Prime Minister to Foreign Secretary   23 Aug 44

Your telegram to Washington [about a suggested meeting of Foreign Ministers on future world organisation].

I certainly agree that such a meeting should be held in London, as it is our turn. I hope the French will not be admitted to such a discussion until they have broadened their Government. The rapid liberation of France should facilitate this and obtain some national authority which we can recognise. This should not be so long delayed now that great parts of France are already liberated.

2. That China is one of the world's four Great Powers is an absolute farce. I have told the President I would be reasonably polite about this American obsession, but I cannot agree that we should take a positive attitude on the matter. The latest information from inside China points to the rise already of a rival Government to supplant Chiang Kai-shek, and now there is always a Communist civil war impending there. While not opposing the President's wish, I should object very much if we adopted other than a perfectly negative line, leaving him to do the needful with the Russians.

Prime Minister to First Lord and First Sea Lord   25 Aug 44

It is very discreditable that the Navy and Merchant Navy mails take so much longer than those of the Army and the Air to reach these short distances. It is a great reflection not only on the Post Office, but mainly on the Admiralty.

Pray let me have a better service on this matter.*

Prime Minister to Secretary of State for India   28 Aug 44

Considering that India will undoubtedly remain in the war picture after Europe is pacified, it is extremely important that a really big move should be set on foot to deal with the comforts and conditions of British troops now in the East or who may be sent there. I cannot tell when transport will be available. It seems to me however that a programme should be worked out first.

Prime Minister to Foreign Secretary, and to General Ismay   29 Aug 44
for C.O.S. Committee

I do not consider that we should turn down the Russian proposal for an International Air Force. The matter raises very large questions of principle and cannot be decided on purely military grounds. I am quite sure the Russian proposal will gain very great acceptance in Great Britain, and it certainly seems by weaving the forces of different countries together to give assurances of permanent peace. My views on this subject were expressed after the last war, and may be read in the opening chapters of *The Aftermath*. I am not aware of any Cabinet decision against an International Air Force.

2. In my view the method should be as follows. Each member of the Peace Council should divide its Air Force into two parts, National and International. The National Air Force should remain as at present. The International Air Force could be similarly trained and organised, but service in it would be for ten years. They would wear a different uniform, and would mingle freely by interchange with the similar dedicated squadrons in the other member countries. No section of the International Air Force would ever be required to take action against its own country.

3. The introduction of an International Air Force is an event of the utmost importance, and cannot possibly be settled on departmental considerations.

SEPTEMBER

Prime Minister to First Lord of the Admiralty   1 Sept 44

Thank you for the statement on merchant ship repairs, which shows that over twenty-five million tons of merchant, troop, and hospital

---

* See minute of July 26 to First Lord.

ships were repaired in the six months January–June 1944, a quantity larger by one-tenth than that of any previous six months. Though the kind of repair now required is less serious than in the past, owing to our victory over the U-boats, this performance is much to be commended, since you have also had to undertake much special work in preparation for "Overlord".

Prime Minister to Colonial Secretary   1 Sept 44

The establishment of the apes on Gibraltar should be twenty-four, and every effort should be made to reach this number as soon as possible and maintain it thereafter.*

Prime Minister to Sir Edward Bridges   2 Sept 44

I am concerned at the position of our prefabricated houses policy. Nearly five months have been wasted exhibiting our designs to the fastidious criticism of housing faddists, local authorities, Members of Parliament, etc., with the result that there is practically no progress at all. I take a most serious view of this, because I have myself made public pledges to the troops. We are being more nice than wise. What does it matter whether the house is the best that can be built or not? The great thing is to have some kind of a house for a soldier returning who wishes to marry. If we are not careful we shall be so busy over design that we shall finish up with Nissen huts.

2. I wish to be informed when I come back [from Quebec] exactly what the position is, and to have a plan for the accelerated production of the largest number of prefabricated houses possible. For this purpose I appoint a Committee of the Cabinet under Lord Beaverbrook, with Lord Portal and Mr. Bracken as members. This committee will have the advantage, as it requires, of the assistance of the President of the Board of Trade and the Minister of Health. Mr. Peck will act as secretary.

3. I do not consider that it would be necessary for Lord Woolton, who is concerned with the general layout of policy over the whole field of reconstruction, to involve himself in these details. Anyhow, I have got to have a plan for action when I come back in about a fortnight. It can then be brought before the Cabinet.

4. Will you please convey the substance of this to the different Ministers concerned, and make it quite clear that I have made up my mind about it because of my own personal commitment to the troops, which I regard as sacred so far as I am concerned.

5. At the same time I wish the President of the Board of Trade to furnish me with an effective report of what he has done in regard to the necessary commodities, crockery, furniture, and the like, and what

* On June 30, 1950, they numbered thirty.

progress has been made with civilian suits, etc. The rapid movement of the war has invested all this with great urgency, and it is probable that very considerable changes of priority as between munitions and domestic articles will have to be made.

Prime Minister to Minister of Reconstruction　　　　　3 Sept 44
There must be no delay in ordering jigs and tools for the temporary houses, or in proceeding with the erection of such houses as are necessary for the provision of needed shelter for people whose homes have been destroyed. You have my authority to anticipate the sanction of Parliament for such measures. Pray let there be no delay.

Prime Minister to Minister of Reconstruction　　　　　3 Sept 44
Please let me have within a fortnight a complete statement of the controls and "Orders" that should be repealed (*a*) immediately hostilities in Europe end, (*b*) within two months after the end of hostilities.

Prime Minister to Minister of Reconstruction　　　　　3 Sept 44
There will be great need, when the war ends, for the provision of clothing and domestic furniture and utensils for the civil population.
Please report to me what steps are being taken to meet this need generously.

Prime Minister to Sir Edward Bridges　　　　　　　　4 Sept 44
The reference should be as follows:
To make the best plan possible, with programme attached, for the largest possible construction in the shortest time of prefabricated houses of all types, without trenching notably on the normal building trade. In this inquiry the Committee should assume that the necessary impingement will be made upon munitions production, provided the scheme is approved by the Cabinet. Reports should be completed by the 20th for Cabinet decision on my return.*

Prime Minister to C.I.G.S.　　　　　　　　　　　　10 Sept 44
I should be glad if this [table of divisions in the West, dated 1.9.44] could be brought up to date.
2. You do not seem to have credited the British, or perhaps the Americans, with the number of Army tank brigades, two of which together should count as the equivalent of a division.
3. Even without this it would seem that, taking Italy and France together, the British Empire has thirty-four divisions and the United States thirty divisions. This figure may have been altered by new arrivals, but in any case it shows the strong basis on which we approach this Conference [at Quebec].

* See minute of September 2.

Prime Minister to General Ismay, for C.O.S. Committee   14 Sept 44

Pray see this.* Although this matter is being settled at home it would be best that the Chiefs of Staff themselves should express a view.

2. I have drafted the following for Mr. Lyttelton, on which I should also like their opinion:

"I am very doubtful myself whether the German war will end by the end of the year. It may struggle over in a reduced condition into 1945. Nevertheless I approve the practical steps you propose to take, subject to Cabinet and Defence Committee agreement. You may therefore submit your paper to the Cabinet as you wish."

OCTOBER

Prime Minister to Secretary of State for War   23 Oct 44

A serious appeal was made to me by General Alexander for more beer for the troops in Italy. The Americans are said to have four bottles a week, and the British rarely get one. You should make an immediate effort, and come to me for support in case other departments are involved. Let me have a plan, with time schedule, for this beer. The question of importing ingredients should also be considered. The priority in issue is to go to the fighting troops at the front, and only worked back to the rear as and when supplies open out.

2. The question of leave is also pressed seriously. If only a small proportion could have leave it would be much appreciated. Try to work out a thousand a month plan. Could these men come back across France? I am aware that Marseilles is greatly congested, but are there not other routes which could be used? In this case also priority should be given to the troops who have been engaged in the fighting.

Prime Minister to Minister of War Transport   28 Oct 44

During the winter it is most important that increased travel facilities should be given to the public, especially the London public, while suffering from the black-out. Could you let me have a report on bus queues, which, as far as I can see, seem to be getting longer, and make me your proposals for alleviating conditions.†

Prime Minister to First Sea Lord   31 Oct 44

I asked Mr. Geoffrey Lloyd for a short report on the progress of "Pluto",‡ and he informs me that a cross-Channel pumping capacity of 1,000,000 gallons of petrol a day is aimed at. A figure of this

---

* Telegram from the Minister of Production about the effect on production of the Cabinet decision that for man-power purposes it could now be assumed that the war against Germany would not continue beyond the end of 1944.

† See Vol. V, pp. 586, 590, 591.

‡ The cross-Channel pipe-line.

magnitude must effect a large saving of tankers and man-power, and I hope that you will make every effort to ensure the prompt laying of the necessary pipe-lines. Pray keep me informed of your progress in this vital matter.

NOVEMBER

Prime Minister to Secretary of State for War   3 Nov 44
and C.I.G.S.

It is certain that we shall have to replace Field-Marshal Dill in Washington.* I can see no other officer of sufficient status to fill this gap except General Maitland Wilson. He would head our delegation, and would, I believe, by his personality and record obtain access to the President and intimacy with General Marshall.

2. At the same time the Mediterranean Command, which was formed at the beginning of this year, has been greatly affected in scope and scale. The Levant is quiet; Greece is in rapid process of liberation; the Germans in the islands will fall in like rotten plums. The Riviera front is taken over by the Americans, and General Devers's army has become a part of General Eisenhower's command. The French are in full charge of Tunis and Algeria. There remains only the great campaign in Italy, which is in General Alexander's province, and any movement we may make across the Adriatic, which again is an offshoot of the Italian campaign.

3. Therefore I consider the time is approaching and will soon come when General Maitland Wilson should be appointed to fill Field-Marshal Dill's place in Washington and when General Alexander should become the Supreme Commander in the Mediterranean, or what is left of it. General Alexander's assumption of the Supreme Command would undoubtedly lead to telescoping in quantities in the two Staffs at present in Italy. There will of course be subsidiary reactions, on which we can discourse later. Pray let me have your views.

Prime Minister to General Ismay, for C.O.S. Committee   5 Nov 44

Just look at the vast numbers of staff units—ancillary and tail—which have been poured into Greece, and the very small proportion of fighting troops. Can we not send a few battalions of infantry? There seem to be only two there, and 1,500 ordinary British infantry out of 22,600 all told.

Prime Minister to General Ismay, for C.O.S. Committee   16 Nov 44

General Eisenhower mentioned to me the possibility that should we

---

* It was clear that Field-Marshal Dill would not survive his illness. He died on November 4.

reach the Rhine opposite the Ruhr in the course of the present operations the American long-range artillery would be able to dominate and destroy at least half of that area. Cannot some of our heavy batteries erected in the Dover area be of use for this, especially those on railway mountings? A range of 30,000 yards is achieved by the medium American guns up to 240 mm., but might not the intervention of our 12-inch and 13.5-inch, and even 15-inch, be accepted? What are the ranges of the principal guns that can be mounted on railway mountings? I am afraid our 18-inch howitzers would be judged too short-range.

2. Anyhow, let the whole matter be examined with care, and a plan made out that can be offered to General Eisenhower for transporting, probably through the port of Antwerp when it is open, about twenty of these long-range, very heavy guns. Every dog has his day, and I have kept these for a quarter of a century in the hope that they would have their chance.

Prime Minister to Minister of Production and   16 Nov 44
Minister of Food

CANNED MEAT FOR THE U.S.S.R.

I do not understand why the Minister of Food did not mention the fact that this would involve us in the payment of twenty million dollars, nor how it was he had authority to authorise his representative in Washington to agree to our providing the 45,000 tons. The Cabinet and the Chancellor of the Exchequer should certainly have been consulted on this matter. Were they? If not, the subject must be brought before them now.

2. What I said was of course based upon the Minister of Food's message. But it was said in relation to a jocular remark of M. Stalin's about the advance made since the Stone Age by the discontinuance of cannibalism, and followed by another from him in an equally non-serious vein. I cannot consider that this mere statement of our intentions involved an obligation or bargain, and this can be seen if read in its context in the secret records.

3. What other negotiations have there been on this subject? Have the Russians in fact made any demand for this food? What have we said to them officially up to the present? The above does not mean that I am not anxious to supply the Russians with this meat if it can be done without loss to Britain. I certainly think however that it should be made to play its part in some larger negotiations, and I do not consider that any action is called for at the present moment. We could quite well await some demand from the Russians and then refer this to the Americans. The matter should be considered at the next

Cabinet, when I hope to receive the Minister of Food's explanation and to have the matter cleared up.

Prime Minister to Colonial Secretary   17 Nov 44

I have been thinking about this matter [my proposed statement to the House of Commons on Palestine and the assassination of Lord Moyne] all day, and the following points have occurred to me. Will not suspension of immigration or a threat of suspension simply play into the hands of the extremists? At present the Jews generally seem to have been shocked by Lord Moyne's death into a mood in which they are more likely to listen to Dr. Weizmann's counsels of moderation. The proposed announcement would come as a shock of a different kind, and, so far from increasing their penitence, may well provide a not unwelcome diversion and excite bitter outcry against the Government. Dr. Weizmann will no doubt join in the protests (saying that the whole community are being punished for the acts of a small minority), but the initiative in such a situation will pass to the extremists. Thus those responsible for the murder will be themselves the gainers. It may well unite the whole forces of Zionism, and even Jewry throughout the world, against us instead of against the terrorist bands.

2. Certainly the situation calls for signal action; but should it not be more clearly directed against that section of the community with whom the responsibility lies—*e.g.*, by *enforcing* even more drastic penalties in the case of those found in possession of firearms or belonging to proscribed societies? In particular, might not action be taken against the nominally respectable leaders of the party, whose extremist wing are the authors of these political crimes? If their national status is non-Palestinian they might be deported; if Palestinian they should be banished.

Prime Minister to Home Secretary   19 Nov 44

I do not feel I could agree to the setting up of a non-Ministerial chairman to the committee. The matter is essentially Ministerial. I thought that we were all agreed that the soldiers must have the best possible chance of exercising their right to vote. For this the ballot for candidates and a reasonable knowledge of the issues in their constituencies at the election is unquestionably superior to the vote by proxy, though this must be used if nothing better serves in the most remote districts. Since we are agreed on the principle there is no need to have an umpire. I understood that your committee was to look into details and clear difficulties out of the way, and I am waiting to send my telegrams to the Commanders-in-Chief asking them to look at the matter from their standpoints. I consider myself specially responsible for securing proper facilities to the soldiers to cast an intelligent vote. If necessary the War Cabinet will have to go into the details themselves, calling such witnesses as they may think fit. General Alexander will be here very soon, and I expect Field-Marshal Montgomery could fly over almost any fine day. These two Commanders-in-Chief cover four-fifths of the whole military vote. Any outstanding issues must be decided by the War Cabinet itself. I cannot see that since we are agreed on the principle of the soldiers having the best show possible there should be any serious divergences between us.

Prime Minister to Secretary of State for War　　　　19 Nov 44

We arranged with Montgomery that he was to disband some battered war-time-raised units, rather than drain white ancient regiments of modernised British Yeomanry. The attached letter [from General O'Connor, about the replacement of the 91st (Argyll and Sutherland Highlanders) Anti-Tank Regiment, R.A., by the Oxfordshire Hussars] however shows that the axe has fallen upon a very fine permanent territorial unit. Surely some better arrangement than this can be made to make a place for the Oxfordshire Hussars. I will, if necessary, telegraph to Field-Marshal Montgomery on the subject. Meanwhile let me have a report on the subject, as well as on the special points mentioned in the attached letter.

2. With regard to the attached letter, it was brought me by an officer, and I take full responsibility for his bringing it to me, and also for its having been written by General O'Connor. No kind of victimisation is to be made in this case, as the communication must be regarded as privileged, and I am responsible for the use of it.

Prime Minister to Secretary of State for War　　　　20 Nov 44

Good. Press on. Make sure that the beer—four pints a week—goes to the troops under the fire of the enemy before any of the parties in the rear get a drop.*

Prime Minister to Foreign Secretary　　　　20 Nov 44

I cordially agree with this minute of the Paymaster-General's about food relief in France and the Low Countries. I believe that Western Europe has eaten better meals than Britain in the last four years, and that the real help should go to solving transportation difficulties.

Prime Minister to Foreign Secretary　　　　26 Nov 44

I consider that at the end of the winter campaign it would be appropriate for the City of London to confer its freedom upon this remarkable American general [General Eisenhower]. It is obvious that no one else should be included at the time.

* See minute of October 23.

613

Prime Minister to First Lord   27 Nov 44

There is no immediate urgency, but the Navy cannot be deprived of their prize money. I will not consent to this. I think we had a fight about it last time.

Prime Minister to Foreign Secretary   30 Nov 44

Please never forget that we can always say to the United States, "Will you take over a mandate for Ethiopia? We certainly will not." You will find they will recoil most meekly and with great rapidity.

DECEMBER

Prime Minister to Foreign Secretary   2 Dec 44

The first thing is arms in the hands of the French. The second step is to improve the quality and type as fast as possible. I cannot conceive how a decision to arm a number of French divisions with captured German weapons for the next two or three years could possibly affect the general treatment of the disarmament of Germany and the destruction of German munitions factories. One lot of German arms would be used while there was nothing else, and fade out as better and more agreeable weapons came in. Your suggestion starts out on a basis which we have by no means reached yet. It is almost like saying, "Don't shoot that German with a German-made pistol now. Far better be shot yourself by him, and have a thoroughly harmonious type of armament developed on a scientific scale a few years after your funeral."

Prime Minister to First Lord, Secretary of State for War,   3 Dec 44
Secretary of State for Air, and to General Ismay for
C.I.G.S. and C.O.S. Committee
(Minister of Labour to be informed of what is passing)

I am much distressed by the impending cannibalisation of the 50th Division. We cannot afford at this stage to reduce our stake in the Western line of battle. We must examine all possibilities which are open. For instance, there are nearly 80,000 Royal Marines. These are no doubt needed as far as ships marked for the East are concerned, but they are not needed in anything like their present strength in vessels in non-Japanese-infected waters. The Admiralty should furnish a paper showing the exact location of all Marines, what ships they are in, what theatres these ships are marked for, how many Marines are ashore, how many are in the training establishments, etc. I should expect at least 10,000 men to be found from this quarter alone. These must be held available for service in France and Italy. The 50th Division might close its three brigades into two and add a third Royal Marine brigade, the rest of the Marines going into a common pool; or, alternatively, they could all be put into the pool.

2. Again, let me have precise figures for the intake to the Navy in the next six months, and the numbers employed in all the schools for recruits, both staff and pupils. In my opinion, at least 5,000 recruits who opted for the naval training and who are now in the Navy training schools should be transferred to the Army.

3. It will be possible to reconsider these matters when the war with Germany is over, which may well be in six months. But now is the time when we require to keep the largest number of British military units in the field.

4. I recognise that the R.A.F. is more closely engaged than the Navy at the present time. But none the less I will ask for a further comb out from the R.A.F. Regiment for the common pool.

5. We should also not hesitate to take in these months younger men who are withheld by pledges to Parliament, and I am prepared at any time, as Minister of Defence, to ask from Parliament the necessary release from our obligations.

6. The staffs should be carefully combed for able-bodied fighting officers and men. We are hoping for the release of several thousand from the Caserta [H.Q.] population. Every training or exceptional establishment in the Army should be examined, not only from the point of view of reducing its size, but also of substituting elderly men or recovered wounded. It is a painful reflection that probably not one in four or five men who wear the King's uniform ever hears a bullet whistle, or is likely to hear one. The vast majority run no more risk than the civil population in Southern England. It is my unpleasant duty to dwell on these facts. One set of men are sent back again and again to the front, while the great majority are kept out of all the fighting, to their regret.

7. This is a moment for supreme effort on the Army front in Europe, and I earnestly beg my Service advisers and their Parliamentary chiefs to do their very utmost to meet the national need.

8. Finally, on no account is the breaking up of the 50th Division to proceed any further until these matters have been thrashed out between us and decisions upon them given by the War Cabinet.

Prime Minister to Secretary of State for War   3 Dec 44

I propose to speak to the King on Tuesday about this [letter from a commanding officer on the Western Front complaining of the delay in men receiving their decorations, especially the "Immediate" awards], but I should first like to have your comments. I am indignant that men should perish without ever receiving decorations awarded them months before. I need scarcely say that no inquiries are to be made about the writer or the regiment, as I take responsibility.

Prime Minister to Foreign Secretary   3 Dec 44

I put this down for record. Of all the neutrals Switzerland has the greatest right to distinction. She has been the sole international force linking the hideously sundered nations and ourselves. What does it matter whether she has been able to give us the commercial advantages we desire or has given too many to the Germans, to keep herself alive? She has been a democratic State, standing for freedom in self-defence among her mountains, and in thought, in spite of race, largely on our side.

2. I was astonished at U.J.'s savageness against her, and, much though I respect that great and good man, I was entirely uninfluenced by his attitude. He called them "swine", and he does not use that sort of language without meaning it. I am sure we ought to stand by Switzerland, and we ought to explain to U.J. why it is we do so. The moment for sending such a message should be carefully chosen. . . .

Prime Minister to Chancellor of the Exchequer   11 Dec 44

I should be glad if you would reconvene the committee on man-power (yourself, Mr. Bevin, Mr. Lyttelton, and Lord Cherwell), for the purpose of preparing for the War Cabinet proposals for allocations to the Services and industry for the first half of 1945, in the light of the Minister of Labour and National Service's memorandum setting out the man-power position for 1945 and the demands for 1945 from the Service and other Ministers.

One of the questions that will arise is whether transfers should be made from the Navy and the Royal Air Force on a considerable scale to meet the Army's man-power difficulties. Decisions on this matter must of course be reserved for the War Cabinet, but it would be helpful to the Cabinet if your committee were to focus the issues—without delaying their proposals for allocating the new intake.

Your committee should proceed on the assumption that the German war will end about June 30, 1945, but the scheme should be sufficiently flexible to allow of adjustments from time to time, without undue disturbance to the war effort, if it becomes necessary to change the assumption as to the date on which the war with Germany is likely to end.

Prime Minister to Foreign Secretary   11 Dec 44

I do not think the balance of help and hindrance given us by Spain in the war is fairly stated [in a certain letter to Franco]. The supreme services of not intervening in 1940 or interfering with the use of the airfield and Algeciras Bay in the months before "Torch" in 1942 outweigh the minor irritations which are so meticulously set forth. Therefore I should like to see the passages reciting our many grievances somewhat reduced. . . . A little alteration in the wording would be compatible with justice and consistency, and I should be glad if you would look them over from that point of view. . . .

2. I am in agreement with the rest of the draft, which as a whole I like very much. I thought however the Cabinet wished some reference to the Falange and to dictatorship to be made. Will you consider whether this should not be included?

3. I should be glad if these points could be settled to-night or to-morrow, and the document then printed and circulated to the Cabinet. . . . The reason why I have become anxious to have this matter cleared up, having myself been rather slow in dealing with it, is . . . because I think it would be very useful now to send a copy of this letter, when agreed, by special messenger or telegraph to Stalin. I believe it would give him great satisfaction, and also help to clear away any doubts that may have been engendered by de Gaulle during his visit that we have desired to build up a Western *bloc* against Russia. I have no doubt de Gaulle endeavoured to acquire as much merit as possible by stating his opposition to such an organism.

4. I am increasingly impressed, up to date, with the loyalty with which, under much temptation and very likely pressure, Stalin has kept off Greece in accordance with our agreement, and I believe that we shall gain in influence with him and strengthen a moderate policy for the Soviets by showing them how our mind works.

5. Finally, I may say I think the letter has been very well drafted, and says the most freezing things with suitable diplomatic restraint.

Prime Minister to Minister of Production   18 Dec 44
Your report on penicillin, showing that we are only to get about one-tenth of the expected output this year, is very disappointing. It is discouraging to find that, although this is a British discovery, the Americans are already so far ahead of us, not only in output but in technique. I hope you are satisfied that we have the right people in charge and that labour and material difficulties are being tackled early enough and energetically enough.

Pray let me have a realistic estimate of 1945 production.

Prime Minister to Chancellor of the Exchequer, Minister   19 Dec 44
of Labour, Minister of Production, Secretary of State for
War, and to General Ismay for C.O.S. Committee
In accordance with my verbal instructions the Commander-in-Chief Middle East Forces has prepared a memorandum showing the present man-power position in the Middle East.

From this report it appears that there are about 662,000 on the ration strength, plus some 242,000 directly employed in civil labour, a total of 904,000 in all. The analysis of the ration strength is as follows:

United Kingdom military personnel	154,000
Royal Air Force	66,000
Dominion forces	29,000
Indian forces .. .. 42,000 ⎫	
Locally enlisted and Colonial forces .. 130,000 ⎭	172,000
Allies, including United States Army forces in Middle East	78,000
Refugees, prisoners of war, and para-military personnel ..	163,000
	————
	662,000

Of the 154,000 U.K. military personnel 88,000 are in formations, garrison units, base depots, etc., 50,000 at the base in Egypt and Palestine, and 16,000 in headquarters and administrative units.

Although the Middle East is an important theatre it cannot be allowed to absorb such large numbers. The need to reduce them is twofold. At the present time of man-power stringency it is essential to bring as many British soldiers as possible into action against the enemy. Therefore I consider that the Middle East should give up to other theatres some 50,000 Service personnel from their total. In addition, there must be a reduction in local labour.

Also, on account of our financial position, our expenditure in the Middle East will have to be curtailed. At June 30 Egypt and Palestine, the main areas of the Middle East base, had amassed sterling balances of some £320,000,000 and £95,000,000 respectively. At present our net military expenditure in these two countries is still running at £6,000,000 per month, against an average of about £8,250,000 per month in both 1942 and 1943. This drain on our sterling resources cannot be allowed to go on. I am of opinion that the time has now come to reduce the man-power in the Middle East base in the first instance by at least one-quarter; *i.e.*, the 662,000 on the ration strength and the 242,000 directly employed civil labour, totalling 904,000, should be reduced to 680,000.

The Secretary of State for War should put forward proposals to bring these reductions into effect.

Prime Minister to Foreign Secretary   19 Dec 44

Why do we want to withdraw our garrison from Persia? I should have thought it was the greatest mistake. It is small, competent, and mainly Indian. There are so many questions unsettled that we had much better stay. It is easy to go and hard to return. I am not pressing on military grounds for any such withdrawal at the present time.

Prime Minister to Foreign Secretary   28 Dec 44

Surely it would be possible to reassure these men who have fought in the Polish divisions that whatever else happens to them the British Empire will find them a home. This is a sad letter from the son of Aubrey Herbert. He was found medically unfit for the British Army, so has been fighting for the last two years with the Poles.

JANUARY 1945

Prime Minister to Secretary of State for War   1 Jan 45

It is not possible for me, on the information available, to determine how far the high rates for sickness in the Burma campaign have been due to failure of unit discipline in enforcing anti-malarial orders and how far to any inadequacy in our medical service; but certain questions present themselves, on which I should be glad to have your observations.

1. To what extent have specialists on tropical hygiene been attached as consultants to the Army at home and overseas?

2. Have the specialists been given independent civilian status, or are they posted in subordinate ranks in the Service hierarchy?

3. What civilian professional advice is available to the Secretary of State and the Director-General in considering medical matters? I understand that there are two committees for this purpose. How often have these met in each of the last five years, and what sort of problems have been referred to them?

These questions are not intended to imply any condemnation of the Royal Army Medical Corps. I have no reason to criticise their work. I am sure you will agree however that the drainage of man-power due to tropical diseases has been of such magnitude that every aspect of the matter requires most searching examination.

Prime Minister to Chief of Air Staff   2 Jan 45

German air attack on our airfields in Belgium. It is very easy to disperse aircraft, and especially to separate large ones by sandbags, etc. I doubt very much whether the question of dispersion was examined as such and rejected. It seems more probable that the aircraft congestion was allowed to grow and no preparations were made. After all, we have had many of these airfields in our possession for months. No doubt it was their crowded state that led the enemy to attack. What will be done now? Will the airfields be left in their congested state, or will they be properly dispersed? I should have thought dispersion would be part of the drill the moment there was time to spare. I cannot consider the incident satisfactory, and I should like to have a report upon the points I have mentioned.

2. There is another incident which requires your attention, and that is the surrender at Kafissia of about 700 R.A.F. ground forces. These men were mostly of the non-combatant variety, but, in spite of several warnings, were left out at this detached station seven miles away from Athens. They had, I believe, two platoons of infantry to defend them. There appears to be a bad arrangement between the military and the Air. In a matter like this the military [command] should have recalled this party instead of letting them linger on, with the inhabitants, on a precarious footing. I fear the sufferings of the prisoners may have been very severe. Naturally I asked Field-Marshal Alexander to have·a searching inquiry made. This is now taking place, but I wish particularly to know how many of these men had rifles and what training they had in rifle fire. One airman told the Field-Marshal, while we were together at one of the advanced posts, that they were only allowed five rounds for practice per annum. Everybody—I repeat, everybody—who wears the King's uniform should be capable of fighting, if it be only with a pistol or a tommy gun.

Prime Minister to Minister of Production   8 Jan 45
Supplies and Relief for Europe
I am not satisfied that enough weight is being given to our present shipping difficulties in considering relief and supplies for Europe.

Pray arrange for the Supplies to Liberated Areas Committee to make an early report on, *inter alia*, the following points. Special regard should be paid to the demands for shipping under each head.

1. The supply of food and raw materials for which they now ask.

2. The repercussions on the general supply situation of any recommendations which may be made by the committee on Belgian food needs.

3. Demands likely to be made for the Dutch if their dykes are blown up and the sea is let in.

4. Demands for Italy—in the light of the 300-gram ration—and the Balkans.

5. The handing over of ships to the Belgians and the French to import for themselves, particularly from their colonies.

6. Programmes put forward by U.N.R.R.A.

7. Can (1) to (6) be achieved without impinging on the British import programme?

8. Would not delay be avoided if matters of detail concerning European economic affairs were handled in London, leaving only questions of principle for discussion in Washington?

You should make definite proposals to cover the situation for the next six months, giving the tonnage required and suggestions as to where it can be got.

Prime Minister to General Ismay and to   14 Jan 45
Sir Edward Bridges

I have been worrying about the following. Before we went to the Quebec Conference in September very sanguine views were expressed about the termination of the German war, and Staff opinion is on record that it would end before Christmas 1944. In consequence of this a great many alterations in departmental plans were made. The same process took place in the United States on a far wider scale.

2. Now however quite a different picture opens, and I am of opinion that we should be prudent to fix October 1, 1945, as the probable target date. Are you sure that the change in the military situation has been properly reflected throughout all the military and other departments concerned, and that no special action is called for from me? At present the expectation is, very hard fighting all through the summer on land, a recrudescence of U-boat activity on a serious scale from February or March, and the revived challenge of the German Air Force implied in their leadership in jet-propelled aircraft and their multiplication of fighter aircraft.

3. Pray propose to me, after consulting with the Chiefs of Staff, any measures which should be taken about the target date for the end of the German war. We can have a Staff meeting and a meeting of the War Cabinet if necessary. Anyhow, we must not be caught short. I expect a great many of the departments have already adjusted their views to the new position. The fact that they are always loth to cut down will be helpful in this matter. The last date officially fixed for the end of the German war was December 31, 1944. Have we fixed a new date? If not, must we not now do so, and how should it be announced to departments?

Prime Minister to General Ismay, for C.O.S. Committee   14 Jan 45

In view of the great operations which impend before the Rhine is reached, the question arises whether we could not destroy the permanent Rhine bridges. If the Air cannot do it, what about fluvial mines? We went far in this in the early days, but aimed then mainly at destroying the traffic on the Rhine.* The destruction of the permanent Rhine bridges seems a great objective as long as the main German armies are engaged west of the Rhine. Destroying them will not hamper us after we have passed that river because they will most certainly be destroyed by the Germans themselves once they are driven across it. It may well be that the present type of fluvial mine is much

---

* See Vol. I (Cassell's 2nd edition), p. 456.

too small to do any good; but the principle, upon which much labour was expended, is perhaps worth considering. Perhaps however the Air Force could oblige. At any rate, the fluvial mine would be very effective against floating bridges, of which surely the Germans must have a number.

2. Do not hesitate to remit the detailed study of this matter to subordinate bodies.

PRIORITIES FOR RESEARCH AND DEVELOPMENT

NOTE BY THE PRIME MINISTER   15 Jan 45

At the present stage of the war research and development projects likely to be effectively used in operations before the end of 1946 must have the highest priority.

Research workers and draughtsmen are scanty, and are needed also by industry in preparation for the change-over to peace-time production and for the development of civil air transport.

All Service research and development projects now in hand must therefore be reviewed forthwith in the light of current hypothesis about the end of the German war and the duration of the Japanese war. Those which are not likely to be used in operations on a considerable scale in the second half of 1946 should be slowed down or temporarily abandoned so as to permit the maximum concentration upon the remainder and some release of man-power to civilian production.

Departments must also review their present practice in making modifications, particularly to obsolescent weapons and equipment (including aircraft), so as to cut out all but those which are essential for operational purposes or to save life.

Prime Minister to General Ismay   19 Jan 45

Arrangements should be made to leave a large number of static anti-aircraft guns which are not needed in their positions for care and maintenance. I do not like breaking up batteries planted with so much care. The personnel can be removed and a few caretakers kept. Otherwise I am sure we shall find that in a few months we have simply stripped the western and northern parts of the country of every form of defence, and should the situation change we should have to begin almost from the very beginning. It is the personnel that we want.

2. I must take exception to the expression "low-grade infantry brigades". This should never be used again. If it is necessary to differentiate between them in any way by name they could be called "Reserve brigades".

Prime Minister to Ministry of Agriculture   22 Jan 45

I am much concerned at the potato shortage, because we had been taught to rely so much upon this form of food.

As the Minister is absent in America the department should let me have a short report showing why this shortage has come about, what measures are being taken to remedy it, and when the position will be restored. The report should not exceed one page.

Prime Minister to General Ismay   25 Jan 45

Pray give me summaries of what our Intelligence predicted [about German strategy and will to resist] on the two previous dates, and what they predict now. Each summary should be limited to fifteen lines. Nothing must be circulated until I have had an opportunity of reading these summaries. Do not let the officers concerned know that what they said a few months ago is being compared with what they say now. It would only dishearten them.

FEBRUARY

Prime Minister to Major-General Jacob   6 Feb 45

All this seems metaphysical hairsplitting. Whatever happens there must be no diminution of the 26 million tons rate of import into the United Kingdom during the German war. Further, no inroad on British stocks below those at present sanctioned. Finally, the loss, if any, to be borne operationally. Any inroad upon the basic requirements of the United Kingdom on imports must be fought out as a first-class issue between Governments. Report to me in what way this affects your metaphysics.

Prime Minister to Foreign Secretary, First Sea Lord, and   6 Feb 45
General Ismay

The Commander-in-Chief Mediterranean told me he had heard of an impending cut of 97 per cent. in the pay of the Greek Navy—*i.e.*, from about £5 a week to 3s. If this, or anything like it, were to be enforced as a part, no doubt, of most necessary and desirable economies in Greece, it would at this moment have disastrous consequences, and it can in no circumstances be allowed to occur. One way would be for us to pay the difference for three or six months. I am assured that the Greek vessels are most useful to the Commander-in-Chief Mediterranean.

Prime Minister to General Hollis   20 Feb 45

The work of British troops is rarely mentioned in the newspapers; but no one ever sees any reference to the work of English troops. Get me the best figure available of the losses sustained by the English in this war from the beginning. I imagine they amount to at least double those of all the other portions of the United Kingdom and British Empire put together; it might be three times. The civilian casualties should also be included. Let me see what can be done about this. Do not confine yourself to the particular questions I have asked, but give me some illustrations.

2. Another calculation which might be made would refer to the losses of the Cockneys. Would it perhaps be true to say that the citizens of London, military and civil, have lost more than the whole of the British Empire, or the rest of the United Kingdom, or the whole of the British Empire and the rest of the United Kingdom put together? Whether and how I shall use these facts I will myself judge. But let me know what they are.*

Prime Minister to Chancellor of the Exchequer and   23 Feb 45
Sir Edward Bridges

I am not prepared to agree to this costly and to a large extent needless investigation [into the effects of the combined bomber offensive]. The pressure in all the Service departments is of course high at the present time to find new jobs for their vastly expanded personnel. I hope I can rely upon the Treasury to tackle a demand like this in the first instance. I regard the whole process proposed as devoid of justification. I think it safe to say that no lessons of use in bombing Japan will be gathered by this special organisation. If the United States choose to embark upon such inquiries, that is only a part of their wasteful methods, which we in our impoverished condition cannot afford to imitate. As Allies however it is probable that they would let us see the results of their inquiries.

2. I am prepared myself to fight this matter through. I think the first collision should however be taken by the Treasury and that the Chancellor of the Exchequer should oppose the scheme.

MAN-POWER

DIRECTIVE BY THE PRIME MINISTER AND MINISTER OF DEFENCE
26 Feb 45

I. It is evident that we are attempting more than our resources will permit, and some restatement of the priorities which underlie our

---

* I was told that from the beginning of the war to January 31, 1945, about 830,000 people from the United Kingdom were killed, injured, missing, or captured. 664,000 of these were English and one in five were Londoners. Casualties in the rest of the Commonwealth were about 317,000.

The death-rate for Londoners was 1 in 130 and for England 1 in 165. Next came New Zealand, with 1 in 175. The other Dominions averaged 1 in 372, and the United States 1 in 775.

planning is called for in order to effect the necessary easement. Our war effort cannot be maintained unless our civil economy receives a considerable reinforcement. In particular we should aim at a total increase of about 275,000 in the first half of 1945 in the labour force allocated to the Board of Trade.

II. The following principles should guide the allocation of man-power:

(a) First priority must of course continue to be given to what is required to achieve the defeat of Germany at the earliest possible date. Above all, the first-line strength of the Army in Europe must be maintained, and such essential needs as artillery ammunition must be met. Some earlier diminution in our air strength in Europe in the latter half of 1945 can however be accepted. Man-power should not in any case be employed on the manufacture of aircraft or other munitions of war designed for use only in the European theatre which cannot be completed until after the end of 1945.

(b) Subject to (a), every effort must be made to meet the essential requirements for a reasonable expansion of civil production.

(c) I do not propose that there should be any change in the ultimate overall strength of the forces to be deployed against Japan, but there can be some delay in their build-up and equipment, including reserves. Munitions programmes should be reviewed in the light of this general principle.

(d) We should ensure that man-power is not employed in the manufacture in quantity of equipment of new and improved types, where equipment of older but serviceable types is available for the purposes of the Japanese war. Scales of initial equipment, assumed wastage rates, and scales of reserves should also be scrutinised. Items of equipment which are not absolutely essential should be eliminated. Over-insurance in provision is a luxury we cannot afford.

(e) This directive does not mean that where no equipment is available for a specific purpose the new equipment to be manufactured should not be of good quality. Nor is this directive intended to debar the manufacture of limited quantities of equipment of improved type, even where some stocks of older and serviceable types are available. But this should be confined to items of special importance.

III. The Ministerial Committee on Man-power are invited to re-examine the man-power position in the light of this directive, consulting the Chiefs of Staff and the Joint War Production Staff as necessary.

MARCH

Prime Minister to Minister of War Transport and   2 Mar 45
Chiefs of Staff

Now that the Yalta Conference is over we must decide at once how to meet the large British shipping deficit that emerged.

The import programme must not be reduced. Various economies may be possible in other fields, but it seems that the main cut will have to fall on military shipments to the Mediterranean and the Indian Ocean. The Services have demanded 122 sailings a month to these theatres over the next four months. I am thinking in terms of a ceiling of eighty sailings a month.

Pray let me know urgently what easement this would provide to our general shipping position, and what effect it would have on our military plans.

Prime Minister to Foreign Secretary   5 Mar 45

I have every intention of working to the utmost for a Poland free to manage its own affairs and to which Polish soldiers in our service will be glad to return. If this fails we must provide for the Poles in arms inside the British Empire, which can easily accommodate such brave and serviceable men. In the first instance no doubt they would be employed upon garrisoning Germany, with proportionate relief of our military burden.

2. However, there may always be a certain number of individual Poles who will not wish to go back to Poland because of their inveterate hostility to Russia. For these, who will not be many if we are successful, the alternative of British citizenship must be open, even if they are unreasonable in their views about the kind of life open to them in Poland.

Prime Minister to Chancellor of the Exchequer and   5 Mar 45
Sir Edward Bridges

I am notifying the three Service Ministers concerned that I wish the matter [of the British Bombing Research Mission] to be dealt with in the first instance between them and the Treasury, and that it is only after it has been thoroughly examined by the Treasury, as such a proposal would be in peace-time, and we have the Treasury safeguards against this expenditure fully operative, that it should be brought before the Cabinet. In particular, the exact cost of this grandiose proposal would be ascertained, as well as the injury to other aspects of our desired revival of civil life.

Prime Minister to Minister of Aircraft Production   6 Mar 45

At the beginning of the war the buildings of Malvern School were requisitioned by the Admiralty, and were, I understand, subsequently taken over by the Ministry of Aircraft Production for research purposes. The Malvern boys found refuge at Harrow, whose numbers had been depleted owing to the raids on London; but the prospective entrants to Harrow have now greatly increased, and while Malvern remains at Harrow it is impossible to find accommodation for them. The Governors of both schools are very anxious that the existing arrangements should be terminated, and I should be glad if you would let me know whether there is any possibility of your Ministry being able to dispense with the buildings at Malvern, in order that the school may now return there.

Prime Minister to General Ismay, for C.O.S. Committee   7 Mar 45
I thought the Polish divisions in Italy had been equipped with British equipment. Is there no more of this available? It would at any rate be desirable to begin the equipment of these two extra Polish units, and for this purpose we should not hesitate to draw upon reserves.

2. In regard to the exact scale of equipment, a considerable tolerance should be allowed. No rigid rule should apply to formations which will be useful and necessary as the war draws to a close but which may not have to sustain its brunt. The use of these Polish units in occupying Germany will take the strain off our man-power, which will be of great importance to us in the period immediately following the collapse of Germany. For this purpose they would not require the equipment, transport, etc., which has been deemed necessary to full combat readiness.

Prime Minister to Minister of Food and Minister of   9 Mar 45
War Transport
The attached proposals for increased production of pigs and eggs were made to me by the Minister of Agriculture. Certainly they seem to offer great advantages, and should not be impracticable in view of the fact that wheat is more plentiful than almost any other food commodity.

Kindly let me have your joint views about this.

Prime Minister to Chancellor of the Exchequer   9 Mar 45
The following points have occurred to me, on which I should like to hear from you.

The first is about children's allowances. Surely these sums should be free from income tax, and should be considered the property of the children? Would not this save a great many complications? How much would it cost?

2. I see to-day in the *Times* that the Bill empowers the Ministers to reduce or withhold allowances payable to the families of Service men

and women. Considering that the object in view is to encourage the birth and extra nourishment of children, I cannot see why this additional benefit should be denied to those classes; in fact, I should think that the prejudice which such a decision would cause would greatly detract from the popularity of the measure, for which nevertheless an immense annual sum is to be paid. . . .

Prime Minister to Foreign Secretary   9 Mar 45
There is no question of the *new* Greek Army and Navy being paid on the British Middle East rates. Their remuneration must be proportionate to the conditions in the Greek State.

2. I am anxious however lest the Navy, which has done so well, and the small military forces, who are to be the key men of the new Greek Army, should be suddenly confronted with such a violent pay cut. Would it not be possible to give them six months' notice that they will revert to Greek rates? It would be a great pity to upset them at the moment, especially when we are so anxious to get our troops away. I do not suppose the amount at stake is very large. Could we not pay the difference between our rates and the Greek rates to these men for the next six months? How much would this cost?

Prime Minister to Minister of Labour   10 Mar 45
Thank you for your minute [about labour controls and persons released from the armed forces]. I confess I was not aware that the universal, overriding power of direction was to be applied after the defeat of Germany. Would you, for instance, in theory be entitled to take an author, a playwright, or an artist from his studies, although these yielded no immediate practical results, and direct him down a coal-pit, the said person having previously been released under Class A from the Army? Would you be entitled to take a man of the above class who is looking after his own farm, although not the actual farmer, and send him into a steel works? Are you entitled to take an officer or a soldier who has had four years' fighting service overseas, and says he wants a year of rest and leisure and is making no claim upon the State for maintenance, and put him in a stone quarry?

2. It seems to me that these dire and overwhelming powers can only be asserted by the State in times of mortal peril. I always understood from you in our very brief talks about these matters that the Class A men and those released in their proper turn from the Army would be free to seek what work they chose, but that those who were released as a privilege before their turn on account of being key men would of course be subject to direction into particular industries.

3. You speak of "an essential job of the highest urgency". What kind of job have you in mind? It is impossible to compel a person to

discharge extremely high-grade functions. You cannot compel a man, for instance, to be a scientist. You can however compel him to sweep out the laboratory. But surely the latter is not "an essential job of the highest urgency for which it is impossible to find a suitable candidate". Clearly you cannot compel a doctor to doctor if he does not wish to do so. Do you then claim the power to order him to grow mangel-wurzels? Will you not allow a gipsy to go back to his caravan after serving his time? May not an ex-soldier with decorations for gallantry become a bookmaker if he chooses? Are you prepared to take a man who is fully employed at some special work of his own and transfer him to a basic industry against his will? If not, what classes of occupation do you regard as reserved for this purpose? What happens to the Bevin Boys after the German war is over? Are they to be compelled to go down the pits when some of them would like to go to the universities or go and fight against Japan?

4. It would be easy to multiply the instances where the claim would be to direct persons who have paid their full debt to war-time compulsion against their will into high-class functions. To direct them into ordinary manual labour may only be to displace those who have need of the job. According to my principles, only mortal peril can justify these extreme invasions of individual liberty.

5. I do not myself consider that the power of universal direction can be maintained over all the men and women in the country for very long after the German war is at an end. I was not aware that the Cabinet had lent itself to such far-reaching decisions of principle. So much is going on, and I have much to do, but I certainly claim to have the matter discussed or reviewed.

6. I may say that the above is all based on principle and theory. I am sure that in practice while Minister of Labour you would never allow any of these hard cases to arise. Nevertheless a Briton is supposed to be a free man except in times of national emergency.

Prime Minister to Minister of Production   10 Mar 45

Thank you for your minute of February 19 about penicillin. I am glad that we are at last producing really substantial quantities here. I am disturbed however to hear that the quality of penicillin produced in this country is said to be inferior to that from the United States. If so, I trust that this will soon be remedied. We must not sacrifice quality for quantity.

No doubt you will let me have a further report when supplies become clearly surplus to Service needs, outlining proposals for increased distribution to the civilian population.

Prime Minister to Lord President of the Council   10 Mar 45

The expression "mandate system" was only used at Yalta to limit the territories which would come within the scope of discussions affecting "territorial trusteeship". This is necessary in view of the disappearance of the old League of Nations, on whose authority the mandates were held. It in no way governs any arrangement that may be made for the future. We are certainly not committed to the maintenance of the mandate system; but there is no question of subjecting any non-mandated British territories to any form of territorial trusteeship unless we choose to do so of our own accord. I should myself oppose such a departure, which might well be pressed upon nations like Britain, France, Holland, and Belgium, who have great colonial possessions, by the United States, Russia, and China, who have none.

Prime Minister to First Lord   10 Mar 45

I was shocked to hear your Civil Lord say in the debate [on the Navy Estimates] that now while the war was on was the best time to procure the enlargement of the Royal dockyards, etc. You must realise that after the war the Air will take a very large part of the duties hitherto discharged by the Royal Navy. In any case, decisions affecting post-war policy ought not to be taken, apart from the ordinary restrictions of peace-time Treasury examination (as long as that system prevails). Pray let me have a statement on your policy and plans for extending the Royal dockyards at the present time, together with estimates of the cost involved.

Prime Minister to Chancellor of the Exchequer, President   14 Mar 45
of the Board of Trade, and Minister of Agriculture

I learn that considerable dissatisfaction exists about the Government decision regarding war gratuities for the Women's Land Army. If a scheme could be devised which would not have wide and costly repercussions I think it would be wise and just to adopt it. I should be glad if you would examine the following scheme which has been put to me, and let me know whether it falls within the definition just mentioned.

The proposal is that women on leaving the Land Army should receive a special allocation of coupons and a proportionate money grant designed to take account of the fact that they have surrendered coupons in excess of the special industrial allowance for their uniforms.

Pray have this examined urgently. Meanwhile I am in no way committed.

Prime Minister to Secretary of State for War   18 Mar 45

I have been studying the latest figures of battle casualties in Europe as contained in the C.I.G.S.'s summary for March 10.

At the middle of February, I understand, the American Army was $2\frac{1}{2}$ times as large as ours, while on D Day the numbers were roughly equal. Taking the average bearing of the strength figures from June 6 to the middle of February, there were roughly twice as many Americans exposed to risk as British and Canadian troops.

Since D Day the Americans have lost 71,000 killed to our 33,000, a proportion of $2\frac{1}{8}$ to 1. Hence the proportion of killed has been very much the same between the two Allies per man on the strength, with a slightly heavier loss to the Americans.

For every American killed they report $4\frac{1}{4}$ wounded; for every British and Canadian soldier killed we find $3\frac{1}{4}$ wounded. In Italy also the American ratio of wounded to killed is higher than the British, though admittedly the variation is not so large as in Northern Europe—3.5 to 3.1. Unless we are to assume that the British soldier succumbs to wounds more easily than the American it seems very difficult to account for the discrepancy, save by the definition of "Wounded", as both Armies are exposed to the same risks by the same weapons. I should be interested to know what is the War Office explanation of this point.

Prime Minister to General Ismay, for C.O.S. Committee,   18 Mar 45
Secretary of State for Air, and C.A.S.

This complaint [from the Netherlands Foreign Minister] reflects upon the Air Ministry and Royal Air Force in two ways. First, it shows how feeble have been our efforts to interfere with the rockets, and, secondly, the extraordinarily bad aiming which has led to this slaughter of Dutchmen. The matter requires a thorough explanation. We have had numerous accounts of the pin-point bombing of suspected Gestapo houses in Holland and of other specialised points; but good indications are given in this account of the wood where the rockets are stored, and of the railway lines which, if interrupted, would hamper the supply of rockets. All this ought to have been available from Air Intelligence. Instead of attacking these points with precision and regularity, all that has been done is to scatter bombs about this unfortunate city [The Hague], without the slightest effect on their rocket sites, but much on innocent human lives and the sentiments of a friendly people.

Prime Minister to Chancellor of the Exchequer   18 Mar 45

I see that the Foreign Office salaries are up by £666,893. What are the principal causes of this very large increase? How was this settled with the Treasury? Did your officials go into it in detail, or is it simply done under war powers?

Surely Treasury control should be resumed in an effective manner now that the peak of the war has been passed in Europe.

Prime Minister to Home Secretary   18 Mar 45

I see no reason why your leaflets should not be dropped now, and you should talk to the Minister of Information about some broadcast features for the Channel Islands. I doubt if it will be possible for me to introduce the subject into my broadcasts. These have to be conceived as a whole, and not as a catalogue of favourable notices. I have not made a broadcast to the nation for over fifteen months.

Prime Minister to Chief Whip   18 Mar 45

Look me up the motion or Bill moved about 1888 by Curzon, Brodrick, and Wolmer asking for relief from going to the Lords on their fathers' decease.

It is a terrible thing for a father to doom his son to political extinction, which must happen to many if they have not had time to make their way in the House of Commons.

Prime Minister to Field-Marshal Wilson   19 Mar 45

I should be glad if you would in private conversation convey the following to General Marshall: The Prime Minister feels that it would look very bad in history if we were to let the French force in Indo-China be cut to pieces by the Japanese through shortage of ammunition, if there is anything we can do to save them. He hopes therefore that we shall be agreed in not standing on punctilio in this emergency. You should also express my good wishes and compliments.*

Prime Minister to Mr. Assheton   19 Mar 45

I notice in the newspapers that the Central Office or Party Chiefs have issued instructions that no one over seventy should be tolerated as a candidate at the forthcoming election. I naturally wish to know at the earliest moment whether this ban applies to me.

Prime Minister to First Lord and First Sea Lord   25 Mar 45

It seems to me that a case has been made out (a) for the better and quicker outfitting of released trawlers; (b) for the release of more trawlers, especially the high-grade ones. Pray let me know what you will do. If you are unable to do anything we shall have to bring it before the Cabinet.

Prime Minister to Mr. Norman Brook   27 Mar 45

The Housing Squad meeting should be at 10 p.m. on Wednesday.

I wish to raise particularly the supply of labour in the first year after Germany is beaten, including of course releases from the Army, formation of special units for housing construction, releases from munitions works, etc. I cannot accept a ceiling of 800,000 or an average bearing of 500,000 for the first year after the German defeat. I am prepared

* General Marshall acted the next day.

to go into the whole question of releases from the Army from the moment that organised resistance in Germany has ceased. All men brought home out of turn from the Army will naturally be in the directed class. I am also considering the formation of special mobile units, on a two or three years' contract, to prepare sites or erect bungalows. One would naturally draw in the main upon Army engineers for this purpose.

We must clear up the position of the local authorities. The strength of the national hand must be more predominant while the emergency lasts.

Prime Minister to Secretary of State for Air　　　　　　　28 Mar 45

You have no grounds to claim that the Royal Air Force frustrated the attacks by the "V" weapons. The R.A.F. took their part, but in my opinion their effort ranks definitely below that of the anti-aircraft artillery, and still farther below the achievements of the Army in cleaning out all the establishments in the Pas de Calais, which so soon would have opened a new devastating attack upon us in spite of all the Air Force could do.

As to V2, nothing has been done or can be done by the R.A.F.

I thought it a pity to mar the glories of the Battle of Britain by trying to claim overweening credit in this business of the "V" weapons. It only leads to scoffing comments by very large bodies of people.

Prime Minister to Chancellor of the Exchequer　　　　　　29 Mar 45

Our progress on the Continent seems to justify a revision of the Yalta dates for the end of the European war. I have referred this to the Chiefs of Staff, who have asked for time to consider the question. No doubt we shall know more in a fortnight; but we cannot delay the necessary studies for a single day. Your Man-power Committee should therefore plan now on the assumption that the European war will end not later than May 31, bearing in mind that if we should suffer an unexpected reverse in Germany we might be forced to revert to the Yalta dates, and conversely that an earlier collapse is also a possibility.

President Roosevelt to Marshal Stalin*　　　　　　　　　29 Mar 45

I cannot conceal from you the concern with which I view the development of events of mutual interest since our fruitful meeting at Yalta. The decisions we reached there were good ones, and have for the most part been welcomed with enthusiasm by the peoples of the world, who saw in your ability to find a common basis of understanding the best pledge for a secure and peaceful world after this war. Precisely because of the hopes and expectations that these decisions raised, their fulfilment is being followed with the closest attention.

*See Chapter XXV, p. 382.

We have no right to let them be disappointed. So far there has been a discouraging lack of progress made in the carrying out, which the world expects, of the political decisions which we reached at the Conference, particularly those relating to the Polish question. I am frankly puzzled as to why this should be, and must tell you that I do not fully understand in many respects the apparent indifferent attitude of your Government. Having understood each other so well at Yalta, I am convinced that the three of us can and will clear away any obstacles which have developed since then. I intend therefore in this message to lay before you with complete frankness the problem as I see it.

Although I have in mind primarily the difficulties which the Polish negotiations have encountered, I must make a brief mention of our agreement embodied in the Declaration on Liberated Europe. I frankly cannot understand why the recent developments in Roumania should be regarded as not falling within the terms of that agreement. I hope you will find time personally to examine the correspondence between our Governments on this subject.

However, the part of our agreement at Yalta which has aroused the greatest popular interest and is the most urgent relates to the Polish question. You are aware of course that the Commission which we set up has made no progress. I feel this is due to the interpretation which your Government is placing upon the Crimean decisions. In order that there shall be no misunderstanding I set forth below my interpretation of the points of the agreement which are pertinent to the difficulties encountered by the Commission in Moscow.

In the discussions that have taken place so far your Government appears to take the position that the new Polish Provisional Government of National Unity which we agreed should be formed should be little more than a continuation of the present Warsaw Government. I cannot reconcile this either with our agreement or our discussions. While it is true that the Lublin Government is to be reorganised and its members play a prominent rôle, it is to be done in such a fashion as to bring into being a new Government. This point is clearly brought out in several places in the text of the agreement. I must make it quite plain to you that any such solution which would result in a thinly disguised continuance of the present Warsaw régime would be unacceptable and would cause the people of the United States to regard the Yalta agreement as having failed. It is equally apparent that for the same reason the Warsaw Government cannot under the agreement claim the right to select or reject what Poles are to be brought to Moscow by the Commission for consultation. Can we not agree that it is up to the Commission to select the Polish leaders to come to Moscow to consult in the first instance and invitations be sent out accordingly? If this could be done I see no great objection to having the Lublin group come first in order that they may be fully acquainted with the agreed interpretation of the Yalta decisions on this point. In order to facilitate the agreement the Commission might first of all select a small but representative group of Polish leaders who could suggest other names for the consideration of the Commission. We have not and would not bar or veto any candidate for consultation whom Mr. Molotov might propose, being confident that he would not suggest any Poles who would be inimical to the intent of the Crimean decision. I feel that it is not too much to ask that my Ambassador be accorded the same confidence. It is obvious to me that if the right of the Commission to select these Poles is limited or shared with the Warsaw Government the very foundation on which our agreement rests would be destroyed.

While the foregoing are the immediate obstacles which in my opinion have prevented the Commission from making any progress in this vital matter, there are two other suggestions which were not in the agreement but nevertheless have a very important bearing on the result we all seek. Neither of these suggestions has been as yet accepted by your Government. I refer to:

(1) That there should be the maximum of political tranquillity in Poland and that dissident groups should cease any measures and counter-measures against each other. That we should respectively use our influence to that end seems to me so eminently reasonable.

(2) It would also seem entirely natural, in view of the responsibilities placed upon them by the agreement, that representatives of the American and British members of the Commission should be permitted to visit Poland.

I wish I could convey to you how important it is for the successful development of our programme of international collaboration that this Polish question be settled fairly and speedily. If this is not done all of the difficulties and dangers to Allied unity which we had so much in mind in reaching our decision at the Crimea will face us in an even more acute form. You are, I am sure, aware that genuine popular support in the United States is required to carry out any Government policy, foreign or domestic. The American people make up their own mind, and no Government action can change it. I mention this fact because the last sentence of your message about Mr. Molotov's attendance at San Francisco made me wonder whether you give full weight to this factor.

Prime Minister to Chancellor of the Exchequer   30 Mar 45

I am greatly disturbed by your minute [attaching the report of the Committee on Public Relations Branches of Government Departments], and most certainly, should the matter be brought to Cabinet, I shall urge that an arbitrary 25 per cent. cut should be made throughout the Public Relations services, and that a Ministerial committee be set up to share it between the different departments. Perhaps in the meantime you will let me know what were the charges for Public Relations work the year before the war, and what are proposed for the present year, or what were actually charged in the last completed year.

I always do my best to help in coping with expenditure, and you are right in thinking I am very disappointed.

Prime Minister to First Lord and First Sea Lord   30 Mar 45

In view of the progress on the Western Front and the fact that the Admiralty forecast of shipping losses by U-boats has been proved to be so far much beyond the actual results, I must ask that 10 per cent. of the high-grade trawlers are released in April and 10 per cent. in May. We can discuss the numbers of the June releases later on in the light of the war situation. I trust you will be able to conform to this direction without my having to bring the matter before the War Cabinet as a matter of confidence. The food supply of Britain has a military as well as a national significance.

2. Report also whether it is possible to send these ships to sea for fishing while they retain their guns, and what is the estimated time for conversion to shipping purposes in each case. When I asked for more speed in converting the older trawlers which had been released I noted that a large proportion of them were so worn out as to be unfit to move from port to port. This was therefore not a great sacrifice.

APRIL

Prime Minister to Dominions Secretary   3 Apr 45
  (*Copy to Foreign Secretary*)

You and I are both determined to make the strongest possible stand against Russia's behaviour since Yalta about Poland. The only question is how best to do it. Obviously we cannot change the course to which we and the United States are committed, namely, to urge Russia to send the strongest possible delegation to San Francisco. Anyhow, Russia's attitude must be judged in relation to the very serious telegrams dispatched by the President and by me in full and perfect accord. No one can form any opinion on this matter till we receive replies to these.

2. If the replies are wholly hostile I think it most unlikely that Russia will come to San Francisco. She will prefer to fight it out on the side of the Lublin Poles. The question will then arise whether the San

Francisco Conference should be held or not. We have not yet reached this point. But, looking ahead at it, Anthony and I both consider that it would be a great blow to our cause and prestige and also to the cause of a free Poland if the sulkiness of Russia prevented this World Conference from being held. The Russians would feel that their mere abstention paralysed world action. Although I have never been at all keen on this Conference, I should then in that event become very keen upon it. It would be an admission of weakness on the part of the whole of the rest of the friendly world which would far exceed any inconvenience of holding the Conference at the present time. I am quite sure that in the event of a Russian refusal we should confront her with the blunt fact that the civilised world is not afraid of her and is not dependent upon her, and that their organisation would proceed with her outside unless she will come inside. No doubt we should make suitable modifications if she definitely remains aloof.

3. The picture of all the United Nations, with Britain and the United States at their head, being put off their stroke by a mere gesture of insolence from Stalin and Molotov is a bad one. The picture of the United States and Britain holding a Conference without Russia which all the United Nations attend is an immense rebuke to Russia. Moreover, the military power of Britain and the United States is at this moment greater than that of Russia and comprises practically the whole world outside Russian territory and the conquered satellite Powers. There is no doubt on which side the hopes of humanity will rest.

4. Therefore should events take the unhappy turn assumed in this note of mine I should without the slightest hesitation proceed to carry forward the Conference. *Les absents ont toujours tort.* Nothing would show the Soviets where they get off more clearly. Thus I consider that this is the best tactical path to follow, and also the rightful moral path for the great mass of the world to follow, headed by the English-speaking armies and forces of all kinds.

5. I give you these reasons, which will determine my action irrespective of any other consequences which may arise. Pray give them careful thought and see whether they do not carry out your purposes in the most effectual manner. Also always remember, please, that there are various large matters in which we cannot go farther than the United States are willing to go. Anyhow, we have to wait (*a*) until the Russian response to the Notes we have sent Stalin is received; (*b*) till we know whether the Russians will reject their San Francisco observations; and (*c*) till we know whether in that event the Americans wish to persist or not. If they do I shall certainly urge that we back them up.

6. You say you can "only trust to your own instincts". I have offered you reasons at the end of a long day.

Prime Minister to Paymaster-General, General Jacob,　　　3 Apr 45
and Private Office

I have not been able to read these papers [about the rocket sites in Northern France], but I am quite clear that the services rendered by the Army in sweeping up the coast were the prime cause of our deliverance. Next in order comes the anti-aircraft artillery, and the Air may divide the credit with them; but anything more foolish than to leave out the great services, alone decisive, of Field-Marshal Montgomery's army, including the Canadians, in sweeping the coast can scarcely be imagined.

Prime Minister to First Lord and First Sea Lord　　　3 Apr 45

Further to my minute about release of trawlers, it may interest you to know that the figure to which my 10 per cent. cut referred was about 425 requisitioned British trawlers in home waters which are the cream of the fishing fleet. I will accept for April forty-two of these, but I shall require, unless there is a change in the war situation, 20 per cent. for the month of May. We can consider June later. I am expecting the Admiralty to render this great service to the British nation by handing over these vessels with extreme promptitude and seeing that they are made good for fishing, so far as the Admiralty is concerned, in the full spirit of a naval evolution. Perhaps you do not realise that "British-caught white fish" in 1938 was 750,000 tons, but in 1944 only 240,000 tons.

Prime Minister to Minister of Agriculture, Minister of　　　3 Apr 45
Food, and Minister of War Transport

Your minute about increasing our home production of pork and eggs.

Everything depends upon the progress of the German war. I have given as a guesswork figure towards which we should now work the end of May 1945, but it may well be that the end will come before this. At any rate, before the end of April we should be in a position to take a much more sure and precise view. There is no reason why preparations should not be made and shipping assigned on the basis of the war ending on April 30, but let me see these in detail, so that the Chiefs of Staff can consider whether it is an undue risk to run. It is a good cause.

2. Do not, I pray you, give up the egg scheme and the chicks necessary for full-scale production in the spring of 1946.

3. On no account reduce the barley for whisky. This takes years to mature, and is an invaluable export and dollar producer. Having regard to all our other difficulties about exports, it would be most improvident not to preserve this characteristic British element of ascendancy.

4. The Minister of War Transport must be bold and not let himself be overlaid by the sombre deadweight of military demands. The people of this country are entitled to have their minimum food supplies. He will run a terrible risk if a lot of shipping is thrown on his hands before the end of April as a result of the German submission.

5. The Treasury should certainly be asked to sanction the purchase of 200,000 tons of cereals from the Plate on the assumptions mentioned. How much would it cost?

6. I hope you will meet together and take an altogether bolder line on the very considerable lead I have given you, which I will do my best to support. The revised paper could be circulated to the Cabinet. Let me see it first.

Prime Minister to Sir Edward Bridges   4 Apr 45

Arrange with the Admiralty to bring up both cases of the transfer of warships to Canada and Australia at some Cabinet meeting to which the Dominion Ministers are summoned. Then make them a full and free presentation there and then across the table. The Admiralty should propose this. No financial considerations should be adduced. We owe too much to Canada in money alone, and the effect of gestures like this upon both Dominions concerned will be achieved far better than by arguments about trading off the value of the ships against certain financial considerations. This is not a moment for a "penny-wise, pound-foolish" policy. We must either keep the ships or give them. If the Admiralty think they can be given, now is the time to make the presentation in the most friendly form. Cast your bread upon the waters; it will return to you in not so many days.

Bring this before the parties involved.

Prime Minister to Foreign Office   4 Apr 45

Attention should be drawn to the misspelling "inadmissable". I have noticed this several times before in Foreign Office telegrams.

Prime Minister to General Ismay, for C.O.S. Committee   7 Apr 45

My view [about future operations in South-East Asia] is the following, and, *if necessary*, I will communicate with General Marshall through Field-Marshal Wilson on the subject on this sort of line:

"In the last war and this we have found it in all our affairs most detrimental to give absolute overriding priorities to any one set of operations or supplies. Once this is agreed to those who own the absolute priority take the last ounce of their requirements, of which they become the sole judges, irrespective of the ruin of the lesser priorities. For instance, a department that needs five tons of a commodity and has the absolute priority will not hesitate to take the five tons without consideration of other vital and important priorities which all together may need no more than a hundredweight of the said commodity. Widespread havoc is therefore wrought without a due sense of proportion.

"We have always in our dealings in this war, and to a large extent on both sides of the Atlantic, accompanied and modified first priorities by assignments. We could certainly not be bound to accept absolute priorities for the main effort without agreements being reached as to the very much smaller, but none the less essential, supplies required for other operations. We should resent and resist to the utmost of our capacity any proposal which proceeded without the slightest regard to operations to which we attach intrinsic importance, though undoubtedly on a much smaller scale than the main operation. We therefore hope that these matters may be settled by reasonable discussion."

Prime Minister to General Ismay, for C.O.S. Committee   8 Apr 45

Proposed British Bombing Research Mission. I much regret that I do not find myself in agreement with your proposal to spend such very large sums of money and use so much highly skilled personnel, required in the starting up again of our own country, upon what I judge to be a sterile task. You have asked for 1,000 persons, of whom one-half are high-grade experts. I cannot conceive that any results can be yielded commensurate with such an expenditure of our remaining resources.

I offer you thirty experts, who, with the large number of air groundsmen who will be scattered about Germany during the next few months, and who could be drawn upon temporarily, in any locality, should be quite sufficient to find out the particular points about which you are interested.

Prime Minister to Foreign Office   8 Apr 45

This war would never have come unless, under American and modernising pressure, we had driven the Habsburgs out of Austria and Hungary and the Hohenzollerns out of Germany. By making these vacuums we gave the opening for the Hitlerite monster to crawl out of its sewer on to the vacant thrones. No doubt these views are very unfashionable. . . .

Prime Minister to First Lord   8 Apr 45

I will certainly try to fit in an opportunity for seeing you and the First Sea Lord this week, but I cannot hold out any expectation that my most modest request [for the release of trawlers] should not be met now, so that our fishermen can get to work and relieve the great strain on British food. There should be no delay in taking action.

Prime Minister to President of the Board of Trade   14 Apr 45

It is absolutely essential to increase the supply of civilian clothes. The

suggestions I have seen that there will be a critical shortage after V.E.
Day are intolerable, and it would be a grave reflection on the Board of
Trade if such a thing occurred. You should be getting a considerable
increase of labour as a result of my directive of February 26, and of
this I trust that the clothing trades will receive an early and adequate
quota.

If this is not enough to ensure sufficient supplies of civilian clothing
in the autumn I am quite willing to consider diverting from making
military clothing up to 20 per cent. of the man-power engaged, even if
this entails delay in fulfilling Service demands.

MAN-POWER IN 1945

DIRECTIVE BY THE PRIME MINISTER AND MINISTER OF DEFENCE

14 April 45

At the Yalta Conference it was agreed that for the purpose of
planning production and allocating man-power the earliest date for
the end of the German war should be assumed to be July 1, 1945, and
the latest December 31, 1945. The deterioration in the German situation in the few weeks that have elapsed since the Conference has been
far more rapid than was foreseen. The disruption of the Western
Front, the annihilation of a great part of his armies in the West, and
the virtual exhaustion of his oil stocks have presented the enemy with
a situation which he cannot retrieve. The time has now come when
we can look forward with confidence to the end of organised resistance
this summer. I therefore propose that we should now adopt as a firm
date for all plans May 31. If our preparations are related to that date
we shall not be materially at fault if the end of the war comes a week
or two earlier or later.

2. It will be the duty of the Chancellor of the Exchequer's Man-
power Committee to discharge the same tasks in the period of national
demobilisation as they have undertaken in the period of national
mobilisation. They should first call for a report from the Service
departments on the state of readiness of their machinery for release.
It is of the utmost importance that releases from all three Services should
begin not more than six weeks after the end of the German war.

3. The Chiefs of Staff should present without delay to the Defence
Committee a statement of the size of the forces which they plan to
deploy against Japan. These should be related strictly to what can be
brought into action in time to play a part within the assumed duration
of the Japanese war. When this calculation has been completed the
level of munitions required to sustain these forces should be worked
out, and administrative preparations in India and the Far East should
be adjusted as may be found necessary.

4. The Man-power Committee should then draw up a balance-sheet to cover the period June 1 to December 31, 1945. I call upon the Service departments and the Ministry of Labour to ensure that much greater progress is made during this period in reducing the Services and munitions industries to the ultimate State II levels than has hitherto been suggested.

5. I trust that the processes I have outlined above may be brought to a conclusion so that the results can be presented to the War Cabinet by the middle of May.

Prime Minister to Major Lloyd George and   15 Apr 45
Mr. Geoffrey Lloyd

I understand that the position with regard to easements of petrol rationing after the end of the German war was provided for during the discussions in Washington.

I attach importance to easing these restrictions on the civilian population as soon as possible after V. Day. If the German war ends soon and the supply position permits it should be possible to restore the basic ration on June 1. Please let me have a report on the administrative preparations.

Prime Minister to Secretary of State for Air and C.A.S. 16 Apr 45

I never understood why all this British Commonwealth Air Training Plan in Canada was broken up so abruptly. Many people have been distressed thereby. The nucleus should have been preserved. I have however agreed to your various telegrams, but please press me no further.

Prime Minister to Minister of Works   23 Apr 45

I saw a paragraph in Saturday's newspapers that no preferential treatment of any kind would be given to the repairs at Buckingham Palace, and that the King and Queen would be treated exactly like ordinary persons. I deprecate this kind of propaganda, and do not consider it expresses the feelings of the people of this loyal country.

Moreover, the Royal palaces are conserved to the nation and serve public purposes. It may well be that in the next six months many of the principal personages in Europe may have to be received and entertained in Buckingham Palace or St. James's, and the public service would suffer if these were not kept in a reasonable state of repair.

Prime Minister to Foreign Office   23 Apr 45

I do not consider that names that have been familiar for generations in England should be altered to study the whims of foreigners living in those parts. Where the name has not particular significance the local custom should be followed. However, Constantinople should never be abandoned, though for stupid people Istanbul may be written in brackets after it. As for Angora, long familiar with us through the Angora cats, I will resist to the utmost of my power its degradation to Ankara.

2. You should note, by the way, the bad luck which always pursues people who change the names of their cities. Fortune is rightly malignant to those who break with the traditions and customs of the past. As long as I have a word to say in the matter Ankara is banned, unless in brackets afterwards. If we do not make a stand we shall in a few weeks be asked to call Leghorn Livorno, and the B.B.C. will be pronouncing Paris "Paree". Foreign names were made for Englishmen, not Englishmen for foreign names. I date this minute from St. George's Day.

Prime Minister to C.I.G.S. 25 Apr '45

What does the Intelligence Department of the War Office think the casualties inflicted by the Russians on the Germans actually amount to? Is it possible that they could have killed or captured ten million Germans, leaving a few for us and the Americans? Personally I think about half would meet the case.*

Prime Minister to Sir H. Knatchbull-Hugessen (Brussels) 26 Apr 45

It is no part of the policy of His Majesty's Government to hunt down the Archduke Otto of Habsburg or to treat as if it were a criminal organisation the loyalty which many Austrians friendly to Britain cherish for their ancient monarchy. We should not actively intervene on their behalf, being at all times resolved that in any case where we are forced for the time being to depart from the ideal of non-intervention our guide is the will of the people, expressed by the vote of a free, unfettered, secret ballot, universal suffrage election. The principle of a constitutional monarchy, provided it is based on the will of the people, is not, oddly enough, abhorrent to the British mind.

2. Personally, having lived through all these European disturbances and studied carefully their causes, I am of opinion that if the Allies at the peace table at Versailles had not imagined that the sweeping away of long-established dynasties was a form of progress, and if they had allowed a Hohenzollern, a Wittelsbach, and a Habsburg to return to their thrones, there would have been no Hitler To Germany a symbolic point on which the loyalties of the military classes could centre would have been found, and a democratic basis of society might have been preserved by a crowned Weimar in contact with the victorious Allies. This is a personal view, but perhaps you would meditate upon it.

* The War Office estimated that the Russians had inflicted seven million casualties on the Germans.

Prime Minister to Sir Edward Bridges   30 Apr 45

Ministers are entitled to keep all telegrams, minutes, or documents circulated to the Cabinet which they wrote and signed themselves. Many of the Ministers have copies of these documents, of which usually a good many were struck. These must be regarded as their personal property, except that they will be bound by the rules governing the use of official papers, which are well established. To these should be added, in the case of the Prime Minister, correspondence with heads of Governments. All other papers should be available to the departmental Ministers concerned, and they should have free access to them, although they must be deposited in the Government strong room.

With regard to the use of Cabinet or other papers in controversial disputes, this is strictly prohibited by the Privy Counsellors' oath, which requires the consent of the Crown for any use or misuse for controversial purposes of work for which all the Ministers must take collective responsibility. Ministers below Cabinet rank must return all their papers. Quotations by the Ministers of a future Government from any of the documents used in the Coalition Government must be settled by the party leaders concerned, with final reference by the Prime Minister of the day to the King. It may be observed that the use by one side may necessitate use by the other, if the public interest is to be served. . . .

Prime Minister to Secretary of State for War   30 Apr 45

You should let me see your estimate of casualties on the assumption that peace is reached in Europe by May 31, and also your man-power requirements for the occupation of Germany in a simple form, showing the broad basis you have taken and the system upon which it is proposed to act. I have not yet given a final decision on any of these points while the war was going on.

2. I have some ideas which have occurred to me which I should like tested. First, the conversion of a large number of our occupying force into mobile military police with armoured cars, jeeps, etc. Second, the formation of mobile columns capable of street warfare, with flame-throwers and artillery, including mortars, suitable to this particular task. Third, the movement to Germany at a very early stage of all the young troops training in this country in order that they may continue their training and relieve older men. Fourth, the definite transference to our zone in Germany of a certain number of our training establishments in this country, if suitable buildings can be found. Fifth, the Air Force proposals which must be linked with these. I have no doubt the Air Force proposals will be most excessive and will need most careful scrutiny. Any ideas you may have on these points I should welcome, as I am proposing to write a more lengthy paper after having made various inquiries.

Of course we must make the Germans govern themselves and face their own future, instead of lying down and being fondled by us and the United States.

Prime Minister to Minister of Fuel and Power　　　　30 Apr 45
Last Friday the War Cabinet took certain decisions on the release of miners from the forces. I should be grateful if you would report to me what this means in terms of coal, and to what extent the proposals made will meet the deficit. We must make quite sure that there is no coal shortage next winter.

MAY

Prime Minister to Secretary of State for War　　　　2 May 45
Can anything be done with military bands during the celebrations [for V.E. Day], when they occur, in London and the country?

Prime Minister to General Ismay　　　　3 May 45
I should be quite ready to address a minute to Lord Leathers if I saw the slightest sign in military quarters of curtailing their vast and unbridled use of our limited available shipping. Here is the real evil we have to face—the whole world being strangled in its development by demands for the war on Japan which have absolutely no relation to the number of warships or troops or aircraft which can be engaged there. The idea that everyone is going to suffer without the military departments making any real effort to search their own pools is one which cannot be preserved. Before I sign any minute I must have a very full statement showing stringent treatment by the three fighting Services of their own problems.

Prime Minister to Monsieur Herriot (France)　　　　4 May 45
I am delighted to learn that you and Madame Herriot have escaped safe and sound from the long ordeal which you have faced so courageously. Please accept my warm congratulations.

Prime Minister to Field-Marshal Montgomery (Germany)　　　　6 May 45
The formidable mass of helpless [German civilians] and wounded in this area must be a great problem to you. Do not hesitate to address me direct if ordinary channels work too slowly.

2. Why is it necessary to put the commanding generals into prisoners of war cages? Have we no facilities for observing the ordinary distinctions of military rank pending any war charges that may be afterwards formulated against individuals?

Prime Minister to M. Reynaud, M. Daladier, and　　　　9 May 45
M. Blum (France)
I send you my warmest congratulations on your liberation. I need

not tell you how often my thoughts were with you during the long years of your captivity, nor how glad I am to be able to rejoice with you on this day of victory.

Prime Minister to Foreign Office   14 May 45

It is of high importance that the surrender of the German people should be completed through agencies which have authority over them. I neither know nor care about Doenitz. He may be a war criminal. He used submarines to sink ships, though with nothing like the success of the First Sea Lord or Admiral King. The question for us is, has he any power to get the Germans to lay down their arms and hand them over quickly without any more loss of life? We cannot go running round into every German slum and argue with every German that it is his duty to surrender or we will shoot him. There must be some kind of force which will give orders which they will obey. Once they obey we can do what we like to carry through unconditional surrender.

2. I deprecate the raising of these grave constitutional issues at a time when the only question is to avoid sheer chaos. You seem startled at General Busch giving orders. The orders seem to be to get the Germans to do exactly what we want them to do. We will never be able to rule Germany apart from the Germans unless you are prepared to let every miserable little German school-child lay its weary head upon your already overburdened lap. Sometimes there are great advantages in letting things slide for a while. In a few days, when we have arrived at solutions to the more important questions requiring action and possibly gunfire, we will find a great many things will settle down. We can then lay down the great principles applicable to the qualities of vast communities.

3. It must of course be remembered that if Doenitz is a useful tool to us that will have to be written off against his war atrocities for being in command of submarines. Do you want to have a handle with which to manipulate this conquered people, or just to have to thrust your hands into an agitated ant-heap?

Prime Minister to First Lord and First Sea Lord   15 May 45

Let me have your proposals forthwith for returning to the fishermen the largest number of the trawlers which are in your possession, and also for 'doing your utmost to repair and help them to get to sea at the earliest moment.

We need another three or four hundred thousand tons of fish, which are all there waiting, to help us through the hard years which are coming.

Prime Minister to Minister of Agriculture   16 May 45

If all the Germans are put to work to grow food on the land what are the crops that can be started if they begin to dig on June 1?

Have you any reports on the state of their tilth?

Prime Minister to General Ismay   20 May 45

What is known about the number of Russians taken prisoners by the Germans and liberated by us? Can you discriminate between those who were merely workers and those who actually fought against us?

Could I have a further report on the 45,000 Cossacks of whom General Eisenhower speaks in his telegram? How did they come into their present plight? Did they fight against us ? *

Prime Minister to Minister of Labour, First Lord, and   20 May 45
Minister of War Transport

I am glad you are releasing the maximum number of trawlers at the highest possible rate. It is above all important to get the largest and most modern ones to sea at once. But releasing them is not enough. They must be converted for immediate service on the long-distance fishing grounds. This conversion should be given every kind of priority over warship repairing and warship building. If more labour is required I hope the Minister of Labour will be able to supply it.

It is also essential that transport be arranged to get the catches of fish away from the landing ports. The Minister of War Transport should make sure that no hauls of fish are wasted. With the threatened shortage of meat no effort should be spared to make the maximum quantity of fish available for consumption.

Prime Minister to Minister of Labour and others concerned   21 May 45

It has not been possible as yet to set out my proposals for the release of doctors in relation to the demobilisation. But the standard of medical attention available to civilians is so low that as a first step 1,600 doctors should be returned to civilian life forthwith from the Services.

Prime Minister to General Ismay, for C.O.S. Committee   27 May 45

Re-deployment and demobilisation. This about the most important thing the Chiefs of Staff Committee have to watch now. They must keep in close touch with me. On the one hand people are looking for release; on the other we must not be found lacking in the number of divisions which we have, or be unable to build them up again if necessary. Pains must be taken to preserve divisional formations. The Russian divisions are only about six or seven thousand strong. I should

---

* I was told that the Western Allies liberated nearly two million Russian prisoners of war and displaced persons. The Germans recruited a cavalry corps of 45,000 Cossacks, and used them against the Partisans in Yugoslavia.

be prepared to keep more overheads and have more divisions and relax as far as individuals are concerned, thus reducing the strength of the division while still preserving the power of remobilisation.

2. You cannot at this moment throw yourselves heartily into the business of demobilisation. I had hoped that this would be so, but I am sure that we had better get some solution in the main field of international relations. Field-Marshal Montgomery told me that he had six divisions which were to be kept for a while and six were to be made occupational. It would be prudent to keep these six occupational divisions on a mobile basis. Let me know what you think about this. Is it being done, and if not how can it be done? I do not wish to be left alone with no troops at all and great Russian masses free to do whatever they choose in Europe.

3. The above applies still more to the Air Forces, which would be our method of striking at the communications of the Russian armies should they decide to advance farther than is agreed. The Chiefs of Staff should occupy their minds with these matters, which may conceivably become of the gravest consequence. We shall know more after the next Big Three meeting.

Prime Minister to Minister of Aircraft Production   28 May 45

Thank you for your minute of March 27 about jet-propelled aircraft.

I note that instead of the sixty Meteor IIIs promised by the end of March only thirty-five have been produced, and that we shall only make fifty Vampires this year, although some 150 of the Goblin engines for these will be available. Can we not get enough jet aircraft to equip a few squadrons to obtain operational experience in the war against Japan?

I hope that the performance of the Rolls-Royce Nene will be equal to its promise. If it is it will be a remarkable engine.

Prime Minister to Lord President of the Council   30 May 45

I regard it as most important to increase supplies of fish, particularly during the coming months, when supplies of meat will be reduced. Accordingly I shall be glad if you will convene a committee consisting of the Ministers to whom this minute is sent and see that speedy continuous action is taken to ensure the following objects:

(a) Quickest possible release, repair, and conversion of trawlers, especially modern trawlers, for fishing; also minesweepers where necessary.

(b) Arrangements to ensure that they are manned and then put to fishing as soon as possible.

(c) Arrangements to ensure that all catches of fish are used, which involves guaranteed purchase of all catches, and the necessary transport and distribution arrangements from the ports. This should, if necessary, be carried out as a military evolution; and the Admiralty should arrange for men of the Royal Navy to assist in the landing of fish at ports when the available civilian labour is insufficient and cannot be supplemented from other sources.

Will you please put this in hand as soon as possible, and keep me informed by fortnightly reports.

Prime Minister to Foreign Office and War Office　　　　　31 May 45
This gallant man [General Anders] has long fought with us. I am not prepared to allow our distribution of military honours to be over-shadowed by Bolshevik prejudices. I should propose that General Anders should receive a decoration for his long fighting services.

Prime Minister to Foreign Secretary and Secretary of　　　　31 May 45
State for War
My views are as follows:
The 128,000 Polish troops who have fought for us and served under us should be formed into a Corps of Occupation in some part of the British Zone [in Germany] which is not limitrophe with the Russian Zone. The question of their recruitment from time to time requires further study. I should have thought it would not present serious difficulties. We need these men desperately, and I cannot see what the Russians have to say to it, any more than we are consulted by them when they deport a few hundred thousand people to Siberia. This is good work for the Poles, and brings them into no political collision with Russia. The War Office should not be diverted from sending over the additional division to the Twenty-first Army Group. We need every man we can get at the present time, in view of the demands for demobilisation, to which we shall have in great measure to succumb. No decision reversing these ideas should be taken until after the impending Conference of the Three.

Prime Minister to Secretary of State for War, and to　　　　31 May 45
General Ismay, for Chiefs of Staff
The Minister of Agriculture tells me that there is no hope of adequate food production next year in Germany unless the present order from the Combined Chiefs of Staff to General Eisenhower to arrest all members of the Reich Food Estate is cancelled. This order was framed before we entered Germany. It was based on an assumption that all German officials concerned were virulent Nazis. Individual officials should be judged on their records, as in the case of other German industries.

I wish you to go into this question as a matter of urgency, consulting Mr. Hudson as necessary, thereafter taking appropriate action with the Combined Chiefs of Staff.

JUNE

Prime Minister to Foreign Secretary   2 June 45

Atrocities in Bulgaria. What is this horror about Petrov's secretary being tortured? Ought we not to tell the Russians through the Ambassador that we shall publish all these facts as they have reached us, and state the facts in their full hideousness?

2. If Dimitrov has been a British agent we should defend him with the whole respectable power of the British flag. Wherever these Bolsheviks think you are afraid of them they will do whatever suits their lust and cruelty. But the Soviet Government has no wish to come out into the world smeared with such tales. Let them then behave, and obey the ordinary decencies of civilisation.

Prime Minister to Field-Marshal Montgomery   5 June 45

I see considerable signs of changes of opinion here on the subject of non-fraternisation. The Russians seem to be following the opposite plan, and gaining thereby.

2. I am alarmed by the winter prospect in Germany. I expect they will do everything you tell them, and hold you responsible that they are fed. I wonder myself whether anything but German responsibility can secure the full German effort. It would not be thought a good ending to the war if you had a Buchenwald in Germany this winter, with millions instead of thousands dying.

3. I did not like to see the German admirals and generals with whom we had recently made arrangements being made to stand with their hands above their heads. Nor did I like to see the infantry component of the 11th Armoured Division used in this particular task. I understand the whole was ordered from S.H.A.E.F.

4. I only send you these as notes, and you are welcome to ask for further information.

Prime Minister to Mr. Bracken, Minister of Information,   9 June 45
Sir Edward Bridges, and Private Office

The rules about Ministers writing for the Press arise from a decision taken in Mr. Baldwin's Government that Ministers may not write for newspapers on any subject connected with their departments, whether for payment or not, nor may they write for payment on any Governmental or political matter. They may of course take part in controversy in defence of Government policy. It would be unsuitable for a Minister to make a particular newspaper his organ.

2. On the other hand, a Minister is entitled to write on literary, historical, scientific, or philosophical topics not connected with current politics at his pleasure, and may receive payment for such work, guarding himself all the time against criticism that he is neglecting his official duties. Many Ministers, myself included, have written books in office under the above limits, and these books have been serialised.

3. In times of election greater freedom prevails. In this present election Ministers may, without payment, write in any newspaper in prosecution of the Government's aims or defence of its policy. They must however beware that favouritism of particular newspapers, if pronounced, will lead to retaliation by others. The matter might be mentioned in Cabinet.

Prime Minister to Admiralty   11 June 45
Please pass the following message:

"*Prime Minister to Commanding Officer H.M.S. 'Kelvin'*
"Please convey my warm thanks to all those stokers under your command who sent me the delightful flowers and kind message on the anniversary of my trip in H.M.S. *Kelvin* to the Normandy beaches. This was the only time I have been in action on one of His Majesty's ships."

Prime Minister to Minister of Food   16 June 45
What is the point of bringing 500 tons of fish per day to London if only one-half of it is edible? To what use is the other half put? If there is no use for it could not the salting be made at the place of delivery? Who pays for the inedible fish?

2. You must do your best to reduce the fish queues by quicker service. At the same time it should be made public that people are coming for fish because they have heard rightly that the supply of fish to London has been greatly increased.

Prime Minister to Secretary of State for Air and   26 June 45
Minister of Labour
What is the truth in the suggestion that the Air Photographic Units at Medmenham are to be re-employed on a larger project, the total survey of Europe? This is no time to make [official] work for people, but, on the contrary, to release as many as possible. It is intolerable that these efforts should be made to find all kinds of sterile jobs for people I had the same trouble with the commission which was to examine the effects of bombing, which was to amount to about a thousand persons. The whole of this proposal must be brought immediately before the Chancellor of the Exchequer. Why are we to burden ourselves with a total survey of Europe at this moment?

2. I expect you to address yourselves to these matters with regard to the public interest and public finance. We shall be tempted from every quarter to keep in being needless appointments. Pray look into this yourselves. I am not going to have a vast mass of women kept at full charge as Government employees when they are needed in many other spheres in private life.

Prime Minister to Sir Alexander Cadogan   29 June 45

Is it possible that we are pursuing a policy of treating the Austrians the same as the Germans in the matter of non-fraternisation? All this matter requires grave and urgent attention. We are dignified and insulting, and the Russians are boon companions and enslavers. I never realised such follies were being committed.

JULY

PROGRAMME FOR JULY 1945

3 July 45

Victory in Europe was closely followed by the break-up of the Coalition Government, and much of the time and attention of Ministers in the last few weeks has necessarily been occupied with the clatter of the General Election. Many problems await decisions which are now required to enable the nation's efforts to be replanned and re-phased to meet the end of the war in Europe.

2. My colleagues have been busy in the last few weeks, but I must call on them for an intensive effort during the period between polling day and the announcement of the result of the election. During this period frequent meetings of the Cabinet, possibly as many as three a week, may be necessary, and much work must also be done by the Standing Committees of the Cabinet, which have now been reconstituted. . . .

Domestic Programme

The Housing Squad, and, as necessary, the Cabinet, will ensure an intensive drive forward with the housing programme as a military operation, in which all controls are to be used and special brigades of demobilised men enlisted for two years at exceptionally favourable terms, to go from one part of the country to another, getting the thing started. Among the Engineers there are large numbers of eminently suitable men. First, five regiments of a thousand each should be formed. The matter of building these houses is to be handled exactly with the energy that would have been put into any of the battles we have won. Nothing is to stand in the way. Apart from this, points to which special attention should be given are:

(a) Supply of labour, both for the building and civil engineering industries, and also for factories producing building materials and components.

(b) Permanent houses: measures to secure an early start with the construction of permanent houses by local authorities and by private enterprise, and to accelerate the production of prefabricated houses and components.

(c) Temporary houses: measures to expedite the preparation of sites and the production of houses.

(d) Emergency shelter: means of increasing, by temporary expedients, the living accommodation available for next winter— e.g., requisitioning, adaptation of large houses, etc.

(e) Measures for establishing proper control over building to ensure that the available labour is concentrated on high-priority work. . . .

8. *Exports.*—Preparations for the export drive should be advanced.

9. *Coal.*—A detailed scheme must be prepared to give effect to the statement of policy made by the Minister of Fuel and Power on May 29 on the future organisation of the coal industry. All possible steps must be taken to ensure that adequate supplies of coal are available to meet the demand of next winter.

10. *Rent Restriction.*—Urgent consideration must be given to the recommendations of the Ridley Report.

11. *National Insurance.*—Further progress should be made with the preparations, both legislative and administrative, for the introduction of this scheme.

12. *National Health Service.*—The form of the legislation required to give effect to the modified scheme should now be considered.

MAN-POWER

NOTE BY THE PRIME MINISTER

5 July 45

Women ought not to be treated the same as men. The reason why we do not bring home the men we want to get the industries started is because of the anger it would cause among Class A. Class A stands first and super-sacred. Any infringement upon their rights might cause the gravest disaster, as it did last time. We must carry them with us at all costs.

2. But women are an entirely different category. They do not mutiny or cause disturbances, and the sooner they are back at their homes the better. This idea of keeping masses of highly paid young women hanging around at the Air Force stations and in Army and Navy work, with a redundant staff finding a job in teaching them to

lead a better life, is one which should be completely cut out of our system. All women should be free to retire as soon as possible from the Services, and those who like to stay will be found sufficient to do the necessary jobs. A regular ramp is growing up to hold on to these women. No one who is not absolutely necessary to the war effort can in any circumstances be retained. Any who are to go to the Far East or India must volunteer from those who wish to continue in the Services. I trust within a few weeks we may be able to declare that every woman who wishes to leave the Services may do so, provided the numbers who wish to leave are not so great as to affect the rate of release of Class A.

3. Pray let all efforts be made to realise this objective.

Prime Minister to Colonial Secretary and Chiefs of Staff   6 July 45
Committee

The whole question of Palestine must be settled at the peace table, though it may be touched upon at the Conference at Potsdam. I do not think we should take the responsibility upon ourselves of managing this very difficult place while the Americans sit back and criticise. Have you ever addressed yourselves to the idea that we should ask them to take it over? I believe we should be the stronger the more they are drawn into the Mediterranean. At any rate, the fact that we show no desire to keep the mandate will be a great help. I am not aware of the slightest advantage which has ever accrued to Great Britain from this painful and thankless task. Somebody else should have their turn now. However, the Chiefs of Staff should examine the matter from the strategic point of view.

Prime Minister to Sir Edward Bridges   7 July 45

All women in the Services or in the war munitions works are to be released if they wish to go, irrespective of the category, corresponding to the male, in which they fall. Direction will however be maintained to women to go into civil manufacturing like textiles, etc. No women are to be kept dawdling about waiting for their turn to come, and no people employed finding occupation for them meanwhile.

2. I am expecting half a million to be out of uniform or out of munitions factories in three months from now.

Prime Minister to First Lord of the Admiralty, Secretary   17 July 45
of State for War, Secretary of State for Air, Minister
of Health, Secretary of State for Dominion Affairs, and
Secretary of State for India

In May I gave directions that 1,600 doctors should be returned to civilian life forthwith from the Services. I presume these are already out, and I should like a report confirming this. The time has now come to make a further cut of doctors in the Services in order to ensure adequate medical attention for civilians in the coming winter. A further 1,600 doctors should therefore be returned to civilian life by October 1. The proportion in which the three Services release these doctors should be the same as applied to the first 1,600.

Prime Minister to General Ismay   23 July 45

What is being done with German rifles? It is a great mistake to destroy rifles. If possible, at least a couple of million should be preserved for Britain.

APPENDIX D

THE ATTACK ON THE SOUTH OF FRANCE*

OPERATIONS IN THE EUROPEAN THEATRES

NOTE BY THE PRIME MINISTER AND MINISTER OF DEFENCE

28 June 44

PART I

I have thought it right to put down a few points which seem to me dominant.

2. At the present stage of the war in Europe our overall strategic concept should be the engagement of the enemy on the largest scale with the greatest violence and continuity. In this way only shall we bring about an early collapse. Here is the prime test.

3. For this purpose sufficient ports must be acquired to allow the direct and speedy deployment in Europe of the thirty or more American divisions which are in the United States.

4. In choosing points of landing or attack regard must be paid, first, to their tactical relation with the main enterprise and battle proceeding under General Eisenhower in Western France, and, secondly, to the strain produced upon the central power of Germany, the O.K.W. The optimum is to combine both.

5. Political considerations, such as the revolt of populations against the enemy or the submission and coming over of his satellites, are a valid and important factor.

6. It is better to have two ventures than three, and there are certainly not enough L.S.T.s, etc., available for more than two major ventures.

7. The various choices now open should be examined in the light of the above requirements.

PART II

8. The supreme priority must naturally be accorded to the support of "Overlord", for it is certain that the number of divisions now assigned to that enterprise up to the end of August, namely, forty plus, are not sufficient to establish mastery over the enemy resources available in Western France (apart from a psychological collapse, which should not be reckoned upon). It was understood that United States divisions would directly reinforce "Overlord" after August at the rate of five per month. The number of divisions which can be provided

* See Chapter IV, "Attack on the South of France?"

to reinforce "Overlord" in this period should be limited only by shipping possibilities and port accommodation on the western shores of France. The fundamental problem for S.H.A.E.F. is the reception of the maximum of divisions from any quarter, together with the necessary tail.

9. For this purpose one ought not to consider only the ports envisaged. There are many small ports besides, as Port-en-Bessin, Courseulles, and Ouistreham, with an aggregate capacity of 4,000 tons per diem, which have already been found, even on the very closely studied beaches of the actual "Overlord" assault. The use of landing-craft enormously increases the discharge from these small ports. For this reason it would seem a mistake to move large quantities of landing-craft from the supreme operation across the Channel to any diversion elsewhere which was not in tactical relation to the battle. The question is how to give General Eisenhower the maximum support directly in the shortest time and without causing needless havoc elsewhere.

10. The whole facilities for reception of troops and vehicles along the French Atlantic coast should be re-examined in the light of newly won experience. Moreover, the gaining of new ports to the north and south of our present "Overlord" objectives is greatly facilitated by the use of shore-based airfields or fuelling grounds now soon to be available in France. The taking of Havre and St. Nazaire is a necessity in far closer relation to the battle than any ports in the Mediterranean. In short, it is the main interest of "Overlord" to receive the great volume of troops who are waiting in the United States and can, if they can take them and if they can come into action sooner, be drawn from the Mediterranean. It would be a great pity to sweep aside all possibilities of broadening the intake direct from the United States or by stages through the United Kingdom into the western coasts of France.

11. Not only should the quantity of the intake be expanded to the utmost limit, but also the quality should be related to the fighting prospects of the next few months. Attached to this paper will be found, in a note prepared for me, the arrivals in the United Kingdom during May and the estimate of arrivals for June, July, and August. From this it will be seen that 553,356 American soldiers have arrived or are to arrive in these four months, but they only constitute seven divisions. The field troops of seven divisions amount to about 20,000 men a division, and with other fighting accessories, such as tank brigades and independent brigades, etc., to, say, 25,000. Total, 175,000. Deducting this from 553,356 leaves 378,356. The question arises whether it might not be possible by severe adjustments, within the limits of existing shipping arrangements, to give a higher priority to at least four or five more fighting divisions at the expense of some

378,356 servicing troops of many details comprised in this immense figure. The battle in France in this period may turn upon the more speedy arrival of these additional fighting units. This would still leave nearly a quarter of a million for the tail. Here also it must be observed that the casualties in France have happily been much less than those provided for in the scale of build-up, and we should be justified on the results of May and June alone in sending in two additional formed divisions instead of 50,000 replacements.

12. There are three French divisions which could be withdrawn from North Africa and a further four French divisions which might be withdrawn from Italy if ports and shipping and tail could be found for them. General Eisenhower plainly foresees this possibility as his second choice.

13. Thus there are possibilities of a considerable increase on the schedule of arrivals in the "Overlord" area in the next three months. Let us be sure that we are right in discarding these possibilities before we turn to more sombre alternatives, for it is certain that in no other way can so great or so timely a reinforcement be given to "Overlord".

PART III

14. We must now consider the application of the axioms set forth in Part I to the Mediterranean in relation to the remarks in Part II about reinforcement of "Overlord" from the west. If there were any way of capturing Bordeaux within the present fighting season by a thrust from the Gulf of Lions, and thus opening Bordeaux and other smaller ports near it to the advance across the ocean of the main United States Army, this would clearly take priority over any purely Mediterranean enterprise which could be launched. Let us therefore examine in this setting the variants of "Anvil" which have for so many months held our thoughts. Two projects have been put forward, to wit, a landing of, say, ten divisions with a three-division lift and a seven-division follow-up at Sète or at Marseilles. Sète has the great advantage of being only 225 miles from Bordeaux and is without any serious mountain obstructions. It is, I understand, admitted by all sides that there is no possibility of any landing on August 1 and the earliest possibility is August 15, and that even this is doubtful. If we attack Sète between August 15 and 30 we are told it would be conceivable to land up to ten divisions by the end of September or the middle of October. There would then be the 225-mile march to accomplish in the face of such opposition as might be offered. If there were any opposition worthy of the name it would be very surprising if a rate of more than five miles a day could be maintained by a substantial force. Thus we could not expect to take Bordeaux from the back

before the beginning or middle of December. Thereafter there would be the need to put the port in order, and therefore the Sète operation, even if the naval objection to the landing-places were overcome, would not influence the war in 1944 except in so far as German troops now on the Riviera or dispatched from O.K.W. were kept out of the "Overlord" theatre. On this plan there could not be any large over-sea intake from the United States. This heavy-footed method of approach to Bordeaux is not to be compared with the results to be obtained by a descent upon Bordeaux either from Bayonne or from neighbouring small landing-points. This might by a *coup de main* give a port and bridgehead into which French troops from Africa and the Mediterranean could enter France, and another great port be opened directly on the Atlantic. Anyhow, in view of the naval objections Sète has been ruled out.

15. We are therefore left with the Toulon-Marseilles operation. The more I have thought about this the more bleak and sterile it appears. It adds another 130 miles to the march upon Bordeaux, making a total of 355 miles in all. This march would present a flank to any German forces to the northward. The landing itself cannot be begun till August 30, and then only if the L.S.T.s, etc., can be spared from "Overlord" by July 10. All that can be said against Sète as a means of access to Bordeaux is reinforced in the case of Marseilles by these facts. Indeed, the march to Bordeaux from Marseilles could not begin in ten-division strength till a month after August 30, and could not be accomplished for probably three months after that. For these reasons I cannot feel convinced that the attack on Bordeaux from the Gulf of Lions is a practical possibility.

16. But the successful capture of Toulon and Marseilles by August 30 and the landing of ten divisions by September 30 would also have as a possible objective a march up the Rhone valley, with Lyons, 160 miles to the north, as its first objective. Here we should have, if successful, the advantage of putting in all the French available and such American divisions as were withdrawn from Italy, from Africa, or diverted from the United States at the cost of "Overlord". We should also be in close contact with the Maquis, who have developed a moderate guerrilla in the mountains. We should have a first-class port through which to pour American troops into this part of France if and as desired. It is as easy to talk of an advance up the Rhone valley as it is of a march from Italy to Vienna. But very great hazards, difficulties, and delays may menace all such projects. Once we are committed to the landing at Marseilles, all the enemy troops along the Riviera, at present seven or eight divisions, can be brought to oppose us. It will always be possible for O.K.W. to move any forces they

have in Italy through the tunnels under the Alps, or till winter comes along the great motoring roads which have been made over them, and intercept our northward advance at any point they choose. The country is most formidable. Without the enemy withdrawing a single division from the "Overlord" battle we could be confronted with superior forces at every step we advance up the Rhone valley. The evacuation by the enemy of Piedmont would not entail more than his guarding the Corniche roads along the Riviera and the mountain passes, which, with the winter coming on, would not be difficult. He can always blow up the tunnels at his discretion. If we blow them up by air action he can always, except in the depth of winter, escape over the top or along the Riviera coast.

17. It seems to me very difficult to prove that either Sète or the Marseilles operation would have any tactical relation to the battle we have to fight *now* and throughout this summer and autumn for "Overlord". The distance, as the crow flies, from Marseilles to Cherbourg is 600 miles, and from Marseilles to Paris 400 miles. It would seem clear that, even with great success, neither of these operations would directly influence the present battle in 1944.

18. Moreover, before we embark upon either of these two forms of "Anvil" in the hopes of helping "Overlord" it would be well to count the cost that must be paid for either of them.

PART IV

19. The telegrams from Wilson, Alexander, and Field-Marshal Smuts put before us the project of an attack eastwards across the Adriatic or/and around its shores, and General Wilson conceives it possible that, on this plan, he and General Alexander could have possession of Trieste by the end of September. This movement is of course as equally unrelated tactically to "Overlord" as are the variants of "Anvil".

20. Whether we should ruin all hopes of a major victory in Italy and all its fronts and condemn ourselves to a passive rôle in that theatre, after having broken up the fine Allied army which is advancing so rapidly through that peninsula, for the sake of "Anvil", with all its limitations, is indeed a grave question for His Majesty's Government and the President, with the Combined Chiefs of Staff, to decide. For my own part, while eager to do everything in human power which will give effective and timely help to "Overlord", I should greatly regret to see General Alexander's army deprived of much of its offensive power in Northern Italy for the sake of a march up the Rhone valley, which the Combined Chiefs of Staff have themselves described as unprofitable, in addition to our prime operation of "Overlord".

ANNEXE

U.S. ARMY, INCLUDING A.A.F. ARRIVALS IN U.K., MAY TO AUGUST 1944

Serial No.	Detail	Arrivals in U.K. during May 1944	U.S. Estimate of Arrivals in U.K. for June, July, and August		
			June	July	August
1	U.S. Army (excluding Air Force) ..	88,432	135,775	107,639	189,541
2	U.S. Army Air Force	16,257	7,196	3,301	5,215
3	Total U.S. Army (including A.A.F.)	104,689	142,971	110,940	194,756
4	Number of infantry divisions ..	1	1	—	2
5	Number of armoured divisions	—	1	—	1
6	Number of airborne divisions	—	—	—	1

21. To sum up:

    (a) Let us reinforce "Overlord" directly, to the utmost limit of landings from the west.

    (b) Let us next do justice to the great opportunities of the Mediterranean commanders, and confine ourselves at this phase to minor diversions and threats to hold the enemy around the Gulf of Lions.

    (c) Let us leave General Eisenhower all his landing-craft as long as he needs them to magnify his landing capacity.

    (d) Let us make sure of increasing to the maximum extent the port capacity in the "Overlord" battle area.

    (e) Let us resolve not to wreck one great campaign for the sake of winning the other. Both can be won.

President Roosevelt to Prime Minister   29 June 44

I have given careful personal consideration to your memorandum, and I have had our joint Staffs give the whole subject further consideration.

2. I agree with you that our overall strategic concept should be to engage the enemy on the largest scale with the greatest violence and continuity, but I am convinced that it must be based on a main effort, together with closely co-ordinated supporting efforts directed at the heart of Germany.

3. The exploitation of "Overlord", our victorious advances in Italy, an early assault on Southern France, combined with the Soviet drives to the west—all as envisaged at Teheran—will most surely serve to realise our object—the unconditional surrender of Germany. In this connection also I am mindful of our agreement with Stalin as to an operation against the south of France, and his frequently expressed views favouring such an operation and classifying all others in the Mediterranean as of lesser importance to the principal objective of the European campaign.

4. I agree that the political considerations you mention are important factors, but military operations based thereon must be definitely secondary to the primary operations of striking at the heart of Germany.

5. I agree that the "Overlord" build-up must receive continuing attention, but consider this to be definitely Eisenhower's responsibility. The forces we are sending him from the United States are what he has asked for. If he wants divisions ahead of service troops he has but to ask—the divisions will be ready.

6. Until we have exhausted the forces in the United States, or it is proved we cannot get them to Eisenhower when he wants them, I am opposed to the wasteful procedure of transferring forces from the Mediterranean to "Overlord". If we use shipping and port capacity

to shift forces from one combat area (the Mediterranean) to another ("Overlord") it will certainly detract from the build-up of "Overlord" direct from the United States, and the net result is just what we don't want—fewer forces in combat areas.

7. My interest and hopes centre on defeating the Germans in front of Eisenhower and driving on into Germany, rather than on limiting this action for the purpose of staging a full major effort in Italy. I am convinced we will have sufficient forces in Italy, with "Anvil" forces withdrawn, to chase Kesselring north of Pisa–Rimini and maintain heavy pressure against his army at the very least to the extent necessary to contain his present force. I cannot conceive of the Germans paying the price of ten additional divisions, estimated by General Wilson, in order to keep us out of Northern Italy.

8. We can—and Wilson confirms this—immediately withdraw five divisions (three United States and two French) from Italy for "Anvil". The remaining twenty-one divisions, plus numerous separate brigades, will certainly provide Alexander with adequate ground superiority. With our air superiority there is obviously sufficient air in the Mediterranean to furnish support both for operations in Italy and for "Anvil", and to provide overwhelming air support during the critical moments of either operation. We also have virtual mastery of the sea in the Mediterranean.

9. I agree that operations against Bordeaux or Sète with Mediterranean forces are out of the picture. As to Istria, I feel that Alexander and Smuts, for several natural and very human reasons, are inclined to disregard two vital considerations: the grand strategy firmly believed by us to be necessary to the early conclusion of the war, and the time factor as involved in the probable duration of a campaign to debouch from the Ljubljana Gap into Slovenia and Hungary. The difficulties in this advance would seem far to exceed those pictured by you in the Rhone valley, ignoring the effect of organised Resistance groups in France and the proximity to "Overlord" forces. I am informed that for purely logistical reasons it is doubtful if, within a decisive period, it would be possible to put into the fighting beyond the Ljubljana Gap more than six divisions. Meanwhile we will be struggling to deploy in France thirty-five United States divisions that are now in continental United States, plus an equivalent of corps and army combat troops, not to mention the necessary complement of service troops. I cannot agree to the employment of United States troops against Istria *and into the Balkans*,* nor can I see the French agreeing to such use of French troops.

10. The beaches, exits, communications, and cover in the Toulon

* My italics.—W. S. C.

area are most suitable. The Rhone corridor has its limitations, but is better than Ljubljana, and is certainly far better than the terrain over which we have been fighting in Italy.

11. I am impressed by Eisenhower's statement that "Anvil" is of transcendent importance and that he can and will furnish the required additional means to Wilson without undue detriment to "Overlord", and by Wilson's statement that he can conduct the operation if given an immediate directive.

12. Wilson's plans for "Anvil" are well developed, and hence the operation can be launched with no delay.

13. Since the agreement was made at Teheran to mount an "Anvil", I cannot accept, without consultation with Stalin, any course of action which abandons this operation. In the event that you and I are unable to agree to issue a directive to General Wilson by July 1 to launch "Anvil" at the earliest possible date, we must communicate with Stalin immediately. Furthermore, I feel that if we are to abandon "Anvil" we must at once discuss with the French the use of their forces, which might by this decision be kept out of the battle in France while taking losses in a secondary effort in Italy or the Balkans.

14. I again urge that the directive proposed by the United States Chiefs of Staff be issued to General Wilson immediately. It is evident that the drawing out of this discussion, if continued, will effectively kill the prospects of "Anvil" in time to be of major benefit to "Overlord".

15. At Teheran we agreed upon a definite plan of attack. That plan has gone well so far. Nothing has occurred to require any change. Now that we are fully involved in our major blow history will never forgive us if we lost precious time and lives in indecision and debate. My dear friend, I beg you to let us go ahead with our plan.

16. Finally, for purely political considerations over here, I should never survive even a slight setback in "Overlord" *if it were known that fairly large forces had been diverted to the Balkans.**

* My italics.—W. S. C.

APPENDIX B

MONTHLY TOTALS OF SHIPPING LOSSES, BRITISH, ALLIED, AND NEUTRAL, BY ENEMY ACTION

(Corrected to June 1952)

MONTH	BRITISH		ALLIED		NEUTRAL		TOTAL	
	No. of Ships	Gross Tons	No. of Ships	Gross Tons	No. of Ships	Gross Tons	No. of Ships	Gross Tons
June 1944 ..	17	54,665	8	47,382	1	2,037	26	104,084
July ..	11	40,539	6	38,217	—	—	17	78,756
August ..	17	80,590	5	37,661	1	53	23	118,304
September..	4	20,407	3	16,961	1	1,437	8	44,805
October ..	2	1,722	2	9,946	—	—	4	11,668
November ..	4	11,254	3	24,621	2	2,105	9	37,980
December ..	11	46,876	15	88,037	—	—	26	134,913
TOTALS ..	66	262,053	42	262,825	5	5,632	113	530,510
January 1945	9	45,691	9	37,206	—	—	18	82,897
February ..	13	43,636	12	50,116	1	1,564	26	95,316
March.. ..	13	46,653	13	63,406	1	1,145	27	111,204
April	11	52,496	11	52,016	—	—	22	104,512
May	1	2,878	3	14,320	—	—	4	17,198
June	—	—	2	18,615	—	—	2	18,615
July	—	—	3	7,237	—	—	3	7,237
August ..	—	—	1	36	—	—	1	36
September..	—	—	—	—	—	—	—	—
Not known	2	1,806	—	—	—	—	2	1,806
TOTALS ..	49	193,160	54	242,952	2	2,709	105	438,821

APPENDIX F
PRIME MINISTER'S VICTORY BROADCAST,
MAY 13, 1945

It was five years ago on Thursday last that His Majesty the King commissioned me to form a National Government of all parties to carry on our affairs. Five years is a long time in human life, especially when there is no remission for good conduct. However, this National Government was sustained by Parliament and by the entire British nation at home and by all our fighting men abroad, and by the unswerving co-operation of the Dominions far across the oceans and of our Empire in every quarter of the globe. After various episodes had occurred it became clear last week that so far things have worked out pretty well, and that the British Commonwealth and Empire stands more united and more effectively powerful than at any time in its long romantic history. Certainly we are—this is what may well, I think, be admitted by any fair-minded person—in a far better state to cope with the problems and perils of the future than we were five years ago.

For a while our prime enemy, our mighty enemy, Germany, overran almost all Europe. France, who bore such a frightful strain in the last great war, was beaten to the ground and took some time to recover. The Low Countries, fighting to the best of their strength, were subjugated. Norway was overrun. Mussolini's Italy stabbed us in the back when we were, as he thought, at our last gasp. But for ourselves —our lot, I mean, the British Commonwealth and Empire—we were absolutely alone.

In July, August, and September 1940 forty or fifty squadrons of British fighter aircraft in the Battle of Britain broke the teeth of the German air fleet at odds of seven or eight to one. May I repeat again the words I used at that momentous hour: "Never in the field of human conflict was so much owed by so many to so few." The name of Air Chief Marshal Lord Dowding will always be linked with this splendid event. But conjoined with the Royal Air Force lay the Royal Navy, ever ready to tear to pieces the barges, gathered from the canals of Holland and Belgium, in which a German invading army could alone have been transported. I was never one to believe that the invasion of Britain, with the tackle that the enemy had at that time, was a very easy task to accomplish. With the autumn storms the immediate danger of invasion in 1940 passed.

Then began the Blitz, when Hitler said he would "rub out our cities". That's what he said, "rub out our cities." This Blitz was borne without a word of complaint or the slightest sign of flinching, while a very large number of people—honour to them all—proved that London could "take it", and so could our other ravaged centres. But the dawn of 1941 revealed us still in jeopardy. The hostile aircraft could fly across the approaches to our Island, where forty-six millions of people had to import half their daily bread and all the materials they needed for peace or war. These hostile aircraft could fly across the approaches from Brest to Norway and back again in a single flight. They could observe all the movements of our shipping in and out of the Clyde and Mersey, and could direct upon our convoys the large and increasing numbers of U-boats with which the enemy bespattered the Atlantic—the survivors or successors of which U-boats are now being collected in British harbours.

The sense of envelopment, which might at any moment turn to strangulation, lay heavy upon us. We had only the North-Western Approach between Ulster and Scotland through which to bring in the means of life and to send out the forces of war. Owing to the action of the Dublin Government, so much at variance with the temper and instinct of thousands of Southern Irishmen who hastened to the battle-front to prove their ancient valour, the approaches which the Southern Irish ports and airfields could so easily have guarded were closed by the hostile aircraft and U-boats. This was indeed a deadly moment in our life, and if it had not been for the loyalty and friendship of Northern Ireland we should have been forced to come to close quarters or perish for ever from the earth. However, with a restraint and poise to which, I say, history will find few parallels, His Majesty's Government never laid a violent hand upon them, though at times it would have been quite easy and quite natural, and we left the Dublin Government to frolic with the Germans and later with the Japanese representatives to their hearts' content.

When I think of these days I think also of other episodes and personalities. I think of Lieutenant-Commander Esmonde, V.C., of Lance-Corporal Kenneally, V.C., and Captain Fegen, V.C., and other Irish heroes whose names I could easily recite, and then I must confess that bitterness by Britain against the Irish race dies in my heart. I can only pray that in years which I shall not see the shame will be forgotten and the glories will endure, and that the peoples of the British Isles, as of the British Commonwealth of Nations, will walk together in mutual comprehension and forgiveness.

My friends, when our minds turn to the North-Western Approaches we will not forget the devotion of our merchant seamen, and our minesweepers out every night, and so rarely mentioned in the headlines. Nor will we forget the vast, inventive, adaptive, all-embracing, and, in the end, all-controlling power of the Royal Navy, with its ever more potent new ally, the Air. These have kept the life-line open. We were able to breathe; we were able to live; we were able to strike. Dire deeds we had to do. We had to destroy or capture the French Fleet, which, had it ever passed undamaged into German hands, would, together with the Italian Fleet, have perhaps enabled the German Navy to face us on the high seas. This we did. We had to make the dispatch to General Wavell all round the Cape, at our darkest hour, of the tanks—practically all we had in the Island—and this enabled us as far back as November 1940 to defend Egypt against invasion and hurl back with the loss of a quarter of a million captives and with heavy slaughter the Italian armies at whose tail Mussolini had already planned to ride into Cairo or Alexandria.

Great anxiety was felt by President Roosevelt, and indeed by thinking men throughout the United States, about what would happen to us in the early part of 1941. The President felt to the depths of his being that the destruction of Britain would not only be an event fearful in itself, but that it would expose to mortal danger the vast and as yet largely unarmed potentialities and the future destiny of the United States. He feared greatly that we should be invaded in that spring of 1941, and no doubt he had behind him military advice as good as any that is known in the world, and he sent his recent Presidential opponent, the late Mr. Wendell Willkie, to me with a letter in which he had written in his own hand the famous lines of Longfellow which I quoted in the House of Commons the other day.

We were, however, in a fairly tough condition by the early months of 1941, and felt very much better about ourselves than in those months immediately after the collapse of France. Our Dunkirk army and field force troops in Britain, almost a million strong, were nearly all equipped or re-equipped. We had ferried over the Atlantic a million rifles and a thousand cannon from the United States, with all their ammunition, since the previous June. In our munitions works, which were becoming very powerful, men and women had worked at their machines till they dropped senseless from fatigue. Nearly one million of men, growing to two millions at the peak, although working all day, had been formed into the Home Guard. They were armed at least with rifles, and armed also with the spirit "Conquer or die".

Later in 1941, when we were still alone, we sacrificed unwillingly, to some extent unwittingly, our conquests of the winter in Cyrenaica and Libya in order to stand by Greece; and Greece will never forget how much we gave, albeit unavailingly, of the little we had. We did this for honour. We repressed the German-instigated rising in Iraq. We defended Palestine. With the assistance of General de Gaulle's indomitable Free French we cleared Syria and the Lebanon of Vichyites and of German aviators and intriguers. And then in June 1941 another tremendous world event occurred.

You have no doubt noticed in your reading of British history—and I hope you will take pains to read it, for it is only from the past that one can judge the future, and it is only from reading the story of the British nation, of the British Empire, that you can feel a well-grounded sense of pride to dwell in these islands—you have sometimes noticed in your reading of British history that we have had to hold out from time to time all alone, or to be the mainspring of coalitions, against a Continental tyrant or dictator, and we have had to hold out for quite a long time: against the Spanish Armada, against the might of Louis XIV, when we led Europe for nearly twenty-five years under William III and Marlborough, and a hundred and fifty years ago, when Nelson, Pitt, and Wellington broke Napoleon, not without assistance from the heroic Russians of 1812. In all these world wars our Island kept the lead of Europe or else held out alone.

And if you hold out alone long enough there always comes a time when the tyrant makes some ghastly mistake which alters the whole balance of the struggle. On June 22, 1941, Hitler, master as he thought himself of all Europe—nay, indeed, soon to be master of the world, so he thought—treacherously, without warning, without the slightest provocation, hurled himself on Russia and came face to face with Marshal Stalin and the numberless millions of the Russian people. And then at the end of the year Japan struck a felon blow at the United States at Pearl Harbour, and at the same time attacked us in Malaya and Singapore. Thereupon Hitler and Mussolini declared war on the Republic of the United States.

Years have passed since then. Indeed, every year seems to me almost a decade. But never since the United States entered the war have I had the slightest doubt that we should be saved, and that we only had to do our duty in order to win. We have played our part in all this process by which the evildoers have been overthrown—and I hope I do not speak vain or boastful words, but from Alamein in October 1942, through the Anglo-American invasion of North Africa, of Sicily, of Italy, with the capture of Rome, we marched many miles and never knew defeat. And then last year, after two years' patient preparation and marvellous devices of amphibious warfare—and mark you, our scientists are not surpassed in any nation in the world, especially when their thought is applied to naval matters—last year, on June 6, we seized a carefully selected little toe of German-occupied France and poured millions in from this Island and from across the Atlantic, until the Seine, the Somme, and the Rhine all fell behind the advancing Anglo-American spearheads. France was liberated. She produced a fine army of gallant men to aid her own liberation. Germany lay open.

Now from the other side the mighty military achievements of the Russian people, always holding many more German troops on their front than we could do, rolled forward to meet us in the heart and centre of Germany. At the same time, in Italy, Field-Marshal Alexander's army of so many nations, the largest part of which was British or British Empire, struck their final blow and compelled more than a million enemy troops to surrender. This Fifteenth Army Group, as we call it, British and Americans joined together in almost equal numbers, are now deep in Austria, joining their right hand with the Russians and their left with the United States armies of General Eisenhower's command. It happened, as you may remember—but memories are short—that in the space of three days we received the news of the unlamented departures of Mussolini and Hitler, and in three days also surrenders were made to Field-Marshal Alexander and Field-Marshal Montgomery of over two million five hundred thousand soldiers of this terrible warlike German Army.

I shall make it clear at this moment that we never failed to recognise the immense superiority of the power used by the United States in the rescue of France and the defeat of Germany. For our part, British and Canadians, we have had about one-third as many men over there as the Americans, but we have taken our full share of the fighting, as the scale of our losses shows. Our Navy has borne incomparably the heaviest burden in the Atlantic Ocean, in the narrow seas and the Arctic convoys to Russia, while the United States Navy has had to use its immense strength mainly against Japan. We made a fair division of the labour, and we can each report that our work is either done or going to be done. It is right and natural that we should extol the virtues and glorious services of our own most famous commanders, Alexander and Montgomery, neither of whom was ever defeated since they began together at Alamein. Both of them have conducted in Africa, in Italy, in Normandy, and in Germany battles of the first magnitude and of decisive consequence. At the same time we know how great is our debt to the combining and unifying command and high strategic direction of General Eisenhower.

And here is the moment when I pay my personal tribute to the British Chiefs of Staff, with whom I worked in the closest intimacy throughout these heavy, stormy years. There have been very few changes in this small, powerful, and capable body of men, who, sinking all Service differences and judging the problems of the war as a whole, have worked together in perfect harmony with each other. In Field-

Marshal Brooke, in Admiral Pound, succeeded after his death by Admiral Andrew Cunningham, and in Marshal of the Royal Air Force Portal, a team was formed who deserved the highest honour in the direction of the whole British war strategy and in its relations with that of our Allies.

It may well be said that our strategy was conducted so that the best combinations, the closest concert, were imparted into the operations by the combined Staffs of Britain and the United States, with whom, from Teheran onwards, the war leaders of Russia were joined. And it may also be said that never have the forces of two nations fought side by side and intermingled in the lines of battle with so much unity, comradeship, and brotherhood, as in the great Anglo-American Armies. Some people say, "Well, what would you expect, if both nations speak the same language, have the same laws, have a great part of their history in common, and have very much the same outlook upon life, with all its hope and glory? Isn't it just the sort of thing that would happen?" And others may say, "It would be an ill day for all the world and for the pair of them if they did not go on working together and marching together and sailing together and flying together, whenever something has to be done for the sake of freedom and fair play all over the world. That is the great hope of the future."

There was one final danger from which the collapse of Germany has saved us. In London and the south-eastern counties we have suffered for a year from various forms of flying bombs—perhaps you have heard about this—and rockets, and our Air Force and our ack-ack batteries have done wonders against them. In particular the Air Force, turned on in good time on what then seemed very slight and doubtful evidence, hampered and vastly delayed all German preparations. But it was only when our armies cleaned up the coast and overran all the points of discharge, and when the Americans captured vast stores of rockets of all kinds near Leipzig, which only the other day added to the information we had, and when all the preparations being made on the coasts of France and Holland could be examined in detail, in scientific detail, that we knew how grave had been the peril, not only from rockets and flying bombs, but from multiple long-range artillery which was being prepared against London. Only just in time did the Allied armies blast the viper in his nest. Otherwise the autumn of 1944, to say nothing of 1945, might well have seen London as shattered as Berlin.

For the same period the Germans had prepared a new U-boat fleet and novel tactics which, though we should have eventually destroyed them, might well have carried anti-U-boat warfare back to the high peak days of 1942. Therefore we must rejoice and give thanks, not only for our preservation when we were all alone, but for our timely deliverance from new suffering, new perils not easily to be measured.

I wish I could tell you to-night that all our toils and troubles were over. Then indeed I could end my five years' service happily, and if you thought that you had had enough of me and that I ought to be put out to grass I tell you I would take it with the best of grace. But, on the contrary, I must warn you, as I did when I began this five years' task—and no one knew then that it would last so long—that there is still a lot to do, and that you must be prepared for further efforts of mind and body and further sacrifices to great causes if you are not to fall back into the rut of inertia, the confusion of aim, and "the craven fear of being great". You must not weaken in any way in your alert and vigilant frame of mind. Though holiday rejoicing is necessary to the human spirit, yet it must add to the strength and resilience with which every man and woman turns again to the work they have to do, and also to the outlook and watch they have to keep on public affairs.

On the continent of Europe we have yet to make sure that the simple and honourable purposes for which we entered the war are not brushed aside or overlooked in the months following our success, and that the words "freedom", "democracy", and "liberation" are not distorted from their true meaning as we have understood them. There would be little use in punishing the Hitlerites for their crimes if law and justice did not rule, and if totalitarian or police Governments were to take the place of the German invaders. We seek nothing for ourselves. But we must make sure that those causes which we fought for find recognition at the peace table in facts as well as words, and above all we must labour that the World Organisation which the United Nations are creating at San Francisco does not become an idle name, does not become a shield for the strong and a mockery for the weak. It is the victors who must search their hearts in their glowing hours, and be worthy by their nobility of the immense forces that they wield.

We must never forget that beyond all lurks Japan, harassed and failing but still a people of a hundred millions, for whose warriors death has few terrors. I cannot tell you to-night how much time or what exertions will be required to compel the Japanese to make amends for their odious treachery and cruelty. We—like China, so long undaunted —have received horrible injuries from them ourselves, and we are bound by the ties of honour and fraternal loyalty to the United States to fight this great war at the other end of the world at their side without flagging or failing. We must remember that Australia and New Zealand and Canada were and are all directly menaced by this evil Power. They came to our aid in our dark times, and we must not leave unfinished any task which concerns their safety and their future.

I told you hard things at the beginning of these last five years; you did not shrink, and I should be unworthy of your confidence and generosity if I did not still cry: Forward, unflinching, unswerving, indomitable, till the whole task is done and the whole world is safe and clean.

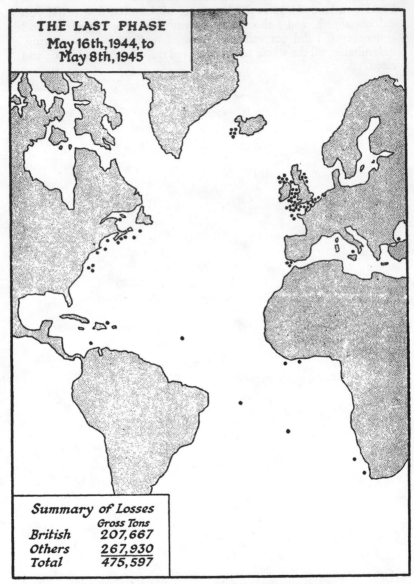

THE LAST PHASE
May 16th, 1944, to
May 8th, 1945

Summary of Losses

	Gross Tons
British	207,667
Others	267,930
Total	475,597

THE BATTLE OF THE ATLANTIC
MERCHANT SHIPS SUNK BY U-BOAT IN THE ATLANTIC

* See Chapter XXXII. "The German Surrender."

APPENDIX H

MINISTERIAL APPOINTMENTS, JUNE 1944 – MAY 1945

(Members of the War Cabinet are shown in italics)

Prime Minister and First Lord of the Treasury, Minister of Defence	*Mr. Winston S. Churchill*
Admiralty, First Lord of the	Mr. A. V. Alexander
Agriculture and Fisheries, Minister of	Mr. R. S. Hudson
Air, Secretary of State for	Sir Archibald Sinclair
Aircraft Production, Minister of	Sir Stafford Cripps
Burma, Secretary of State for	Mr. L. S. Amery
Chancellor of the Duchy of Lancaster	Mr. Ernest Brown
Chancellor of the Exchequer	*Sir John Anderson*
Civil Aviation, Minister of	Viscount Swinton (appointed October 9, 1944)
Colonies, Secretary of State for	Colonel Oliver Stanley
Dominion Affairs, Secretary of State for	Viscount Cranborne
Economic Warfare, Minister of	Earl of Selborne
Education, President of the Board of (became Minister by Education Act, 1944)	Mr. R. A. Butler
Food, Minister of	Colonel J. J. Llewellin
Foreign Affairs, Secretary of State for	*Mr. Anthony Eden*
Fuel and Power, Minister of	Major G. Lloyd George
Health, Minister of	Mr. H. U. Willink
Home Department, Secretary of State for / Home Security, Minister of	*Mr. Herbert Morrison*

INDIA, SECRETARY OF STATE FOR	MR. L. S. AMERY
INFORMATION, MINISTER OF	MR. BRENDAN BRACKEN
LABOUR AND NATIONAL SERVICE, MINISTER OF	*Mr. Ernest Bevin*
LAW OFFICERS:	
ATTORNEY-GENERAL	SIR DONALD SOMERVELL
LORD ADVOCATE	MR. J. S. C. REID
SOLICITOR-GENERAL	SIR DAVID MAXWELL FYFE
SOLICITOR-GENERAL FOR SCOTLAND	SIR DAVID KING MURRAY
LORD CHANCELLOR	VISCOUNT SIMON
LORD PRESIDENT OF THE COUNCIL	*Mr. Clement Attlee*
LORD PRIVY SEAL	LORD BEAVERBROOK
MINISTER OF STATE	MR. R. K. LAW
MINISTER WITHOUT PORTFOLIO (until November 18, 1944) MINISTER OF NATIONAL INSURANCE	} SIR WILLIAM JOWITT
PAYMASTER-GENERAL	LORD CHERWELL
PENSIONS, MINISTER OF	SIR WALTER WOMERSLEY
POSTMASTER-GENERAL	CAPTAIN H. F. C. CROOKSHANK
PRODUCTION, MINISTER OF	*Mr. Oliver Lyttelton*
RECONSTRUCTION, MINISTER OF	*Lord Woolton*
SCOTLAND, SECRETARY OF STATE FOR	MR. THOMAS JOHNSTON
SUPPLY, MINISTER OF	SIR ANDREW DUNCAN
TOWN AND COUNTRY PLANNING, MINISTER OF	MR. W. S. MORRISON
TRADE, PRESIDENT OF THE BOARD OF	MR. HUGH DALTON
WAR, SECRETARY OF STATE FOR	SIR JAMES GRIGG
WAR TRANSPORT, MINISTER OF	LORD LEATHERS
WORKS, MINISTER OF	LORD PORTAL (until November 22, 1944) MR. DUNCAN SANDYS
MINISTERS OVERSEAS:	
MIDDLE EAST, MINISTER OF STATE RESIDENT IN THE	LORD MOYNE (until November 22, 1944) SIR EDWARD GRIGG
WASHINGTON, MINISTER RESIDENT FOR SUPPLY IN	MR. BEN SMITH

ALLIED FORCE HEADQUARTERS, MR. HAROLD MACMILLAN
MEDITERRANEAN COMMAND,
MINISTER RESIDENT AT

WEST AFRICA, MINISTER RESI- VISCOUNT SWINTON
DENT IN (until November 22, 1944)
CAPTAIN H. BALFOUR

HOUSE OF LORDS, LEADER OF THE VISCOUNT CRANBORNE
HOUSE OF COMMONS, LEADER OF *Mr. Anthony Eden*
THE

THE FRONTIERS OF CENTRAL EUROPE

INDEX

INDEX